Inorganic Chemistry

Concepts and Main Group of Elements

Inorganic Chemistry

Concepts and Main Group of Elements

Dr. PL Soni

Formerly Associate Professor
Department of Chemistry
ARSD College
University of Delhi

Dr. Vandna Soni

Assistant Professor
Department of Chemistry
Maharaja Agrasen College
University of Delhi

Manakin
PRESS

First published 2025
by CRC Press
4 Park Square, Milton Park, Abingdon, Oxon, OX14 4RN

and by CRC Press
2385 NW Executive Center Drive, Suite 320, Boca Raton FL 33431

CRC Press is an imprint of Informa UK Limited

British Library Cataloguing-in-Publication Data
A catalogue record for this book is available from the British Library

ISBN13: 9781041023166 (hbk)
ISBN13: 9781041023173 (pbk)
ISBN13: 9781003618706 (ebk)

DOI: 10.1201/9781003618706

Typeset in Times New Roman
by Manakin Press, Delhi

Manakin
PRESS

Detailed Contents

Unit I

Concepts in Inorganic Chemistry

0

Broad Division between Inorganic, Organic and Physical Chemistry

The word organic is derived from living organisms. In the early days the source of organic compounds was mostly from living organisms as the synthesis of organic compounds was not advanced whereas the source of inorganic compounds was from all non-living things such as rocks, soil and mineral deposits. The inanimate source gave the name Inorganic (non-living).

Inorganic chemistry is the oldest branch of chemistry. It includes the study of all the elements. Out of all the elements the element carbon shows uniqueness in forming very strong bonds with other carbon atoms. This property of making strong bonds between like atoms is called Catenation. This property has led to the formation of a very wide variety of organic compounds. Rough comparison indicates that there are approximately 100 million organic compounds known comprising of C, H, O, S halogens and nitrogen (4–6 elements) in comparison to 40 million inorganic compounds embracing roughly 106 or more elements.

Inorganic compounds behave as molecular substances in a wider capacity to organic compounds. Inorganic chemistry includes atomic gases, solids that are air and water sensitive. Inorganic compounds are water soluble and soluble in other polar solvents. There are other inorganic compounds soluble in nonpolar solvents as well. Therefore inorganic chemistry covers greater variety of substances than does organic chemistry.

Further comparison between organic and inorganic chemistry lies regarding their coordination number. As organic compounds are carbon based therefore, their coordination number rarely exceeds 4 (*e.g.* CH_4, NR_4^+) but in inorganic chemistry the coordination number frequently exceeds 4 and goes upto 14. *e.g.* PF_5, $Nd(H_2O)_9$ etc.

This book deals with structures, properties and reactions of inorganic compounds. Inorganic compounds offer a wide variety of complex structures and reactions. There are so many methods of synthesis and their characterization. The properties of inorganic substances are dependent on the study of physical aspects for example thermodynamics, electronic structure of atoms, molecular bonding and reaction kinetics. In brief the book highlights the periodicity in properties, types of structures and their reactivities.

Classifications

Inorganic substances are classified into (*a*) the study of elements (*b*) ionic compounds (*c*) molecular substances and (*d*) polymeric compounds.

(*a*) Elements have different type of structures and properties:

(*i*) Some are monoatomic molecules *e.g.* Ar, Kr and Xe and molecular gases O_2, N_2, Cl_2, H_2 etc.

(*ii*) Molecular solids *e.g.* P_4, S_8 etc.

(*iii*) Large molecular solids like graphite and diamond.

(*iv*) Solid metals W, Ni, Cu or liquid Hg, Ga etc.

(*b*) Simple ionic solids

(*i*) Compounds of the type MX (NaCl) soluble in H_2O and polar solvents.

(*ii*) Ionic oxides insoluble in H_2O *e.g.* ZnO_2 and mixed oxides $MgAl_2O_4$ and silicates $CaMg\ (SiO_3)_2$.

(*iii*) Binary halides, sulphides arsenides and nitrides AgCl, GaAs, BN etc.

(*iv*) Compounds with complex ions *e.g.* $[SiF_6]^{2-}$ $[Co(NH_3)_6]^{3+}$, $[Fe(CN)_6]^{4-}$.

(*c*) Molecular compounds for example:

(*i*) Binary compounds PF_5, UF_6, OsO_4.

(*ii*) Complex compounds of metals $PtCl_2(PMe_3)_3$.

(*iii*) Organo-metallic compounds with metal-to-carbon bonds $[Ni(CO)_4]$, $Fe(C_5H_5)_2$.

(*iv*) Inorganic polymers of the type Boron nitrides, Boranes and silicates. These include super conductors of the formula $YBa_2Cu_3O_7$.

Structures of Inorganic Compounds

Organic compounds are carbon based. The structures of organic compounds are *sp*, *sp*^{2} and *sp*^{3} hybridised *i.e.* linear, triangular and tetrahedral (4 coordinate structures). Inorganic structures can form more than 4 bonds. Inorganic substances can give five, six, seven and more bonds and hence their geometries are more complicated. The five polyhedra are illustrated below:

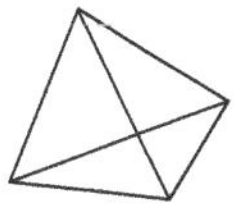

Tetrahedron
Faces: 4 equilateral triangles
Vertices: 4
Edges: 6

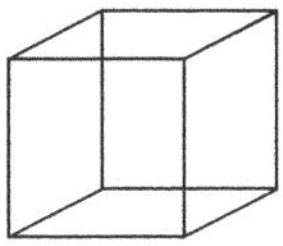

Cube
Faces: 6 Squares
Vertices: 8
Edges: 12

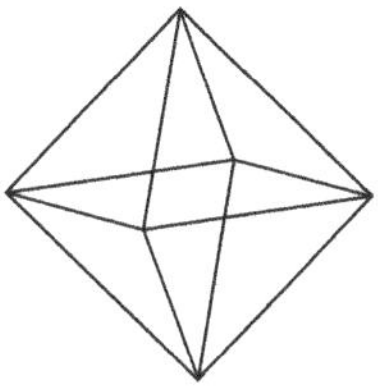

Octahedron
Faces: 8 equilateral triangles
Vertices: 6
Edges: 12

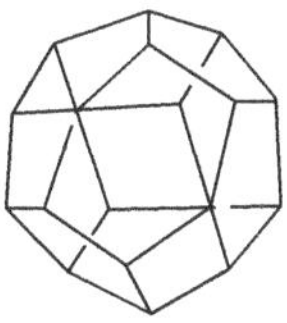

Dodecahedron
Faces: 12 regular pentagons
Vertices: 20
Edges: 30

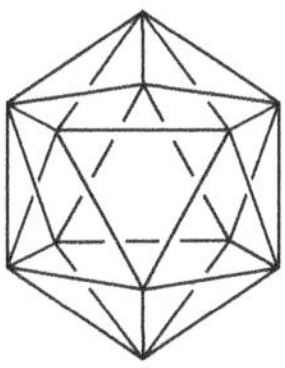

Icosahedron
Faces: 20 equilateral triangles
Vertices: 12
Edges: 30

The structures of inorganic substances are less symmetrical polyhedra *e.g.* trigonal bipyramid, trigonal prisms, square bipyramid or octahedral and so on.

Reactions

Contrary to organic reactions the exact reaction mechanism of inorganic reactions cannot be properly understood. This is because of two factors (*i*) bonds in inorganic compounds are labile. Therefore lot of bond making and bond-breaking is taking place at all stages of reactions giving numerous products. Inorganic reactions requires vigorous stirring, high temperature and pressure due to heterogeneous nature of reactants. This makes reaction mechanism impractical to find. Inorganic reactions are best, described by the overall outcome of the reaction. Inorganic compounds have ionic character therefore the reactions are fast and instantaneous while organic reactions are slow due to semi-ionic and non-ionic nature of the reactions. Organic reactions therefore take days and sometimes weeks to complete. Still the inorganic reactions may be classified in the following categories:

1. Acid-base reactions
2. Addition
3. Elimination reactions
4. Redox
5. Insertion
6. Substitution
7. Rearrangement
8. Exchange
9. Solvolysis
10. Chelation
11. Cyclization and condensation
12. Nuclear

For understanding of inorganic reactions the different stages from reactants, intermediates to products be detailed. Therefore thorough knowledge of thermodynamics, kinetics, structure and bonding is very essential.

1

Atomic Structure

Bohr's atomic model is based on Planck's theory of radiation (1900) and Rutherford; model (1911). According to Bohr (1913) electron keeps revolving round the nucleus only in certain fixed orbits where there is no loss or gain in energy called stationary orbits. The angular momentum *mvr* can acquire only certain permitted values as expressed mathematically by the relation:

$$mvr = \frac{nh}{2\pi}$$

where h is Planck's constant (6.6×10^{-27} ergs-sec), n = principal quantum number 1, 2, 3, … n.

According to Planck's quantum theory energy absorbed or emitted by black body such as heat and light is in multiples of small packet of energy called 'Quantum'. The energy of one quantum is given by:

$$\nu = \text{cm}^{-1}$$
$$\lambda = \text{cm}$$
$$c = \text{cm/sec.}$$

$E = h\nu$ or $E = h\frac{c}{\lambda}$ because $hc\nu$ as wave number $\bar{\nu} = \frac{1}{\lambda}$. In the case of light instead of quantum the word photon is used, each photon has the characteristics of a wave as well as a particle.

In case a photon has mass 'm' and velocity of light 'c', the relation between m and energy is given by $E = mc^2$.

This Einstein's mass-energy relationship.

Postulates of Bohr's Energy

(*i*) Out of infinite number of orbits round the nucleus electron chooses to stay only in those orbits in which there is no loss or gain in energy. Each orbit carriers a definite amount of energy and are therefore called energy levels. They are numbered 1, 2, 3, …, n starting from the nucleus or designated by capital letters

K, L, M ..., shells. The energy of the shells increases as we move outwards from the nucleus. E_1, E_2 and E_3 denote the energies for K, L, M shells in the order:

$$E_1 < E_2 < E_3 \ldots$$

There is no loss or gain in energy for electron in these shells's therefore they are called stationary state because the energy remains constant.

(*ii*) The electrons can jump from one energy level to the other

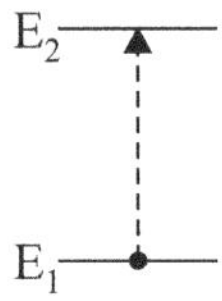

The amount of energy absorbed or emitted is given by $E_2 - E_1 = \Delta E = h\upsilon$ where υ is the frequency emitted or absorbed.

(*iii*) There are infinite number of concentric orbits, out of which orbit with quantised angular momentum (mvr) is chosen by the electron. Angular momentum is a whole number multiple of $\frac{h}{2\pi}$

$$mvr = n\frac{h}{2\pi}$$

where 'm' is mass of an electron 'v' is tangential velocity; 'r' is the distance of electron from the nucleus and 'n' is the principal quantum number.

Bohr's Theory of hydrogen atom with the help of concept of quantisation of energy and the laws of classical mechanics Bohr worked out a mathematical derivation for radius of an electron and energy of electron in the nth orbit.

In hydrogen atom nucleus has charge of one proton and the electron with equal and opposite charge revolves round the nucleus in an orbit of radius r. Let v be the tangential velocity and m the mass of moving electron.

Consequently two types of forces act upon the nucleus (*i*) centrifugal force takes the electron away from the circular path and (*ii*) centripetal force tends to pull the electron towards the nucleus due to electrostatic attraction. The two forces are equal and opposite and thus balance each other.

(*a*) Radius of electron in n^{th} orbit:

$$\text{Coulombic forces} = e \times \frac{e}{r^2} = \frac{e^2}{r^2}$$

$$\frac{e^2}{r^2} = \frac{mv^2}{r}$$

$$v^2 = \frac{e^2}{mr}$$

Due to quantisation the revolving electron is expressed by $mvr = \frac{nh}{2\pi}$

Therefore $$v^2 = \frac{n^2h^2}{4\pi^2m^2r^2}$$

Comparing the two equations

$$\frac{e^2}{mr} = \frac{n^2h^2}{4\pi^2m^2r^2}$$

or $$r = \frac{n^2h^2}{4\pi^2me^2}$$

for n^{th} orbit it becomes

$$r_n = \frac{n^2h^2}{4\pi^2me^2}$$

If $n = 1$ as for hydrogen atom the radius of first orbit of hydrogen atom we can express Bohr's radius as

$$r_1(r_H \text{ or } a_0) = \frac{h^2}{4\pi^2me^2}$$

Comparing r_n and r_1

$$\frac{r_n}{r_1} = \frac{n^2h^2}{4\pi^2me^2} \times \frac{4\pi^2me^2}{h^2}$$

***(b)* Energy of electron in *n*^th^ orbit:** Energy E_n of an electron moving in n^{th} orbit is equal to the sum of K.E. and P.E.

$$\text{K.E.} = \frac{1}{2}mv^2$$

and P.E. by virtue of electron in x shell is equal to $-\frac{e^2}{r}$ *i.e.* $E_n = \frac{1}{2}mv^2 - \frac{e^2}{r}$.

To keep the electron moving in the orbit both centrifugal force $\frac{1}{2}mv^2$ and centripetal force must balance each other. On putting $mv^2 = \frac{e}{r}$

$$E_n = \frac{e^2}{2r} - \frac{e^2}{r} = \frac{e^2}{2r}$$

Substitution the value of $r = \dfrac{n^2h^2}{4\pi^2 me^2}$

$$E_n = -\frac{e^2}{2} \times \frac{4\pi^2 me^2}{n^2h^2}$$

or

$$E_n = -\frac{2\pi^2 me^4}{n^2h^2}$$

In shell number 1 of H-atom

$$\frac{E_n}{E_1} = \frac{2\pi^2 \cancel{me^4}}{n^2 \cancel{h^2}} \times \frac{\cancel{h^2}}{2\pi^2 \cancel{me^4}}$$

$$E_n = E_1 \times \frac{1}{n^2} = -\frac{13.6}{n^2} \text{ eV/atom}$$

Subsequently value of

$$E_2 = -\frac{13.6}{2^2} = -3.4 \text{ eV/atom}$$

$$E_3 = -\frac{13.6}{3^2} = -1.51 \text{ eV/atom}$$

$$E_4 = -\frac{13.6}{4^2} = -0.85 \text{ eV/atom}$$

from these values we conclude:

1. Energy difference between two successive energy levels decrease as n increases.
2. Electron in a H-atom can have only definite energy and no energies in between the two shells is permissible.
3. The lowest energy state w.r.t. $n = 1$ is called the ground state. The next higher energy state $n = 2$ that is energy level E_2 is called the first excited state and the next higher E_3 is called the second excited state and so on.

Frequency (Ʊ), wave number ($\overline{Ʊ}$) and wave length's of the spectral lines: When an electron jumps from lower level to higher level energy is absorbed and we get absorption spectrum as in Fig. (i) and when the electron jumps from higher to lower level we get emission spectrum as shown in Fig. (ii).

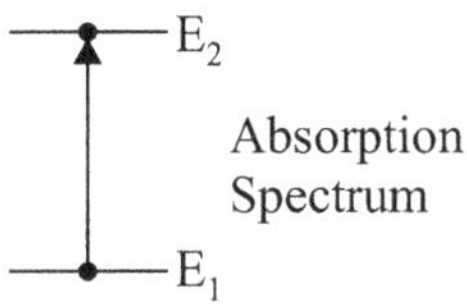

Fig. (i)

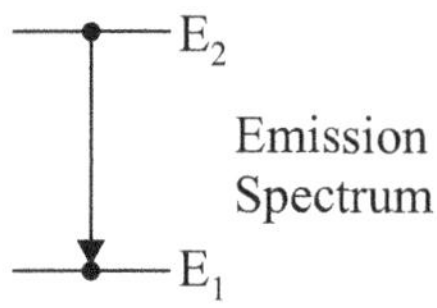

Fig. (ii)

The energy absorbed or emitted is equal and is given by $E_2 - E_1 = \Delta E = h\nu$. If inner energy is n_1 and outer is n_2 then

$$(\Delta E)n_2 - n_1 = -\frac{2\pi^2 e^4 m}{n_2^2 h^2} - \left(-\frac{2\pi^2 e^4 m}{n_1^2 h^2}\right) = h\bar{\upsilon}$$

or $$h\nu = \frac{2\pi^2 e^4 m}{h^2}\left(\frac{1}{n_1^2} - \frac{1}{n_2^2}\right)$$

$\therefore$ $$\nu = \frac{2\pi^2 e^4 m}{h^2}\left(\frac{1}{n_1^2} - \frac{1}{n_2^2}\right)$$

or $$\bar{\nu}C = \frac{2\pi^2 e^4 m}{h^2}\left(\frac{1}{n_1^2} - \frac{1}{n_2^2}\right) \qquad \ldots(a)$$

or $$\bar{\nu} = \frac{2\pi^2 e^4 m}{h^2 c}\left(\frac{1}{n_1^2} - \frac{1}{n_2^2}\right)$$

or $$\frac{1}{\lambda} = \frac{2\pi^2 e^4 m}{h^2 c}\left(\frac{1}{n_1^2} - \frac{1}{n_2^2}\right) \quad \text{as} \left[\bar{\upsilon} = \frac{1}{\lambda}\right]$$

or $$\bar{\nu} = R_H\left(\frac{1}{n_1^2} - \frac{1}{n_2^2}\right) \quad \because \quad R_H = \frac{2\pi^2 e^4 m}{h^2 c}$$

$$(R_{H\ cal.}) = 109.679\ \text{cm}^{-1}$$

The equation is reduced to:

$$\bar{\nu} = \frac{1}{\lambda} = 109{,}679\left(\frac{1}{n_1^2} - \frac{1}{n_2^2}\right) \qquad \ldots(b)$$

From equation *a* and *b* we can calculate (*i*) frequency (*ii*) wave number (*iii*) wave length of respective spectral lines observed in the hydrogen spectrum in which values of n_1 and n_2 are fixed.

Spectral Lines in the Hydrogen Spectrum

To define 106 elements we require 7 permitted shells out of infinite number of shells around the atom.

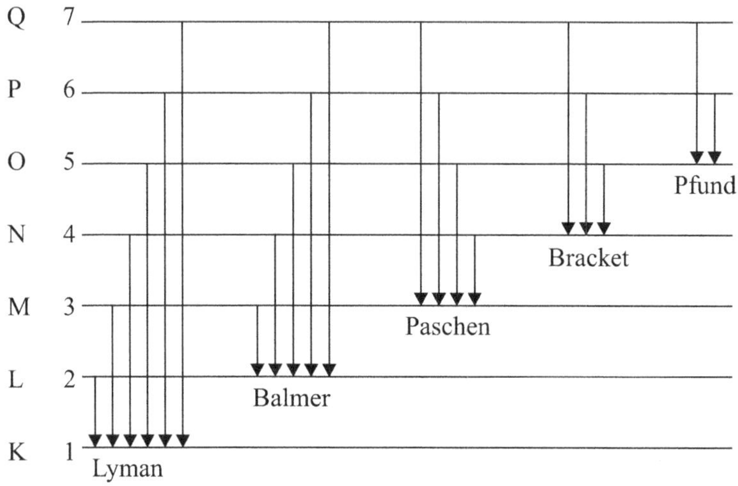

Spectral lines	Wavelength λ in Å	Spectral region
Lyman series	less than 4000 Å to 7000 Å	UV
Balmar series	between 4000 Å to 7000 Å	Visible
Paschen series	more than 7000 Å	Near IR
Bracket series	more than 7000 Å	Far IR
Pfund series	more than 7000 Å	Far IR

The jumps of electron from 7th to 6th shell is of very high frequency not observable by the instruments available.

Facts in Agreement with Bohr's Theory

(*i*) The frequencies of spectral lines of the hydrogen atom as per experimental determination have similar values as calculated by Bohr's theory.

(*ii*) The Rhydberg constant determined experimentally 109678 cm^{-1} is in close agreement with the Bohr value *i.e.* 109679 cm^{-1}.

(*iii*) The emission and absorption spectra of hydrogen atom is explained by the Bohr's concept of stationary states of electrons.

(*iv*) The radii and energy values proposed by Bohr on the basis of stationary states are quite close to the experimental values.

Failure of Bohr's Theory

(*i*) The electron jumps from one energy orbit to another is successfully explained by Bohr but how this jump is related to radiation is not explained.

(*ii*) Bohr could explain the five types of lines in the hydrogen spectrum but with more sophisticated instruments its was observed that each spectral line like Lyman, Balmer, Paschen is made up of a number of fine closely spaced lines which implies that there are several energy levels for each quantum number '*n*'.

(*iii*) Bohr's theory can successfully explain the spectra of simple atoms or ions like He^+, Li^{2+} and Be^{3+} which have very limited electrons but in case an atom has large number of electrons it is unable to explain the spectral lines.

(*iv*) Bohr gave no satisfactory explanation for the assumption that electron can rotate in those orbits in which the angular momentum *mvr* is a whole number multiple of $\frac{h}{2\pi}$ *i.e.* quantisation of angular momentum.

(*v*) According to Bohr electron moves in an orbit which is at definite distance from the nucleus and moving with a definite velocity. This is against the Heisenberg uncertainty principle which says that it is impossible to determine the position and momentum of a fast moving particle correctly simultaneously.

(*vi*) Bohr's theory is unable to explain Zeeman effect which means that the substance for which gives a line emission spectrum when place under the influence of magnetic field then lines in the emission spectra split into a number of closely spaced lines.

(*vii*) If the substance that gives line emission spectra is placed under external electro-magnetic field even then the lines get split into a number of closely spaced lines. This is known as Stark's effect. Bohr was not able to explain.

Improvement in Bohr Model or Sommerfeld Concept

In the hydrogen spectrum when an electron moves from one orbit to another it should give single sharp into corresponding to the energy difference but when the hydrogen spectrum is observed under high resolution spectro photo meter it shows a bunch of closely spaced lines. Sommerfeld explained this by assuming elliptical orbits precessing round the nucleus. These elliptical orbits have major axis '*n*' and minor axis *k*.

Thus, $$\frac{\text{major axis}}{\text{minor axis}} = \frac{n}{k}$$

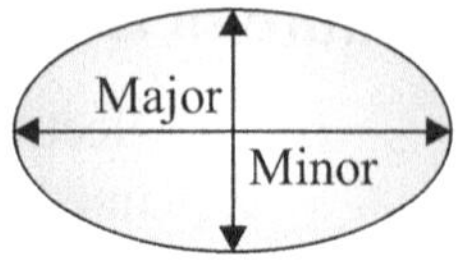

k is called the azimuthal quantum number. n may have values 1, 2, 3, ..., n for $n = 2, \frac{n}{k} = \frac{2}{2}$ = circular orbit. Then $n = 3 = \frac{n}{k} = \frac{3}{3}$ = circular $\frac{3}{2}$ elliptical and $\frac{3}{1}$ narrow ellipse.

Weakness of Sommerfeld Concept

It could not give correct number of lines in the structure, no information regarding the intensities of five lines, the exact position of the nucleus is against the uncertainty principle.

Quantum numbers

Bohr could not explain the number of five lines observed in the hydrogen spectrum. The appearance of these five lines cannot be explained by one quantum number 'n'. There are other quantum numbers which are required to explain the spectra. These quantum numbers are associated with position and energy of any electron in an atom.

1. **Principal quantum number '*n*':** This is the main or principal shell in which the electron is present. It has integral values from 1 to 7 or it is denoted by letters K, L, M, N, ... These letters or digits stand for major energy level and are denoted by 'n'. 'n' gives the idea of size of the shell as given by Bohr in $mvr = n\frac{h}{2\pi}$.
2. **Azimuthal quantum number (l):** It is denoted by 'l' and is also called subsidiary quantum number, angular quantum number or orbital quantum number. It gives the idea of the shape of the orbital. It can have value from 0 to $n - 1$.

   For principal quantum number $n = 1$, $l = 0$; which proposes an orbital which is non-directional or a sphere or s-orbital principal quantum number having value $n = 2$; l will have values 0, 1 and s and p-orbital. The value one means that the orbital is lying along an axis. Therefore, it is dump bell shaped lying along x, y and z axes.

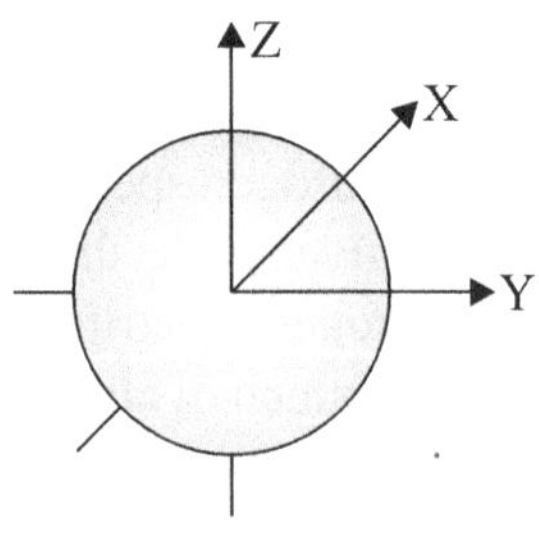

s-orbital

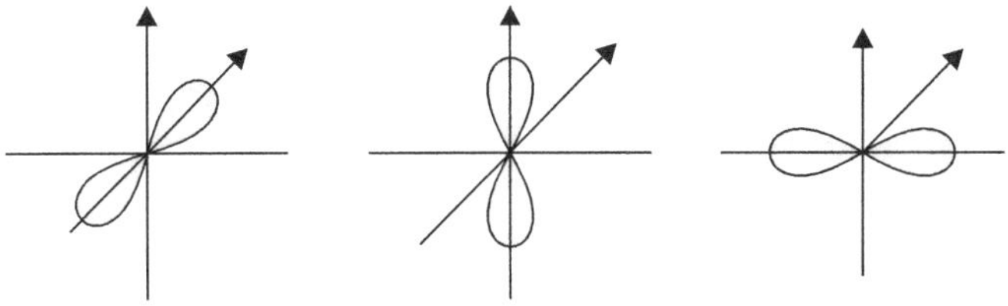

Shapes of *p*-orbital

for $n = 3$ the values of l invoked are 0, 1, 2 *i.e. s, p* and *d*-orbital. The value 2 indicates that the orbital is lying along all the 3 axes, therefore, it may be double dumpbell shaped. It is given the name *d*-orbital. It is visualised for $n = 4$; $l = 0$, 1, 2, 3, *i.e. s, p, d* and *f* orbital.

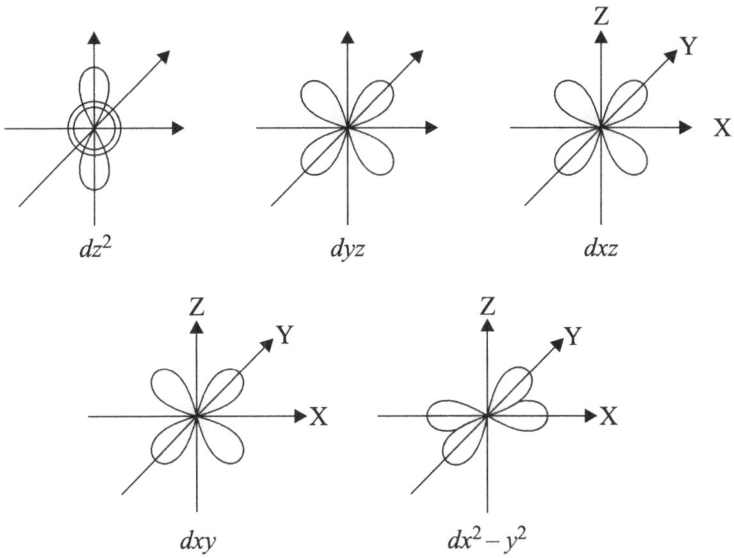

Shapes of *d*-orbital

It implies that *f*-orbital is spread in three directions *i.e.* it is 3 dimensional, therefore, its shape is difficult to draw.

Magnetic Quantum Number '*m*'

Bohr could not explain the splitting of original line in the spectrum into a number of fine closely spaced lines under the influence of magnetic field (Zeeman effect) or electric field called Stark effect, these fine lines indicate energy levels that are sub divided by the application of magnetic or electric fields. These lines represent the same value of '*l*' but different space orientations. This phenomena is represented by the magnetic quantum number '*m*' along which is orbital angular moment along an axis.

Values of *m*

Magnetic quantum number values depend on the values of azimuthal quantum number 'l'. Therefore the permitted values of 'm' are 0, ±1, ±2, ±3, ..., ±l. Mathematically $m = (2l + 1)$. *e.g.*

(*i*) $n = 1, l = 0$ (one orbital) or no related sub-levels *i.e.* s sub-shell is present only.

(*ii*) $n = 2; l = 0$. If $m = 2l + 1$ then $m = 2 \times 1 + 1 = 3$ related sub-levels *i.e.* p_x, p_y and p_z. It can be shown as 0, ±1 or 0 s sub-shell –1, 0, +1, p sub-shells along x, y and z axes. It has s and p sub-shells.

(*iii*) $n = 3, l = 2; 0, \pm 1, \pm 2, l = 2$ stands for related d-orbtails which according to formula $m = 2e + 1 = 5$. This indicates the presence of 5 related energy levels of d sub-shell. Therefore s, p and d sub-shells are present.

(*iv*) $n = 4, l = 0, \pm 1, \pm 2, \pm 3$. The value of $e = 3$ stands for f sub-shells which have –3, –2, –1, 0, +1, +2, +3 *i.e.* 7 related f sub-shells. In all $n = 4$ has s, p, d and f sub-shells.

Spin Quantum Number and Pauli Principle

The spatial distribution of electron in an atom is described by the three quantum number that have been discussed already. To completely describe an electron in an atom spin quantum number is involved owing to the spin motion of an electron, spin angular momentum is given by $\sqrt{s(s+1)} \cdot \frac{h}{2\pi}$ where it stands for absolute value of spin quantum number without positive and negative sign. The permitted values of spin are $\pm\frac{1}{2}$. According to Pauli's exclusion principle an orbital can contain no more than two electrons. Two electrons with parallel spins has $s = 1$ and opposite spins has $s = 0$. In later case spins are paired or cancelled. Atoms with $s = 0$ are slightly repelled by a magnet and atoms where $s \neq 0$ are attracted by the magnet (unpaired). If number of unpaired electrons are more then the atoms are strongly attracted by a magnet and are called paramagnetic. The atoms with paired electrons are called diamagnetic.

Pauli principle says that the two electrons must be spin paired in an orbital to remain at low energy. For parallel spins the repulsive forces come into play which means unacceptable state or high energy state.

A maximum of two electrons with opposite spins can be accommodated in a sub-shell or an orbital indicated by the sign ↑↓ or $\pm\frac{1}{2}$.

Dual nature of an electron

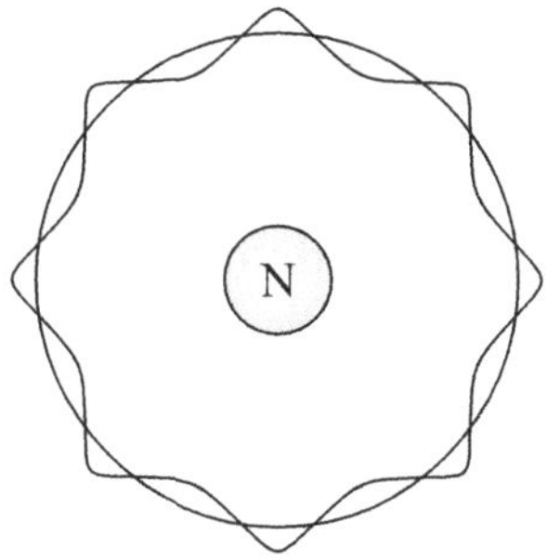

It is assumed that electron is a material particle revolving round the nucleus in circular orbits. De-broglie pointed out in 1924 that electron behaves as a particle as well as a wave like light. Its exact behaviour can be derived after the experiment but cannot be predicted before. For calculation the wave length λ is associated with the electron. Let us assume particle with mass *m* going round the nucleus with velocity *c*.

With the help of Plank's equation $E = h\nu$ and Einstein's equation $E = mc^2$ we conclude that

$$\lambda = \frac{h}{mv} = \frac{h}{\text{Momentum of electron}} \quad \text{or} \quad \lambda = \frac{h}{p}.$$

De-Broglie's equation was later verified by Davison and Germer's experiment in 1927. They focussed a beam of electrons on the nickel surface. The pattern obtained was similar to X-ray diffraction. This represents the dual nature.

Heisenberg's Uncertainty Principle

According to Heisenberg it is impossible to measure simultaneously and accurately both the position and momentum (mv) of a fast moving sub-atomic particle like electron, proton etc.

Mathematically it can be expressed by:

$$\Delta x \times \Delta p \geq \frac{h}{4\pi}$$

where Δp is uncertainty in momentum, h in Planck's constant sign $\geq$ indicates that the product of $\Delta x \cdot \Delta p$ is either equal or greater than $\frac{h}{4\pi}$ but never less than $\frac{h}{4\pi}$. Δx and Δp are inversely related to each other. This implies that if Δx is determined accurately than uncertainty in Δp will be large and vice-versa.

Generally the relation is expressed as:

$$\Delta x \cdot \Delta p = \frac{h}{4\pi}$$

or

$$\Delta x \cdot m\Delta v = \frac{h}{4\pi}$$

Schrodinger's Wave Equation

De-Broglie unfolded both the material particle as well as wave nature of an electron. To describe the wave motion of an electron there must exist a wave equation. Schrodinger proposed a wave equation describing the wave motion of an electron wave propagating in three dimensions (x, y and z axes) in space.

The derivation of the Schrodinger wave equation is described below:

For a vibrating string of wave length λ whose amplitude along x axis is given by $f(x)$ is shown by:

$$\frac{d^2 f(x)}{dx^2} = -\frac{4\pi^2}{\lambda z} f(x)$$

If electron is taken as a wave moving in only one direction.

Then, $$\frac{d^2\psi}{dx^2} = -\frac{4\pi^2}{\lambda^2}\psi$$

An electron may move along x, y and z axes simplifies to:

$$\frac{\partial^2\psi}{dx^2} + \frac{\partial^2\psi}{dy^2} + \frac{\partial^2\psi}{dz^2} = -\frac{4\pi^2}{\lambda^2}\psi$$

Using the symbol ∇ for three partial differentials

$$\nabla^2\psi = -\frac{4\pi^2}{\lambda^2}\psi$$

According to de-Broglie's relation:

$$\lambda = \frac{h}{mv}$$

where h is Planck's constant, m is the mass and v is the velocity of an electron, hence:

$$\nabla^2\psi = -\frac{4\pi^2 m^2 v^2}{h^2}\psi$$

or $$\nabla^2\psi + \frac{4\pi^2 m^2 v^2}{h^2}\psi = 0 \qquad \ldots(a)$$

Total energy E of the system is made of kinetic energy 'K' and potential energy V. *i.e.*

$$E = K + V$$

or $$K = E - V$$

but kinetic energy is $\frac{1}{2}mv^2$ so

$$\frac{1}{2}mv^2 = E - V$$

or

$$v^2 = \frac{2}{m}(E - V)$$

Substituting v^2 in the equation A above, Schordinger equation becomes:

$$\nabla^2\psi + \frac{8\pi^2 m}{h^2}(E - V)\psi = 0$$

The above equation will be applicable if:

(*a*) ψ is continuous.

(*b*) ψ must be finite.

(*c*) ψ must have single value.

(*d*) The probability of finding the electron in space from $+\infty$ and $-\infty$ is equal to one.

Therefore, probability of finding electron at a point x, y, z is ψ^2;

$$\int_{-\infty}^{+\infty} \psi^2\, dx\, dy\, dz = 1$$

The value of ψ which satisfy above four, conditions are called Eigen-functions.

Significance of ψ and ψ^2

Heisenberg gave us the idea that the exact location of electron from the nucleus is uncertain therefore the probable region in which we can find an electron became the next option. The probability of finding an electron in a region round the nucleus is proportional to ψ^2. Hence ψ^2 is a real quantity. The wave function ψ is an imaginary quantity. If a and b are real functions of the coordinates then probability $\psi^2 = (a + ib)^2$ or $a^2 + 2ab + b^2$ is an imaginary quantity, since probability is a real quantity then it must be equal to $\psi\psi^*$ and ψ^2 since ψ^* is a complex conjugate of ψ (*i.e.* $\psi^* = a - ib$). Under the situation:

$$\psi\psi^* = (a + ib)(a - ib)$$

$$= a^2 + b^2 \ldots \qquad \text{(realquantity)}$$

Thus conjugate complex ψ^* must be real function ψ hence probability ψ^2 is:

$$\psi\psi^* = \psi^2$$

or ψ^* and ψ are the same.

The probability of finding electron round the nucleus may be Angular and Radial.

Relative energies of principle quantum numbers and sub-levels (orbitals) in multi-electron atoms.

In constructing the poly electron atoms the energies of various orbitals belonging to principle quantum numbers have been found in the order of increasing energies *i.e.*

$1s < 2s < 2p < 3s < 3p < 4s < 3d < 4p < 5s < 4d < 5p < 6s < 4f < 5d < 6p$ … and so on.

This is called energy-level diagram. The electron has a tendency to fill the orbital with the lowest energy first and if that is already filled it goes to the orbital that is next higher in energy. Thus the orbitals are filled according to increasing energy order. The energy-level diagram as given below, is the best way to remember the relative energies of orbitals:

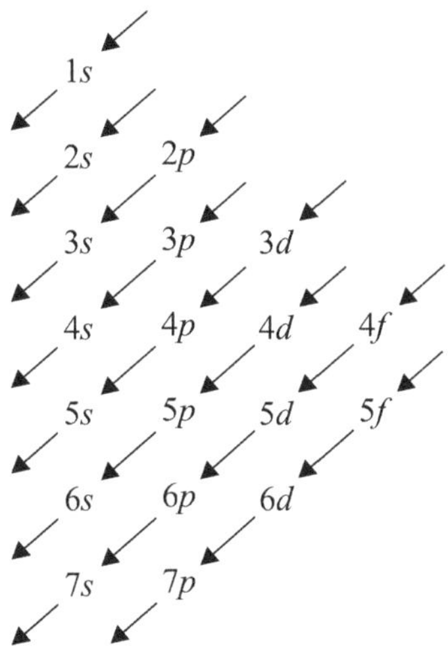

The rules governing filling of electrons in main shells and sub-shells are like this:

(*i*) The maximum electrons in principle quantum number are governed by the formula $2n^2$.

(*ii*) The maximum electrons in the sub-shells or orbitals *s, p, d, f* is equal to $2l(l + 1)$; $l = 0, 1, 2, 3, \ldots, n - 1$.

(*iii*) The building up of electrons or construction of atoms is according to the energy-level diagram as given in Figure in the increasing order of energy.

(*iv*) No more than two electrons can be present in an orbital according to pauling.

(*v*) **Hund's Rule:** Related energy levels of the same energy are first singly occupied before getting paired *i.e.* such an orbital must have maximum number of unpaired electrons. This rule gives maximum stability or lowest energy to half-filled and completely filled configuration *e.g.*

Half filled d-orbitals	↑	↑	↑	↑	↑	Extra stability
Completely filled d-orbitals	↑↓	↑↓	↑↓	↑↓	↑↓	Extra stability

Half filled

In half filled *d*-configuration lowest energy is because of maximum exchange forces.

Completely filled

In this configuration lowest energy is represented because of neutralisation of electron spins.

Electronic Configurations of Elements

Table 1.1 Ground State Electronic Configuration of the Elements

Period number	*Elements with atomic number*	*K* *n = 1*	*L* *n = 2*	*M* *n = 3*	*N* *n = 4*	*O* *n = 5*	*P* *n = 6*	*O* *n = 7*
		1*s*	2*s* 2*p*	3*s* 3*p* 3*d*	4*s* 4*p* 4*d* 4*f*	5*s* 5*p* 5*d* 5*f*	6*s* 6*p* 6*d* 6*f*	7*s* 7*p*
1	H_1	1						
	He_2	2						
	Li_3	2	1					
	Be_4	2	2					
2	B_5	2	2 1					
	C_6	2	2 2					
	N_7	2	2 3					
	O_8	2	2 4					
	F_9	2	2 5					
	Ne_{10}	2	2 6					
	Na_{11}	2	2 6	1				
	Mg_{12}	2	2 6	2				
	Al_{13}	2	2 6	2 1				
3	Si_{14}	2	2 6	2 2				
	P_{15}	2	2 6	2 3				
	S_{16}	2	2 6	2 4				
	Cl_{17}	2	2 6	2 5				
	Ar_{18}	2	2 6	2 6				
	K_{19}	2	2 6	2 6	1			
	Ca_{20}	2	2 6	2 6	2			
	SC_{21}	2	2 6	2 6 1	2			
	Ti_{22}	2	2 6	2 6 2	2			
	V_{23}	2	2 6	2 6 3	2			
	Cr_{24}	2	2 6	2 6 5	1			
	Mn_{25}	2	2 6	2 6 5	2			
	Fe_{26}	2	2 6	2 6 6	2			
4	CO_{27}	2	2 6	2 6 7	2			
	Ni_{28}	2	2 6	2 6 8	2			
	Cu_{29}	2	2 6	2 6 10	1			
	Zn_{30}	2	2 6	2 6 10	2			
	Ga_{31}	2	2 6	2 6 10	2 1			
	Ge_{32}	2	2 6	2 6 10	2 2			
	As_{33}	2	2 6	2 6 10	2 3			
	Se_{34}	2	2 6	2 6 10	2 4			
	Br_{35}	2	2 6	2 6 10	2 5			
	Kr_{36}	2	2 6	2 6 10	2 6			
	Rb_{37}	2	2 6	2 6 10	2 6	1		
	Sr_{38}	2	2 6	2 6 10	2 6 1	2		
	Y_{39}	2	2 6	2 6 10	2 6 2	2		
	Zr_{40}	2	2 6	2 6 10	2 6 4	2		
	Nb_{41}	2	2 6	2 6 10	2 6 5	1		
	Mo_{42}	2	2 6	2 6 10	2 6 5	1		
	Te_{43}	2	2 6	2 6 10	2 6 7	2		
5	Ru_{44}	2	2 6	2 6 10	2 6 8	1		
	Rh_{45}	2	2 6	2 6 10	2 6 10	1		
	Pd_{46}	2	2 6	2 6 10	2 6 10	0		
	Ag_{47}	2	2 6	2 6 10	2 6 10	1		
	Cd_{48}	2	2 6	2 6 10	2 6 10	2		
	In_{49}	2	2 6	2 6 10	2 6 10	2 1		
	Sn_{50}	2	2 6	2 6 10	2 6 10	2 2		
	Sb_{51}	2	2 6	2 6 10	2 6 10	2 3		
	Tc_{52}	2	2 6	2 6 10	2 6 10	2 4		
	I_{53}	2	2 6	2 6 10	2 6 10	2 5		
	Xe_{54}	2	2 6	2 6 10	2 6 10	2 6		

Period number	Elements with atomic number	*K* *n* = *1*	*L* *n* = *2*	*M* *n* = *3*	*N* *n* = *4*	*O* *n* = *5*	*P* *n* = *6*	*O* *n* = *7*
		1*s*	2*s* 2*p*	3*s* 3*p* 3*d*	4*s* 4*p* 4*d* 4*f*	5*s* 5*p* 5*d* 5*f*	6*s* 6*p* 6*d* 6*f*	7*s* 7*p*
	Cs_{55}	2	2 6	2 6 10	2 6 10	2 6	1	
	Ba_{56}	2	2 6	2 6 10	2 6 10	2 6	2	
	La_{57}	2	2 6	2 6 10	2 6 10	2 6 1	2	
	Ce_{58}	2	2 6	2 6 10	2 6 10 2	2 6 0	2	
	Pr_{59}	2	2 6	2 6 10	2 6 10 3	2 6 0	2	
	Nd_{60}	2	2 6	2 6 10	2 6 10 4	2 6 0	2	
	Pm_{61}	2	2 6	2 6 10	2 6 10 5	2 6 0	2	
	Sm_{62}	2	2 6	2 6 10	2 6 10 6	2 6 0	2	
	Eu_{63}	2	2 6	2 6 10	2 6 10 7	2 6 0	2	
	Gd_{64}	2	2 6	2 6 10	2 6 10 7	2 6 0	2	
	Tb_{65}	2	2 6	2 6 10	2 6 10 10	2 6 1	2	
	Dy_{66}	2	2 6	2 6 10	2 6 10 11	2 6 0	2	
	Ho_{67}	2	2 6	2 6 10	2 6 10 12	2 6 0	2	
	Er_{68}	2	2 6	2 6 10	2 6 10 13	2 6 0	2	
	Tm_{69}	2	2 6	2 6 10	2 6 10 14	2 6 0	2	
	Yb_{70}	2	2 6	2 6 10	2 6 10 14	2 6 0	2	
	Lu_{71}	2	2 6	2 6 10	2 6 10 14	2 6 1	2	
	Hf_{72}	2	2 6	2 6 10	2 6 10 14	2 6 2	2	
	Ta_{73}	2	2 6	2 6 10	2 6 10 14	2 6 3	2	
6	W_{74}	2	2 6	2 6 10	2 6 10 14	2 6 4	2	
	Re_{75}	2	2 6	2 6 10	2 6 10 14	2 6 5	2	
	Os_{76}	2	2 6	2 6 10	2 6 10 14	2 6 6	2	
	Ir_{77}	2	2 6	2 6 10	2 6 10 14	2 6 7	2	
	Pt_{78}	2	2 6	2 6 10	2 6 10 14	2 6 9	1	
	Au_{79}	2	2 6	2 6 10	2 6 10 14	2 6 10	1	
	Hg_{80}	2	2 6	2 6 10	2 6 10 14	2 6 10	2	
	Ti_{81}	2	2 6	2 6 10	2 6 10 14	2 6 10	2 1	
	Pb_{82}	2	2 6	2 6 10	2 6 10 14	2 6 10	2 2	
	Bi_{83}	2	2 6	2 6 10	2 6 10 14	2 6 10	2 3	
	Po_{84}	2	2 6	2 6 10	2 6 10 14	2 6 10	2 4	
	At_{85}	2	2 6	2 6 10	2 6 10 14	2 6 10	2 5	
	Rn_{86}	2	2 6	2 6 10	2 6 10 14	2 6 10	2 6	
	Fr_{87}	2	2 6	2 6 10	2 6 10 14	2 6 10	2 6	1
	Ra_{88}	2	2 6	2 6 10	2 6 10 14	2 6 10	2 6	2
	Ac_{89}	2	2 6	2 6 10	2 6 10 14	2 6 10	2 6 1	2
	Th_{90}	2	2 6	2 6 10	2 6 10 14	2 6 10 0	2 6 2	2
	Pa_{91}	2	2 6	2 6 10	2 6 10 14	2 6 10 2	2 6 1	2
	U_{92}	2	2 6	2 6 10	2 6 10 14	2 6 10 3	2 6 1	2
7	Np_{93}	2	2 6	2 6 10	2 6 10 14	2 6 10 4	2 6 1	2
	Pu_{94}	2	2 6	2 6 10	2 6 10 14	2 6 10 6	2 6 0	2
	Am_{95}	2	2 6	2 6 10	2 6 10 14	2 6 10 7	2 6 0	2
	Cm_{96}	2	2 6	2 6 10	2 6 10 14	2 6 10 7	2 6 1	2
	Bk_{97}	2	2 6	2 6 10	2 6 10 14	2 6 10 9	2 6 0	2
	Cf_{98}	2	2 6	2 6 10	2 6 10 14	2 6 10 10	2 6 0	2
	Es_{99}	2	2 6	2 6 10	2 6 10 14	2 6 10 11	2 6 0	2
	Fm_{100}	2	2 6	2 6 10	2 6 10 14	2 6 10 12	2 6 0	2
	Md_{101}	2	2 6	2 6 10	2 6 10 14	2 6 10 13	2 6 0	2
	No_{102}	2	2 6	2 6 10	2 6 10 14	2 6 10 14	2 6 0	2
	Lw_{103}	2	2 6	2 6 10	2 6 10 14	2 6 10 14	2 6 1	2
	Ku_{104}					14		

EXERCISES

1. What is the difference between an orbit and an orbital?
2. What is the physical significance of ψ and ψ^2?
3. Explain radial wave functions of hydrogen atom.
4. What is the justification of Schrodinger wave equation?

5. Give the shapes of *d*-sub-shell orbitals state the maximum electron capacity of this shell and why?
6. How *n*, *l*, *m* explain the relative energies of the subshell and give the number of orbitals in a given sub-shell?
7. What do you understand by the effective nuclear charge?
8. Write the wave equation based on the assumption that electron is standing wave. How its wavelength 'λ' is given by de Broglie relationship?
9. (*a*) Give the value of all the four quantum numbers for the two electrons in the He configuration.

   (*b*) Give the value of all the four quantum number for 3*d*-electrons.
10. Write the configuration of lanthanum ($z = 57$), Cerium ($z = 58$) and Gold (79)
11. Discuss with example the limitation of Aufbau principle.
12. Explain, giving examples, the factors governing extra stability of half-filled and fully filled sub-shells.
13. Define electro negativity and discuss Pauling's scale of electronegativity.
14. Calculate electronegativity of C in C—H bond if $E_{C—H}$, $E_{H—H}$ and $E_{C—C}$ are 98.8, 104, 83 Kcal/mol respectively.
15. What is Pauli's exclusion principle? Discuss its application.
16. (*a*) Explain why 4*s* subshell is filled first than 3*d* while 2*p* subshell is filled first than 3p.

    (*b*) Write the electric configuration of Pd ($z = 46$), Cu^{2+} ($z = 29$) ion.
17. Discuss in brief the important factors affecting electronegativity of an atom.
18. Give the possible values of quantum number of an electron in 4*d* orbitals.
19. Write the ground state configuration of the elements Cd ($z = 48$), Sn ($z = 50$), Cu ($z = 29$) and I ($z = 53$)
20. Discuss the factors involved for the low reactivity of noble gases.
21. Arrange the following orbitals in the order of increasing energy for a multi-electron atom.

    1*s*, 3*s*, 5*s*, 2*p*, 4*p*, 5*p*, 3*d*, 4*d* and 4*f*
22. In forming Fe^{2+} from Fe ($z = 26$) the electrons are removed from 4d and not from 3d orbitals. Explain.
23. Which has the higher first ionization energy

    (*i*) Li or Cs^{+} (*ii*) F or Br (*iii*) Sc or Cu (*iv*) Cu or Pt
24. Determine the number of unpaired electrons in the ground state configuration of the following ions:

    (*i*) Ti^{3+} (*ii*) Mn^{2+} (*ii*) Mn^{2+} (*iii*) Cu^{2+} (*iv*) Pd^{2+} (*v*) Cd^{3+}

25. Which has the higher electron affinity.

    Li or Cs; Li or F; Cs or F; F or Cl; Cl or Br; O or S; S or Se

26. What are Lyman, Balmer and Paschen series in the atomic spectra of hydrogen?

27. Which of the following does the Bohr theory apply to?

    (*a*) H (*b*) H^+ (*c*) He (*d*) He^+ (*e*) Li

    (*f*) Li^+ (*g*) Li^{2+} (*h*) Be (*i*) Be^+ (*j*) Be^{2+}

28. Write the electronic configuration for the elements with atomic number 6, 11, 17 and 25 and name there elements.

29. Give the name and symbol for each atom which have the ground state configuration in the outer shells.

    (*a*) $2s^2$ (*b*) $3s^2 3p^5$ (*c*) $3s^2 3p^6 4s^2$

    (*d*) $3s^2 3p^6 3d^6 4s^2$ (*e*) $5s^2 5p^2$ (*f*) $5s^2 5p^6$

30. Calculate the radii of the first three Bohr orbits for hydrogen.

    Planck's constant $h = 6.36262 \times 10^{-24}$ Joules

    Mass of electron 9.09×10^{-34} kg

    Charge of an electron 1.6×10^{-19} C

    Permittivity if vacuum 8.8×10^{-12} $kg^{-1}m^{-1}A^2$

2

Periodic Table and Periodic Properties (Periodicity)

PERIODIC TABLE

Though to date more than 106 elements are known but only 106 have been authenticated. It is almost impossible to study and remember the properties and behaviour of all the elements individually. Therefore it become's imperative to classify all the 106 elements in order to systematise their properties. Attempts were made by various chemists to classify them. In 1861 Mendeleev made a serious attempt which was noted by the chemists all over the world. He gave the famous Periodic Law which states that physical and chemical properties of the elements are a periodic function of their atomic weights. This means that if the elements are arranged according to their rising atomic weights then elements with similar properties recur after a regular interval. Accordingly every 1st and 8th elements show similar properties. Some anomalies cropped up after Mendeleev in the arrangement of these pairs such as Te and I ; K and Ar and Ni and Co. These pairs are not arranged according to rising atomic weights. If atomic weights are the basis then isotopes of elements and rare-earths must find a place in the Periodic Table which is not so.

Moseley in 1911 proposed Modern Periodic Law which was able to remove some of the defects and anomalies of the Mendeleev's Table. Moseley proposed the change in the basis of classification of elements. His Modern Periodic Law states that physical and chemical properties of the elements are the periodic function of their atomic numbers. Atomic number is a more fundamental property of elements as it deals with sub-atomic particles like protons, neutrons and electrons therefore, some of the defects of Mendeleev's Table automatically disappeared.

A more scientific table was proposed later which had minimal defects or anomalies. The new Periodic Table was called Extended or long-form Table. It is also known by the name Bohr's Table. This table has horizontal columns known as periods and vertical columns called groups. All the elements in the same group show similar or graded properties. It is based on Modern Periodic Law. It is most widely used Periodic Table. It was evolved by various chemists over a period of years therefore no single person can be given the credit.

Long Form Periodic Table

This table is an extended form of the Mendeleev's Table. Sub-group A and sub-group B elements in the Mendeleev's Table find a separate place in this table. Therefore group I to VII have been elongated to 14 groups ; Fe, Co, Ni; Ru, Rh, Pd; and Os Ir, Pt have been split into three groups. The noble gases have been assigned group zero. In all, there are 18 groups in the Long-Form table.

All the elements are divided into four categories according to the following portions :

1. **Left Portion** comprise of group I_A and II_A of the Mendeleev's Table. These elements are extremely positive in character. (Alkali and alkaline earth metals).
2. **Right Portion :** Elements in the category include III_A, IV_A, V_A, VI_A, VII_A and zero group or noble gases. This portion can have metals, metalloids and non-metals including noble gases on extreme right. Except these noble gases, elements are strongly electro negative elements are on left of noble gases. Elements in the gas form are essentially non-metals.

Middle Stretch : This portion houses III_B, IV_B, V_B, VI_B and VII_B elements of the Mendeleev's Table. These elements are called Transition elements. There are 4 transition series called *3d*, 4d, *5d* and *6d* series of elements. Each series contains 10 elements except *6d* series as *6d* series is not complete with elements *3d*, *4d* and *5d* series contain 10 elements each. New elements as and when discovered will fill the empty space provided in *6d* elements series.

Bottom Portion : It has two series of elements each series *4f* and *5f* contains 14 elements. They are called Inner Transition elements. These elements are very closely related to each other but have very little resemblence with rest of the elements in the periodic table. Therefore they cannot be placed in the main body of the Periodic Table as they do not resemble with any of these elements. Hence the best place for them is at the bottom of the table separated from the main body of table. *4f* elements are also called lathanides after the parent Lanthanum and *5f* elements Actinides after the parent Actinium. There are 28 elements distributed between *4f* and *5f* series (14 elements in each). They are also known as Inner transition element as they are inner to the transition elements.

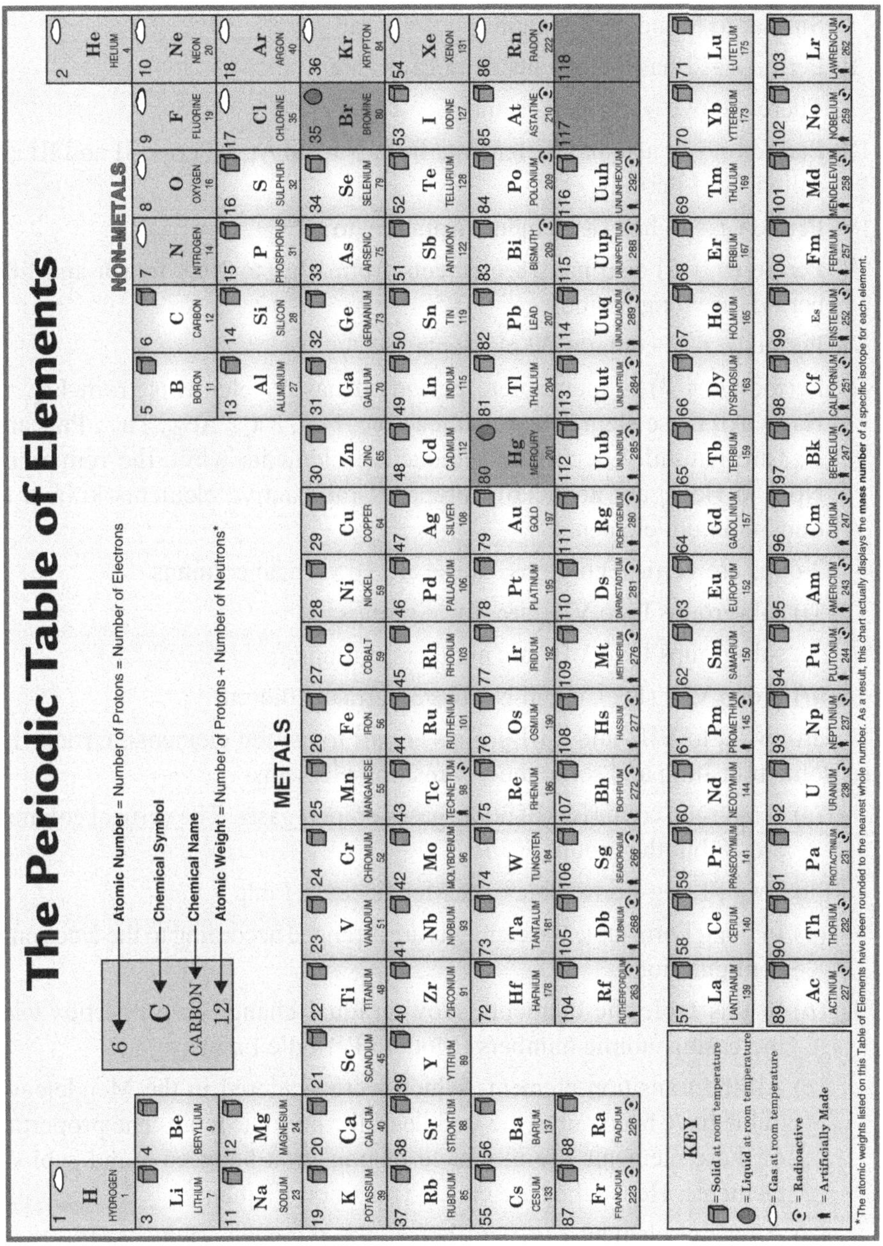
The Periodic Table of Elements
6 C CARBON 12
Atomic Number = Number of Protons = Number of Electrons
Chemical Symbol
Chemical Name
Atomic Weight = Number of Protons + Number of Neutrons*
METALS
NON-METALS
1 H HYDROGEN 1
2 He HELIUM 4
3 Li LITHIUM 7
4 Be BERYLLIUM 9
5 B BORON 11
6 C CARBON 12
7 N NITROGEN 14
8 O OXYGEN 16
9 F FLUORINE 19
10 Ne NEON 20
11 Na SODIUM 23
12 Mg MAGNESIUM 24
13 Al ALUMINIUM 27
14 Si SILICON 28
15 P PHOSPHORUS 31
16 S SULPHUR 32
17 Cl CHLORINE 35
18 Ar ARGON 40
19 K POTASSIUM 39
20 Ca CALCIUM 40
21 Sc SCANDIUM 45
22 Ti TITANIUM 48
23 V VANADIUM 51
24 Cr CHROMIUM 52
25 Mn MANGANESE 55
26 Fe IRON 56
27 Co COBALT 59
28 Ni NICKEL 59
29 Cu COPPER 64
30 Zn ZINC 65
31 Ga GALLIUM 70
32 Ge GERMANIUM 73
33 As ARSENIC 75
34 Se SELENIUM 79
35 Br BROMINE 80
36 Kr KRYPTON 84
37 Rb RUBIDIUM 85
38 Sr STRONTIUM 88
39 Y YTTRIUM 89
40 Zr ZIRCONIUM 91
41 Nb NIOBIUM 93
42 Mo MOLYBDENUM 96
43 Tc TECHNETIUM 98
44 Ru RUTHENIUM 101
45 Rh RHODIUM 103
46 Pd PALLADIUM 106
47 Ag SILVER 108
48 Cd CADMIUM 112
49 In INDIUM 115
50 Sn TIN 119
51 Sb ANTIMONY 122
52 Te TELLURIUM 128
53 I IODINE 127
54 Xe XENON 131
55 Cs CESIUM 133
56 Ba BARIUM 137
72 Hf HAFNIUM 178
73 Ta TANTALUM 181
74 W TUNGSTEN 184
75 Re RHENIUM 186
76 Os OSMIUM 190
77 Ir IRIDIUM 192
78 Pt PLATINUM 195
79 Au GOLD 197
80 Hg MERCURY 201
81 Tl THALLIUM 204
82 Pb LEAD 207
83 Bi BISMUTH 209
84 Po POLONIUM 209
85 At ASTATINE 210
86 Rn RADON 222
87 Fr FRANCIUM 223
88 Ra RADIUM 226
104 Rf RUTHERFORDIUM 263
105 Db DUBNIUM 268
106 Sg SEABORGIUM 266
107 Bh BOHRIUM 272
108 Hs HASSIUM 277
109 Mt MEITNERIUM 276
110 Ds DARMSTADTIUM 281
111 Rg ROENTGENIUM 280
112 Uub UNUNBIUM 285
113 Uut UNUNTRIUM 284
114 Uuq UNUNQUADIUM 289
115 Uup UNUNPENTIUM 288
116 Uuh UNUNHEXIUM 292
117
118
57 La LANTHANUM 139
58 Ce CERIUM 140
59 Pr PRASEODYMIUM 141
60 Nd NEODYMIUM 144
61 Pm PROMETHIUM 145
62 Sm SAMARIUM 150
63 Eu EUROPIUM 152
64 Gd GADOLINIUM 157
65 Tb TERBIUM 159
66 Dy DYSPROSIUM 163
67 Ho HOLMIUM 165
68 Er ERBIUM 167
69 Tm THULIUM 169
70 Yb YTTERBIUM 173
71 Lu LUTETIUM 175
89 Ac ACTINIUM 227
90 Th THORIUM 232
91 Pa PROTACTINIUM 231
92 U URANIUM 238
93 Np NEPTUNIUM 237
94 Pu PLUTONIUM 244
95 Am AMERICIUM 243
96 Cm CURIUM 247
97 Bk BERKELIUM 247
98 Cf CALIFORNIUM 251
99 Es EINSTEINIUM 252
100 Fm FERMIUM 257
101 Md MENDELEVIUM 258
102 No NOBELIUM 259
103 Lr LAWRENCIUM 262
KEY
= Solid at room temperature
= Liquid at room temperature
= Gas at room temperature
= Radioactive
= Artificially Made
* The atomic weights listed on this Table of Elements have been rounded to the nearest whole number. As a result, this chart actually displays the mass number of a specific isotope for each element.

Long-Form Table

The periods are horizontal columns.

I Period $n = 1$ has 2 elements *i.e.* H_1 and He_2

Similar to Mendeleev's Table

II Period $n = 2$ has 8 elements from Li_3 to Ne_{10}

There are two such shells s and p.

III Period ($n = 3$) also has 8 elements from Na_{11} to Ar_{18}. Period II and III are called short Periods.

IV Period ($n = 4$) has 18 elements from K_{19} to Kr_{36}

V Period ($n = 5$) also has 18 elements from Rh_{37} to Xe_{54} fourth and fifth periods are long periods.

VI Period ($n = 6$) contains 32 elements

VII Period ($n = 7$) is incomplete period. It has 19 elements from Fr_{87} to Ha_{105}. All these elements are radioactive. Fr_{87}, Ra_{88}, Ac_{89}, Th_{90}, Pa_{91} and U_{92} are naturally occurring radioactive elements while the remaining Np_{93} to Ha_{105} are artificially prepared radioactive elements known as trans-uranium elements.

Groups are vertical columns. These are 18 vertical columns

(*i*) sub-groups I_A to VII_A are seven groups

(*ii*) sub-groups I_B to VII_B are another 7 groups.

(*iii*) group VIII (Fe, Co and Ni) has 3 vertical columns.

Groups I_B to VII_B and VIII groups are all transition elements. Group III_B has some transition and some inner transition elements.

(*iv*) zero group comprises of noble gases or inert gases. The vertical columns are 18 but the groups are 16.

Plusses of Long-Form table over Mendeleev's Table

(*a*) In Long Form table the elements are arranged according to the electronic configuration.

(*b*) In this table the elements show gradual change in properties with increasing atomic numbers (Modern Periodic Law).

(*c*) All the transition elements which were scattered in the Mendeleev's Table have been put in the middle of the periodic table. The properties of these elements are in a state of transition between *s* and *p* block elements. Hence they are called Transition Elements.

(*d*) After the element La_{57} and before the transition element Hf_{72} there is interruption by 14 elements called Inner transition elements or Lanthanides. There is a similar interruption after the element $Actinium_{89}$

and before the element Ku (104) *i.e.* elements with atomic number 90 to 103 (14 elements) known as Inner Transition or Actinides. The two sets of 14 elements (28 elements) have been placed at the bottom of the table. These set of elements show no resemblance with any of the elements in the main body of the Periodic Table. These elements are very closely related to each other in properties. The best option for these elements is the bottom position of the table.

(*e*) In the Long Form table sub-group A and sub-group B elements having dissimilar properties have been separated *i.e.* Alkali metals belonging to I_A and coinage metals Cu, Ag and Au belonging to I_B find separate place or groups.

(*f*) Broadly it is an attempt to separate metals from non-metals. Elements on the left of the table are all metals and those on the right of the table are non-metals. Elements I_A, II_A, III_B to VII_B, VIII, I_B and II_B represent all metals on the right III_A to VII_A and noble gases are non-metals although they do contain some metalloids and metals.

The thick line shown in the Periodic Table shows separation of metals from non-metals.

(*g*) The distribution of elements in the Long Form table is easy to remember.

(*h*) The advantage of the Long Form table lies with the fact that all the hundred odd elements are divided broadly into four main blocks of elements *i.e.*, *s*-block, *p*-block, *d*-block and *f*-block of elements. Along a period *s*-block has a capacity of 2 elements *p*-block 6 elements *d* had 10 elements and *f*-block has 14 elements depending on the *s*, *p*, *d* and *f* orbitals to hold maximum electrons.

Defects in the Long Form Table

(*i*) **Position of Hydrogen.** The position of hydrogen is no better than the Mendeleev's table. Only on the basis of one electron in the valency shell ns' it is placed on top of Alkali metals but the properties of hydrogen and alkali metals are dissimilar.

(*ii*) Position of He Helium. He has the configuration $1s^2$ similar to alkaline-earth metals but its properties are not similar. Noble gases ns^2np^6 and He $1s^2$ show similar properties but their configurations do not match as except He, all other noble gases have ns^2np^6.

(*iii*) Position of lanthanides (4*f*) and configuration. Actinides (5*f*) in this table shows no improvement, they find no place in the main body of the table.

(*iv*) The Periodic Table does not show detailed and exact configuration of all the elements.

Classification of elements on the basis of electronic configuration

This classification divides the elements in *s*, *p*, *d* and *f*-blocks depending upon the electrons present in the atomic orbital of the outermost shell. These electrons are named differentiating electrons.

s-block elements. These elements contain electrons in the outermost s-orbital. These elements are called alkali and alkaline-earth metals and they hold ns^1 and ns^2 electrons in the valency shell orbitals. These elements occupy extreme left position in the periodic table.

***p*-block elements:** The elements have p-orbitals in the process of filling np^1 to np^6. These elements comprise of III_A (Boron family), IV_A (Carbon family), V_A (nitrogen family), VI_A (oxygen family), VII_A (Halogen family) and noble gases. The differentiating or extra electron enters 2*P*, 3*P*, 4*P*, 5*P* and 6*P* orbitals. The electronic configuration of atoms in these elements $ns^2\ np^{1\text{-}6}$. They are placed at the extreme right of the table. They compose of metals, metalloids and non-metals.

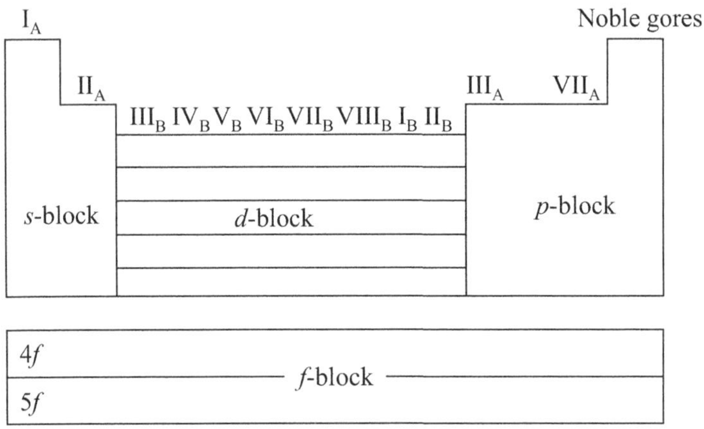

***d*-block elements.** These elements have the defining electron entering $(n-1)$ *d* orbital. They are called *d*-block elements and are represented by the general formula $(n-1)\ d^{1-10}\ ns^{0,1,2}$. They find the place in the middle of the periodic table flanked by *s*-block elements of their left and *p*-block elements on their right. Their properties are in a state of transition from *s*-block to *p*-block elements. The transition elements are classified in 3*d*, 4*d*, 5*d* and 6*d*-series of elements. Each series contains 10 elements except the 6*d*-series which is incomplete. The new elements as and when discovered are expected to be fill the vacant spaces left in the 6*d*-series.

3*d*-series starts from element Sc_{21} and ends with Zn_{30}. It includes all the elements such as Sc_{21}, Ti_{22}, V_{23}, Cr_{24}, Mn_{25}, Fe_{26}, Co_{27}, Ni_{28}, Cu_{29} and Zn_{30}. The elements Cr and Mn have half-filled d-orbital configuration *i.e.* d^5. In element Cr one electron shifts from the outer most *s*-orbital to inner *d*-orbital to make it half-filled (d^5).

4*d*-series of element contain 10 elements such as Y_{39}, Zr_{40}, Nb_{41}, Mo_{42}, Tc_{43}, Ru_{44}, Rh_{45}, Pd_{46}, Ag_{47} and Cd_{48}. Mo and Tc show half-filled configuration (d^5).

5*d*-series contain 10 elements of the type La_{57}, Hf_{72}, Ta_{73}, W_{74}, Re_{75}, Os_{76}, Tr_{77}, Pt_{78}, Au_{79}, Hg_{80}. This series shows a similar bias for half-filled configuration.

6*d*-series is incomplete. Presently it contains these elements Ac_{89}, Ku_{104}, Ha_{105}. It is hoped that elements up to atomic numbers 106 to 112 will be part of the series.

***f*-block elements.** The elements in which the new incoming electron fills the $(n - 2)$ *f*-orbitals are called *f*-block elements. The general configuration of these elements in the ground state $(n-2)f^{0,1-14}\,(n-1)\,d^{0,1,2}\,ns^2$ *f*-block elements are further divided into two series 4*f* and 5*f*-series. Each series comprises of 14 elements.

4*f* contains Ce_{58}, Pr_{59}, Nd_{60}, Pm_{61}, Sm_{62}, Eu_{63}, Gd_{64}, Tb_{65}, Dy_{66}, Ho_{67}, Er_{68}, Tm_{69}, Yb_{70}, Lu_{71}. Similarly 5*f*-series has 14 elements starting from Th_{90} to Lw_{103}. Both the series show a prejudice for half-filled configuration (f^7).

Bohr's classification. Bohr divided the elements into four classes depending upon the unfilled shells in the atom.

(*i*) **Normal or representative elements.** In these atoms the outer most nth shell is incompletely filled. The shells beneath are all fully filled. These elements have some metals, non-metals and metalloids. They are divided into two blocks namely *s*-block and *p*-block elements.

***s*-block elements.** The valency shell configuration for these elements is ns^1 or ns^2. These elements are placed on the extreme left of the periodic table. They consist of alkali metals (I_A group) and alkaline earth metals (II_A group). They are 13 in number. since the incoming electron goes to the outermost *s*-orbital therefore they are called *s*-block elements.

***p*-block elements.** These elements are represented by the configuration ns^2np^{1-5}. They are placed on the extreme right of the table. They occupy the space in between transition elements and noble gases. There are 25 elements in the group. Since the incoming electron enters *p*-orbitals hence they are known as *p*-block elements. Noble gases are also referred to as *p*-block elements.

(*ii*) **Noble gases:** In these elements both *s* and *p*-orbitals in the *n*th shell are completely filled. The general configuration for these elements is ns^2np^6 except He which is $1s^2$. On account of highly stable configuration these elements have very little tendency to react and form chemical compounds. They form fewer compounds.

(*iii*) **Transition elements :** These elements have the outermost and next to outermost shell (last but one shell) incomplete. They are placed between *s* and *p*-block elements in the periodic table. Atoms in these elements show a general configuration $(n-1)\,d^{1\text{-}10}\,ns^{0,1,2}$. In these the penultimate *d*-orbital is in the process of filling the extra electron therefore they are called *d*-block elements.

They comprise of four series 3*d*, 4*d*, 5*d* and 6*d*-series.

3*d*-series comprises of 10 elements from Sc_{21} to Zn_{30} in which the electron enters 3*d*-orbital 4*d*-series contain 10 elements from Y_{39} to Cd_{48}. The addition electron fills 4*d*-orbital.

5*d*-series also has ten elements. They include La_{57}, Hf_{72} to Hg_{80}. Here electron enters 5*d*-orbital 6*d*-series is incomplete. At present they contain only three elements *e.g.* Ac_{89}, Ku_{104} and Ha_{105}. It is hoped that elements from atomic numbers 106 to 112 will become members of this series.

Inner Transition or *f*-block Elements

The atoms of these elements have three outer most shells ns, $(n-1)d$ and $(n-2)f$-orbital. The general configuration is represented as $(n-2)f^{1-14}$, $(n-1)$ $d^{0or1or2}\,ns^2$. The electrons fill orbital inner to the transition elements therefore the name Inner Transition elements. This is also true that f has overlying *d* and *s*-orbitals. It is buried deep inside an atom. If we consider La which has the configuration $4f^0 5f^1 6s^2$ should not be member of 4*f*-block elements but owing to the similarity in physical and chemical properties with 4*f*-series it is justified to include La in the inner transition series. There are 14 elements in the series. They are called Lanthanides or Lanthanones. 5*f*-series also have 14 elements Th_{90} to Lw_{103}. They are called Actinides after the parent actinium. In these 5*f*-orbital is in the process of filling. They are also called Actinones.

In Bohr's classification the properties of the elements depend upon the complete configuration but the differentiating classification not only stresses on the role of entire configuration but emphasis is on the effect of successive electron addition in modifying the properties. This is now most commonly accepted that the differentiating electron configuration is more rational.

General Properties of the Elements

The elements consist of both physical and chemical properties. Some of the physical properties to an extent control the chemical properties. To understand the trend of physical properties in the periodic table *i.e.* in a group or a period it becomes important to understand shielding or screening effect which in turn controls most of the physical properties namely atomic size, ionization energy, electronegativity and electron affinity. Let us focus the mental faculties to clearly see through the property of shielding or screening.

Shielding or screening: The energy of electron in an atom is guided by the relation z^2/n^2 where 'z' is the atomic number and 'n' is the main quantum member. Atomic number increases more rapidly than the principal quantum number, in the building up of elements. The energy required to remove an electron from the outermost shell of an atom increases with rising atomic number. Hydrogen ($1s'$) and lithium ($2s'$) with atomic numbers $z = 1$ and $z = 3$ respectively are 1312 kJ mol^{-1} and 520 kJ mol^{-1}. The ionization energy of Li is lower because $2s^1$ is repelled by inner $1s^2$ so that so that the hold of nucleus over $2s^1$ decreases. In other words the inner electrons offer 'shielding' or 'screening' to the nucleus. In fact, the effective nuclear charge or the net charge operative on the $2s^1$ electron shows a decrease. The ionization energy of Li is lower because of two reasons (*i*) the radius of $2s$ electron is greater than that of $1s$ electron (*ii*) $2s^1$ electron is repelled by inner $1s^2$ electrons so that the $2s^1$ electron is easily removed. Looking at the phenomenon from another angle it means inner core electrons are 'shielding' or 'screening' the nucleus from the outermost electron. Therefore ionization energy is related to the effective nuclear charge between nucleus and the valency shall electron. The $2s^1$ electron penetrates the cage of $1s^2$ and experiences the nuclear charge equivalent to $z = 1.0$. If the nuclear charge is unshielded then the nuclear charge attraction is $z = 3.0$. Hence the effective nuclear charge z^* becomes somewhat less than the actual nuclear charge *i.e.*

$$z^* = z - s \qquad \text{('}s\text{' stands for shielding)}$$

The penetration power of orbitals depends upon the size of an orbital therefore, the order of penetration is $s > p > d > f$ in the broad approximation. As far as shielding effect is concerned s orbitals are very penetrating and are less shielded by inner orbital electrons, d and f orbitals are poorer in penetration and are less shielding. These rules are rough approximation of the extent of shielding. To estimate the shielding constant for an electron in ns and np orbital:

(*i*) Write the electronic configuration in the order. ($1s$), ($2s$, $2p$), ($3s$, $3p$, $3d$), ($4s$, $4p$), ($4d$) ($4f$), ($5s$, $5p$) etc.

(*ii*) Electrons to the right of (*ns*, *np*) do not contribute much to the shielding.

(*iii*) Electrons in the (*ns*, *np*) group shield the valency electron by 0.35 each.

(*iv*) Electrons in $n - 1$ shell shield by 0.85 each.

(*v*) Electrons in $n - 2$ or lower shell shield completely by 1.00 each.

(*vi*) Electrons lying to the left of *nd* or *nf* group contribute 1.00 to shielding

Periodic Properties (periodicity)

Most of the chemical properties owe their dependence on physical properties namely Atomic or ionic radii, ionization energy, electronegativity, electron affinity and metallic and non-metallic character of elements.

Atomic and Ionic radii: The radii is a measure of distance between the nucleus and the outer most shell of electrons belonging to atoms or ions. It is not possible to isolate an atom or an ion. The radii can be measured by taking two bonded atoms in the gaseous molecule or nuclei of two ions in crystalline state. The internuclear distance between two atoms or ions is measured. Half of the internuclear distance is called the radii. The internuclear and inter-ionic distance are determined by X-ray diffraction, electron diffraction, NMR spectroscopy techniques etc.

Atomic radii are further subdivided into (*a*) covalent radii (*b*) metallic radii and vander Waal's radii.

Covalent radii: It is half of the internuclear distance between two like atoms bonded by a single covalent bond.

It can be measured in a homonuclear diatomic molecule and a heteronuclear diatomic molecule

$$\gamma_A = d_{A-A}/2 \text{ in case of single bond}$$

or

$$\gamma_{A-B} = \gamma_A + \gamma_B$$

for ionic bond $A - B$ schomaker and stevenson (*i*) interatomic distance d_{A-B} for a bond $A - B$ which is sum of radii $\gamma_A + \gamma_B$ plus a correction factor equal to $-0.09\ (\chi_A - \chi_B)$.

$$d_{A-B} = \gamma_A + \gamma_B - 0.09(\chi_A - \chi_B)$$

$\chi_A - \chi_B$ is the difference between absolute values of electronegativity of A and B single covalent bond Radii of *s*- and *p*-block elements are given in the Table 2.1. Covalent bond radii may be single bond, double bond or triple bond respectively.

Table 2.1 Covalent Radii (Single-bond Covalent radii) for *s*- and *p*-block Elements (Å)

s-block elements				***p-block elements***			
I A	***II A***	***III A***	***IV A***	***V A***	***VI A***	***VII A***	***Zero***
H 0.32							**He 0.93**
Li 1.23	Be 0.90	B 0.82	C 0.77	N 0.75	O 0.73	F 0.72	Ne 0.71
Na 1.54	Mg 1.36	Al 1.18	Si 1.11	P 1.16	S 1.02	Cl 0.99	Ar 0.98
K 2.03	Ca 1.74	Ga 1.26	Ge 1.22	As 1.20	Se 1.16	Br 1.14	Kr 1.12
Rb 2.16	Sr 1.91	In 1.44	Sn 1.41	Sb 1.40	Te 1.36	I 1.33	Xe 1.31
Cs 2.35	Ba 1.98	Tl 1.48	Pb 1.47	Bi 1.46	Po 1.46	At (1.45)	Rn —
Fr —	Ra —						

Metallic radii: It is one half of the distance between the nuclei of two adjacent metal atoms in close packed crystal lattice.

Metallic bond controls the nearness between two metal atoms. It is not localised between the bonded atoms as in the case of a covalent bond. The metallic radii is smaller than VanderWaals radii since the metallic bond is stronger than van der Waal's forces.

Table 2.2 Metallic Radii of *s*- and *p*-block elements (in Å). These Values are for Coordination Number of 12.

s-block elements			***p-blocks elements***		
I A	***II A***	***III A***	***IV A***	***V A***	***VI A***
H —					
Li 1.55	Be 1.12	B 0.98	C 0.91	N 0.92	O —
Na 1.90	Mg 1.60	Al 1.43	Si 1.32	P 1.28	S 1.21
K 2.35	Ca 1.97	Ga 1.41	Ge 1.37	As 1.39	Se 1.40
Rb 2.48	Sr 2.15	In 1.66	Sn 1.62	Sb 1.59	Te 1.60
Cs 2.67	Ba 2.22	Tl 1.71	Pb 1.75	Bi 1.70	Po 1.76

Van der Waal's Radii: In the solid form non-metallic elements exist as aggregation of molecules. In this case the bonding is largely covalent. The individual molecules are held together by weak van der Waal's forces (residual forces). Half of the internuclear distance between two atoms within two molecules may be termed as van der Waal's radius. It is defined as half of the distance between the nuclei of two non-bonded neighbouring atoms of two adjacent molecules. Therefore, the radii are non-bonded. When two molecules come together without forming a chemical bond due to slight attraction between them due to distortion of their electron clouds then this force is called van der Waal's forces. These force are absent in the gaseous state best can be felt in liquid and crystal lattice.

Table 2.3 van der Waal's radii* of some elements (in Å)

V A	*VI A*	*VII A*	*Zero*
		H 1.2	He 1.20
N 1.5	O 1.40	F 1.36	Ne 1.60
P 1.9	S 1.85	Cl 1.80	Ar 1.91
As 2.0	Se 2.0	Br 1.95	Kr 2.0
Sb 2.2	Te 2.20	I 2.15	Xe 2.20

* The values of van der Waals radii are obtained from X-ray study of the elements in soild state.

The sizes of atoms or ions are related to the ionization energies and electron affinities. The effective nuclear charge pulls the electrons towards the centre of the atom. With the increase in the nuclear charge the size of any particular orbital decreases. On the other hand when the nuclear charge increases more electrons are added and due to mutual repulsion keep the outer orbitals large. The interaction of these two effects results in the gradual decrease in atomic size across the period Table 2.4 in case of non-polar covalent radii. There are other factors to be considered such as van der Waal's radius, chemical structure, physical state of molecules and polarity which contribute to the atomic size.

Table 2.4 Non-polar Covalent radii (pm)

1	2	3	4	5	6	7	8	9	10	11	12	13	14	15	16	17	18
H 32																	He 31
Li 123	Be 89											B 82	C 77	N 75	O 73	F 71	Ne 69
Na 154	Mg 136											Al 118	Si 111	P 106	S 102	Cl 99	Ar 98
K 203	Ca 174	Sc 144	Ti 132	V 122	Cr 118	Mn 117	Fe 117	Co 116	Ni 115	Cu 117	Zn 125	Ga 126	Ge 122	As 120	Se 117	Br 114	Kr 111
Rb 216	Sr 191	Y 162	Zr 145	Nb 134	Mo 130	Tc 127	Ru 125	Rh 125	Pd 128	Ag 134	Cd 148	In 144	Sn 140	Sb 140	Te 136	I 133	Xe 126
Cs 235	Ba 198	La 169	Hf 144	Ta 134	W 130	Re 128	Os 126	Ir 127	Pt 130	Au 134	Hg 149	Tl 148	Pb 147	Bi 146	Po (146)	Al (145)	Ra

Source: R.T. Sanderson, *Inorganic Chemistry,* Reinhold, New York, 1967, p. 74; and E.C.M. Chen, J.G. Dojahn, and W.E. Wentworth, *J.Phys. Chem. A,* 1997, *101,* 3088.

Ionic Radii

Similar problems arise in the case of the size of ions. The stable ions of different elements have different charges, different number of electrons and varying crystal structures of the compounds. Therefore it is difficult to find a reasonable comparison. According to Paulings approach ratio of radii of isoelectronic ions was found to be equal to their effective ions was found to be equal to their effective nuclear charges. Electron derived desired from X-ray data shows larger cations and smaller anions than previously found Table 2.5. Crystal radii is different from ionic radii by +14 pm for cations and –14 pm for anions.

Table 2.5 Crystal Radius and Nuclear Charge

Ion	*Protons*	*Electrons*	*Radius (pm)*
O^{2-}	8	10	126
F^{-}	9	10	119
Na^{+}	11	10	116
Mg^{2+}	12	10	86

Factors influencing ionic size are the coordination number of the ion, the covalent character, delocalisation of electrons and distortion of the crystal geometries.

Table 2.6 Crystal Radius and Total Number of Electrons

Ion	*Protons*	*Electrons*	*Radius (pm)*
O^{2-}	8	10	126
S^{2-}	16	18	170
Se^{2-}	34	36	184
Te^{2-}	52	54	207

Radius of anion is influenced by the size of cation, by the charge and the coordination number. The values in the Table show that anion are generally larger than cations with the similar charge *e.g.* F^{-} and Na^{+} (radius of F^{-} is 37% bigger). The radius decreases as nuclear charge increases for ions with same electronic structure such that O^{2-}, F^{-}, Na^{+} and Mg^{2+}. The charge is larger for cation with change in nuclear charge. In a vertical column or a group radius increases as Z (atomic number) increases because of more number of electrons for the ions. For the same element radius decreases with increasing charge on the cation. This trend is well illustrated in the Table 2.7.

Table 2.7 Crystal Radius and Ionic Charge

Ion	*Protons*	*Electrons*	*Radius (pm)*
Ti^{2+}	22	20	100
Ti^{3+}	22	19	81
Ti^{4+}	22	18	75

Ionization energy: Ionization energy or ionization potential is the energy to remove an electron from the outermost shell of a neutral gaseous atom or ion in its ground state of configuration *e.g.*

$Ag^{\circ}(g) \longrightarrow Ag^{+} + e$ Ionization energy $= \Delta U$

where n is zero it is called the first ionization energy, when $n = 1, 2$ it is second and third ionization energy. Ionization energy varies with different nuclei and different number of electrons. The first ionization energy values of the early elements in the periodic table are shown in Fig. 2.1. Along a period it increases with increase in nuclear charges. The ionization energy is measured in kJ mol^{-1}. A graph has been shown between electron and the shielded nucleus is almost

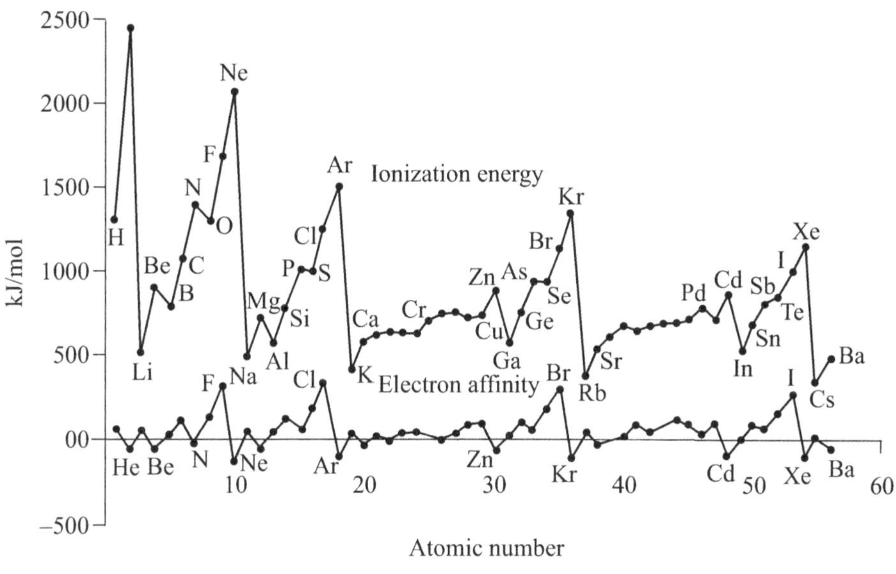

Fig. 2.1 Ionization energies and electron affinities. Ionization energy.

a straight line shown in the Fig. 2.1 Z^*/r where Z^* effective nuclear charge and r the radii. In Fig. 2.1 the expcrimcntal values show a break in the trend at boron electron in B goes to p-orbital therefore its electron density is farther away from the nucleus than the 2s electrons. Hence its ionization energy is smaller than that of $2s^2$ electrons of Be. At the stage of oxygen the fourth p-electron has a higher energy than indicated by the trend because it must be paired in the same p-orbital. The pairing energy or repulsion between two electrons in the same region would reduce the ionization energy. The transition elements have smaller differences in the ionization energies with a lower value for heavier atoms in the same family because of increased shielding by inner electrons and large difference between nucleus and the outer electrons.

At the start of the period larger decrease in the ionization energy occur because the next quantum number requires that the new s-electron have a higher energy. The maxima for the noble gases decrease with increase in Z because outer electrons are far removed from the nucleus in the heavier elements. The overall trends towards higher ionization energy from left to right in the period and lower ionization energy from top to bottom in a group is observed. The differences mentioned earlier are super imposed on these general changes.

Electron affinity: In ionization, energy is supplied to remove an electron from gaseous atom while in electron affinity energy is associated with the neutral atom. The amount of energy released when an electron is added to an isolated, neutral gaseous atom in its ground state of configuration is called electron affinity. It is represented by EA and is measured in electron volts eV or kilo calories per gram atom. It can be shown by the equation:

$$A(g) + e(g) \longrightarrow A^{-}(g) + \text{energy released}$$

The above process is exothermic since energy is released. It is given the negative sign (–). Actually it represents first electron affinity since it pertains to addition of one electron only. If two electrons are added, they do so in two steps.

$$A(g) + e^{-}(g) \longrightarrow A^{-}(g) + \text{energy released} \quad \ldots(I)$$

$$A^{-}(g) + e^{-}(g) + \text{Energy supplied} \longrightarrow A^{2-}(g) \quad \ldots(II)$$

In Step I energy is released but in Step II energy is supplied. In Step II due to electrostatic repulsion between two negative entities energy is given to convert $A^{-}(g)$ to $A^{2-}(g)$. This is energy absorbed and thus given a positive sign.

Table 2.8 Electron Affinities (in eV) of Some Elements. Values given in Brackets are Second Electron Affinity Values

I A	*II A*	*III A*	*IV A*	*V A*	*VI A*	*VII A*	*Zero*
H –0.75							He 0.0
Li –0.61	Be 0.0	B –0.30	C –1.25	N –1.20	O –1.48 + 7.3	F –3.6	Ne 0.0
Na –0.82	Mg 0.0	Al –0.52	Si –1.39	P –0.80	S –2.0 + 3.4	Cl –3.8	Ar 0.0
K –0.69						Br –3.5	Kr 0.0
						I –3.2	Xe 0.0

Electron affinity is influenced by (*i*) Atomic size, (*ii*) Effective nuclear charge and (*iii*) Shielding effect.

In general EA value decreases with increasing atomic radius and increases with decrease in screening effect by inner electrons. Besides, EA value also depends upon the type of orbital to which the electron is added *i.e.* *s*-, *p*-, *d*- and *f*-orbitals.

Periodic variations: (*i*) **Group:** On moving down the group there is gradual increase in atomic radius and consequent decrease in electron affinity *e.g.* EA of Cl > Br > I.

In halogens the energy of dissociation of molecules also comes into play which leads to unexpected values in certain elements.

Period: Electron affinity values generally increase when we move left to right in a period in the periodic Table.

This rule has some exceptions *e.g.* Be and Mg have zero *EA* value. Be has $1s^2$ and Mg has $3s^2$ electrons in the valence shell. Similarly noble gases are ns^2np^6 configuration. The *ns* and *np* orbitals are completely filled and have no tendency to accept an additional electron therefore *EA* value is zero.

Nitrogen and phosphorus have low *EA* values due to half filled configuration of $2p$ and $3p$ orbitals. This is very stable configuration and it is difficult to put an additional electron to it.

Effect of *EA* values on properties of elements:

(*i*) The higher *EA* value would mean a strong tendency to accept on electron or the element will have strong oxidizing power.

(*ii*) If ionization energy valves and *EA* values have greater difference than it is an ionic bond and if less it is covalent bond.

Electronegativity: In a covalent bond the electron pair between two similar or dissimilar atoms is shared mutually. In dissimilar atoms electronegativity is defined as the relative power of the atoms to pull the shared electron paired to itself. Electronegativity is symbollised by χ_A in case of atom A and χ_B in case of atom B.

It is measured by different scales proposed by different chemists. Some of these scales are Pauling's scale (1923), Mulliken's scale (1934). The Alfred Rochow (1958) etc. The most commonly used scale is Pauling's scale.

Table 2.9 Electronegativity Value of *s*-and *p*-block Elements as Determined by Pauling (Scale F = 4.0)

***s*-block elements**		***p*-block elements**				
IA	IIA	IIIA	IVA	VA	VIA	VIIA
H 2.1						
Li 1.0	Be 1.5	B 2.0	C 2.5	N 3.0	O 3.5	F4.0
Na 0.9	Mg 1.2	Al 1.5	Si 1.8	P 2.1	S 2.5	Cl 3.0
K 0.8	Ca 1.0	Ga 1.6	Ge 1.8	As 2.0	Se 2.4	Br 2.8
Rb 0.8	Sr 1.0	In 1.7	Sn 1.8	Sb 1.9	Te 2.1	I 2.5
Cs 0.7	Be 0.9	Ti 1.8	Pb 1.8	Bi 1.9	Po 2.0	At 2.2
Fr 0.7	Ra 0.9					

Periodic Variations

Period: In a period left to right in the periodic table there is increase in the electronegativity value due to increase in the nuclear charge so that electrons can be more tight. While moving across second period from Li to F electronegativities increase because electrons add to the same outer shell.

Second Period

Electronic configuration:

Li	Be	B	C	N	O	F
$2s^1$	$2s^2$	$2s^22p^1$	$2s^22p^2$	$2s^2p^3$	$2s^22p^4$	$2s^22p^5$

Electronegativity value:

1.0 1.5 2.0 2.5 3.0 3.5 4.0

$\longrightarrow$ increasing order

Group

On moving down the group in the periodic total the nuclear charge increase with increase in the number of electrons and the number of shells. The increasing atomic radii and the shielding effect more than compensate the increase in nuclear charge, consequently the element lower in the group has lesser electronegative value than the element upper in the group *i.e.*, small atoms attract electrons strongly than larger ones. Hence most electronegative element is top element in the group and less electronegative or electropositive is the bottom element in the group.

Ionisation Potential Versus Electron Affinity and Electronegativity: Higher ionisation potential means that it is very difficult to remove an electron from an atom while conversely it has greater tendency to attract an electron or higher electron affinity. Greater ionisation potential value will have greater electron affinity and electronegativity of an atom.

Non-metals for example halogens have high ionisation potential and electron affinity. This means non-metal atoms have little tendency to lose electrons. Contrary to this they have strong tendency to accept electrons are strongly electronegative. On the other hand metals lose electrons very easily. They are strongly electropositive. They have low electro affinity value *e.g.*, alkali and alkaline earth metals.

Metallic and Non-metallic Elements

The elements having higher value of electronegativity will be invariably non-metals while those with low value of electronegativity are metals. In a period electronegativity value shows an increase and in a group top to bottom shows a decrease. Thus trend is well illustrated in elements belonging to IV_A and V_A group.

Group IVA	*Group VA*
C $\longrightarrow$ non-metal	N $\longrightarrow$ non-metal
Si $\longrightarrow$ non-metal	P $\longrightarrow$ non-metal
Ge $\longrightarrow$ metalloid	As $\longrightarrow$ metalloid
Sn $\longrightarrow$ metal	Sl $\longrightarrow$ metalloid
Pb $\longrightarrow$ metal	Bi $\longrightarrow$ metal

The groups begin with non-metals and end up with metals. The oxides of the metals are basic and non-metals acidic. Fr at the bottom left hand corner of the table is the most reactive metal and F at the top right hand corner of the table in the most reactive non-metal. This can be properly understood on the basis of ionisation potential and electronegativity values of these two elements.

EXERCISES

1. In what way modern periodic is more fundamental to Mendeleev's periodic law?
2. 'Mosley arranged all the elements into groups and periods' elaborate.
3. What is periodicity? Explain.
4. What is the relation between groups and periods with electronic configuration? Explain *s*-block, *p*-block, *d*-block and *f*-block of elements.
5. What are three anomalous pairs of elements; K, Ar; Te, I and Co, Ni and how the anomaly is rectified?
6. Discuss lanthanide contraction and its effect on the properties of elements.
8. How does the atomic radius vary along a period an in the group?
9. In ionic radii which of the ions Mg^{2+} and Al^{3+} is smaller? Give reasons.
10. What is ionization potential? How does it vary in a group or a period?
11. Explain why ionization energy of N is greater than O.
12. Why subsequent ionization potentials are greater than the first I.P.?
13. Noble gas elements have the highest I.P. Explain.
14. Cation is always smaller than an Anion. Discuss.
15. What is the trend of electron affinity and electronegativity along a period and down the group?

16. $AlCl_3$ is a covalent compound but it ionizes in solution. Why?
17. What is the cause of diagonal relationship of elements?
18. $Ca(OH)_2$ is less soluble than $Ba(OH)_2$. Explain.
19. Why an electron prefers to enter 45 orbital than 3*d* orbital?
20. Ca is more reactive than Mg why?
21. Among H_2O, HCl and NH_3; H_2O has the highest boiling point. Why?
22. I_2 dissolves readily in KI. Explain.
23. Blue litmus paper first turns red and gets decolorized in HOCl. Why?
24. Fluorine has less electron affinity than chlorine. Give reason.
25. Li, Mg; Be, Al; and C, Si show similar properties. Explain.
26. Valence shell configuration of Cr(24) is $3d^5 4s^1$ and not $3d^4 s^2$. Given reason.
27. What are magic numbers? Discuss.
28. Electron affinity of nitrogen is negative. Give reason.
29. Bond energy of F_2 is less than that of chlorine. Why?
30. NH_3 is polar but BF_3 is non-polar. Give reason.
31. Structure of H_2O is angular. Explain.
32. Ionic radius of P^{3+} is greater than P^{5+} ion. Why?
33. Why bond angles $\angle$ONO decreases along NO_2^+, NO_2, NO_2^-?
34. Metals have a high m.p. with respect to non-metals. Why?
35. What is effective atomic number?
36. What is screening effect? How does it vary?
37. Why the colour of AgCl is white while AgBr is pale yellow and AgI is yellow?
38. Why $PbCl_2$ is more stable than $PbCl_4$. Elaborate.
39. Cl has higher electron affinity than F. Explain.

3

Chemical Bonding

Chemical Bond

Literally bond is an agreement between two persons or parties while chemical bond is an arrangement between any two or more atoms to stay together. For any two atoms to stay together there ought to be close distance and attraction between them. Atoms have a central nucleus having a positive charge containing protons and neutrons and electrons with negative charge on the periphery. When two atoms are quite close these should be forces of repulsion between the two nuclei and the two adjacent electrons, simultaneously.

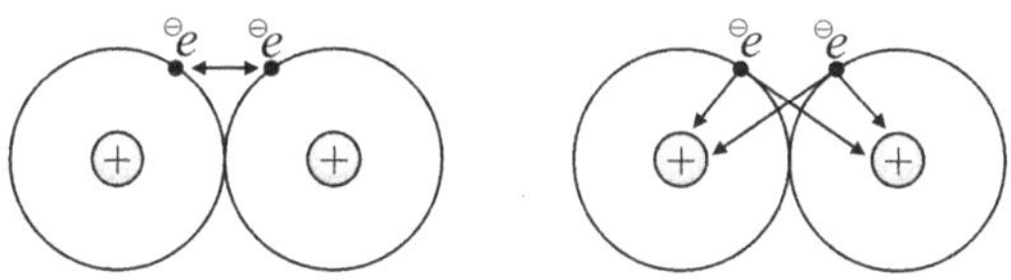

(*a*) Forces of Repulsion (*b*) Forces of attraction

Obviously, if a molecule or compound is formed the forces of attraction would be more than the forces of repulsion otherwise they will fall apart. Nucleus of one atom is attracting its own electrons and the electrons from the other nucleus of the other atom. Forces of repulsion arise out of repulsion between the two nuclei and repulsion between an electron belonging to one atom with the electron belonging to the other atom. When the two atoms decide to stay together indeed the forces of attraction between them outdo the forces of repulsion.

Octet Rule: Lewis, Kossel and Langmuir in 1916 found that inertness in the character of atoms or elements is related to the noble gas configuration. Except the noble gas He which has the configuration $1s^2$ all other noble gases have ns^2np^6 (8 electrons) in the outermost shell. This is the fully complete or the stablest configuration. All other atoms try to acquire 8 electrons in their outermost (valency) shell. Since He atom has only two electrons therefore some atoms that acquire He configuration are governed by the rule of two. Rule of eight is more commonly applicable.

Atom has two categories of electron:

(*i*) **Core electrons:** These are the electrons present in the inner shell. They are very stable and do not take part in the bond formation.

(*ii*) **Valence electrons:** The electrons present in the outermost shell determine the valency or combining power of the atom. These are called valence electrons. They take part in the bond formation and are guided by the Octet theory.

The assumptions of Octet theory are given below:

(*i*) Atoms having 8 electrons in the outermost shell are most stable and do not easily form a chemical bond.

(*ii*) Atoms having less than 8 electrons are chemically active and are capable to form a chemical bond.

(*iii*) Atoms gain, lose or mutually share electrons to form a chemical bond.

(*iv*) Atoms having less than 8 electrons show chemical activity.

(*v*) Sharing of electron pairs, loosing and gaining electron determines the chemical activity.

Energy Concept

When the combining atoms approach each other closely then there is either lowering of energy or increase in energy. In case of increase in energy, no bond can be formed. The energy of bounded atoms must be lower than the uncombined atoms only then the bond can be formed. The bonded atoms have lower enthalpy.

Energy change:

$$E_2 - E_1 = \Delta E$$

Enthalpy change:

$$\Delta H_2 - \Delta H_1 = \Delta H$$

molecule or compound $A + B \longrightarrow A - B$ compound atoms.

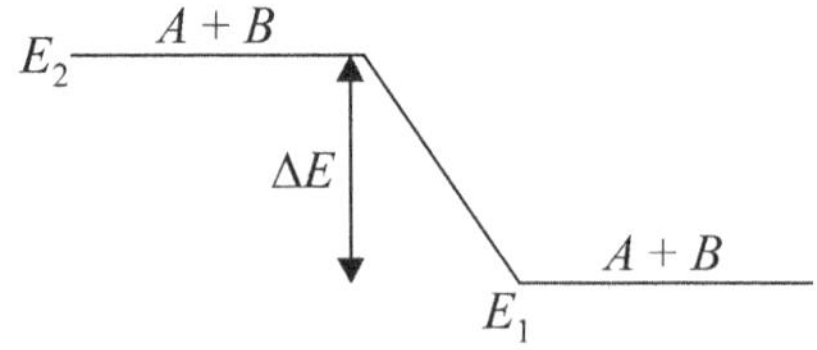

The internuclear distance is called bond lengths.

Chemical Bond

The final definition that has emerged out of the discussion so far is that a chemical bond is the force of attraction between atoms which hold the constituent atoms together in a molecule or a compound. Atoms combine to acquire eight electron in their outermost shell because 8 electrons configuration is the stablest form. This process can be brought about by:

(*i*) Gain or loss of valence shell electrons.

(*ii*) Mutual sharing of electrons between bonded atoms.

(*iii*) One sided sharing of electron pairs between bonding atoms.

The bond thus formed may be classified as:

(*a*) Electrovalent bond

(*b*) Covalent bond

(*c*) Coordinate covalent bond

(*d*) Metallic bond

(*e*) Hydrogen bond

(*f*) Bond due to van der Waals forces

Electrovalent Bond or Ionic Bond

This bond is formed by the transference of one or more electrons from one atom to another *e.g.*

$$A^{\bullet} + B \longrightarrow A^{+}B^{-}$$

It atom *A* has lost one electron it will acquire a positive charge and atom *B* has gained one electron therefore it acquires a negative charge. As a result the formation of ⊕ and ⊖ poles gives the formation of ionic or polar bond. The atoms are held together due to electrostatic force of attraction (attraction between unlike charges)

Example:

$$Na^0 \longrightarrow 1s^2, 2s^22p^6, 3s^1 \text{ or } 2, 8, 1$$
$$Cl^0 \longrightarrow 1s^2, 2s^22p^6, 3s^23p^2 \text{ or } 2, 8, 7$$

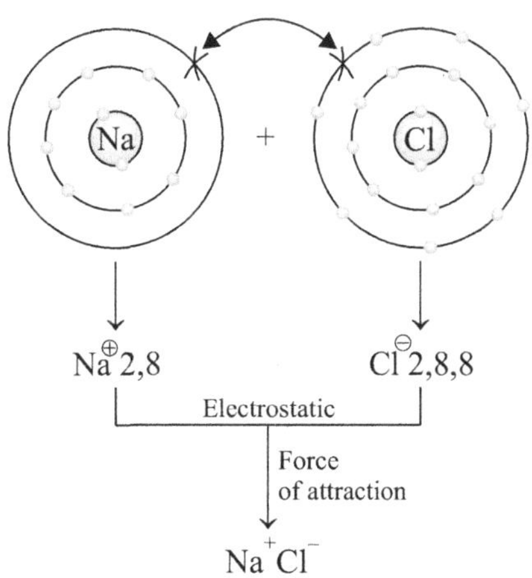

CaF_2

$$Ca_{20} \longrightarrow 1s^2, 2s^22p^6, 3s^23p^6, 4s^2$$
$$F_{(9)} \longrightarrow 1s^2, 2s^22p^5$$

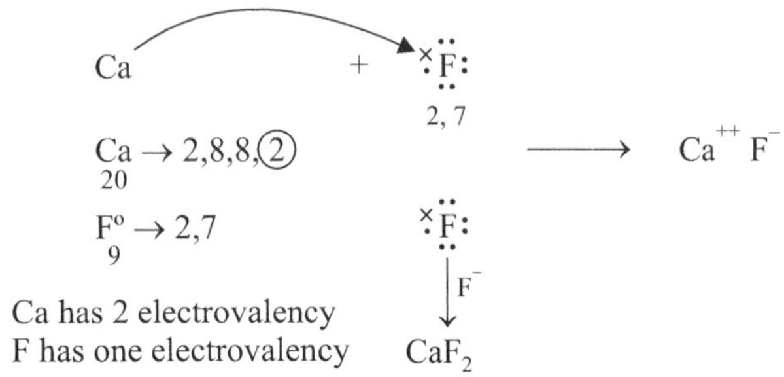

To neutralise the charges one Ca and 2F are required, Octet of both Ca and F is satisfied.

Al_2O_3

$$Al_{(13)} \longrightarrow 1s^2, 2s^22p^6, 3s^2, 3p^1$$
$$O_{(8)} \longrightarrow 1s^2, 2s^22p^4$$

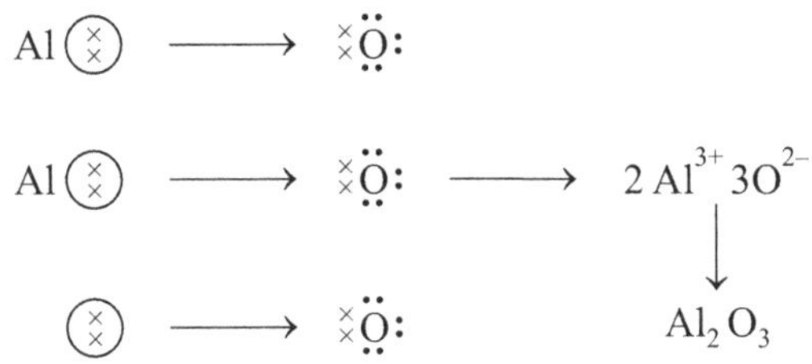

Electrovalency of Al is 3 positive and O is 2 negative.

Octet of both Al and O must be satisfied and charge neutralised.

Characteristics of Electrovalent Bond

Electrovalent or ionic bond is formed by transference of valency electron or gain or loss of valence electron. By this cations and anions are formed and due to Coulombic force of attraction well-ordered geometrical shape show lattice structure. In the solid state these compounds are bad conductors but they conduct electricity in molten state. (electron gets a medium to travel in molten state). Electrovalent compounds being ionic dissolve in water (ionic solvent) because solvation energy being greater than lattice energy.

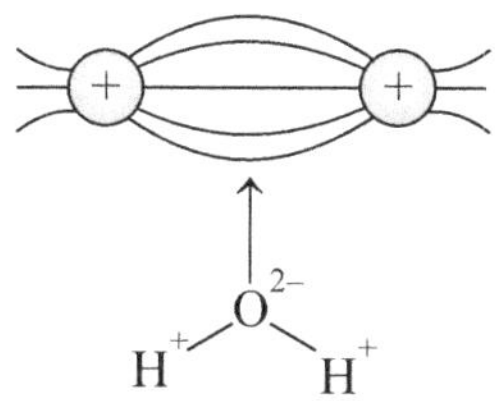

(*i*) The cations and anions are held together by Coulombic forces of attraction (opposite charges attract each other).

(*ii*) These compounds are polar therefore they are soluble in polar solvents like H_2O, NH_3 etc. but insoluble in non-polar solvents like C_6H_6, CCl_4 etc.

(*iii*) These compounds split into ions in fused state.

(*iv*) Solution of these compounds is a good conductor of electricity.

(*v*) The linkages in these compounds are non-directional.

(*vi*) Due to lattice structure and strong electrostatic force of attraction between ions they are brittle when hammered.

(*vii*) Due to ionic nature of the compounds they are quite hard and have low volatility.

(*viii*) These compounds have high melting and boiling points due to strong electrostatic force of attraction in the crystal lattice.

(*ix*) These compounds are quite stable.

Born-Haber Cycle

The enthalpy of formation of ionic solid is determined by the Born-Haber cycle. Born-Haber applied Hess's law which means that the enthalpy of reaction is the same whether the reaction takes place in one step or several steps. According to the first law of thermodynamics that means that energy is conserved in a reaction. For example the enthalpy of formation of NaCl is given by Born Haber cycle:

$$\left.\begin{array}{l} \text{Na(metal)} \xrightarrow{\text{Sublimation}} \text{Na}^0(g) \xrightarrow[-e]{\text{IE}} \text{Na}^+(g) \\ \frac{1}{2}\text{Cl}_2(g) \xrightarrow{\text{Dissociation}} \text{Cl}^0(g) \xrightarrow{\text{EA}} \text{Cl}^-(g) \end{array}\right] \text{Lattic energy NaCl}(s)$$

or

$$\underset{\text{Formation}}{\Delta H} = \Delta H_{\text{sub}} + \frac{1}{2}D + \text{IE} + \text{EA} + U_{\text{NaCl}}$$

Therefore heat of formation is the summation of all other terms in the formula mentioned along with its sign (positive or negative). For many ionic solids the calculated and the experimental values are much different.

For NaCl solid

$$-Q = S + \frac{1}{2}D + IE - EA - U$$

or

$$U = S + \frac{1}{2}D + IE - EA + Q$$

By substituting the values on the right hand side we get the value of lattice energy (U).

Lattice energy (U): It is the energy released which when gaseous cations and anions form ionic solids of fixed geometrical shape *e.g.*

–	–	–	–	–	–
+	+	+	+	+	+
–	–	–	–	–	–
+	+	+	+	+	+
–	–	–	–	–	–
+	+	+	+	+	+

Fig. Arrangement of ions.

Mathematically lattice energy U is given by the relation:

$$U = \frac{NAe^2z^2}{r}\left(1 - \frac{1}{n}\right)$$

where,

N = Avogardro's number

A = Madelung constant

n = Born exponent.

The higher the negative value of heat of formation, greater will be stability of the ionic solid produced.

Advantages of Born-Haber cycle:

(*i*) It is the best and the easiest method for the determination of electron affinity which is so difficult to find.

(*ii*) Born-Haber cycle is the best method in analysing and correcting the stability of various ionic compounds. Large energy is required to vapourise MgO to Mg^{2+} ion and large energy to dissociate O_2 to O^{2-} ion. Both are endothermic process the heat of formation of MgO is –143.8 Kcal/mole which is very high value. High negative energy is due to high lattice energy which compensates the energy expenditure.

(*iii*) Born-Haber cycle also helps us to understand metals fails to form stable ionic compounds in low oxidation states *e.g.* MnCl, CaCl and AlO.

Formation of MCl_2 is favoured over MCl although energy required to form $M^+(g)$ and $Cl^-(g)$ is lower than the required for the formation of $M^{2+}(g)$ and $2Cl^-(g)$. The high lattice energy of MCl_2 over MCl more than compensates the advantage of MCl over MCl_2.

Lattice Energy (U)

Energy of the crystal lattice of an ionic compound is the energy released when ions get lodged in a crystal from infinite distance:

$$M^+(g) + X^-(g) \longrightarrow MX(s) \text{ solid}$$

Besides electrostatic force which contributes 90% of the bonding energies. Other factors included in the lattice energy are repulsion of closed shells, dispersion forces and zero-point energy. Energy curve for the ion pair is shown below:

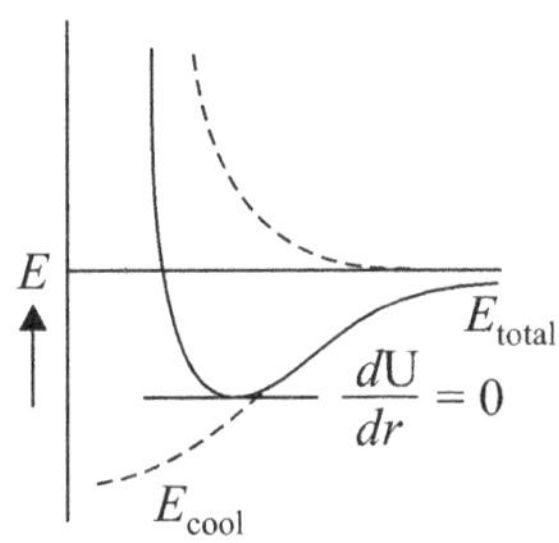

Fig. 3.1

Theoretical discussion regarding lattice energy was initiated by Born and Lande. Derivation of lattice energy is as mentioned.

Ions M^+, X^- separated by distance r, then electrostatic attraction given by Coulomb's law is:

$$E_c = \frac{Z^+Z^-}{4\pi E_0 r}$$

since there is a negative charge the energy is negative with respect to energy at infinite distance. The effect is increasing in crystals as inter ionic distance decreases. The dotted line shows the Coulombic energy (Fig. 3.1). Z^+ and Z^- are multiples of electronic charge $e = 1.6 \times 10^{-19}$ Coulombs therefore

$$E_c = \frac{Z^+Z^-e^2}{4\pi E_0 r}$$

In crystal lattice there are more interactions than in simple ion-pair *e.g.* in NaCl there are six immediate neighbours of opposite charge therefore attractions are more. Similarly there is repulsion by 12 neighbours. The summation of these interactions is called Madelung constant A. The energy of ion pair in crystal becomes

$$E_c = \frac{AZ^+Z^-e^2}{4\pi E_0 r}$$

Madelung constant for common crystal lattices are given in the table

Structure	Coordination Number	Geometrical factor, A	Conventional factor, Aa
Sodium chloride	6 : 6	1.74756	1.74756
Cesium chloride	8 : 8	1.76267	1.76267
Zinc blende	4 : 4	1.63806	1.63806
Wurtzite	4 : 4	1.64132	1.64132
Fluorite	8 : 4	2.51939	5.03878
Rutile	6 : 3	2.408[b]	4.816[b]
β-Cristobalite	4 : 2	2.298	4.597
Corundum	6 : 4	4.1719[b]	25.0312[b]

[a] Use $Z_{\pm}$ = highest common factor

[b] Exact value depend upon details of structure.

Polarisation of Ions: When a cation comes closer to the anion then the positive charge on the cations attract the electron cloud on the anion due to the electrostatic force.

At the same time cation repels the positively charged nucleus of the anion. Due to these two forces anion no longer remains symmetrical. The anion is elongated towards the cation.

This is deformation or polarisation of the anion.

Fig. 3.2 Polarisation of anion B^- by cation A^+.

Similarly cation is also polarised by the anion. Cation has a small size therefore polarization of cation by the anion is by lesser extent or negligible in some cases.

In brief, when cation and anion come closer then polarisation of anion by the cation is appreciable while polarisation of cation by the anion is negligible.

The amount of polarisation of an ion is governed by Fajan's Rules:

(*i*) **Charge on the cation or anion:** Increase in the positive charge on the cation increases the ability of cation to polarise the anion, the polarising power is in the order $Na^+ < Mg^{2+} < Al^{3+}$ same rule applies to anions also *e.g.* in F^- and O^{2-}; O^{2-} is more polarised.

(*ii*) **Size of the cation or anion:** Cation with a smaller size is able to polarise the anion to a greater extent. The order of polarising power is $Be^{2+} > Mg^{2+} > Ca^{2+} > Sr^{2+} > Ba^{2+} > Ra^{2+}$.

The polarisability of an anion increases with increase in size *e.g.* $F^- < Cl^- < Br^- < I^-$ CaF_2 has the melting point of 1392°C therefore it is the most electrovalent out of $CaCl_2$, $CaBr_2$ and CaI_2.

(*iii*) **Electronic configuration:** If the cation has a configuration containing 18 electrons in the outermost shall it has greater polarising power to polarise an anion than the cation having 8 electrons in the outermost shell. This is because the *d* electrons shield the nuclear charge less effectively than *s* and *p* electrons. Therefore Cu^+ ns^2np^6 np^63d^{10} has greater polarising power to polarise Cl^- anion than Na^+ (anhydrous or dry condition) although both cations have the same size $Cu^+ = 0.96$ Å $Na^+ = 0.95$ Å. Hence chlorides of Cu^+, Ag^+ and Au^+ are more covalent than those of Na^+, K^+ and Ru^+. The decrease in character can be judged by studying their melting points *e.g.* CuCl = 442°C and NaCl = 800°C.

Fajan's rules

Molecules having high positive charge on the cation, high negative charge on the anion, small cation and big anions are covalent in nature while those with less positive charge on the cation, low negative charge on the anion, large cation or small anion have ionic nature. Molecules having 18 electron valence shell configuration are more covalent than those having 8 electron valence shell configuration.

Structure of Crystalline solids: The ionic compounds due to strong electrostatic attraction give crystal structures out of these four structures the number of counter ions grouped about or given ion may he four, six and eight.

For example sodium chloride crystallises to give face-centered cubic structure. Fig. 3.3(*a*)

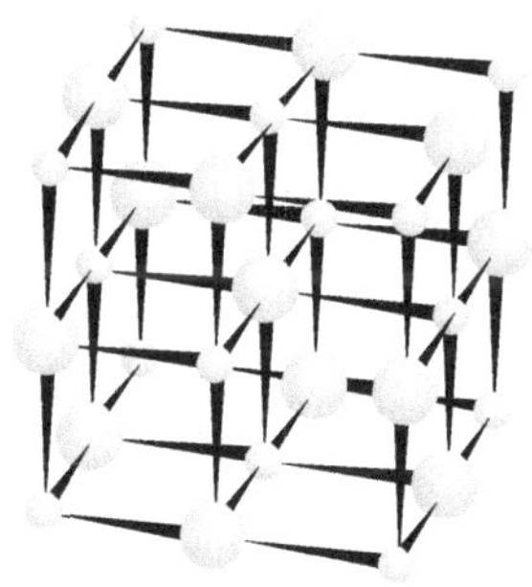

Fig. 3.3 (a)

Eight Na^+ ions form corner of a cube and six Cl^- ions occupy centre of the faces. The coordination number of both ions in the NaCl lattice is 6 *i.e.* there are $6Cl^-$ ions surrounding each Na^+ and six Na^+ ions round each Cl^- ion.

NaCl type structure is acquired by most of the alkali metals. This structure is also found in the oxides of magnesium, calcium, strontium, basium and cadmium.

Cesium chloride crystallizes in the cubic arrangement Fig. 3.3(*b*). Cs or Cl^- ion occupy the 8 corners of a cube and the counter ion occupies the centre of the cube. The coordination number of both ions in CsCl is 8 *i.e.* there are 8 anions round a cation and 8 cations round an anion. At high pressure all the alkali halides except lithum can be forced to acquire CsCl structure.

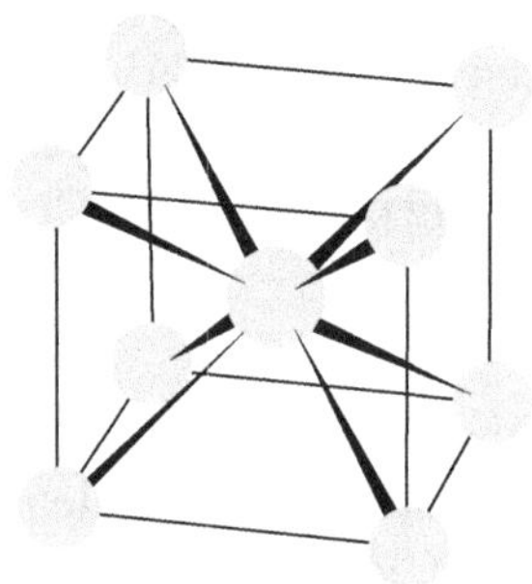

Fig. 3.3 (b)

Zinc blende (Z*n*S) crystallises into two distinct lattices *i.e.* hexagaonal wurtizite. Figure 3.3(*c*) and cubic zinc blende Fig. 3.3(*d*). Huheey. Many divalent metal oxides and sulphides *e.g.* BeO, ZnO, BeS, MnS, ZnS, MnS, CdS and H_gS acquire these structures of zinc blende and wurtzite. Some other compounds included against these structures are AgI, NH_4F and SiC.

Fig. 3.3 (c)

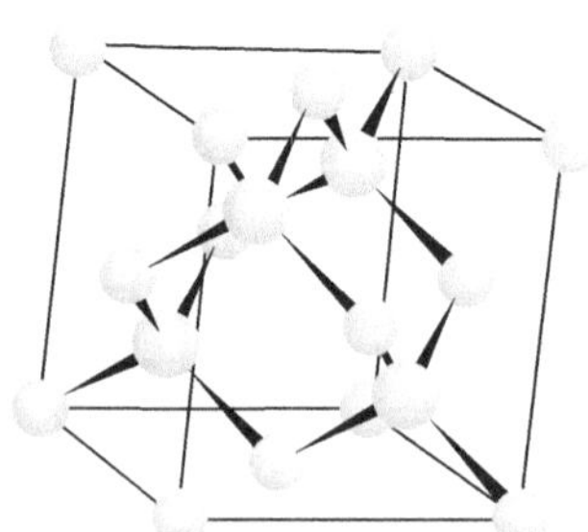

Fig. 3.3 (d)

The fluorite structure is adopted by CaF_2 Fig. 3.3(*e*). The coordination numbers are 8 for the cation *i.e.* $8F^-$ ions form a cube round the Ca^{2+} ion and $4Ca^{2+}$ ions about each F^- ion *e.g.* fluorides of Ca, Sr, Ba, Cd, Hg and Pb. Dioxides of Zn and Hf and some other lanthanides and actinides. If cations and anions, are reversed then antifluorite structure is obtained. The examples are oxides and sulphide of Li, Na, K and Pb. Fig. 3.3 (e).

Fig. 3.3 (e)

Rutile Structure: TiO_2 crystallizes in these three at atmospheric pressure such as anatase, brookite and sectile Fig. 3.3(*f*). Only tetragonal will be considered. The coordination number is 6 for cation and 3 for anion *e.g.* fluorides and dioxies of Cr, Mn, Ge, Rh, Sn, Os, Ir, Pt and Pb.

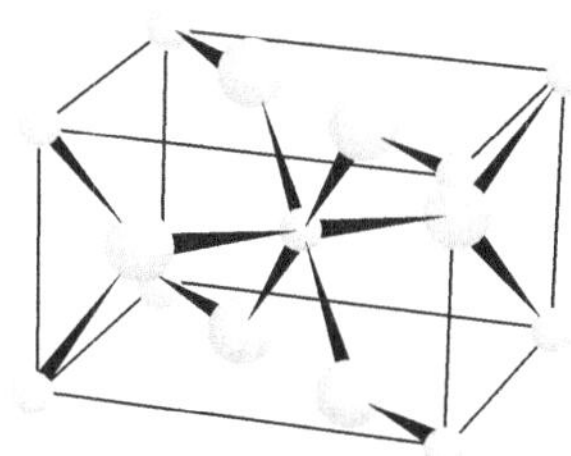

Fig. 3.3 (f)

The β-crystobalite structure: SiO_2 crystallises in several kinds which are sometimes stabilised by foreign atoms. β-crystobalite is related to zinc blende Fig. 3.3 (*d*) having silicion atom in place of zinc and sulphur in zinc blende with oxygen atoms between silicon atoms. The compounds with this structure are $Be(OH)_2$, $Zn(OH)_2$, BeF_2, $ZnCL_2$ and SiS_2 are formed at high pressures. Tridymite SiO_2 is related to wurtzite structure. The coordination number in both β-crysobalite and tridymite is 4 for Si and 2 for oxygen.

Calcite and aragonite structures: In this case crystals have polyatomic cations and anions where the situation is more complicated. It contains CO_3^{2-} ion and had two structures vis-a-vis calcite and aragonite. They contains $CaCO_3$ in addition to $MgCO_3$, $FeCO_3$, $LiNO_3$, $NaNO_3$ and YBO_3.

Calcite represents rhombohedral structure. The coordination number of metal ion is 6. Large metal ions imbibe aragonite structure with 9 oxygen atoms around the metal ion *e.g.* $CaCO_3$ in addition to $SrCO_3$, KNO_3 and $LaBO_3$.

Close-packed structures: Atoms or ions are considered as hard spheres. The closest packed spheres may be arranged in the form of touching spheres. If hard spheres are of the same size and are arranged in two dimensional lattice then a coordination number of 6 is obtained. If they are of the same size but of opposite charge then six coordinate structure will experience too many repulsions. The most stable arrangement will prefer 4 coordinate square lattice of alternate charges Fig. 3.4.

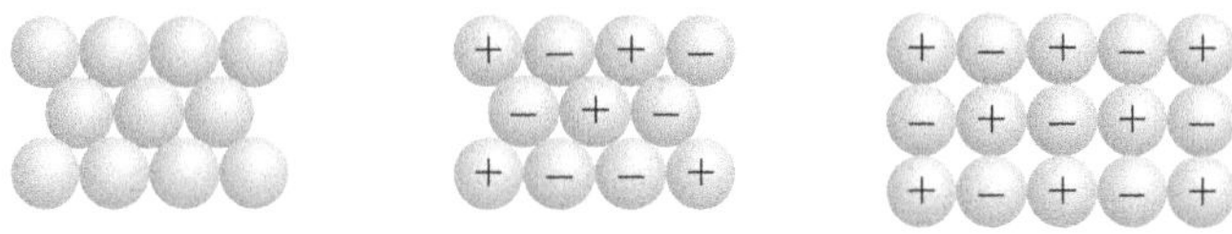

(*a*) (*b*) (*c*)

Fig. 3.4

This arrangement can be extended to three-dimensional lattices. In the first instance consider the lattice of uncharged atoms *e.g.* metal or noble gas atoms. The second layer is centered over depressions Fig. 3.4(*a*). Therefore *n* atoms will have 2*n* sites to accept atoms but when the sites accept an atom than the remainder of the layer continue to accept atoms like wise Fig. 3.4(*b*). It becomes easy to visualise hexagon therefore the term hexagonal closest packed Fig. 3.4(*a*) in one layer. There can be similar arrangement in layers *B* and *C*.

The metals and noble gas atoms crystallise in *hcp* or the *ccp* structure, hexagonal closest packed (*hcp*) and cubic closest packed (*ccp*). The alkali metals, caesium and a few transition elements crystallise in the body centered cubic system.

If atoms are no longer neutral then they should be half cations and half anions. The closest packed structure are not very stable in two-dimensional arrangement. In LiI, iodide anion is too big than Li cation. The smaller Li ions can fit into interstices between the anions. When anion-anion touch each other it will experience a force of repulsions. On expanding the lattice the repulsion may be reduced and the crystal stabilised. All this will depend upon the relative size of the cation and the anion. The closest-packed arrangement may lead to two types of holes [(*a*) Tetrahedral (*b*) Octahderal Fig. (3.5)].

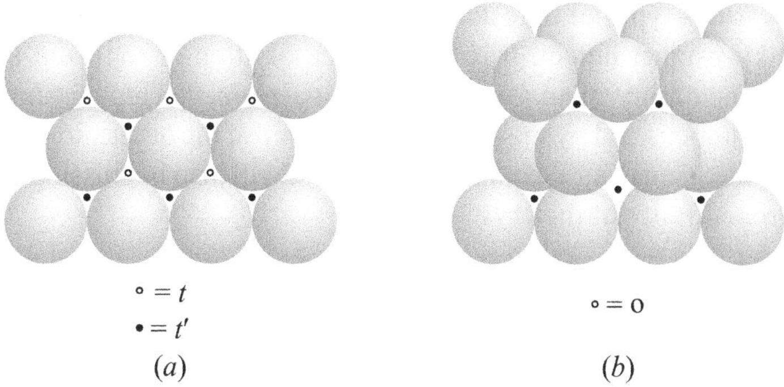

Fig. 3.5

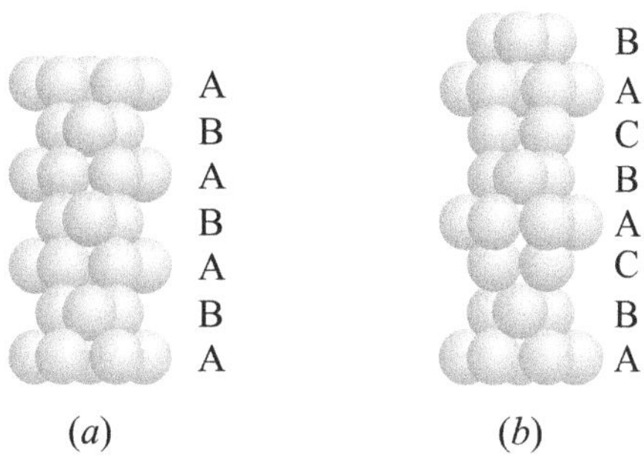

Fig. 3.6

The octaheral holes are larger than the tetrahedral holes hence large cation can be easily accommodated. The structure of LiI is similar to NaCl face-centered cubic (*fcc*) Fig. 3.6.

In zinc blende if the S^{2-} ions form a *ccp* and zinc ion are in tetrahedral holes than the resulting structure is cubic closest-packed and if they form *hcp* then the Wurtzite. In the current discussion size is the only parameter that is considered to determine the occupancy of tetrahedral and octahedral sites. The factor of size will dominate since other factors are of lesser importance.

Imperfect Crystals

The discussion about the crystals so far assumed that the crystals are pure there occurs loss of entropy when random gas state of the compound or molecule changes to the ordered solid. The enthalpy-entropy relationship is tilted in favour of enthalpy because of crystal energy but the entropy factor will result in equilibrium defects at all temperatures above absolute zero (crystals). The defects in the crystals increase cation anion interactions and decreases cation-cation and anion-anion repulsions.

The various types of defects that are observed in the crystals are Schottky or Schottky-wagner defect. It is the absence of atom or ion from the site but the overall neutrality in an ionic crystal is maintained. This missing charge is generally compensated by missing charge created at the other site Fig. 3.7. *i.e.* if a positive site is missing then one negative site will also be missing to maintain the electroneutrality of the crystal.

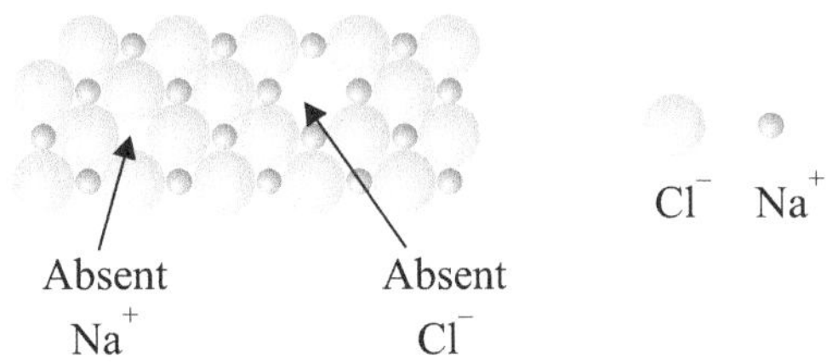

Fig. 3.7

Besides, the missing ion is balanced by the impurity of ion of higher charge. In a crystal of AgCl, the absence of Ag^+ ion is doped with a small amount of Cd^{2+} ion. The depositive charge necessitates another vacancy to balance the change in charge Fig. 3.8 for example NiO aqueous solution contains Ni^{2+} ion which is pale green. If it is doped with a small amount of Li_2O another vacancy to balance +1 charge is created. This induces Ni^{2+} ion to lose electrons to become Ni^{3+} ion to maintain electroneutrality of the crystal. This changes colour of Ni oxide crystal to gray-black. This also acts as a semiconductor. Doping with Li_2O causes a few cation sites to be occupied by +1 Li in place of +2 Nickel. This helps to preserve electrical neutrality of the crystal. The properties of NiO crystal change *i.e.* colour change from pale green to gray-black and from NiO insulator to semiconductor.

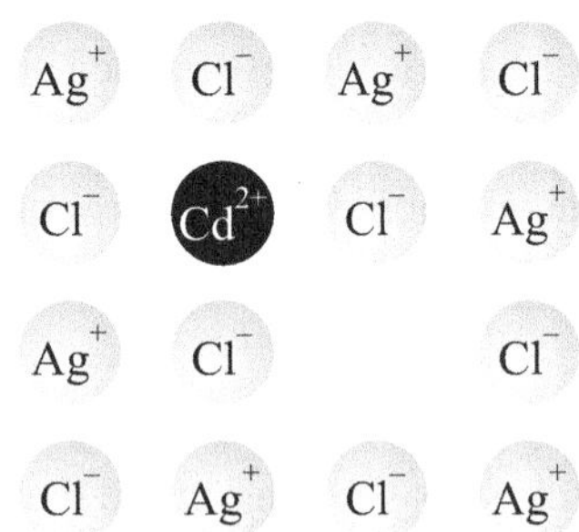

Fig. 3.8

In non-stiochiometric compounds *e.g.* in Cu_2S may not be in the exact ratio of 2 : 1. Some of Cu^+ ions may lie absent and compensated by Cu^{2+} ions. In case the vacancy contains trapped electron the imperfection caused is known as F centre (Farbe, colour). If a small amount of sodium crystal, the energy in the crystal makes the sodium metal ionise $Na - Na^+ + e$ and the electron occupies the site which would otherwise be occupied by Cl^- ion Fig. 3.9. The trapped electron can absorb visible light and give coloured compound. The composition of the crystal may be non-stoichimetric.

Fig. 3.9

If the missing ion does not occupy the site but move in the nearby interstitial site than it is called Frenkel defect Fig. 3.10. Cation has a smaller size than the anion therefore it is easier for cation to enter the small interstitial hole. Theoretically cation and anion can fit into the site depending upon the size.

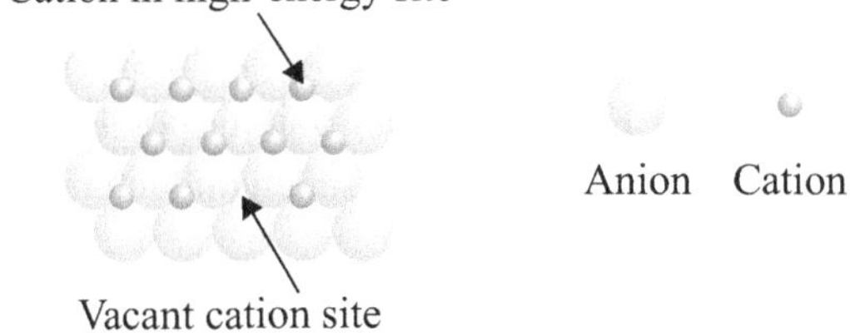

Fig. 3.10

Conductivity of Ionic Solids

Ion Migration: Ordinarily ionic solids have low conductivities *e.g.* NaCl conducts by ions conductance since there are no partially filled bonds or accessible bands for electronic conduction. The conductivities relate to defects in the crystals. The migration of ion is of three types:

(*i*) **Vacancy sites:** Due to the vacancy in the lattice, it may be possible for the adjacent ion that is missing usually a cation to migrate into it. The difficulty in migration is due to the size of the migrating ion and the ion surrounding it may impede it.

(*ii*) **Interstitial site:** This is with regard to Frenkel defects. If the size of the ion is small enough to occupy the interstitial site such tetrahedral hole in an octahedral lattice. It may again move to other interstitial sites.

(*iii*) **Interstitially mechanism:** It is combination of above two. It has one ion moving into the interstitial site and another ion moving into the vacant site so created.

These three mechanisms are shown in Fig. 3.11.

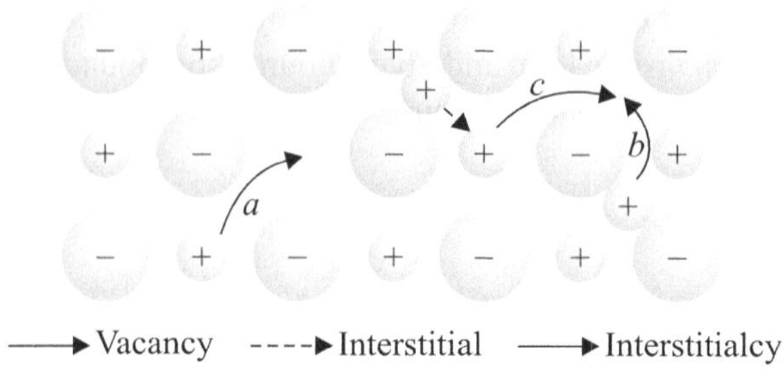

Fig. 3.11

In ionic compound conductivity is because of these three considerations. The conductivity is intrinsic and increases with increase in temperature. Because of the limited defects conductivities are quite low $10^{-6}\ \Omega^{-1}\ cm^{-1}$.

However due to extrinsic vacancies few ionic compounds have high room temperature conductivities $0.27\ \Omega^{-1}\ cm^{-1}$ *e.g.* $Rb\,Ag_4I_5$.

Super conductivity: It was first observed in mercury metal. Below 4.2 K the resistance droups to zero. Presently interest is focussed on high temperature superconductors 100 ± 20 K greater than boiling point. of nitrogen 77 K but lower than the temperature on earth. Earlier superconductors needed to be cooled by liquid helium.

The first superconductor was formulated as $La_{2-x}\ Ba_x\ CuO_{4-\delta}$ ($x < 0.2$), δ is small and have tetragonal, layered K_2NiF_4 perovskite structure. Their critical temperature is about 35 K. At still higher critical temperature Tc, La is substituted by Y and new type of compound $YBa_2Cu_3O_{7-\delta}$ was formed. This is called 1–2–3 superconductor. Recently metals like Tl, Bi and lead have been included in the category of superconductors. *e.g.* Tl, $Ba_2Ca_{n-1}\ Ca_n\ O_{2n+2}$ at higher Tc. At 125 Tc $Tl_2\ Ba_2\ Cu_3$ O has been observed as superconductor.

EXERCISES

1. Tending of atoms to gain or lose electrons is related to the types of bond they form. Explain.
2. Which of the following will conduct electricity.

   (*a*) NaCl (solid) (*b*) NaCl (aqueous solution)

   (*c*) NaCl (fused) (*d*) Cu (solid) and (*e*) CCl_4 (liquid)

   Discuss the mechanism of conduction.
3. Why are ionic compounds generally have high melting point, while covalent compounds have low melting point?
4. Discuss the Born-Haber cycle for the formation of ionic compound NaCl. Define the terms and state how these are measured or calculated. How do enthalpy terms vary throughout the periodic table? How do the properties of NaCl and CuCl differ from each other's?

5. List the type of defects that occur in the solid state giving an example of each. Explain the electrical conduction in each.
6. Calculate the lattice energy of CsCl from the following values:

$Cs(s) \longrightarrow Cs(g)\ \Delta H = +79.9$ kj mol^{-1}

$Cs(g) \longrightarrow Cs^{+}(g)\ \Delta H = +374.05$ kj mol^{-1}

$Cl_2(g) \longrightarrow 2Cl(g)\ \Delta H = +241.84$ kj mol^{-1}

$Cl(g) \longrightarrow 2Cl(g)\ \Delta H = -397.90$ kj mol^{-1}

$Cs(s) + \frac{1}{2}Cl_2 \longrightarrow CsCl(s)\ \Delta H = -623.0$ kj mol^{-1}

7. Give reasons why
   (*i*) $BaSO_4$ is insoluble in H_2O
   (*ii*) Noble gases do not form ionic halides
   (*iii*) KCl_2 is not formed
8. What are ionic crystals? Why the lattice energy is considerably higher in the case of ionic crystals?
9. How do you explain zero dipole moment of BF_3?
10. Discuss Fagan's rules.
11. LiI is soluble in C_2H_5OH and LiF is insoluble in C_2H_5OH. Explain.
12. $ZnCl_2$ is soluble in organic solvents while $MgCl_2$ is insoluble.
13. LiCl has a higher boiling point than HCl.
14. CO_2 has a zero dipole moment though each C—O bond is polar. Give reasons.
15. What are the limitations of radius ratio rule?
16. NF_3 and NH_3 have same structure but dipole moment of NF_3 is less than that of NH_3.
17. What kind of elements combine to form an ionic bond? Why ionic bond has no direction?
18. Why water is a good solvent for ionic compounds and poor solvent for covalent compounds?
19. LiCl is not soluble in water but NaCl is soluble.
20. Solid NaCl does not conduct electricity while fused NaCl does. Explain.
21. What is dipole moment? What is its importance is the structure of molecules?
22. Ionic compounds have high m.p, h.p, high density and brightness. Explain.
23. Why do atoms combine? What are the drawbacks of octet theory?
24. Arrange the follow in the order of increasing ionic character:

BeCl, $CaCl_2$, $MgCl_2$, $BaCl_2$

4

Covalent Bond or Non-ionic Bond

Covalent bond is formed by mutual sharing of electron pair in which both the bonding atoms contribute same number of electrons and later share them equally to complete the octet. This concept is based on Pauli's exclusion principal *i.e.*, the two electrons must have opposite spins to occupy the same region. These electrons are between the two nuclei and consequently both the nuclei attract them. It is also called electron pair bond.

Electron Dot Formula

Here each electron is represented by a dot (·). For most atoms a maximum of eight electrons are shared by the valency shell. This is called Lewis octet structure. The bond sharing atoms follow the octet rule or rule of two (He structure) in the valency shell *e.g.*

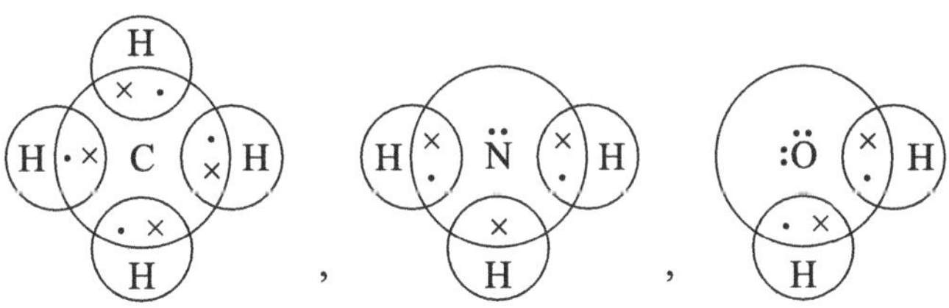

Methane CH_4 , Ammonia NH_2 , Water H_2O

Exceptions to the Octet Rule

There are a number of compounds where the rule of eight is not followed but still the compound is quite stable. *e.g.* for elements with variable *d*-orbitals, shell is expanded beyond the octet.

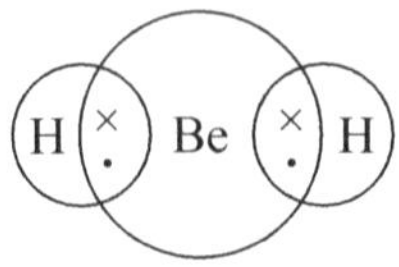

Be shares 4 electrons only

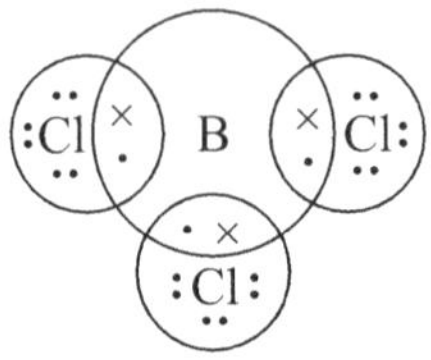

B shares 6 electrons

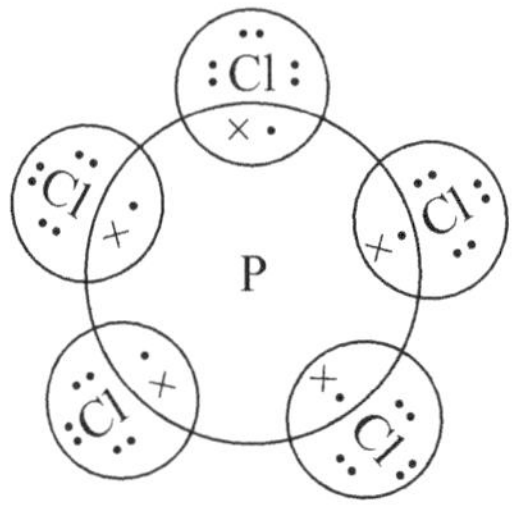

P shares 10 electrons

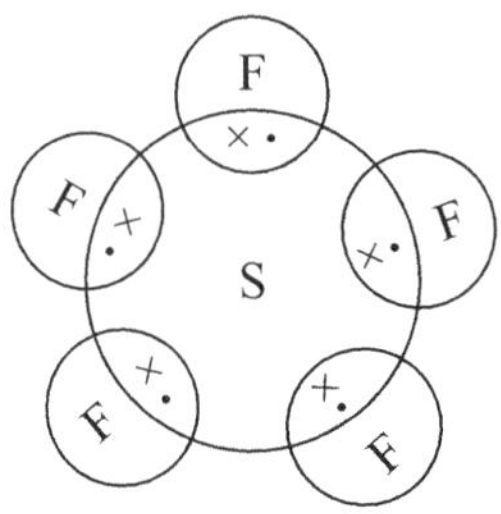

S shares 12 electrons

This can be observed in elements of period three and beyond as given above (PCl_5, SF_6, IF_7 etc.) The same phenomena is observed in transition metal complexes. *e.g.* $[Co(NH_3)_6]^{3+}$

In the case of non-metals the covalency of 5, 6, 7 and 8 are observed in Nitrogen, Oxygen, halogen family and group VIII_A (Noble gases). Covalency and coordination number are not always the same.

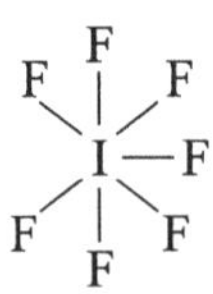

,

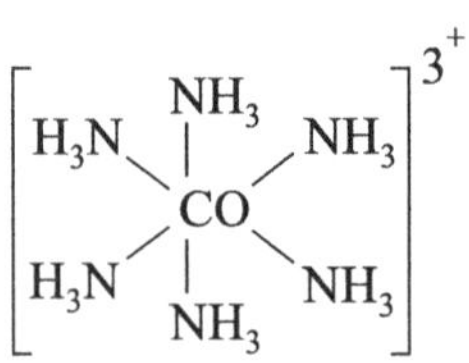

,

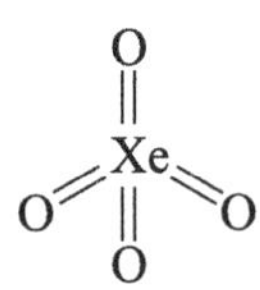

I shares 14 electrons, Co shares 12 electrons and Xe shares 16 electrons.

It is assumed that in all the arrangements of electrons discussed, the molecule should acquire the lowest energy. In all, maximum bonds or strongest bonds will form in such a way that the repulsion energies are at the minimum.

Covalent Bond

If two electrons are mutually shared between two atoms it is represented by a single (—), if 4 electrons are shared it is represented by (=) and in case of 6 electrons by three dashes (≡).

Overlapping of Atomic Orbitals (AO's)

In a covalent bond the central atom is the most important factor in the bond formation. The atomic orbitals present in the valency shell take part in the bond formation whether they have paired electrons or they are vacant. The paired electrons in the atomic orbitals get unpaired for sharing of electrons with the electrons in the atomic orbitals of the combining atoms. For unpairing of electrons and jumping to the atomic orbitals of higher energy, energy is required which is provided by the adjacent or neighbouring atom *e.g.*

In the compound PCl_5, Cl provides the energy, for the P atom to unpair the electrons and secondly go to the orbital with the higher energy. This energy is quantized (fixed amount). Maximum unpaired electrons in the central atom determines the covalency maxima.

	3*s*	3*p*			3*d*				
Pg_{state}	↑↓	↑	↑	↑					
$P_{excited}$	↑	↑	↑	↑	↑				
PCl_5	↑↓	↑↓	↑↓	↑↓	↑↓				
	u	*u*	*u*	*u*	*u*				

Similarly for SF_6

	3*s*	3*p*			3*d*				
Sg_{state}	↑↓	↑↓	↑	↑					
$S_{excited}$	↑	↑	↑	↑	↑	↑			
SF_6	↑↓	↑↓	↑↓	↑↓	↑↓	↑↓			
	F	F	F	F	F				

This determines the maximum covalency for P that is 5 and S is 6.

Polar and Non-polar Covalent Bonds

Non-Polar Bond: A compound containing two similar atoms is called a homodiatomic molecule. Here the sharing of electrons is equal and zero electronegativity difference as well as dipole moment. This can apply to similar atoms. This can also apply to dissimilar atoms with zero electronegativity difference and dipole moment. Such compounds are called heterodiatomic molecules. *e.g.* H_2, O_2, CO_2, CCl_4 etc.

Polar Covalent Bonds: In the event of heterodiatomic molecules, the two bonding atoms have difference in their electronegativities therefore electron-pair shifts towards the more electronegative atom *i.e.* it is farther away from the less electronegative and closer to the more electronegative. By virtue of the drag of the electron-pair towards the more electronegative atom a small δ^- charge is shown and consequently the less electronegative acquires a small δ^+ charge *e.g.*

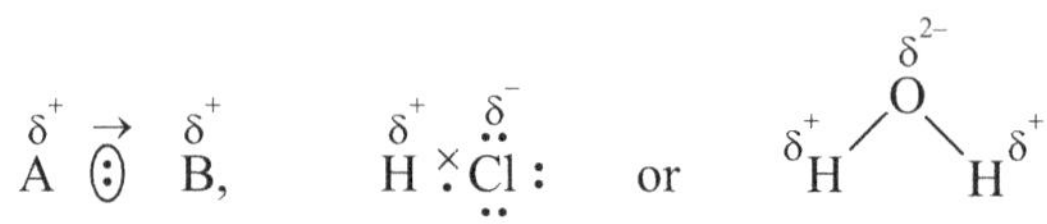

In these Cl and oxygen acquire greater electronegativity and the electron-pair moves towards the more electronegative. The polar covalent compounds show a dipole therefore they show value of dipole moment.

Dipole Moment (μ): The magnitude of polar and non-polar compounds is given by dipole moment. Dipole moment is equal to the product of the electronic charge (e) and the distance between positive and negative centres. Hence,

$$\mu = e \times d \text{ where } \mu \text{ is a vector quantity.}$$

Characteristics of Covalent Compounds

1. Covalent compounds are poor conductors of electricity (absence of ions).
2. These compounds are only slightly soluble in water but highly soluble in non-polar solvents.
3. Most of the covalent compounds are gases or liquid with low melting and boiling points.
4. The rate of reactions of covalent compounds is slow compared to electrovalent compounds.
5. They are rigid and directional.
6. They are soft in the solid form.

Relative Strength of Halogen Acids

The ionic character of covalent compounds depend upon the electronegativity of the bonded atoms *e.g.* A and B are joined by covalent bond, then the polarity of the covalent compounds depend on electronegativity difference X_A and X_B. Roughly, if the difference between the electronegativities is > 1.7 it is largely ionic otherwise it is covalent. (less than 50% ionic).

In case of halogen acids HF, HCl, HBr and HI the most ionic is HF due to electronegativity difference being large. As per the acid strength H—F is dimer therefore it is weak acid while HI is a strong acid. Due to poor electronegativity of iodine it is not able to hold or attract H^+ therefore H^+ are discharged in solution (aqueous) hence HI is the strongest acid.

$$HF < HCl < HBr < HI$$

This the order of acid strength in the above acids.

Coordinate Covalent Bond or Dative Bond

By coordinate covalent bond we coordinate the two atoms in which one of the two atoms donates a lone pair of electrons and the other atom accepts the lone pair. It is a directional bond (dative) with arrow from donor to acceptor. The bond so formed is called a coordinate bond. After donation of lone pair the pair of electrons are mutually shared between the two atoms therefore the covalent (mutual sharing) word is incorporated.

e.g.,

1. NH_4^+

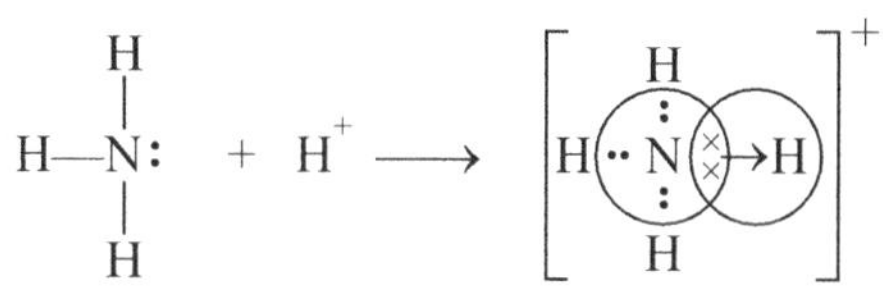

2. H_2SO_4

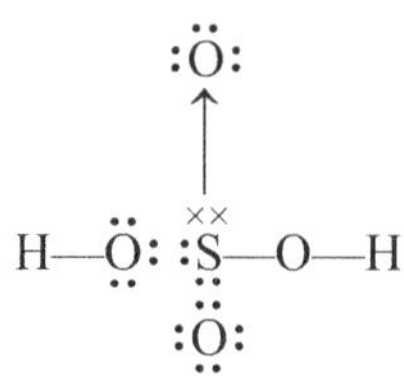

3. $[Cu(NH_3)_4]^{2+}$

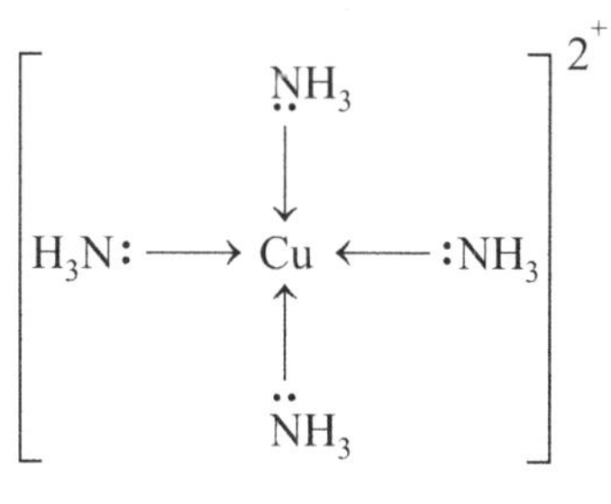

4. Al_2Cl_6

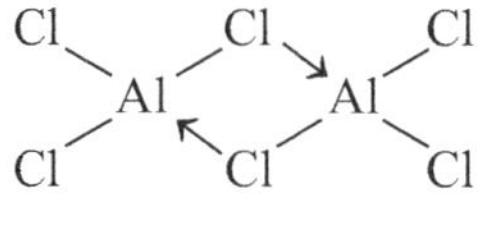

$$A\!: + B \longrightarrow A \rightarrow B$$

$$\overset{+}{A} (:) \overset{-}{B} \longrightarrow A^{+} — B^{-}$$

The bond is neither electrovalent nor covalent *i.e.* it is a semi-polar bond.

Characteristics of Coordinate Compound

1. Formation of coordinate compounds is based on Lewis acid base concept.
2. The nature of these compounds is between electrovalent and covalent compounds.
3. They can have solid, liquid and gaseous states.
4. They are poor conductors of heat and electricity in solid and molten states as they do not easily ionise.
5. The rate of reaction is slow as they react in molecular state.
6. Due to rigid and directional bonds they show isomerism.
7. The stability of compounds is more than covalent and less than electrovalent compounds.

Failure of Electronic Theory

Electronic octet theory put up by Lewis, Kossel and Languir was a very generalized explanation of chemical bonding. There were some shortcomings:

1. It did not explain the role of valency electron in electronic dot formula.
2. There exist a lot of examples where octet theory does not hold *e.g.* $BeCl_2$, BCl_3, PF_5, SF_6.
3. In transition elements due to variable valency the rule of 8 is not followed.
4. The theory could not explain the metallic bond.
5. The energy of double or triple bond is not twice or thrice of a single bond.
6. The magnetic behaviour of molecules could not be explained.
7. The geometry of molecules is not explained.
8. Does not give idea of electron distributed in space and the bond energy.

The Orbital Concept in Covalent Bond

To remove the shortcomings of the octet theory Heitler and London described Valency Bond Theory (VBT) which was later modified into molecular orbital theory proposed by Hund and Muliken in 1951. According to the orbital concept covalent bond is the result of overlapping of atomic orbitals.

ORBITAL is a region in space where the probability of availability of electron is maximum. An orbital can contain a maximum of two electrons with opposite spins (Pauli's exclusion principle). The atomic orbitals are of various shapes.

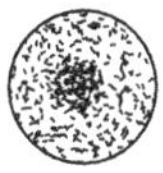

One *s*-orbital (spherical)

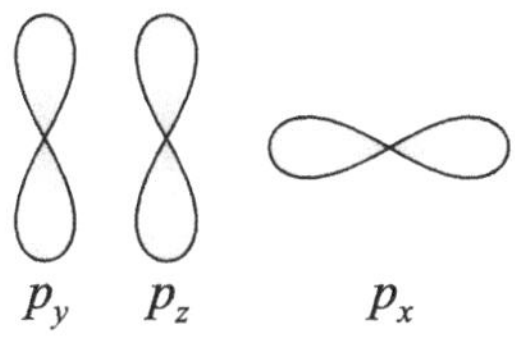

Three *p*-orbital (dump bell shaped) planer along *x*, *y* and *z* axes.
Five *d*-orbitals (double dump bell) along *x*, *y* and *z* planes

dxy, *dyz*, *dzx*

dz^2

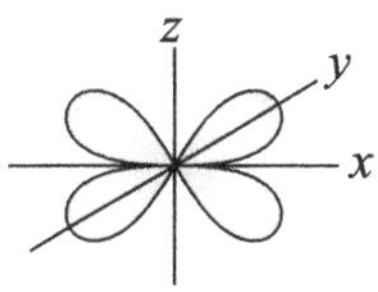

$dx^2 - y^2$

Overlapping of orbitals result in two types of bonds (*i*) sigma (σ) bond and (*ii*) Pi(π) bonds. **Sigma bond (σ)** end to end overlap of atomic orbitals along the same axis results in sigma bond.

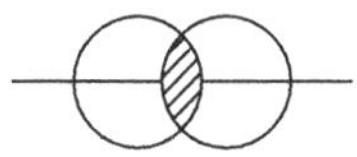

s-s overlap

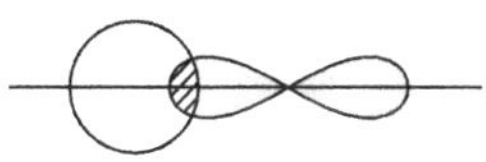

s-p overlap

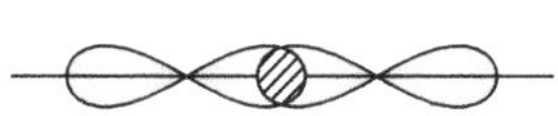

p-p overlap

Pi(π) bond: In case of sideways, head to head and lateral overlap of atomic orbitals a pi (π) bond is formed. The area of overlapping orbitals is small so the bond is weak. p_y and p_z orbital offer a possibility of π bond formation. σ bond is a must for any molecule. π bonds are in addition to a σ bond. A molecule with only π bond cannot exist.

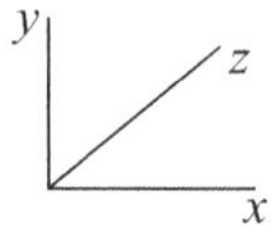

These axes

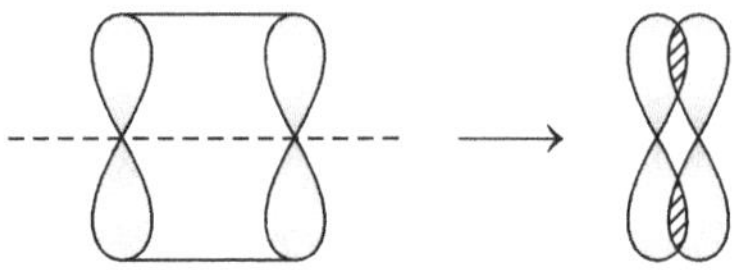

$p_y - p_y$ overlap

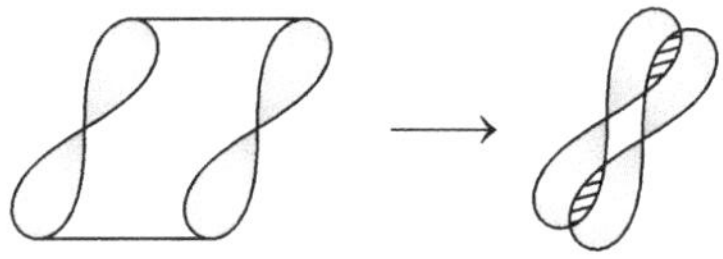

$p_z - p_z$ overlap

p_y–p_y can form both σ bond and π bond

e.g. O_2 molecule:

$$O + O \longrightarrow O_2$$

$$O_8 \longrightarrow 1s^2 \quad 2s^2 \quad 2p^4$$

2s	2p		
↑↓	↑↓	↑	↑

	2s	p_x	p_y	p_z
$O_8 \longrightarrow$	↑↓	↑↓	↑	↑

p_y and p_y orbitals contain unpaired electron which has a tendency to overlap with another p_y and p_z belonging to the other atom.

p_y–p_z overlap results into a σ bond (end-to-end overlap) and p_z–p_z overlap results into a π bond (sideways overlap) as shown in the diagram.

O_2 molecule:

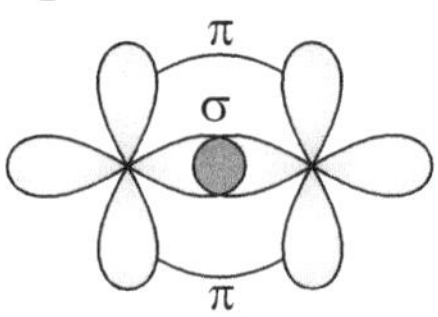

$O\overset{\pi}{\underset{\sigma}{=}}O$ O_2 molecule contain one σ and one π bond.

N_2 molecule:

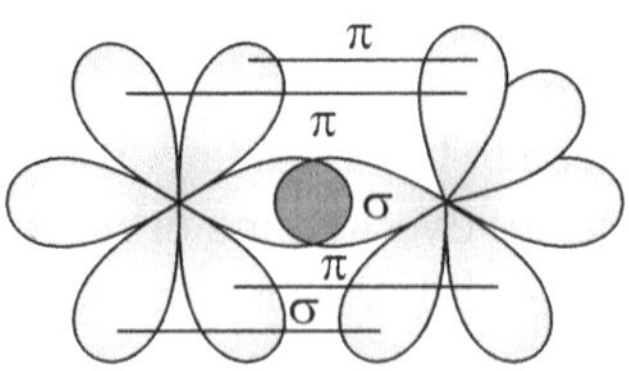

$\longrightarrow$ $N\overset{\pi}{\underset{\pi}{\equiv}}\sigma N$

The area of overlap in the σ bond is large therefore the bond is strong and in the π bond area of overlap is less so the bond is weak.

Objection to the Valence Bond Theory (VBT)

The actual strength of a covalent bond is of much high order than that visualised by the pure overlap of atomic orbitals. This means that something more is happening than mere overlap of atomic orbitals and this is known as Hybridisation.

Hybridisation: It is equalisation of energy of the atomic orbitals belonging to the central atom. This process results into new orbitals having more strength than the atomic orbitals. To equalise the energy of the atomic orbitals energy is required which is known as hybridisation energy. In brief, original energy of the atomic orbitals plus the extra hydrisation energy is divided in atomic orbitals of equivalent energy known as hydrid orbitals. These hybrid orbitals have energy greater than the corresponding atomic orbitals therefore they form bonds which are stronger than those visualized by pure overlap of atomic orbitals.

Hybridisation

1. Atomic orbitals of the same atom undergo hydridisation.
2. Number of hybrid orbitals formed are equal to kind of atomic orbitals mixing together.
3. The number of new hybrid orbitals are of the same energy but different in orientation in space.
4. The hybridisation is named after the type of participating atomic orbitals *e.g.* sp, sp^2, sp^3, sp^3d^2 etc.
5. Hybridisation determines the kind of geometry of the molecule and the bond angle *e.g.*

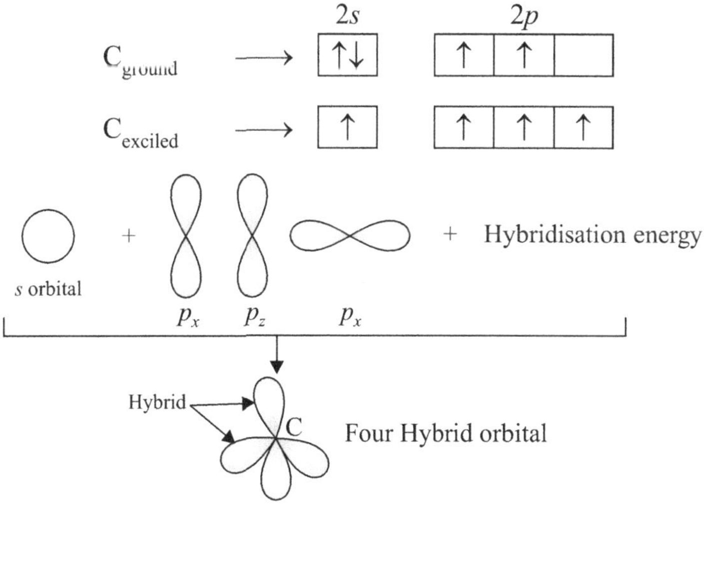

The hybrid orbitals are elongated for better overlap of atomic orbitals *i.e.* area of overlap is large resulting in a strong bond.

e.g. CH_4

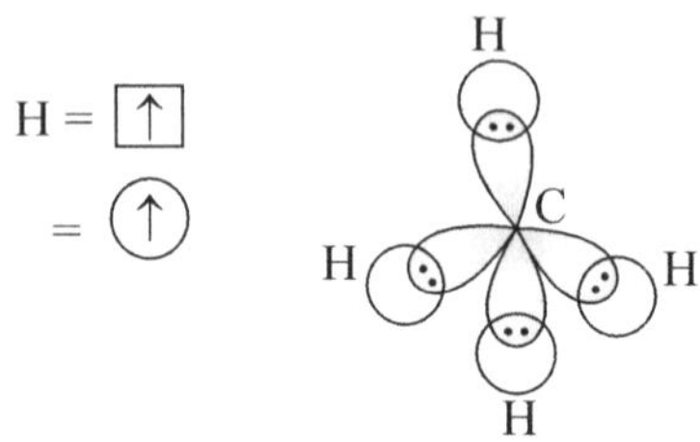

In CH_4 formation carbon has one *s* ad 3*p* orbitals forming four hybrid orbitals of the same energy therefore hybridisation is named sp^3 and CH_4 has four C—H bonds. Due to bond pair-bond pair repulsion of electrons the C—H bonds fall apart till they experience zero or no force of repulsion. At this stage the geometry that emerges is tetrahedral with a bond angle of 109°28′.

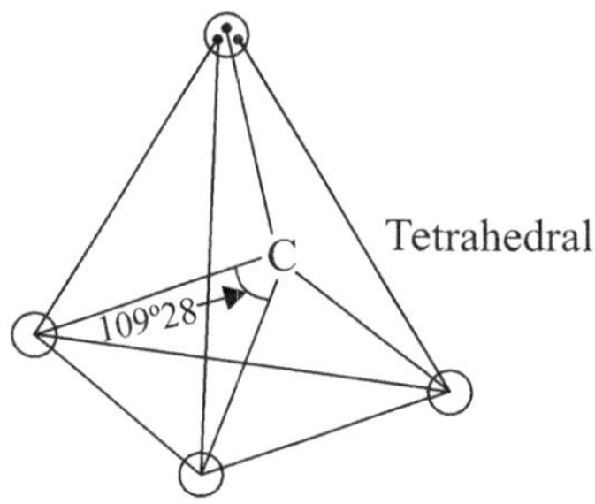

It should be understood that the required hybridisation energy is provided by the approaching atom.

It must be borne in mind that no covalent compound is possible without first undergoing hybridisation and only σ bonds which are strong can participate in hybridisation. Pi (π) bonds are weak due to lateral overlap, these orbitals do not take part in hybridisation. Hybridisation is further linked to the geometry of the molecule. Other examples of sp^3 hybridisation are:

Nitrogen in NH_3

$N \rightarrow 1s\, 2s^2\, 2p\, 2p^3$

$H \rightarrow 1s^1 \quad 2p\, 2p^3$

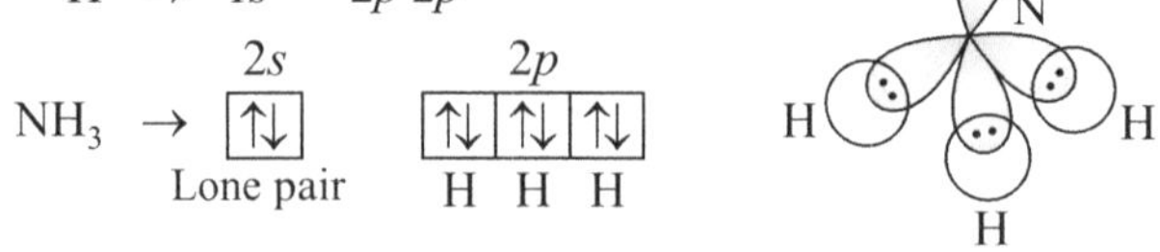

2*s* behave as lone pair of electrons lone pair of electrons are a part of hybridisation. NH_3 accepts sp^3 hybridisation like CH_4 discussed earlier. The lone pair affects the bond angle of tetrahedral geometry due to *lp bp* repulsion being stronger than *bp bp* repulsion. Due to unequal repulsions the angle ∠HNH closes in to 107°48′.

Similarly in H_2O molecule there are two lone pairs but it represents sp^3 hybridisation. The tetrahedral geometry offered is distorted with a bond angle of 104°. The decrease in bond ∠HOH is due to *lp lp* > *lp bp* > *bp bp* as shown in the diagram.

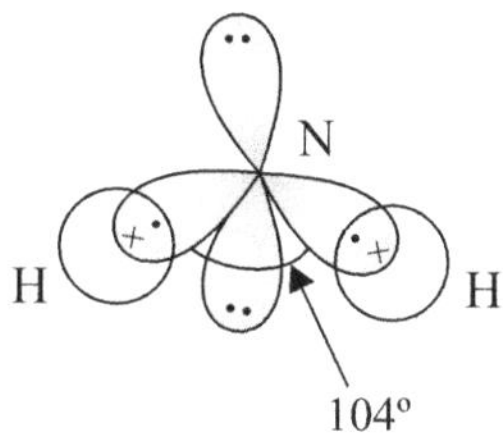

On similar lines in H_2S molecule the bond angle ∠HSH reduces to 91° due to *lp lp* being for more strong than bond pair S—H due to poor electronegativity of S.

Taking examples of C_2H_4 the formation of bonds is shown below:

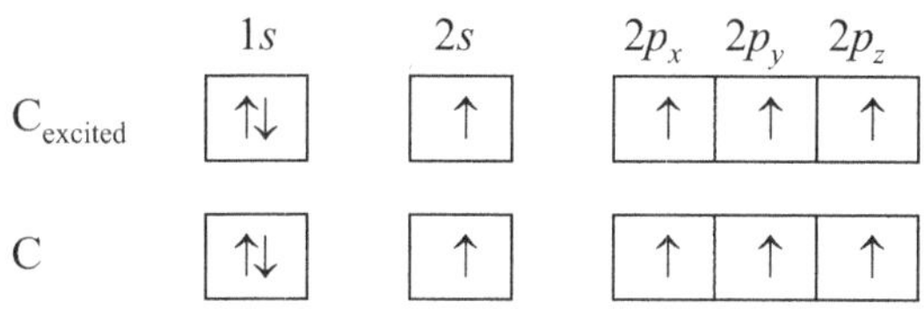

In 2C atoms there is one σ bond and Pi(π) bond.

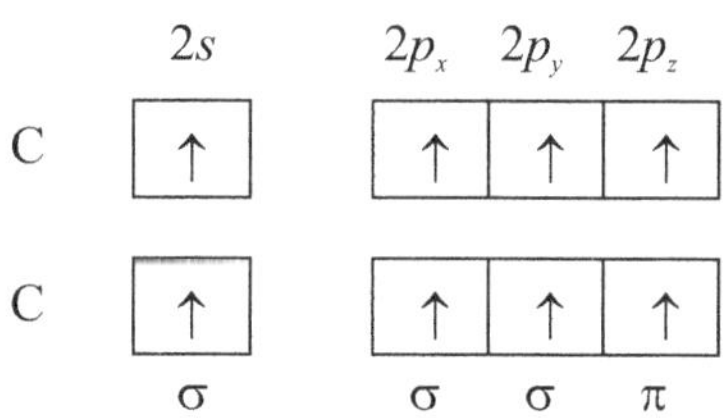

In C_2H_4

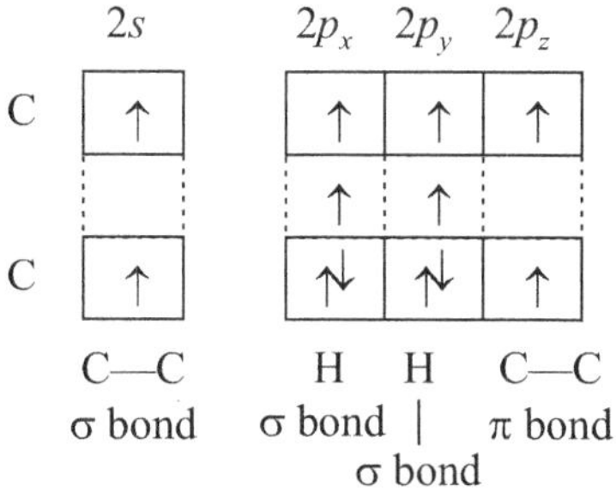

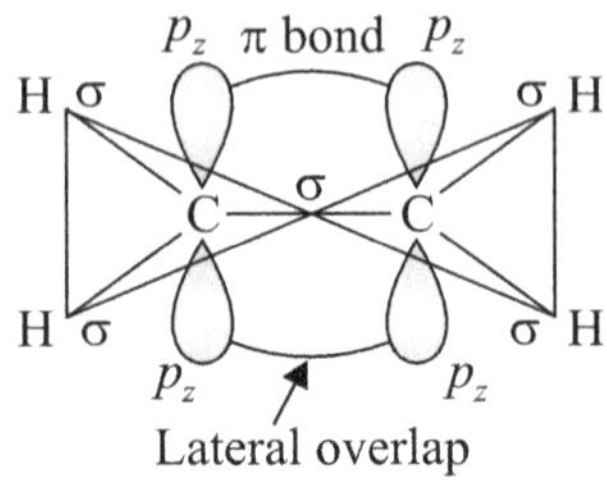

sp^2 hydridisation involves plane triangular geometry of the molecule (equilateral triangle) with bond angle ∠HCH = 60°.

In the case of C_2H_2

	2s	2p			
$C_{excited}$	↑	↑	↑	↑	
$C_{excited}$	↑	↑	↑	↑	all the four unpaired

1s

H → ↑

C_2H_2

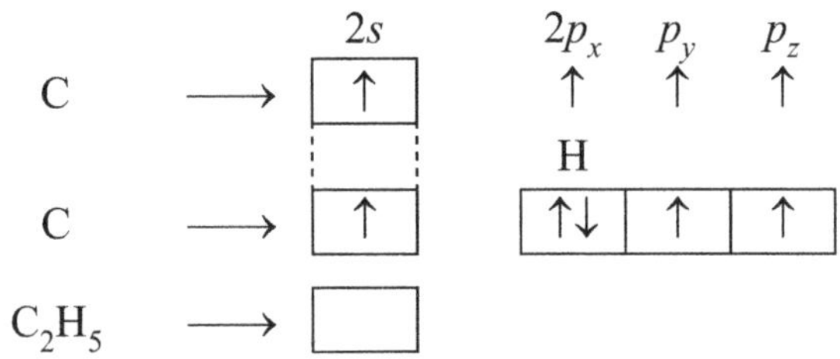

$$2s - 2s \text{ overlap} \longrightarrow \text{C—C } \sigma \text{ bond}$$
$$2(s - 2p_x) \longrightarrow \text{2C—H } \sigma \text{ bond}$$
$$2p_y - 2p_y \text{ overlap} \longrightarrow \text{C—C } \pi \text{ bond}$$
$$2p_z - 2p_z \text{ overlap} \longrightarrow \text{C—C } \pi \text{ bond}$$

The overall molecule represent a bond angle ∠HCH = 180° that is it represents a linear geometry with 2π bonds between the two C-atoms.

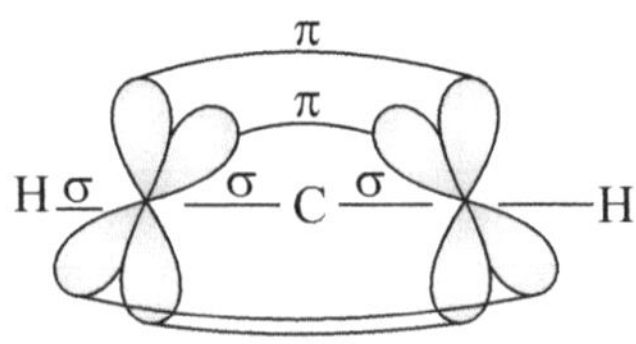

The hybridisation in such a case is *s–p* and the molecule has a straight geometry.

Participation of *d* Orbital in Hybridisation

In the event of compounds where one of atoms is very strongly electronegative like oxygen, chlorine and fluorine maximum energy is provided to the central atom to expose the central atom to maximum valency or to unpair maximum electrons which involves the participation of *d* orbitals *i.e.* lot of energy is provided to the central atom not only to unpair the electrons but also to make them jump to high energy *d* orbitals *e.g.*

PCl_5

	$3s$	$3p$			$3d$				
Valency shell P =	↑↓	↑	↑	↑					
$P_{ex.}$ =	↑	↑	↑	↑	↑				
PCl_5 =	↑↓	↑↓	↑↓	↑↓	↑↓				
	Cl	Cl	Cl	Cl	Cl				

$s - p^3 - d$ hybridisation

This hybridisation represents trigonal bipyramidal geometry. Trigonal pyramidal geometry is less symmetrical with the angle of the triangular base equal to 60° and that of apex chlorines having 90° angle.

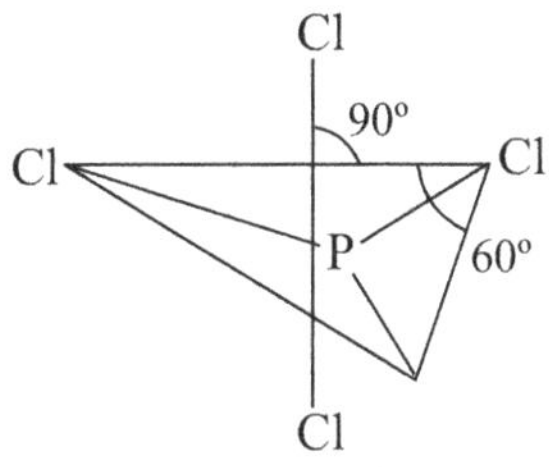

Exhibits

Similarly SF_6 molecule exhibits sp^3d^2 hybridisation with octahedral geometry as show:

	$3s$	$3p$			$3d$				
$S_{ex.}$	↑	↑	↑	↑	↑	↑			
SF_6	↑↓	↑↓	↑↓	↑↓	↑↓				
	F	F	F	F	F				

$S — P^3 — d^2$

The octahedral geometry is also called square bipyramid geometry. This is one of the most symmetrical geometry in which all the angles are of 90° and the sides are equal.

IF_7 is another example where *d* orbital gets involved in hybridisation. It exhibits sp^3d^3 on lines similar to PCl_5 and SF_6. This hybridisation gives pentagonal bipyramidal geometry which is also a less symmetrical geometry where bond angle of pentagon base is 72° and orthogonal bond is 90°.

***dsp*²** hydridisation is involved only in coordinate complexes with strong field ligands. It offers square planar geometry on *xy*-plane. For example

$$[Ni(CN)_4]^{2-}$$
$$2^+ + 4^- = 2^-$$

Here Ni is 2^+ and $4CN^-$ *i.e.,* the sum of the two charges is $2^{\ominus}$

	3*d*					4*s*	4*p*		
Ni^{2+}_{2s}	↑↓	↑↓	↑↓	↑	↑				

CN^- being strong field ligand provides enough energy to pair the two unpaired electrons in 3*d* orbitals leaving one orbital in *d* vacant

3*d*					4*s*	4*p*	
↑↓	↑↓	↑↓	↑↓	×	×	×	×
				CN^-	CN^-	CN^-	CN^-
				d –	*s* –	p^2	

The hybridisation invokes square Planar geometry. Similarly $[Cu(NH_3)_4]^{2+}$ adopts sp^3 hybridisation and a tetrahedral geometry.

Hybridisation Involved in Important Ions

Ammonium ion NH_4^+: In this cation the Lewis dot formula is as shown:

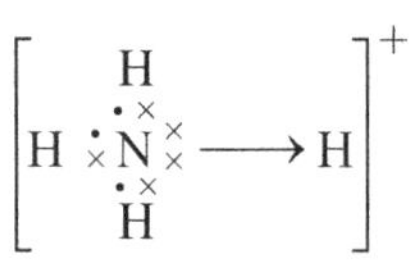

It entails sp^3 hybridisation:

	3*s*	3*p*			3*d*				
N → G. state	↑↓	↑	↑	↑					
NH_3 →	↑↓	↑↓	↑↓	↑↓					
	lone pair	H	H	H					

S—P^3 hybridisation propose tetrahedral structure.

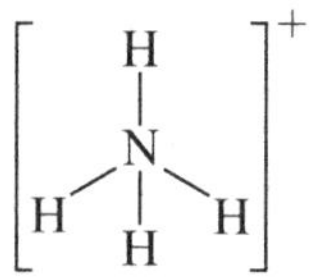

BO_3^{3-} ion Here, boron is the central atom. The Lewis structure is

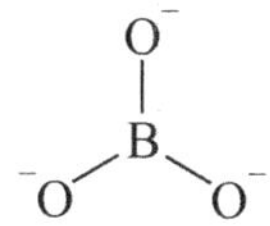

B	ground state	$1s^2$	$2s^2$	$2p^1$
(5)		$2s$		$2p$
B	excited	↑		↑ ↑

After forming hybrid orbitals it overlaps with *p*-orbitals of oxygen to give trigonal geometry with a bond angle 120°.

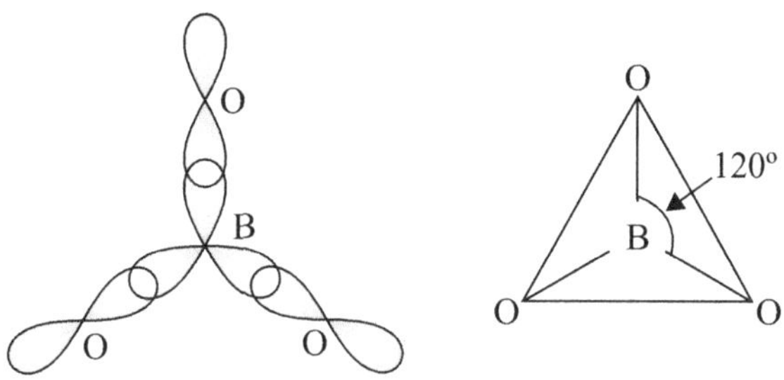

NO_3^- The Lewis structure is

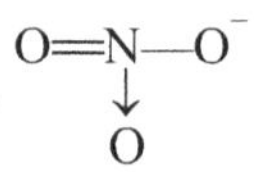

It has two σ bonds and a coordinate bond. The hybridisation involved is sp^2

$$N_{(7)} \longrightarrow 1s^2\, 2s^2\, 2s^3$$

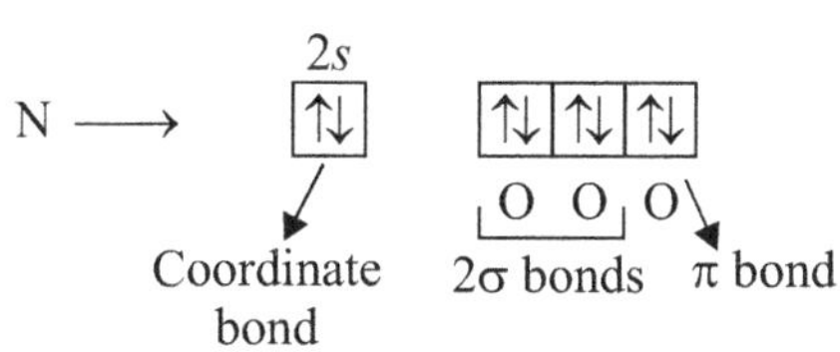

s—p^2

π bond does not participate in hybridisation therefore it shows sp^2 hybridisation

CO_3^{2-}

The dot formula forms the structure

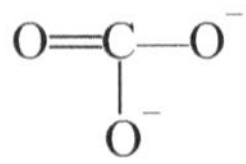

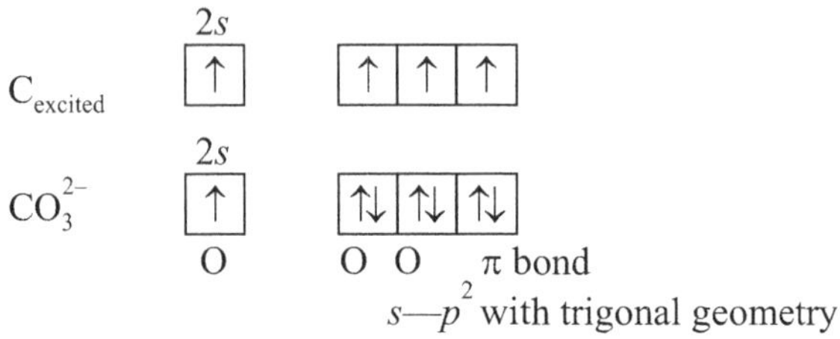

SO_4^{2-} Similarly sp^3 hybridisation is proposed for this ion with a tetrahedral structure

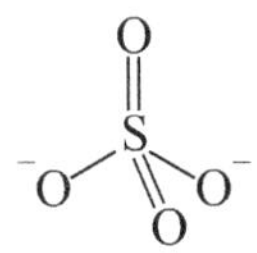

Chlorate ion ClO_3^-: The dot formula for this ion is

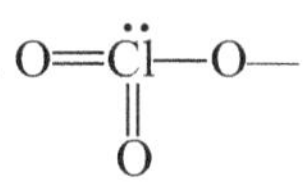

It involves sp^3 hybridisation with a tetrahedral structure.

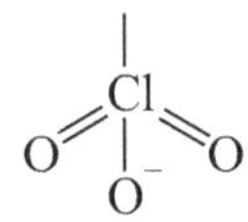

It has 3σ bonds and 2π bonds with one lobe having a lone pair.

Perchlorate in ClO_4^- It represents sp^3 hybridisation like ClO_3^- with the possible Lewis structure

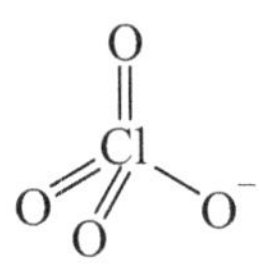

It has one σ bond with O^-

3 σ bonds with the rest of 3 oxygen atoms

3 π bonds with 3 oxygen atoms.

Valency Shell Electron Pair Repulsion (VSEPR)

Broadly, the type of geometry of a molecule is proposed by the type of hybridisation and in the same geometry there is variation in the bond angle or the geometry becomes irregular is explained thoroughly by the VSEPR theory. The molecule may have lone pair of electrons and bond pair of electrons. There is unequal force of repulsion between lone pair and bond pair electrons. The order of repulsion is *lp lp* > *lp bp* > *bp bp*. Due to unequal force of repulsion the geometry becomes irregular or distorted and the bond angle may vary.

1. If the central atom is surrounded only by bond pair electrons then the repulsion between bond pairs of electrons is equal and a regular geometry emerges.
2. In case the central atom is surrounded by *lp* and *bp* electrons. Due to unequal repulsion between *lp*'s and *bp*'s will go on till they experience zero or no repulsion . the geometry that the molecule then acquires is irregular or distorted. This varies the bond angle of the regular geometry. For example a perfect-tetrahedral in CH_4 has a bond angle of 109°, 28°. The molecule NH_3 cantain a lone pair on N-atom

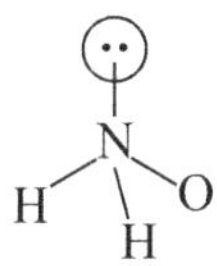

$\angle HNH = 107.5°$

H_2O: Oxygen the central atom contains 2 lone pairs therefore due to strong lone pair, lone pairs repulsion the angle $\angle HOH$ closes in to 105.5°

The tetrahedral with a bond angle of 105.5° would be distorted.

H_2S: The central S atom has two lone pairs like oxygen. There is wide difference in the electronegative value of O = 3.5 and S = 1.9 therefore *bp-bp* repulsion is very weak. In comparison *lp bp* repulsion becomes strong so the angle $\angle HSH = 92.5°$.

Which is just above the right angle hence the tetrahedral geometry is highly distorted and less stable.

The trend of the bond angle is decreasing in the order: H_2O (105.5°) > H_2S (92.5°) > H_2Se (91°) > H_2Te (89.5°).

The bond angle decrease is due to two opposing repulsions. If *bp-bp* repulsion is weak then *lp lp* or *lp bp* becomes all the more strong.

Limit of VSEPR Theory

1. It does not explain the shape of polar molecules Polar Li_2O is linear while angular H_2O is tetrahedral.
2. It could not explain the structure of certain molecules where lone pair becomes inert.
3. Some molecules which have delocalised π-electrons could not be given any geometrical shape.
4. It could not explain the shapes of molecules of transition metals. The Nickel d^8 molecules are square planar and not tetrahedral as predicted by Gillespie and Nyholm theory.

Strength and Stability of a Covalent-Bend

The strength of a covalent is only partly explained by the Valency Bond Theory (VBT). In fact covalent bond is much stronger than it is visualised by VBT theory.

The actual strength of covalent compounds is further accentuated by Molecular Orbital Theory (MOT), resonance, hydrogen bond and metallic bond .

Some of the shortcomings of VBT are listed below

1. In N_2 molecule, according to VB theory there exist one σ and two π bonds. Existence of *pi* bond in a molecule is synonymous with instability but N_2 molecule inspite of having 2 *pi* bonds is very stable or almost inert.
2. VBT is not able to explain the paramagentic behaviour of O_2 molecule when all the electrons (16) are paired.
3. VBT is based an atomic orbitals while in a molecule formation the atoms or atomic orbitals lose their identity *i.e.* the change is from monocentric to polycentric.
4. VBT could not why all the filled atomic orbitals do not form a covalent bond.

To explain the shortcomings of VBT a new theory known as Molecular Orbital Theory was proposed (MOT).

This theory advocated that when atomic orbitals coalesce a new orbital known as molecular orbital is born. This theory was put forward by Hund and Mulike. It proposed a new concept in the formation of a covalent bond. The concept is based on linear Combination of Atomic Orbitals (LCAO).

Based on LCAO method the molecular orbital theory has certain rules of its own

1. There are as many molecular orbitals as the number of combining atomic orbitals.
2. The molecular orbitals are bigger than atomic orbitals and cover the whole of the molecule.
3. The molecular orbitals are of varying energy.
4. The molecular orbitals are filled energy wise as in case of atomic orbitals. The molecular orbitals with lowest energy is filled first and so on.
5. Pauli's exclusion principle is applicable to molecular orbitals as well. Two electrons with opposite spins can occupy one molecular orbital.
6. In VBT atomic orbitals concentrate round one nucleus but molecular orbitals concentrate round the combination of atomic orbitals are polycentric.
7. Molecular orbital theory does not explain the existence of gases.

Bonding and Anti-Bonding Molecular Orbitals

When two atomic orbitals of hydrogen atoms combine according to LCAO method there are two possibilities. In case the atomic orbitals have the same direction or orientation (shown by the positive sign) then they coalesce to form a bonding molecular orbital of lower energy state.

In the other possibility the two atomic orbitals are not having the same direction or orientation then they repel one another or go to a higher energy state such a molecular orbital is given the name antibonding molecular orbital indicated by a star. Both these orbitals obey Aufbau's principle, Hund's rule and Pauli's exclusion principle, Diagramatically it is represented as:

1st Possibility

2nd Possibility

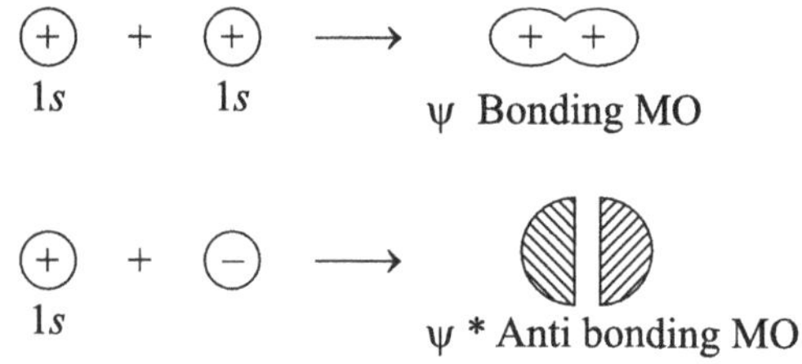

For H_2 molecule

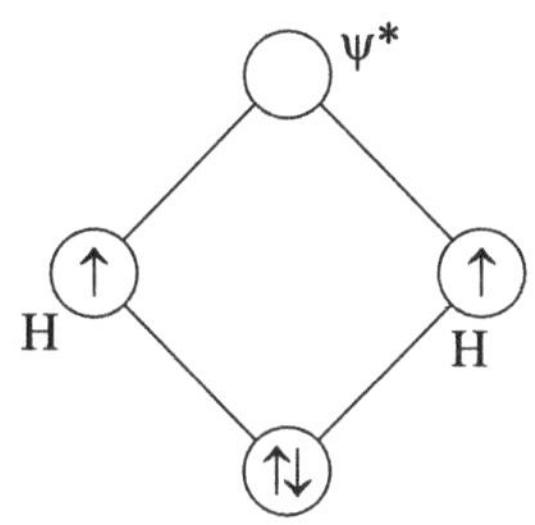

for determination of Bond order the formula $\frac{1}{2}$(B – A) electrons is applicable.

For a stable bond formation the number of electrons in the bonding MO should be more than the number of electrons in the antibonding MO.

In H_2 molecule the Bond order is $\frac{1}{2}(2-0) = 1$. We conclude that there exist a single bond between the two H atoms.

The molecule is stable because B.O. has a positive value. The positive value indicates that the molecule H_2 does exist and is quite stable.

In H_2^+ B.O. is $\frac{1}{2}$ (half) meaning thereby that H_2^+ ion does exist but stability is not of high order.

Similarly He_2 does not exist because.

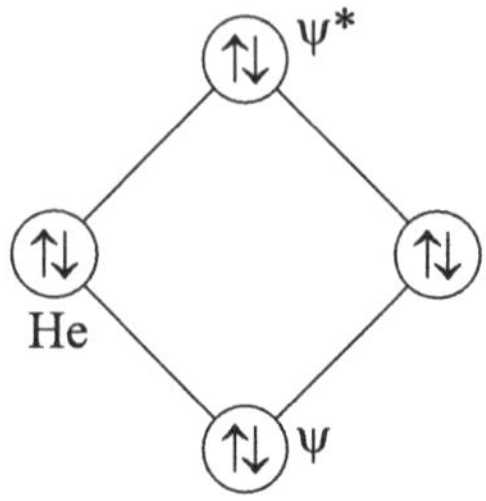

$$B.O. = \frac{1}{2}(2-2)$$
$$= O$$

Therefore He molecule does not exist as He_2 or He exists as a monoatomic molecule noble gas. This generalises the fact that all noble gases are monoatomic gases.

N_2 molecule

$$N_7 = 1s^2\, 2s^2\, 2p^3$$
$$N + N \rightarrow N_2$$
$$7 + 7 \rightarrow 14 \text{ electrons}$$

The energy level diagram up to 20 electron system is in the order:

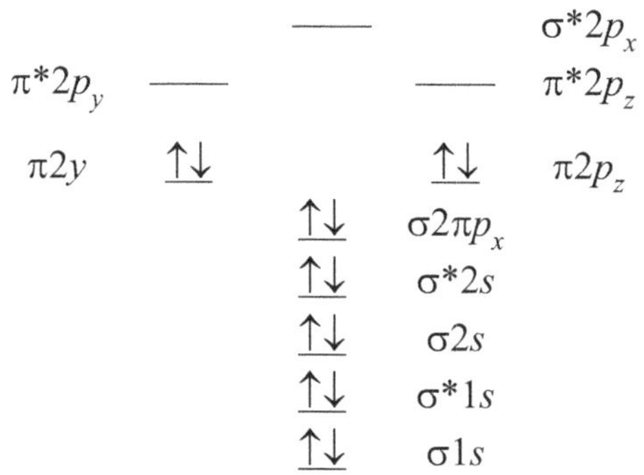

$$B.O. = \frac{1}{2}(B-A) = \frac{1}{2}(10-4) = 3$$

Therefore there exists a ≡ bond between the two nitrogen atoms N≡N.

The extra stability of N_2 molecule is due to the fact that low energy molecular orbitals have much larger number of electrons than the high energy antibonding molecular orbitals. Hence overall energy of the molecular becomes low or the stability is of high order which makes N_2 molecule of almost inert inspite of existence of two π bonds.

Another abnormality of O_2 molecule showing paramagnetic behaviour despite having even number of electrons (16).

MOT in case of Homodiatomic O_2 molecule:

$$O_8 + O_8 \rightarrow O_2\,(16)$$
$$O_8 \rightarrow 1s^2\,2s^2\,zp^4$$
$$8 + 8 \rightarrow 16 \text{ electrons in } O_2 \text{ molecule}$$

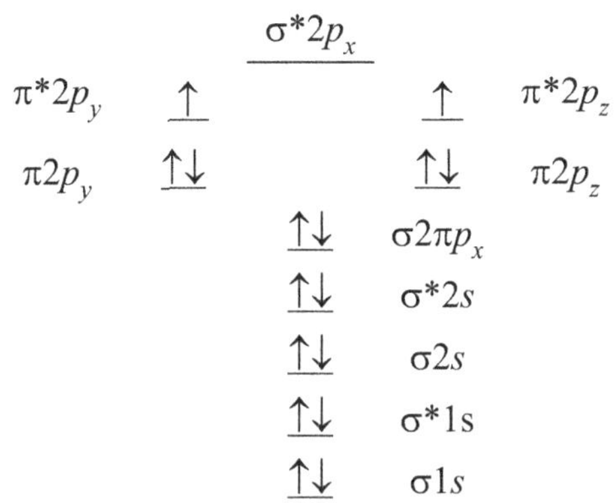

Ultraviolet light removes outer electrons from gases X-ray being more energetic is able to remove inner electrons as well. Photo electron spectra for N_2 and O_2 show lower energy peaks are meant for higher energy orbitals because less energy is required to remove these electrons. The energy levels in N_2 spectrum are closer than in the O_2 spectrum. The energy difference between first and third major peaks in N_2 is 3 eV while in O_2 it is 6 eV. The photo electron spectrum π_μ gives lower peak. The ionisation energy of the orbitals shows interaction between electronic and vibrational energy of the molecule. Vibrational energy levels are closer than electronic energy levels, Any collection of molecules has an energy distribution between various vibrational levels. Therefore, transition between electronic levels also includes vibrational levels, resulting in number of peaks for a single electron transition. Orbitals that are strongly involved in bonding have a vibrational multiple fine peaks. Orbitals that are less involved in bonding have only a few individual peaks at each energy level. N_2 spectrum indicates that more π_u orbitals are involved in bonding than the σ orbitals, The O_2 Photo electron spectrum has much more vibrational fine structure for all the energy levels with π_u levels more involved than other orbitals, The vibrational fine structure shows that all the orbitals are important in bonding in the molecule .

In the energy level diagram of O_2 two electrons remain unpaired in π^* antibonding orbitals (following Hund's Rule) which proves its paramagnetic behaviour.

B_2 molecule is found in gas phase. solid boron is found in hard form with complex bonding involving B_{12} icosahedra.

B_2 is paramagnetic. The Lewis dot formula Cannot account for the paramagnetic behaviour. The paramagnetic behaviour can be explained only by the MO model.

Energy level shift is caused by mixing of *s* and *p* orbitals. In the absence of mixing $\sigma_g(2p)$ orbital is expected to be lower in energy then $\pi_u(2p)$ orbitals resulting in the diamagnetic B_2. During mixing $\sigma_g(2s)$ orbital with $\sigma_g(2p)$ orbital of higher energy lowers the energy of the π orbital and increases the energy of $\sigma_g(2p)$ orbital to a higher level. As a consequence the last two electrons fill the degenerate π orbitals and the molecule behaves paramagnetic. The B.O. is 1 although the two π electrons are in different orbitals.

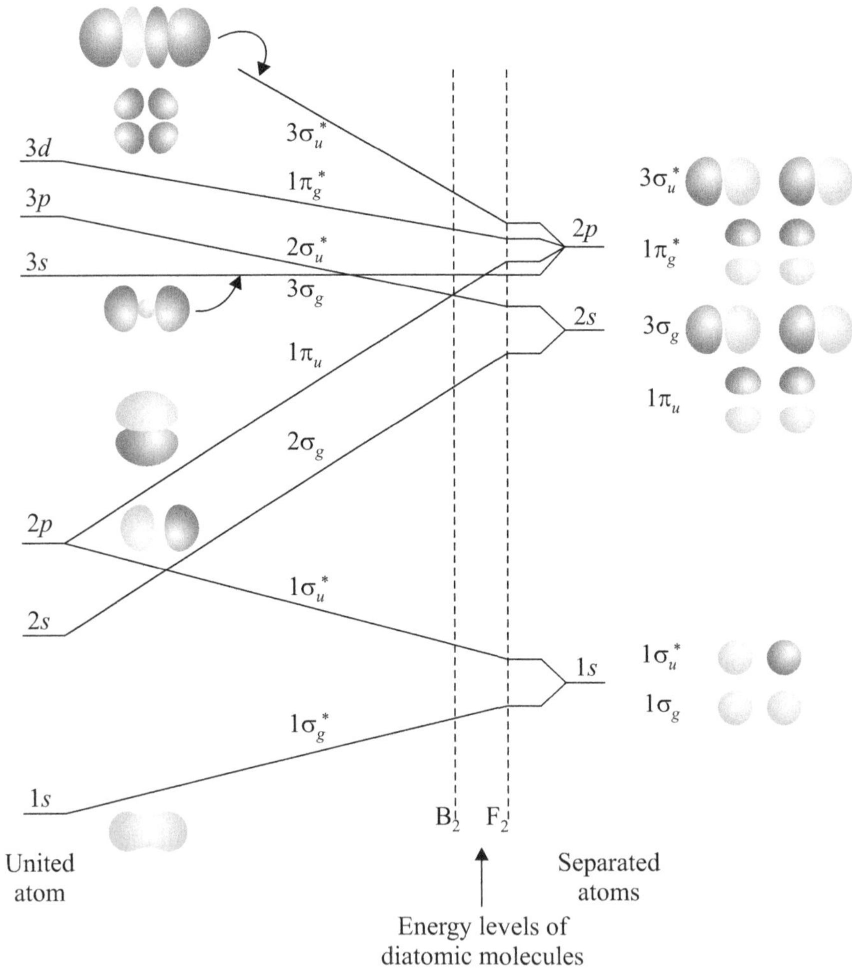

Energy diagram of Homonuclear Diatomic Molecules

$C_2[\pi_u^2\ \pi_u^2\ (2p)]$: MO diagram shows a double bonded molecule with all paired electrons with both highest occupied molecular orbitals (HOMOs) having π symmetry. It has two π bonds and no σ bond. The dissociation energies of B_2, C_2 and N_2 show a steady increase indicating a single, double and triple bond with increasing atomic number. C_2 is commonly encountered in spices like diamond, graphite and fullerenes, C_2^{2-} is well known in acetylides of alkali, alkaline earth and lanthanides, Molecular orbital diagram of C_2^{2-} shows a bond order 3 having configuration $\pi_u^2\ \pi_u^2\ \sigma_g^{\ 2}$. This is supported by —C —C— bond distance in acetylene and calcium acetylide.

Molecule	—C—C— pm
C=C(gas)	132
H—C≡C—H	120.5
CaC_2	119.1

Molecular Orbital in Heterodiatomic Molecules

Molecular Orbitals and Photo electron spectrum of CO: Molecular orbitals of 1*s* orbitals are not shown as they do not matter. These are usual atomic orbitals 1*s*, 2*s* and 2*p* of the two separate atoms. As the atoms approach each other they interact to form bigger molecular orbitals, 1*s* orbitals form $1\sigma_g$ and $1\sigma_u$*, 2*s* form $2\sigma_g$ and $2\sigma_u^*$ and 2*p* form $3\sigma_g$, $1\pi_u$, $1\pi_g^*$ and $3\sigma_u^*$. As they move closer the bonding molecular orbitals decrease in energy while antibonding MOs increase in energy shown in the figure.

Energy level diagram of CO Molecule

Molecular orbital for NO is similar to CO except that it has one unpaired electron in π^*. Hence NO is paramagnetic fig. (*p*.83).

$$B.O = \frac{8-3}{2} = 2.5$$

This agrees with the value 115 pm for bond length and Bond Energy 668 kJ/mol. Removal of electron gives NO^+ mitrosonium ion,

The electron is removed from π^* antibonding molecular orbital, It resembles the configuration of N_2 (14 electrons) and NO^+ (14 electrons). They are iso electronic.

$$B.O.\ in\ NO^- = \frac{8-4}{2} = 2$$

Out of NO, NO^+ and NO^-, NO^- has the largest bond and is also paramagnetic

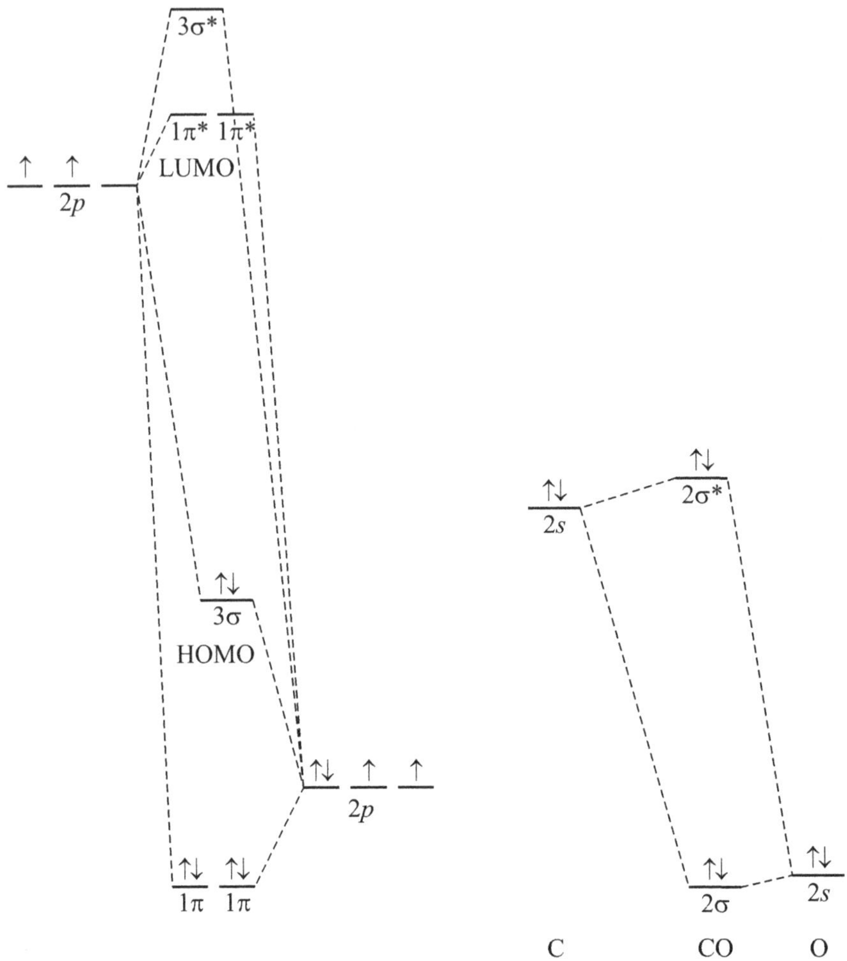

Molecular Orbitals CO. Molecular orbitals 1σ and 1σ* are from the 1*s* orbitals and are not shown.

Molecular Orbitals for Bigger Molecules

CO_2 has a more complicated orbital description. The central atom in CO_2 *i.e.* carbon atom has both *s* and *p* orbitals interacting with 2*p* orbitals on the oxygen atoms figure (*p* 83). The group orbitals of the oxygen atoms are shown in the figure. To determine atomic orbitals of carbon of correct symmetry to interact with the group orbitals, consider the combinations shown in the figure (83).

Group orbitals 1 and 2 in figure formed by adding and subtracting oxygen 2*s* orbitals have Ag and B_{1u} symmetry respectively Group 1 is of correct symmetry to interact with 2*s* orbital of carbon having Ag symmetry. Group orbital 2 has appropriate symmetry to interact with $2p_z$ orbital of carbon (B_{1u} symmetry)

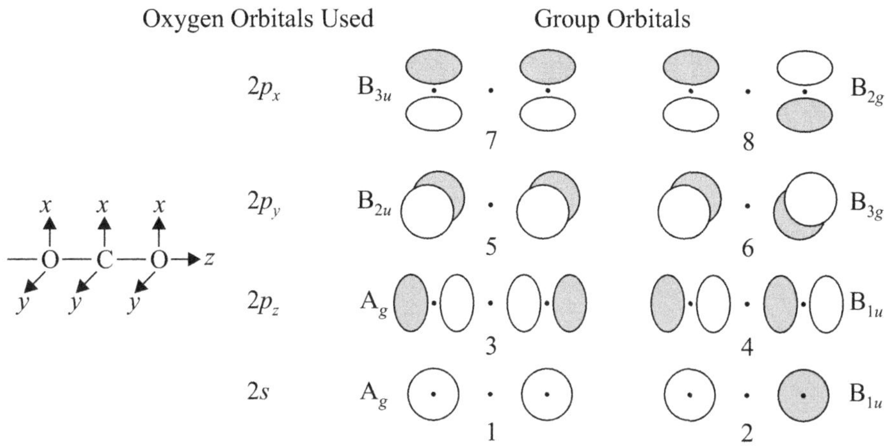

Group Orbital Symmetry in CO_2.

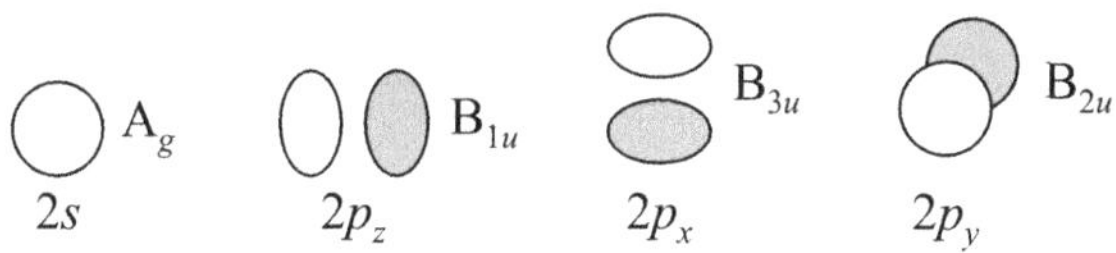

Symmetry of the Carbon Atomic Orbitals in the D_{2h} Point Group.

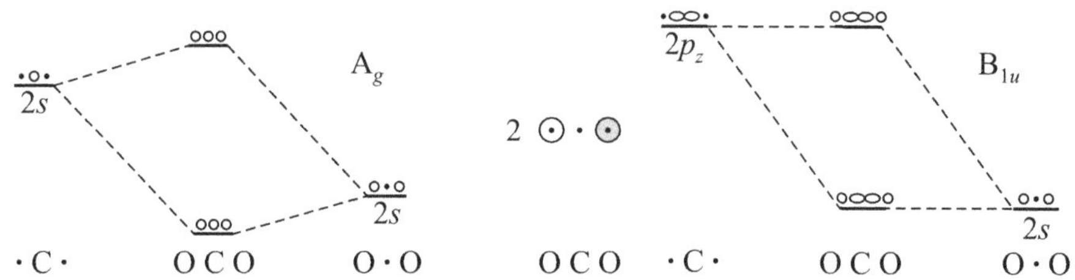

Group Orbitals 1 and for CO_2.

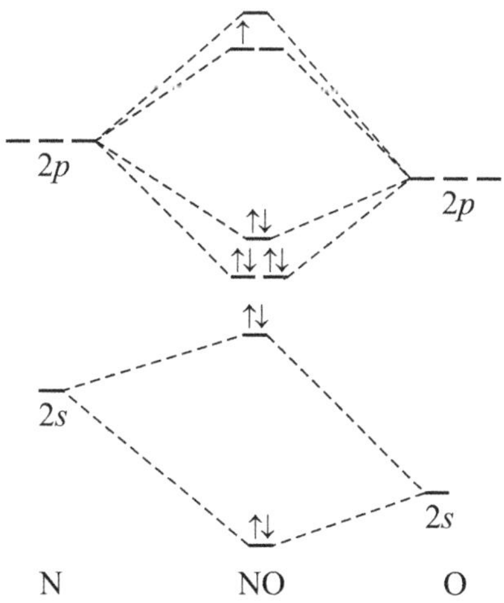

Molecular Orbitals of NO.

Group orbitals 3 and 4 in figure formed by adding and subtraction oxygen $2p_z$ orbitals have orbitals have Ag and B_{1u} symmetries. Group orbital 3 can interact with 2*s* orbital of carbon and group orbital 4 can interact with $2p_z$ of carbon.

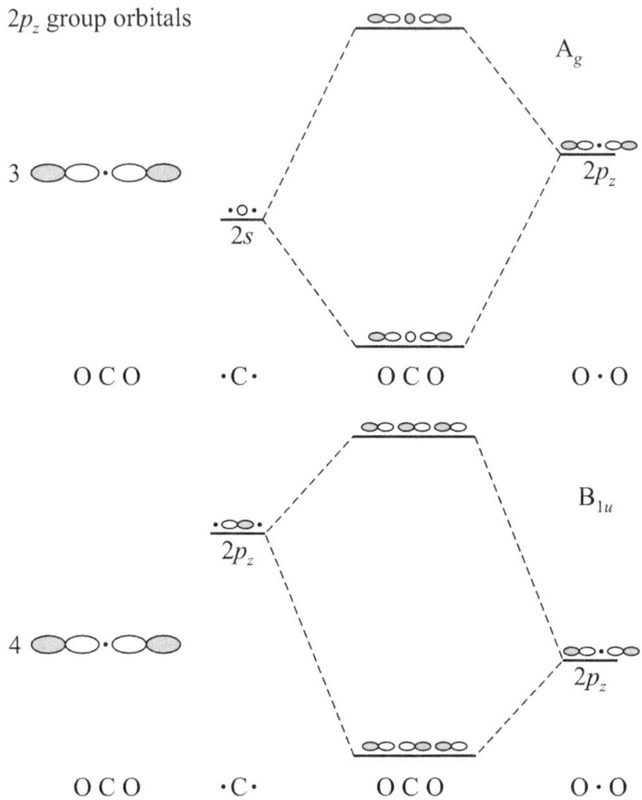

Group Orbitals 3 and 4 for CO_2.

The group 2*s* and $2p_z$ orbitals of carbon have two possible sets of group orbitals with which they may interact, therefore all the four interactions in figure (146) and 5.22 can occur and all the four symmetries are allowed.

Then it is estimated which interactions are the strongest from the potential energies of the 2*s* and 2*p* orbitals of carbon and oxygen figure. Orbitals having similar energies give the strongest interaction. The group orbital 1, from the 2*s* orbitals of oxygen and group orbital 3 from the $2p_z$ orbitals have the appropriate symmetry to interact with 2*s* orbital of carbon. The energy gap between group orbital 3 and 2*s* orbital of carbon is 3.6 eV than the energy gap between group 1 and 2*s* orbital of carbon (12.9 eV) therefore the interactives between $2p_z$ orbital of oxygen and 2*s* orbital of carbon is suggested. Group orbital 2 has energy difference 21.7 eV with $2p_z$ orbital of carbon to suggest low or no interactions.

Orbital	2s	2p
Carbon	–19.4 eV	–10.7 eV
Oxygen	–32.4 eV	–15.9 eV

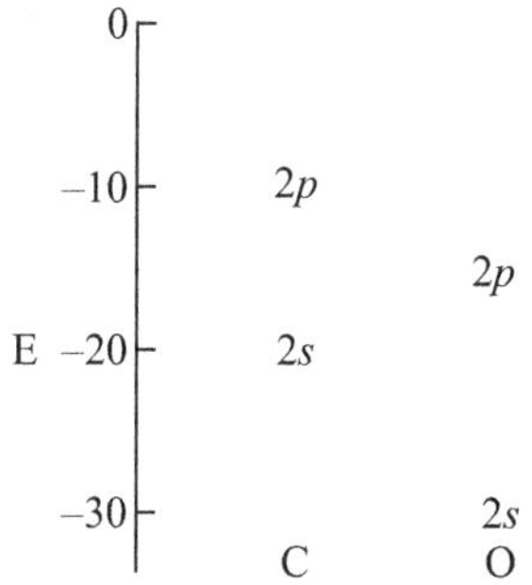

Orbital Potential energies

Orbital Potential Energies of Carbon and Oxygen.

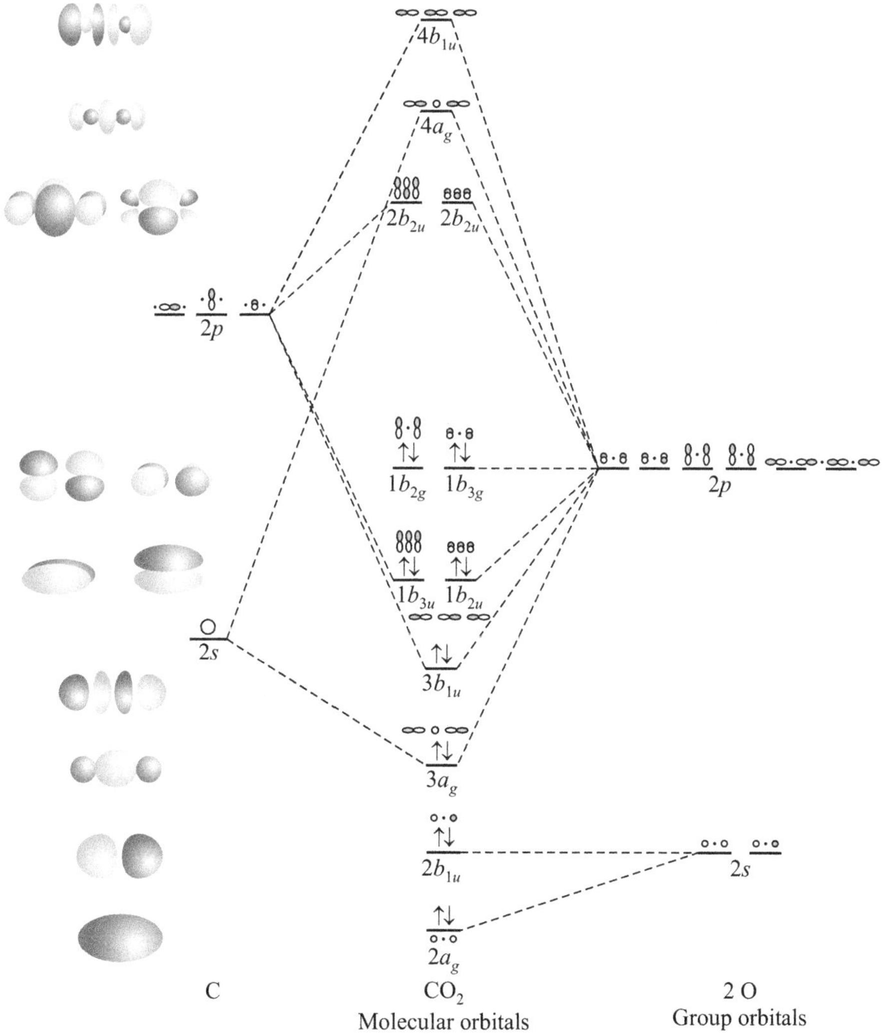

Molecular Orbitals of CO_2.

In the final picture of molecular orbital no interaction with carbon orbitals of group 1 and 2.

The $2p_z$ orbital of carbon interacts with group orbital 5. This interaction results in the formation of π molecular orbitals both bending and antibonding. However, there is no orbital of carbon B_{3g} symmetry to interact with group orbital 6, formed by combining $2p_3$ orbitals of oxygen. Hence group orbital 6 is non-bonding, Interacting $2p_3$ orbitals are akin to $2p_y$ orbitals. Group orbitals 7 with B_{2u} interacts with $2p_z$ orbitals of carbon to form π bonding and antibonding orbitals while group orbital 8 is nonbonding (figure).

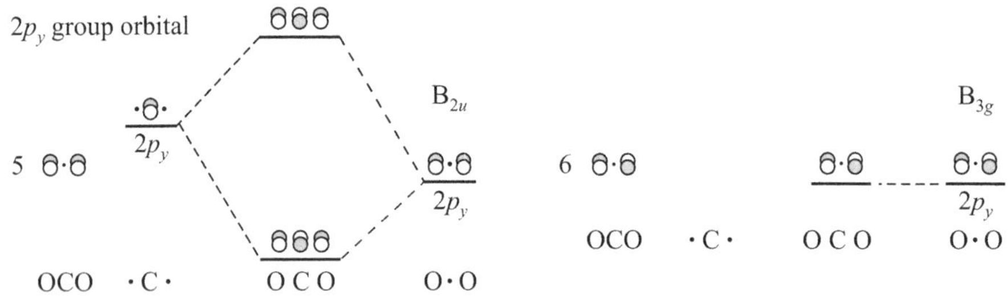

Group Orbitals 5 and 6 for CO_2.

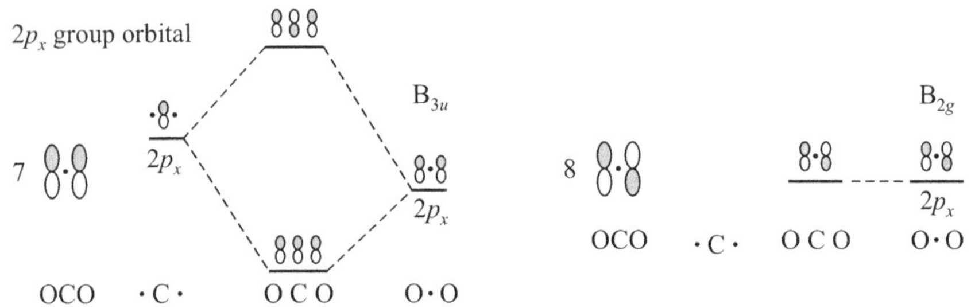

Group Orbitals 7 and 8 for CO_2.

The overall molecular orbital diagram of CO_2 is given in (page 85) According to diagram 16 valence electrons occupy from the bottom, two nonbonding σ orbitals, two bonding σ orbitals, two bonding π orbitals and two non-bonding π orbitals. In a way there are two banding electron pairs in σ orbitals and two are in π orbitals and in all there are four bands in the O=C=O molecule . The molecular orbital picture of linear triatomic molecules such as N_3^-, CS_2 and OCN^- can be determined like wise.

H_2O: H_2O is a non-linear or bent molecule. Molecular orbitals involved can be determined by the procedures adopted in CO_2 molecule. The various steps involved are:

1. Water is triatomic bent molecule with a z-axis through the oxygen atom and two mirror planes intersecting is this axis.

2. The z-axis and xz plane is seen as the plane of the molecule. Hydrogen has non-directional $1s$ orbital therefore it is not necessary to assign axes to the hydrogen.
3. The two hydrogen atoms determine the symmetry of the molecule. The identity operation leaves both hydrogen atoms unchanged with a character of 2. Two fold rotation inter changes the orbitals so that each orbital contributes O to the total character O. Reflection in the plane of the molecule (σv) leaves both hydrogens unchanged reflection perpendicular to the plane of the molecule ($\pi v'$) makes the two orbitals to a character of O.

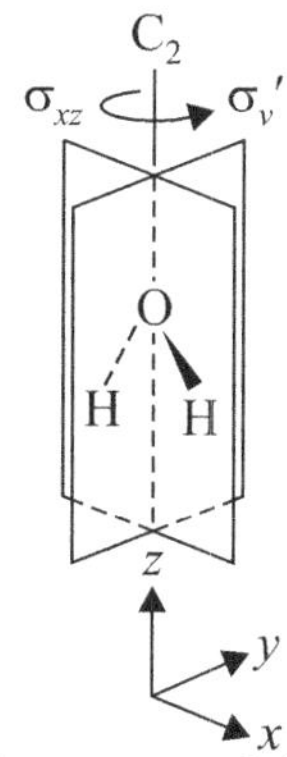

Symmetry of the Water Molecule.

Representations for C_{2v} Symmetry Operations for Hydrogen Atoms in Water

C_{2v} Character Table

C_{2v}	E	C_2	$\sigma_v(xz)$	$\sigma_v'(yz)$		
A_1	1	1	1	1	z	x_2, y_2, z_2
A_2	1	1	– 1	– 1	R_z	xy
B_1	1	– 1	1	– 1	x, R_y	xz
B_2	1	– 1	– 1	1	y, R_x	yz

$$\begin{bmatrix} H_a' \\ H_b' \end{bmatrix} = \begin{bmatrix} 1 & 0 \\ 0 & 1 \end{bmatrix} \begin{bmatrix} H_a \\ H_b \end{bmatrix}$$ for the identify operation.

$$\begin{bmatrix} H_a' \\ H_b' \end{bmatrix} = \begin{bmatrix} 0 & 1 \\ 1 & 0 \end{bmatrix} \begin{bmatrix} H_a \\ H_b \end{bmatrix}$$ for the C_{2v} operation

$$\begin{bmatrix} H_a' \\ H_b' \end{bmatrix} = \begin{bmatrix} 1 & 0 \\ 0 & 1 \end{bmatrix} \begin{bmatrix} H_a \\ H_b \end{bmatrix}$$ for the σ_v reflection (xz plane)

$$\begin{bmatrix} H_a' \\ H_b' \end{bmatrix} = \begin{bmatrix} 0 & 1 \\ 1 & 0 \end{bmatrix} \begin{bmatrix} H_a \\ H_b \end{bmatrix}$$ for the σ_v' reflection (yz plane)

The reducible representation $T = A_1 + B_1$:

C_{2v}	E	C_2	$\sigma_v(xz)$	$\sigma_v'(yz)$	
T	2	0	2	0	
A_1	1	1	1	1	z
B_1	1	– 1	1	– 1	x

[21]Some sources use the yz plane as the plane of the molecule. This convention results in $T = A_1 + B_2$ and switches the b_1 and b_2 labels of the molecular orbitals.

4. Irreducible representations $A_1 + B_1$ indicate the symmetrices of the group orbitals. These group orbitals are matched with matchable symmetries on oxygen. 2*s* and 2*p* orbitals have A_1 symmetry and $2p^z$ orbital has B_1 symmetry , so on combining the two hydrogen 1*s* orbitals, the sum of the two $\frac{1}{\sqrt{2}}[\psi(\text{Ha}) + \psi(\text{Hb})]$ has A_1 symmetry and $\frac{1}{\sqrt{2}}[\psi(\text{Ha}) - \psi(\text{Hb})]$ has B_1 symmetry (page 89). In this case the atomic-orbitals are identical and have equal coefficients, so they contribute group orbitals in equal proportions.
5. The same can be applied to oxygen orbitals each orbital remains unchanged in all operations so it has A_1 symmetry.

   The p_z orbital has B_1 symmetry along *x*-axis.

   The p_y orbital has B_2 symmetry along *y*-axis.

   The p_x orbital has A_1 symmetry along *z*-axis.

   The *x*, *y*, and *z* variable are very complex function.
6. The atomic and group orbitals of the same symmetry combine into molecular orbitals as listed in (Table 4.1) and shown in (figure overleaf). They are given the numbers ψ_1 to ψ_6 with ψ_1 of lowest energy and ψ_6 with ψ_1 of lowest energy and ψ_6 highest energy.

   The A_1 group orbital joins with *s* and p_2 orbital of oxygen to form bonding, non-bonding and antibonding molecular orbitals, *i.e.* ψ_1, ψ_3 and ψ_5.

Table 4.1 Molecular Orbitals for Water

Symmetry	Molecular Orbitals		Oxygen Atomic Orbitals		Group Orbitals from Hydrogen Atoms	Description
B_1	ψ_6	=	$c_9\psi_1(p_x)$	+	$c_{10}[\psi_1(H_a) - \psi_1(H_b)]$	antionding (c_{10} is negative)
A_1	ψ_5	=	$c_7\psi(s)$	+	$c_8[\psi_1(H_a) + \psi_1(H_b)]$	antionding (c_8 is negative)
B_2	ψ_4	=	$\psi_1(p_y)$	+		nonbonding
A_1	ψ_3	=	$c_5\psi(p_z)$	+	$c_6[\psi_1(H_a) + \psi(H_b)]$	nearly nonbonding (slightly bonding: c_6 is very small)
B_1	ψ_2	=	$c_3\psi(p_x)$	+	$c_4[\psi(H_a) - \psi(H_b)]$	bonding (c_4 is positive)
A_1	ψ_1	=	$c_1\psi(s)$	+	$c_2[\psi(H_a) + \psi(H_b)]$	bonding (c_2 is positive)

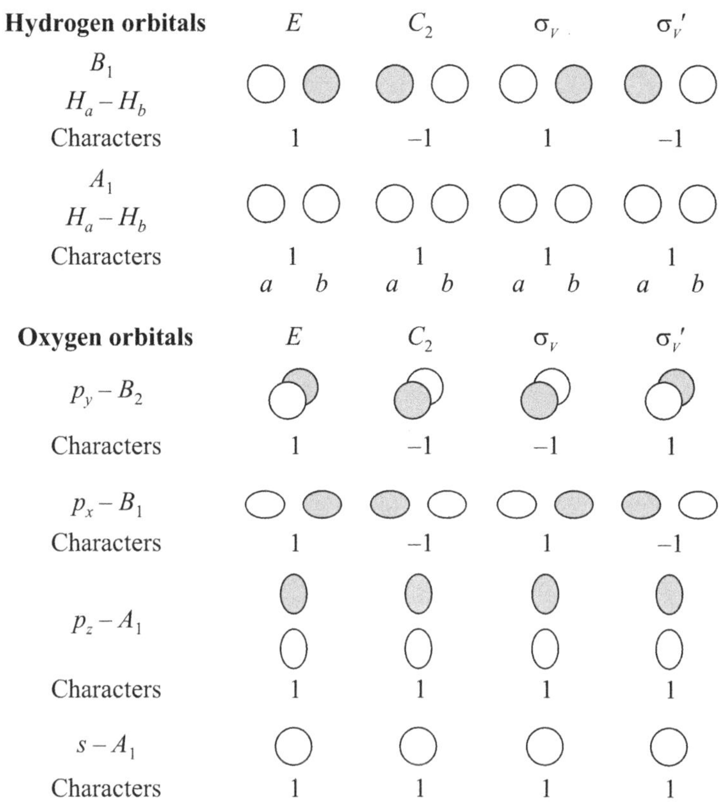

Symmetry of Atomic and Group Orbitals in the Water Molecule.

The p_z orbital of oxygen has only minor contributors from the other orbitals to form weakly bonded ψ_3 orbital, and the oxygen s orbital and the hydrogen group orbitals combine to form bonding ψ_3 and antibonding ψ_5 orbital.

The B_1 group orbital of hydrogen combines with p_z orbital of oxygen bonding, and antibonding molecular orbitals ψ_2 and ψ_6. The p_y oxygen ψ_4 with B_1 symmetry does not have same symmetry as $1s$ hydrogen group orbital therefore it forms non-bonding and two antibonding orbitals, two non-bonding and two antibonding orbitals. The $2s$ oxygen has very little interaction with hydrogen orbitals because of difference in energy. The $2p$ orbitals of oxygen are matchable with $1s$ hydrogen orbital forming bonding b_1 and a_1 molecular orbitals.

There are a total of 8 valence electrons that are added; two pairs in bonding orbitals and two pairs in nonbonding orbitals which are equal to two bonds and two lone pairs of dewis-dot structure. The lone pairs are lodged in molecular orbitals, one b_2 from p_y of oxygen and a_1 from s and p of oxygen and the two $1s$ orbitals of hydrogen (figure above).

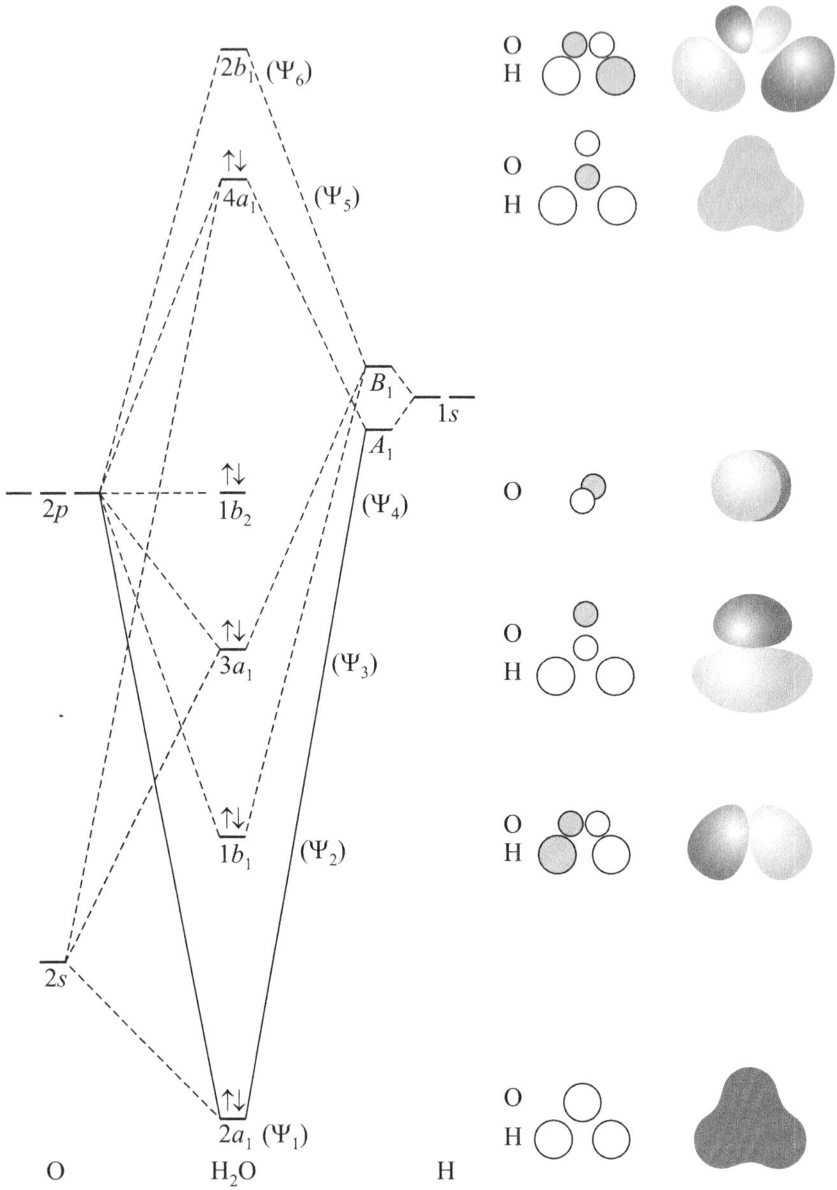

Molecular Orbitals of H_2O.

The molecular orbital diagram is different from two lone pairs and two similar —OH bonds. According to the MO picture the highest energy electron pair is nonbonding. The other two electron pairs are bonding pairs resulting in $2p_z$ and $2p_z$ overlap with $1s$ orbital of the hydrogens.

The lowest energy pair is the lone pair in $2s$ orbital of oxygen. Therefore, all the four occupied molecular orbitals are different.

NH_3: According to the VSEPR theory NH_3 molecular acquires a pyramidal structure with a lone pair of electrons and 3*v* symmetry. In the molecular orbital picture of NH_3 molecule the lone pair down the 3*v* or Z_v axis with the *yz* plane passing through one of the hydrogen atoms. The three hydrogen atoms with 1*s* orbitals is reduced to A_1 and *E* representations with orbital combinations shown in Fig. below. Three hydrogen 1*s* orbitals combines with one A_1 symmetry and two *E* symmetry. The *s* and p_z orbitals of nitorgen have A_1 symmetry and p_z lone pair, p_y has *E* symmetry exactly the same representation holds for three 1*s* hydrogens. In NH_3 it is not possible as A_1 symmetry of the three hydrogen 1*s* orbitals is seen but the two group orbitals of *E* symmetry are not seen.

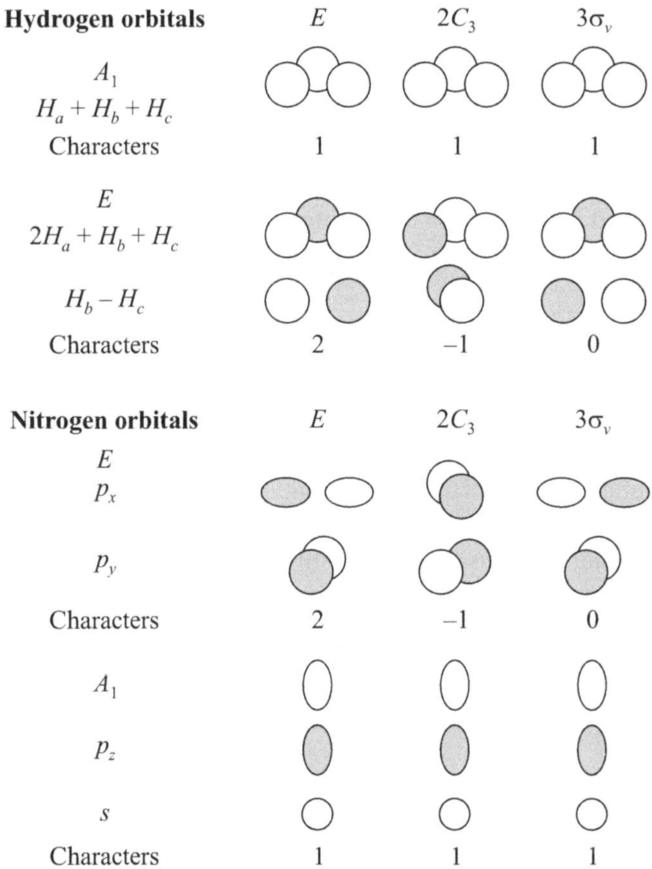

Group Orbitals of NH_3.

The molecular orbitals is the sum of the squares of the coefficients of each of the atomic orbitals in the LCAO equals 1 for each of the atomic orbital. In the second instance symmetry of central atom orbitals matches the symmetry of group orbitals with which they are combined. The *E* symmetry of nitrogen

p_x, p_y group orbitals must match the E symmetry of SALC. This requires one node for each of the E group orbitals with three atomic orbitals. The exact combinations are:

$$\frac{1}{\sqrt{6}}[2\psi(\text{Ha}) - \psi(\text{Hb}) - \psi(\text{Hc})] \text{ and } \frac{1}{\sqrt{2}}[\psi(\text{Hb}) - \psi(\text{Hc})]$$

The coefficients in these group orbitals result in equal contributions by each atomic orbital when each term is squared. The terms for each orbital are summed up:

$$\text{For Ha the contribution is } \left(\frac{2}{\sqrt{6}}\right)^2 = \frac{2}{3}$$

$$\text{Hb and Hc } \left(\frac{1}{\sqrt{6}}\right)^2 + \left(\frac{1}{\sqrt{2}}\right)^2 = \frac{2}{3}$$

Ha, Hb and Hc each has a contributed of $\frac{1}{3}$ in A_1 group orbital.

$$\frac{1}{\sqrt{3}}[\psi(\text{Ha}) + \psi(\text{Hb}) + \psi(\text{Hc})] = \left(\frac{1}{\sqrt{3}}\right)^2 = \frac{1}{3}$$

The total contribution is equal to 1 by each atomic orbital.

Again each group orbital is treated as a single orbital in combining with nitrogen orbitals. The s and p_z orbitals combine with hydrogen A_1 group orbitals to give three a_1 orbitals (one bonding, one non-bonding and one antibonding). The non-bonding orbital is almost p_z, with the nitrogen s orbital combining with hydrogen group orbitals to give bonding and antibonding orbitals.

p_z and p_y orbitals of nitrogen combine with E group orbitals:

$$\frac{1}{\sqrt{6}}[2\psi(\text{Ha}) - \psi(\text{Hb}) - \psi(\text{Hc})] \text{ and } \frac{1}{\sqrt{2}}[\psi(\text{Ha}) - \psi(\text{Hc})]$$

to form 4 orbitals, two bonding and two antibonding (e has a dimension of 2, which requires a pair of degenerate orbitals) when 8 electrons are put into lowest energy levels, three bond and one lone pair are obtained. The 1s orbital energies of hydrogen match well with the energies of the nitrogen 2p orbitals going large energy difference bonding and antibonding orbitals. The nitrogen 2s has a low energy therefore its interaction with hydrogen orbitals is quite small.

The HOMO of NH_3 is slightly bonding because it an electron pair in an orbital resulting from the interaction of 2p_z orbital of nitrogen with the 1s orbitals of hydrogens. (figure overleaf)

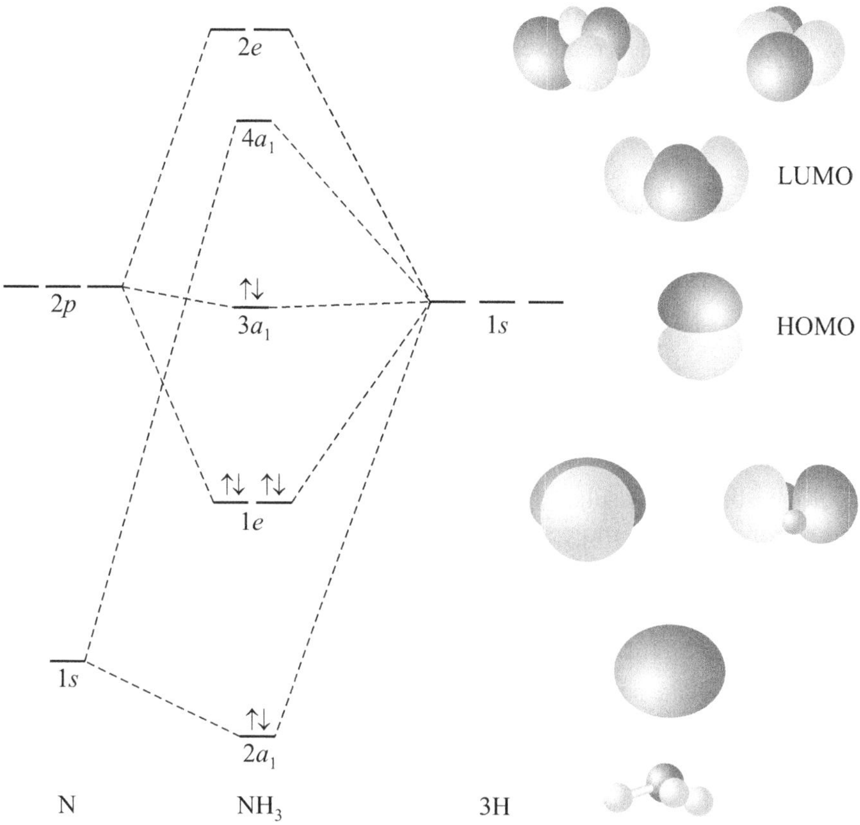

Molecular Orbitals of NH_3. All are shown with the orientation of the molecule at the bottom.

Hydrogen bonding: The concept of hydrogen bonding was introduced by Latener and Rodebush in 1920. It is extreme manifestation of dipole-dipole interactions. When a hydrogen atom is sandwiched between two strongly electronegative atoms like F, O and N having low atomic size then a unique bond is formed in which hydrogen atom bonding to one molecule bonded to electronegative atom such as F, O and N of the other molecule. Due to weak electrostatic forces a weak bond known as hydrogen bond is formed between two or more molecules. For example in molecule AH, A is more electronegative than H atom so A pulls bond electrons towards itself and hydrogen becomes electron deficient and tends to bond with lone pair electron of F, O and N atoms.

$$A^- \leftarrow H \longrightarrow \overset{\delta^-}{A} \rightarrow \overset{\delta^+}{H}$$

$$a_1 \overset{d}{\longleftrightarrow} a_2$$

The bond energy $F = \dfrac{a_1 \times a_2}{d^2}$

It is also called a H-bridge which unites two or more molecules. It is kind of weak covalent bond represented by a dotted line .... Its energy range is between 2 to 10 Kcal/mole while a normal covalent bond is ~ 100 Kcal/mole.

e.g. **water**

(*i*) 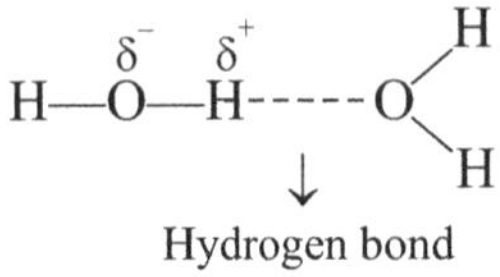

(*ii*) **HF** $H—\overset{\delta^-}{F}------\overset{\delta^+}{H}—\overset{\delta^-}{F}$

(*iii*) 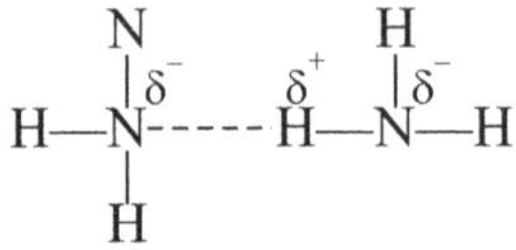

(*iv*) $\overset{\delta^-}{A}—\overset{\delta^+}{H}------\overset{\delta^-}{B}—\overset{\delta^+}{H}$

Generalised example

Hydrogen bonding results in the formation of long chain clusters of a large number of associated molecules.

Existence of hydrogen bond

It exists due to

(*i*) inadequate screening of nucleus by shared pair of electrons.

(*ii*) the small size of hydrogen atom (0.3 Å). A hydrogen atom contains one proton therefore, a strong polarising effect exists due to other electronegative element.

Types of hydrogen bonds

There exist two types of hydrogen bonds

(*i*) Intermolecular hydrogen bond

(*ii*) Intra molecular hydrogen bond

Intermolecular Hydrogen Bond

When a hydrogen bond exists between two or more molecules of the same or different species then it is called inter molecular hydrogen bonding *e.g.*

(*a*) 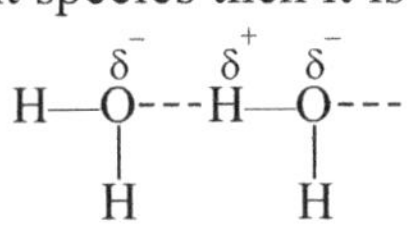

(*b*) 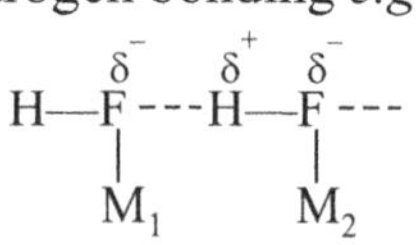

Phenol

(*c*) 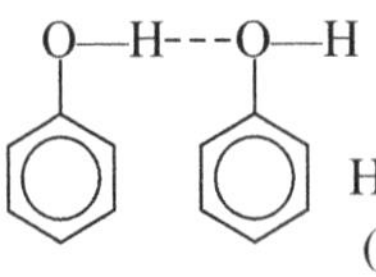

H_2O H_2O

(A) (B)

Inter molecular hydrogen bonding leads to association of large number molecules. This superficially increases the molecular weight and consequently, melting, boiling points, vapourisation and sublimation. All this becomes possible due to weak electrostatic force between the molecules *e.g.* b.pt. of H_2O without hydrogen bonding should have been – 100°C but due to H-bonding b.pt. becomes + 100°C same is applicable in the case of NH_3, AsH and SbH_3.

Intra-Molecular Hydrogen Bond

The hydrogen bond formed within the same molecule where hydrogen atom experiences weak electrostatic force of attraction of strongly electronegative atoms is called intra-molecular hydrogen bond *e.g*:

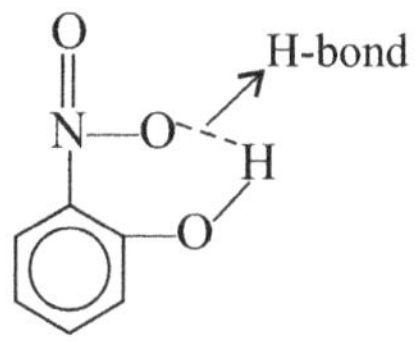

⋆*o*-nitrophenol

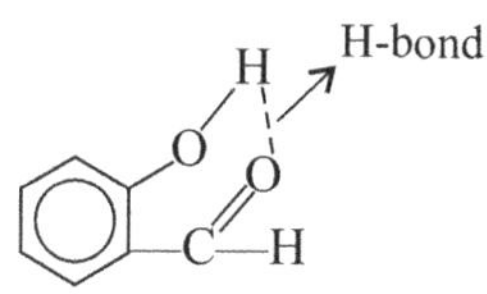

o-hydroxy benzaldehyde

Due to the presence H-bond O-nitrophenol is a liquid, reduces the acidic property. Para-nitrophenol shows inter molecular hydrogen bonding which is shown by high b.pt. of 279°C compared to ortho which is 214°C.

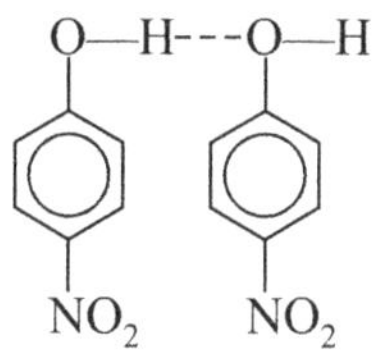

Properties of Hydrogen Bond

(*i*) Due to hydrogen bond H_2O exists as liquid while H_2S is in gas state at room temperature.

(*ii*) Bond energy of a hydrogen bond is between 3 to 10 Kcal/mole while a normal covalent has bond energy in the range of 50–100 Kcal/mole. Therefore, it is inferred that hydrogen bond is weaker than the covalent bond. The strength of H-bond depends upon the electronegativity of atoms with which it is in combination *e.g.*

$$N—H\cdots\cdots O < O—H\cdots\cdots O < F—H\cdots\cdots F$$

Greater size of electronegative atoms project lesser possibility to form a hydrogen bond. Nitrogen forms a H-bond while Cl does not when the electronegativities of N and Cl have similar values.

(*iii*) Hydrogen bond is because of dipole-dipole interactions.

(*iv*) In hydrogen bonds hydrogen atom is nearer to one of the two electronegative atoms.

(*v*) The electronegative atoms forming a hydrogen bond must have small size.

(*vi*) H-bond is directional because it involves *p*-orbital which contain a lone pair of electrons.

Effect of Hydrogen Bonding

It affects daily life of mankind:

1. In the solid state water changes to ice volume of ice is more than water in the liquid form. This happens because of tetrahedral structure of ice with inter molecular H-bonding.

   In this 2H-atoms are covalently bonded and lie close to O-atom at a bond distance 1.0 Å whereas the other two H-atoms form tetrahedral cage structure. The vacant space in the structure is the reason for large volume of ice. The presence of large number of H-bonds in liquid water gives it high density, hence ice floats over liquid water. The density water is maximum at 4°C due to increase in kinetic energy of water at 4°C. Above 4°C density decreases again. The structure of ice is called clathrate compound formation.

2. **Solvation:** Water is miscible with alcohol and milk in all proportions due to the formation of H-bond between H_2O and C_2H_5OH.

3. H-bond controls RNA and DNA. It holds hereditary information. Double helix are held together by hydrogen bonds. Proteins are polymeric due to the presence of H-bonds.

4. Paints and dyes give adhesive properties due to H-bond.

5. The reason for hydrates formation *e.g.* $CuSO_4 \cdot 5H_2O$, $BaCl_2 \cdot 2H_2O$, $FeSO_4 \cdot 8H_2O$ because of H-bond.

6. Wood fibres play an important role due to the presence of H-bond.

   Detection of H-bond spectroscopically:

   (*i*) Infrared spectroscopy: Due to inter-molecular H-bond, the —O—H stretching frequency which appears at 330 cm^{-1} is observed at 3600 cm^{-1}. This variation is because of H-bond.

   (*ii*) **Proton Magnetic Resonance (PMR):** H-bonding modifies electron density round the proton of A—H group where the PMR absorption is displaced down field *i.e.* towards lower magnetic field. This negative shift is because of H-bond.

   (*iii*) *X*-ray and electronic absorption spectroscopy can also detect H-bond.

H-bond Theories

Various theories were put forward to explain the formation of hydrogen bonding:

1. **Pimental theory:** This theory is based on Molecular Orbital Theory (MOT). According to this theory $1s$ orbital of hydrogen and two p-orbitals of atoms A and B form three molecular orbitals:

$$\psi^b \text{ (bonding)} = 2p_A + 2p_B + a_1 s$$
$$\psi^1 \text{ (non-bonding)} = 2p_\text{A} + 2p_B$$
$$\psi^* \text{ (anti-bonding)} = 2p_A + 2p_B - a_2 s$$

where a_1 and a_2 are mixing coefficients. In HF_2^- ion

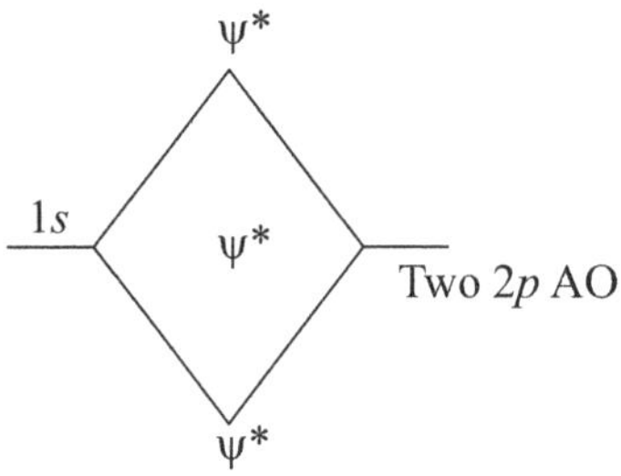

The four electrons will occupy bonding and non-bonding orbitals H-atom and thus give two weak bonds.

2. **Coulson theory:** Coulson is 1957 proposed his theory based on electrostatic force of attraction. According to the theory H-atom develops a weak positive charge as it is linked to strong electronegative small sized F, O and N. H-atom is more towards the electronegative atoms. H-bond is formed with lone pair electrons on the electronegative atom due to electrostatic forces. The delocalisation of electrons on electronegative atoms attracts H-atom due to electrostatic forces *e.g.* $\overset{\delta^+}{H}\longrightarrow\overset{\delta^-}{X}$ when two H—X molecules approach each other then $H^{\delta+}$ will attract extra strong $X^{\delta-}$ off the other molecule. This attractive force between H-atom and $X^{\delta-}$ leads to the formation of a weak bond called H-bridge or H-bond.

$$H{-}\overset{\delta^-}{X}\underset{d}{\text{———}}\overset{\delta^+}{H}{-}X\colon\cdots\cdots H{-}X$$

B.E.

$$F = \frac{\delta^+ \times \delta^-}{d^2}$$

This theory satisfactorily explain the formation of inter molecular hydrogen bonding but does not explain intra molecular hydrogen bond formation.

3. **MC-Clellan theory:** It describes the hydrogen bond formation on the basis of resonance. The molecule AHB contributes to the resonance as explained:

$$\underset{(1)}{A\cdots\cdots HB} \longleftrightarrow \underset{(2)}{A—H\cdots\cdots B} \longleftrightarrow \underset{(3)}{A—H—B} \quad \underset{(4)}{A^{+}H—B} \quad \underset{(5)}{A\cdots\cdots H^{-}B^{+}}$$

These five resonance hybrids form a hydrogen band involving the four types of forces:

(*i*) Electrostatic interaction (*ii*) Vander Waal's forces
(*iii*) Electrosatic attraction (*iv*) Electron-electronrepulsion

Electrostatic repulsion and attraction between original molecules renders polar nature to the species giving rise to dipole interactions.

The polarising forces causes deformation in combining atoms and delocalisation of electron cloud between two nuclei giving rise to covalent nature to the bond.

The van der Waal's forces are very weak and therefore do not contribute much to the interaction. In O—H bond in water it is 1.06 Å and only 1.0 Å in Vander Waal's forces.

Valence Bond Theory (VBT)

This theory explains hydrogen bonding in the compound on the basis of electrostatic force and resonance structures. *e.g.*

$$\underset{(i)}{X—H\cdots\cdots Y} \longleftrightarrow \underset{(ii)}{H \longleftrightarrow X:} \longleftrightarrow \underset{(iii)}{X—H \leftarrow Y}$$

The structure (*iii*) involves bicovalent hydrogen with X—H as a covalent bond and H ← Y as long covalent bond. The exact resonance energy value is not known.

The hydrogen bond energy of H—A—B is given by the following equation:

$$E = k\text{MH} - \text{A}^{1/r}$$

where $k = 2.44 \times 10^{18}\ C^{-1}$, M_{H-A}. Bond moment of the H—A bond, 1 is ionisation energy of the donor atom B and r is the distance H·····B.

Metallic Bond

There are certain properties associated with metals that are given below:

(*i*) Due to large size of metal atoms they loose electrons easily.
(*ii*) Metals are good conductors of heat and electricity.
(*iii*) They have high density and strength.
(*iv*) Metals have a metallic lustre.
(*v*) They forms alloys with other metals.

(*vi*) They have metallic structures:
  (*a*) Face Centred Cubic (FCC)
  (*b*) Body Centred Cubic (BCC)
  (*c*) Hexagonal Close Packed (HCP)

(*vii*) They are opque.

Metals have neither ionic nor covalent bond shows that there are a lot of valence electrons round the metal atom *e.g.* Na atom is bonded to 8 other Na atoms to form a metallic crystal.

Metallic Bond

Metallic bond is defined as that in which a metal atom is embedded in a 'sea' of electrons or electron pool or electron gas. This theory explains qualitatively some of the characteristics of the metals.

In 1916 Lorentz thought of metals as hard spheres or the metal cations with free electrons moving in the interstices. Metals have low ionisation energy to form metal ions and tend to share electrons with all other atoms and this is called a metallic bond.

In a metal atom valence electrons are called mobile electrons, free electrons or lattice electrons. These electrons move from place to place in vacant space, hence electrical conductivity is generated *e.g.* Na metal atom is represented:

$$\begin{array}{ccccccc} Na^+ & — & e & — & Na^+ & — & e \\ | & & | & & | & & | \\ {}^-e & — & Na^+ & — & e^- & — & Na^+ \\ | & & | & & | & & | \\ Na^+ & — & e^- & — & Na^+ & — & e^- \end{array}$$

Metallic Bond and Physical Properties of Metals

1. **Metallic Lustre:** Most of space in the metals is occupied by electrons so when light falls on metal surface the electrons get excited and energy is emitted in the visible light as no light can pass through the opaque metals.
2. **Conductivity of heat and electricity:** When electrical or heat energy is supplied to the metal then mobile electrons from positive pole move towards the negative pole due to which heat and electricity is transferred from one end to the other. This phenomenon give heat and electrical conductivity.
3. **Metals are malleable and ductile:** When metals are heated they can form thin sheets or drawn into wires because atoms of metals can be shifted easily from one place to another. Since the sea of electrons are mobile and move along positive atoms of the metals therefore the new position is permanent and they do not revert back to the original position.

4. **Elasticity:** Elasticity is due to the ease with which the metal ions can move from one lattice to another. When a wire is bent, the pattern of ions and electrons is broken and a new pattern is set up.
5. **Alloy formation:** Metal mix easily with other metal of similar atomic size, mobile electrons of one metal atom easily get bonded with mobile electron of the other metal atom to form metallic structure.
6. **Metals get red when heated:** On supplying heat energy to the metal the metal electron gets excited to form new position or energy and thus radiate visible ray depending upon the wavelength (λ).
7. **Metals are solids with high density:** Metal atoms are bonded with strong electrostatic forces which force metals to have high density, high melting and boiling points. Except Hg and Cs all are solids.
8. **Shining surface:** When a beam of light falls on the metal surface the energy associated with the light is passed on to the electron and the electron shifts from one place to another and the quantised energy of the electron is radiated to get shining property.

Van der Waal's Force

Besides, chemical bonds another type of bond which is very weak is seen to hold molecules, atoms or ions. The force is imaginary and is called van der Waal's force. When particles of a substance are brought nearer than a force which attracts the particles is called van der Waal's force. The modified definition of van der Waal's force are the imaginary cohesive, weak forces of attraction which come into play when all kinds of atoms, molecules and ions are brought close to each other. It is not essentially a valence force but a cohesive force. Between the molecules there is no valence force, the force of attraction that exists is a van der Waal's force. This force is the cause of molecule converted to solid or liquid state. van der Waal's forces consist of three types of attractive interactions in dipolar ions (*i*) Kieson forces (*ii*) Debye forces and (*iii*) London forces.

(*i*) **Kieson Forces:** These forces are exerted between the molecules because of dipole-dipole interactions. In a molecule the positive ion of one is attracted by the negative ion of the other and vice-vessa. This leads to rise in dipole-dipole interactions. Dipole-dipole interactions between the molecules are called Kieson forces they are the strongest of all types of van der Waal's forces.

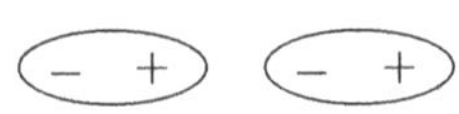

(*ii*) **Debye Forces:** These forces arise out of interaction between polar and non-polar molecules. They are called induced forces and the effect is called induced effect. When a non-polar molecule is brought near a polar molecule then polarity is induced in the non-polar molecule. The amount of polarising effect depends upon the polarising power of the dipolar molecule, and is determined by the polarisibility constants which are independent of temperature but depend upon the distance between polar and non-polar molecules. The forces of attraction between positive end of the polar molecule and the induced non-polar molecule is called Debye forces.

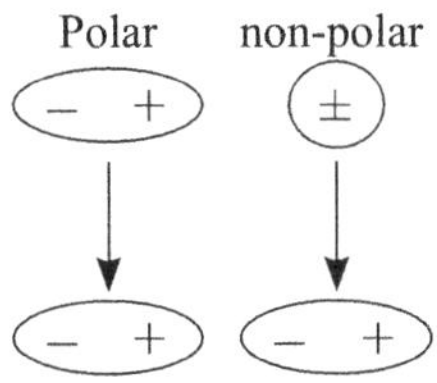

Independent Dipoles

London forces: German physicist Frintz London studies interaction between non-polar molecule of gaseous state attracted by very weak force so that the molecule is broken due to distortion of electron cloud in the molecule. The force is called London forces. This is a very weak force exerted on molecules in the gas state. This may operate in similar kind of molecules.

Van der Waal's forces: According to these considerations van der Waal's forces are weaker than a chemical bond but play significant role to bind the ions when they are brought nearer while the chemical force is absent. When two atoms or molecules approach each other than they experience (*i*) attractive and (*ii*) repulsive forces between the negatively charged electron and positively charged nucleus. The resultant effect favours attractive force.

The attractive and repulsive forces are at equilibrium when internuclear distance equals 4.0 Å. When inter molecular distance is between 4 Å to 10 Å attractive forced predominate and the molecules experience van der Waal's forces.

In molecules having a large number of electrons, the probability of molecule is high and so are van der Waal's forces. Increase in van der Waal's forces increases melting and boiling points as well as the density. The size of the molecule also increases due to increase in polarisation. At low temperature the movement of electrons decreases and van der Waal's force increase. At high pressure also van der Waal's force increases.

Applications

1. van der Waal's forces bind the molecules particularly those in the gaseous state. *e.g.* halogens, inert gases etc. They help in condensation process.
2. The softness and easily broken nature of substances is because of van der Waal's force. Solidification of the substances takes place with the help of van der Waal's force.
3. It helps in the study of kinetic theory of gases.

Comparison with other Bonds

van der Waal's forces are much weaker than ionic or covalent bonds. Like ionic bonds the Vander Waal's forces are additive and cannot be saturated *i.e.* attractive force is independent of neighbouring atoms. Covalent radius is smaller and stronger than van der Waal's radius. Hg is a liquid due to weak van der Waals force.

Resonance: Some molecules or ions are represented by more than one structure without change in the position of the constituent atoms are called resonance or resonance hybrid structure of the molecule. The properties of the molecule are not given by any one structure but all the canonical structures taken together. This is called resonance effect. Each structure is called resonating structure. True structure of H_2 molecule is represented by the four resonating structures shown below. Resonance structures are indicated by the sign ($\longleftrightarrow$) or arrow in both directions.

$$H_2 \rightarrow \underbrace{\underset{\text{I}}{H_A^1\,{}^2H_B} \longleftrightarrow \underset{\text{II}}{H_A^2\,{}^1H_B}}_{\text{Covalent structures}} \longleftrightarrow \underbrace{\underset{\text{III}}{H^{-1}_A\,{}^{2+}H_B} \longleftrightarrow \underset{\text{IV}}{H^{+2}_A\,{}^{1-}H_B}}_{\text{Ionic structures}}$$

i.e. H_2 molecule is resonance hybrid of (I), (II), (III) and (IV) structures. The various structures contributing to resonance are called resonating structures.

As a result of resonance, energy of the molecule is lowered *i.e.* the actual structure has less energy than any of the resonating structures. The amount of energy lowered is called the resonance energy. This energy is determined by taking the theoretical bond energy and the observed energy. More the number of resonating structures of the molecule greater is the resonance energy.

For Example Resonance Structures of CO_2 Molecule

For a good amount of time the CO_2 structure was considered to be O=C=O. The bond distance according to experimental value is 1.15 Å which is in between 1.22 Å for a double bond and 1.10 Å for a triple bond shows that the structure O=C=O does not truly represent the real structure. The real structure is a resonance hybrid of these structures:

$$\underset{\text{I}}{O_A{=}C{=}O_B} \longleftrightarrow \underset{\text{II}}{O_A^{+}{\equiv}C{-}O_B^{-}} \longleftrightarrow \underset{\text{III}}{O_A^{-}{-}C{\equiv}O_B^{+}}$$

Dipole moment of structure I is zero because it is symmetrical. Structure II and III must show a dipole moment as they are unsymmetrical. Actually the dipoles of structures II and III are equal and opposite therefore the resultant dipole moment of the real structure is equal to zero. Structures II and III are equivalent and have the same energy. The resonance energy of the molecule is 33 Kcal/mole. Therefore by 33 Kcal/mole of energy, CO_2 is more stable than that visualised by any of the resonating structures.

CO_3^{2-} ion

Carbonate ion must have 3σ and on π bond. Therefore C═O bond length must be shorter than the C—O bond length. Experimentally all the three C—O bond is CO_3^{2-} ion are of equal bond length. The actual value of C—O bond is 1.31 Å which is intermediate to C═O = 1.22 Å and C—O = 1.43 Å. Thus, structure of CO_2 is a resonance hybrid of:

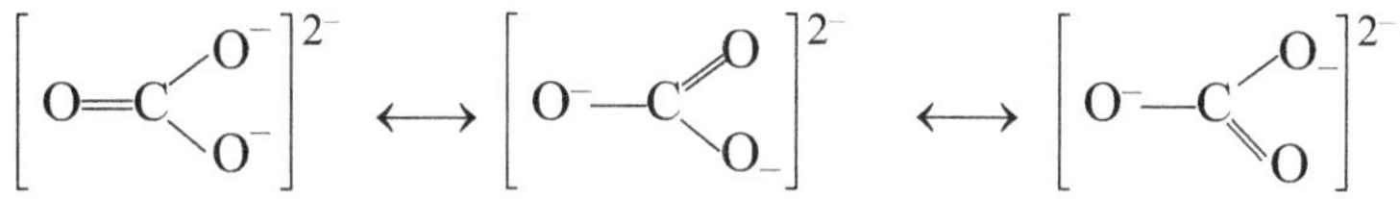

NO_3^- ion

The true structure of NO_3^- ion is a resonance hybrid of these structures:

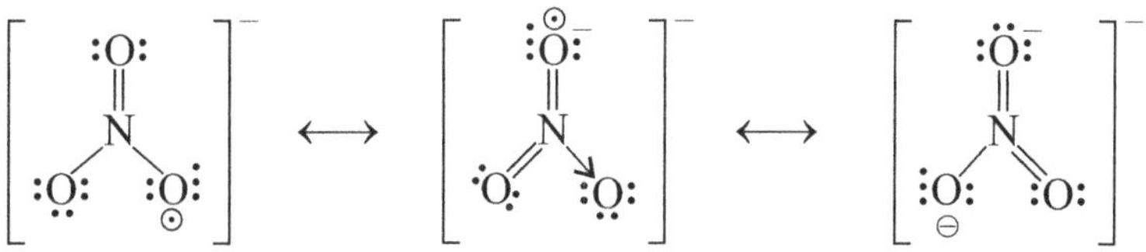

The observed bond length of N—O bond is 1.21 Å. The calculated bond lengths N → O and N═O are 1.36 Å and and 1.15 Å respectively. Thus it is intermediate between two bond lengths. The resonance energy of the NO_3^- ion is 45 Kcal/mole.

Benzene

It has isovalent resonating structures proposed by Kekule's and Dewar. These resonating structures have same number of covalent bonds but they are more significant than heterovalent resonating structures. They can be represented by:

I II III IV V

Structure I and II are called Kekule structures and III, IV and V are Dewar structures. In these structures benzene has a planar regular hexagon having sp^2 hybridisation of each C-atom. All the size C—C bonds are of the same length of (0.139 nm) for C═C bond energy is 31.6 Kcal/mole for 3C═C bonds the enrgy is 31.6 × 3 = 94.8 Kcal/mole.

Experimental bond energy value is 59.0 KCal/mole. Therefore 35.8 or ~ 36 Kcal/mole is the resonance energy of benzene molecule.

Oxygen Molecule (O_2)

The resonating structures of O_2 may be shown as:

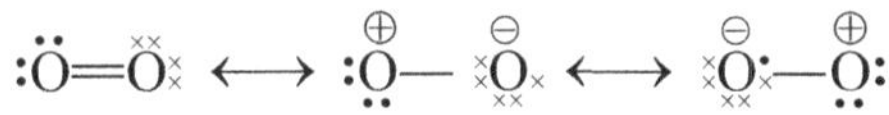

These structures do not account for the paramagnetic behaviour of oxygen molecule. In order to explain this following structures were proposed:

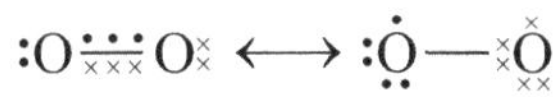

O_2 molecule has one 2-electron and two 3-electron bonds. Each three electron bond has one unpaired electron therefore two unpaired electrons in the O_2 molecule explain the paramagnetism.

The Conditions of Resonance

The following conditions must be satisfied for resonance of molecules or ions to occur:

1. The relative position of all the atoms in each of the resonance forms must be the same while the arrangement of electrons may be different.
2. The number of unpaired and paired electrons in each of the resonating structures must be the same so that a continuous change from one bond type to another may occur:

I and II

In above forms I and II, these structures cannot be contributing structures since the number of paired and unpaired electrons are not the same in the structures.

3. Like charges must not be present in the atoms which are close together and unlike charges should not be greatly separated.
4. Resonating structures should be almost of the same energy.

EXERCISES

1. Define a Covalent bond. Give examples.
2. What is a pseudo inert gas configuration? Give example.

3. Write the lewis dot structures of:
   SO_4^{2-}, CO_3^{2-}, $Cr_2O_7^{2-}$, MnO_4^-,
   NH_4^+, NO^+, N_2O_5, $(NH_4)_2SO_4$
4. What are the postulates of VBT? Explain. In what way VBT is better than lewis theory of covalent bond?
5. What is a singlet bond?
6. What is LCAO theory?
7. What is hybridization of orbitals? Discuss *sp*, sp^2, sp^3d and sp^3d^2 types of hybridization with suitable examples.
8. How do you explain paramagnetic behavior of O_2 molecule with the help of molecular orbital theory?
9. Write the molecular orbital configuration of CO, NO, N_2, C_2H_2.
10. What is a metallic bond? Why metals are malleable, ductile and good conductors of heat and electricity?
11. Boron shows a covalence of four. Why?
12. What do you understanding by a hydrogen bond? How does it affect the properties of the compound?
13. Which of the following has zero dipole moment and why?
    CH_2Cl_2, BF_3, NH_3, CO_2
14. Explain Fagan's rules for covalence.
15. Explain:
    (*i*) Boron has a tendency to form covalent bond.
    (*ii*) MOT is more reliable and acceptable than the VBT.
    (*iii*) AlF_3 has high m.p. while $AlCl_3$ has low m.p.
    (*iv*) LiCl is covalent.
    (*v*) $BeCl_2$ is a linear molecule and the bond angle is 180°.
    (*vi*) σ bond is stronger than π bond.
    (*vii*) CO_2 is non-polar but SiO_2 in angular.
    (*viii*) H_2O, NH_3 and CH_4 have different bond angles.
    (*ix*) S-S overlap is weaker than p-p overlap.
16. Write the hybridization with structure of
    $BeCl_2$, $SnCl_2$, CH_4, NH_3, NO_3^-, SO_4^{2-}, SF_6, IF_7 etc.
17. What is the shape of HF molecule?
18. What is the bond order of H_2^+, He_2, Co, F_2 etc.
19. Which of them has a planer structure? ClO_3^-, ClO_4^-, NO_3^-. Explain with the help of VBT.
20. Explain why the electrical conductivity of a metal decreases as the temperature is increased but the opposite happens in semiconductors.

21. What is super conductivity? What uses and potential use are there for super conductors? What types of materials are superconductors?
22. What do you understanding by the term resonance and resonance energy? Write resonating structure for CO and account for its low dipole moment.
23. Hybridization is based on VBT while the structure depends upon VSEPR theory? Explain.
24. Draw bonding, non-bonding and anti-bonding molecular orbitals. What are their relative energies?
25. Discuss the resonance energy and resonance hybrid in benzene molecule.
26. Arrange the following according to their increasing bond angles:
    (*i*) H_2O, H_2S, H_2Se
    (*ii*) PF_3, PCl_3, PBr_3
27. Phosphorus forms PCl_5 and PCl_3 while N forms only NCl_3. Why?
28. Compare the bond orders in He_2 and He_2^+.
29. Draw MO energy level of NO and predict its (a) bond order and (b) magnetic properties.
30. Explain decreasing bond angle on the basis of VSEPR theory.

5

Recent Concepts of Acids and Bases

History: The concept of acids and bases has been important since very early times. Practical acid-base chemistry developed from time to time. In ancient times acids was described as the one with sour taste and base with a bitter taste. It then developed and acids and bases were considered substances that changed the colour of the indicators. Acids and base react together to form salts and water. Later on acids included the idea of oxygen (oxides of N, P, S and halogen which form acids with water).

Many acids do not contain oxygen, in 1838 Leibig defined acids as compounds containing hydrogen which can be replaced by a metal. Many other definitions of acids were proposed and are useful in particular reactions. Many other definitions were proposed and adopted in certain reactions. The theories put forward were Arrhenius, Bronsted-Lowry, Lewis donor-acceptor, Lux-flood and Usanovich theory which have been described later in the Chapter.

Classical concept: In 1895 Leibig proposed acids as compounds containing hydrogen which could be replaced by a metal. This concept could not be accepted for a long time. From time to time lot of other concepts have been developed about acids and bases.

Arrhenius (1834), Bronsted Lowry in 1923 and G, N Lewis suggested electronic concepts which need to be elaborated now.

Arrhenius concept: Arrhenius in 1903 was awarded Nobel prize for his theory of acids and bases. According to his theory acids form hydrogen ions which are called hydronium ions, H_3O^+ in aqueous solution while bases form hydroxide ions OH^- in aqueous solution. The reaction of hydrogen ions and hydroxide ions form water. This is universal in all acid-based reactions. Besides water formation, the rest of the ions form a salt. The overall Arrhenius acid base reaction is written as:

$$\text{acid} + \text{base} \longrightarrow \text{salt} + \text{water}$$

e.g.

Hydrochloric acid + Sodium hydroxide

↓

Sodium chloride + Water

$$H^+Cl^- + NaOH \longrightarrow Na^+Cl^- + H_2O$$

This reaction works well in aqueous solution but not in non-aqueous solutions. For gaseous and solid phase reactions water may not exist.

Bronsted Lowry Concept: In 1923 Bronsted and Lowry defined 'acids have a tendency to lose a hydrogen ion and base is that which gains a hydrogen ion.' This definition brought in its domain gases such as hydrogen chloride (HCl) and NH_3 along with many other compounds. This concept brought in its arena the concept of conjugate acids and bases. It differs in the presence or absence of hydrogen ion or proton. According to this concept all reactions between a strong base and acid result into a weak base and acid. *e.g.*

$$H^3O^+ + NO_2^- \longrightarrow H_2O + HNO_2$$

$$\text{acid}_1 + \text{base}_2 \longrightarrow \text{base}_1 + \text{acid}_2$$

Cojugate means that they are wedded or linked to each other.

Conjugate acid-base pairs:

Acid	Base
H_3^+O	H_2O
HNO_2	NO_2^-

species differing from one another by a proton are referred to as conjugate acid and base.

In case of strong acid, the base is weak and vice-versa.

In nonaqueous NH_3 solvent *e.g.*

$$NH_4^+ Cl^- + Na^+ NH_2^- \longrightarrow \underset{\text{salt}}{Na^+Cl^-} + \underset{\text{solvent}}{2NH_3}$$

In the above reaction NH_4Cl acts as an acid (NH_4^+) and $NaNH_2$ as a base (NH_2^-). Acid NH_4^+ and base NH_2^- form NH_3 (solvent).

The net reaction is

$$\underset{\text{acid}}{NH_4^+} + \underset{\text{base}}{NH_2^-} \longrightarrow \underset{\text{conjugate base and conjugate acid.}}{2NH_3}$$

In any solvent the reaction favours the formation of weak acids and bases than the reactants.

In the examples studied so far H_3^+O is a stronger acid than HNO_2 and NH_2^- amide is stronger base than NH_3. Therefore formation of HNO_2 and NH_3 is favoured. The usefulness of Bronsted-Lowry theory lies not only in its ability to detect acid-base in protonic solvent but also in non-aqueous solvent system. Bronsted-Lowry theory rationalises acid-base reactions in protonic solvents such as H_2O, NH_3 and CH_3COOH. The theory fails in case of acidic oxides *e.g.* SO_3 which does not have any proton to donate. It reacts with H_2O to give pH < 7.0 *i.e.*, H_2SO_4.

Concept of solvent system: Bronsted gave protonic definition of acid and bases which can be further extended to reactions in non-aqueous solvents, containing hydrogen *e.g.* NH_3, HF, H_2SO_4, CH_3COOH, HCN and alcohols. According to Cady and Elsey (1928) acid is a solute which by dissociation or by reaction with a solvent gives the anion characteristics of the solvent and base is a solute which by direct dissociation or by reaction with the solvent gives the cation characteristic of the solvent. *e.g.*

Water as a solvent in aqueous conditions gives H_3O^+ and OH^- respectively:

$$\underset{\text{acid}}{H_2O} + \underset{\text{base}}{H_2O} \rightleftharpoons \underset{\text{acid}}{H_3O^+} + \underset{\text{base}}{OH^-}$$

Accordingly all those compounds which gives H_3O^+ in H_2O behave as acids and all those compounds which give OH^- ions in water act as bases.

In solvents like N_2O_4, NOCl yields NO^+ ions is an acid and substances like $NaNO_3$ yields NO_3^- ions are bases.

Auto-ionisation of protonic and non-protonic solvents is shown below:

(*a*) **Protonic solvents:**

$$\begin{array}{rcl} \text{Acid Base} & & \text{Acid Base} \\ H_2O + H_2O & \rightleftharpoons & H_3O^+ + OH^- \\ NH_3 + NH_3 & \rightleftharpoons & NH_4^+ + NH_2^- \\ CH_3COOH + CH_3COOH & \rightleftharpoons & CH_3COOH_2^+ + CH_3COO^- \end{array}$$

(*b*) **Non-protonic solvents:**

$$\begin{array}{rcl} \text{Acid Base} & & \text{Acid Base} \\ SO_2 + SO_2 & \rightleftharpoons & SO^{2+} + SO_3^{2-} \\ BrF_3 + BrF_3 & \rightleftharpoons & BrF_2^+ + BrF_4^- \\ N_2O_4 + N_2O_4 & \rightleftharpoons & 2NO^+ + 2NO_3^- \end{array}$$

As per Arrhenius definition reaction between an acid and a base results in salt and water or solvent. Some neutralization reactions in non-aqueous solvents are given below:

Non-aqueous solvents

Acid	Base	Solvent	Acid	Base

$$NH_4Cl + NaNH_2 \xrightleftharpoons{NH_3(s)} NaCl + 2NH_3$$

$$NOCl + NaNO_3 \xrightleftharpoons{N_2O_4(s)} NaCl + N_2O_4$$

$$HCl + CH_3COONa \xrightleftharpoons{CH_3COOH(s)} NaCl + CH_3COOH$$

SOLVOLYTIC AND AMPHOTERIC BEHAVIOUR

Solvolytic Behaviour

In liquid, NH_3

$$AlCl_3 + NH_3 \longrightarrow [Al(NH_2)]^{2+} + H^+ + 3Cl^-$$

in H_2O

$$AlCl_3 + H_2O \longrightarrow [Al(OH)]^{2+} + H^+ + 3Cl^-$$

Amphoteric Behaviour

In liq. NH_3

$$Zn\,(NH_2)_2 + 2NH_2^- \longrightarrow [Zn(NH_2)_4]^{2-}$$

In H_2O

$$Zn\,(OH_2) + 2OH^- \longrightarrow [Zn(OH)_4]^{2-}$$

Utility

Hence the concept of solvent system can be used to explain acid-base reactions in aqueous and non-aqueouis solvents.

Limitations

1. It does not explain a number of acid-base reactions covered in the defintion.
2. It does explain acid-base reactions in the absence of solvent.
3. It does not explain neutralisation reactions without the presence of ions, surely, it is just an extension of Arrhenius concept.

Lewis acid-base concept: This theory explains the structure of acids and bases on the basis of electron pair donation and its acceptance with the formation of a coordination bond, Lewis in (1923) put forward the idea of acid as an electron pair acceptor and base is specy that can donate the electron pair to form a coordinate bond.

The process of neutralisation is the formation of coordinate bond between an acid and a base. The product of neutralisation is a coordinate complex or adduct which may be either non-ionisable or may undergo dissociation or condensation.

For example formation of NH_4^+ ion:

$$\underset{\text{proton acid}}{H^+} + \underset{\text{base}}{:NH_3} \longrightarrow \underset{\text{Adduct}}{[H_3N \rightarrow H^+]}$$

In the above example NH_3 is a proton acceptor (Bronsted base) and the electron pair donor (Lewis base) and H^+ (proton) accepts an electron pair from NH_3 is an acid NH_4^+ ion is the adduct. Bronsted and Lewis theories are similar but the defintion of bases is different. *e.g.* NH_3, H_2O, OH^-, Cl^-, CN^- etc. are all bases on Bronsted and Lewis systems. There are compounds such as amides, ethers, nitriles, C_2H_4, C_2H_2 and C_6H_6 which have no tendency to accept protons but react with Lewis acids.

Types of Lewis acids: Lewis acid must contain one empty orbitals in the valency shell of one of the atoms to accept electron-pair from the Lewis base.

(*i*) Molecules having central atom with incomplete octet *e.g.* halides of Be, B and Al.

$$F_3B + C_2H_5OC_2H_5 \longrightarrow F_3B \leftarrow (O^-C_2H_5)_2$$

$$Cl_3Al + NC_5H_5 \longrightarrow Cl_3Al \leftarrow NC_5H_5$$

(*ii*) **Compounds containing central atom with vacant *d*-orbital:** This includes SiX_4, GeX_4, $TiCl_4$, PCl_3, PCl_5, SF_4 etc. These compounds can accept an electron-pair from Lewis base to form an adduct *e.g.* These halides are hydrolysed by H_2O to form oxides and oxyacids.

(*iii*) **Simple metal cations:** They find good examples in the formation of coordinate complexes,

(*a*) $Ag^+ \; + \; 2:NH_3 \longrightarrow [H_3N—Ag—NH_3]^+$

(*b*) $Cu^{2+} \; + \; 4\,NH_3 \longrightarrow \left[\begin{array}{c} NH_3 \\ \downarrow \\ H_3N \rightarrow Cu \leftarrow NH_3 \\ \uparrow \\ NH_3 \end{array}\right]^{2+}$

(*c*) $Co^{3+} \; + \; 6\,H_2O: \longrightarrow$ 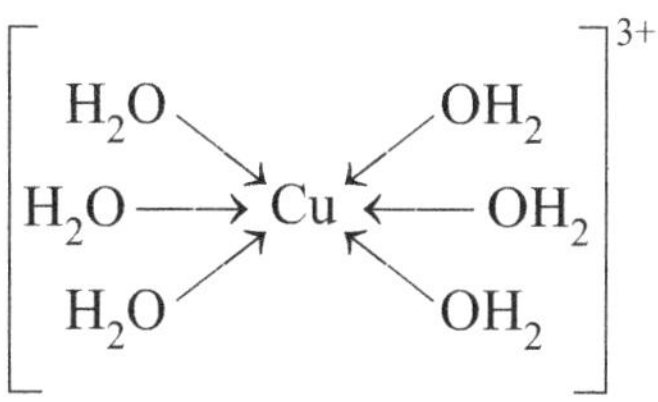

The strength of Lewis acids increawses with (*i*) increase in positive charge of the cation (*ii*) increase in nuclear charge along a period (*iii*) decrease in ionic radius (*iv*) decrease in shielding shells.

(*iv*) **Molecules with multiple bonds between atoms of dissimilar electronegativity:** These include Lewis acids such as CO_2, SO_2, SO_3 and NO_2.

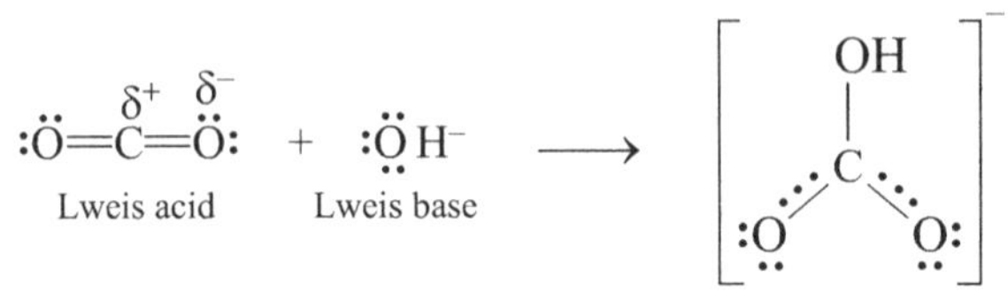

Bicasbonate ion

(*v*) **Elements with electrons sextet:**

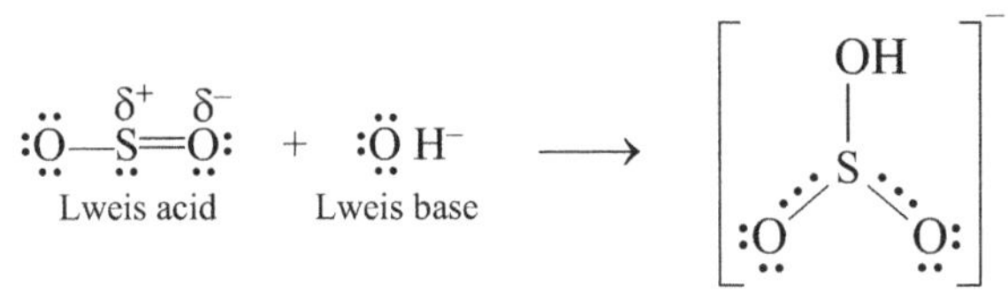

Bisulphite ion

Uses

(*i*) It covers reactions where no proton is involved.

(*ii*) Lewis acid-base concept is more general as it does not require the presence of a solvent.

(*iii*) It includes metallic and non-metallic oxides.

(*iv*) The theory covers gas phase, high temperature and non-solvent reactions.

(*v*) It is applicable where protonic concept is inapplicable.

Limit of the Theory

(*i*) It is not possible to arrange Lewis acids and bases in order of their relative strength.

(*ii*) It is expected that acid-base reaction should be rapid but many acid-base reactions are slow.

Lux flood concept: This concept was proposed by Lux in 1930 and extended by flood in 1947. According to the concept acid is a substance which propose oxide as acceptor and base as a donor oxide.

e.g.

$$\text{Acid} + \text{Base} \longrightarrow \text{Salt}$$

$$SiO_2 + CaO \longrightarrow CaSiO_3$$

$$CO_2 + BaO \longrightarrow BaCO_3$$

$$P_2O_5 + 3Na_2O \longrightarrow 2Na_3PO_4$$

Accordingly an atmospheric substance shows a tendency to gain or loss an oxide ion.

e.g.

$$ZnO \xrightleftharpoons{O_2} ZnO_2^{2-}$$

$$ZnO \xrightleftharpoons{O_2} Zn^{2+} + O_2^-$$

This means that base is oxide donor and acid is oxide acceptor. It is limited to system such as molten oxides. It emphases the acid and base anhydride aspect of acid-base concept.

The Lux-flood base is a basic anhydride:

$$CaO + H_2O \longrightarrow Ca^{2+} + 2OH^-$$

similarly Lux-flood acid is an acid anyhyride.

$$SO_3 + 2H_2O \longrightarrow H_3O^+ + HSO_4^-$$

Usanovich Concept

It is an extension of Lewis acid-base theory. It waives off the concept of donation and acceptance of lone pair. It defines acid as a substance which reacts with bases, gives up cations or accepts anions and electrons and base is a specy which reacts with acids, gives up anions or electrons and combines with cations. This concept applies to all redox acid-base reactions.

$$Fe^{2+} \longrightarrow Fe^{3+} + e$$

$$Sn^{2+} \longrightarrow Sn^{4+} + 2e$$

In the above reactions Fe^{2+} and Sn^{2+} act as bases while Fe^{3+} and Sn^{4+} act as acids.

It gives a very generalised definition of acids and bases.

Limitations

1. It is too generalised a concept.
2. It considers all reactions in the form of redox reactions.
3. It is too difficult a definition of acid-base concept.

Bjerrum in 1951 proposed that acid is a proton donor which may be treated as an anti-base.

Hard and Soft Acids and Bases (HSAB)

In 1958 Arland, Chat and Davis classified the metal ions and the ligands into two categories:

Hard and Soft Acids: Hard acid is difficult to polarise. A **cationic hard acid** like Al^{3+} must have (*i*) small size (*ii*) high positive charge and (*iii*) noble gas configuration.

Soft acids are easily polarised. They must posses (*i*) large size (*ii*) low positive charge or zero charge and (*iii*) do not possess noble gas configuration.

The alkali metal and alkaline earth metal ions are hard acids.

The heavier transition metal ions are branded as soft acids. The outer electrons or orbitals of soft acids like a hard base is difficult to polarise.

In Table 5.1 are given some hard, intermediate and soft acids.

Table 5.1 Hard, Intermediate and Soft Acids

Hard acids	Intermediate acids	Soft acids
H^+, Li^+, Na^+, K^+	Fe^{2+}, Co^{2+}, Ni^{2+}, Cu^{2+}	Cu^+, Ag^+, Au^+, Tl^+, Hg^+
Be^{2+} and Ca^{2+}	Zn^{2+}	Pb^{2+}, Cd^{2+}, Pt^{2+}, Hg^{2+}
Sr^{2+}, Mo^{2+}, Al^{3+}, Ga^{3+}	Pb^{2+}, Sn^{2+}, Sb^{3+}, Br^{3+}	Tl^{3+}, BH_3, $GaCl_3$, $InCl_3$
In^{2+}	Rh^3, SO_2, NO^+	I^+, Br^+, I_2, Br_2
La^{3+}, Lu^{3+}, Cr^{3+}, Co^{3+}	GaH_3	I^+, Br^+, I_2, Br_2 and other zero valent metal atom
Th^{4+}, U^{4+}, Ce^{3+}, Sn^{4+}, VO^{2+}		
VO_2^{2+}, MoO^{3+}, BF_3, SO_3 and Cr^{6+}		
I^{7+}, I^{5+}, CO_2, $(HF)_n$		

Principle of soft and hard acids and bases is represented by SHAB is useful in predicting the stability of the complex. The complex A : B is most stable when A and B are both soft and both hard. The complex is least stable when one of them is very hard acid the other very soft.

This principle implies tha reaction between an acid and two bases or between a base and two acids then a hard acid will choose to combine a hard base and soft acid will prefer to combine with a soft base to form a more stable product.

Hard and soft bases: Hard bases are those in which donor atoms have low polarising power and high electronegativity. On the other hand soft bases contain easily polarised donor atoms having low electronegativity. Softness of bases increase with increase in size of donor atoms. In halide ion softness increases in the order $F^- < Cl^- < Br^- < I^-$. Therefore F^- is the hardest and I^- is the softest.

In Table 5.2 are shown the classification of hard, intermediate and soft bases.

Table 5.2 Bases

Hard bases	Intermediate bases	Soft Bases
H_2O, OH^-, F^-, CH_3COO^- PO_4^{3-}	$C_6H_5NH_2$, C_5H_5N, Br^-	R_2S, RSH, I^-, SCN^-, $S_2O_3^{2-}$, R_3P, R_3As
SO_4^{2-}, Cl^-, CO_3^{2-}, ClO_4^-, NO_3^-	NO_2^-, SO_3^{2-}, N_2	$(RO)_3P$, CN^-, RNC, CO, C_2H_4, C_6H_6, H^-
ROH, RO^-, R_2O, NH_3, RNH_2, N_2H_4		

Application of HSAB

HSAB finds wide applicability in various kinds of chemical reactions.

1. **Relative stablities of compounds and complexes:** In the case of AgI_2^- exists but AgF_2 does not. This is because Ag^+ is a soft acid and F^- is a hard base and I^- is a soft base. Therefore AgI_2^- is a stable complex (combination of soft acid and soft base).

   Likewise $[CoF_6]^{3-}$ (combination of hard acid and soft base) is less stable. Similarly HF is stable while HI is less stable.

   Hydrahalide acts as: HF > HCl > HBr > HI

   $[Cu(NH_3)_4]^{2+}$ is more stable than $[Cd(NH_3)_4]^{2+}$

2. **Occurrence:** Hard acids occurs in the combined state whereas soft acids are available in the free state *e.g.* sulphur in the free state in nature. Intermediate strength metal ions occur in nature as sulphides or carbonates.

3. **Poisoning of metal crystals:** Soft metals *e.g.* Pd^{2+}, At^{2+} act as catalyst but are easily poisoned by soft bases such as Co, P or As ligands because soft acid and soft base form a stable compound. These ligands are absorbed on the surface of the metal and block the negative sites. The soft metal catalyst are not affected by hard base such as F, O, N, etc.

4. The HSAB principle has been correlated to the chemical reactions. The ratio of electrophilic and nucleophilic reactions depend upon hardness and softness of acids and bases.

5. HSAB theory predicts the possibility of a chemical reaction *e.g.*

$$LiI + CsF \longrightarrow CsI + LiF$$

   The reaction takes place easily because (hard acid + hard base) and (soft acid and soft base) prefer to combine to form CsI and LiF.

Limitations of HSAB Concept

1. The concept does not give any quantitative idea of relative strength of acids and bases.
2. The hard and soft factor is in independent of the acidic and basic character of the compound. In some reactions these factors are dependent to acidic and basic character.
3. To explain the reaction, one has to break the compound into acids and bases.

   From the above it becomes clear that there is no fool proof definition of acids and bases but all the concepts have their own advantages and disadvantages but the knowledge of all the concepts is essential to seek the overall picture of acids and bases.

EXERCISES

1. Which of the under mentioned is not a Bronsted acid and Bronsted base:

   (*i*) Ml_3 (*ii*) O_2^- (*iii*) HCO_3^- (*iv*) H_3PO_4
2. Select the answer. Which is the conjugate base of HBr (*i*) H_2Br^+ (*ii*) Br^+ (*iii*) Br^- (*iv*) H^+
3. Explain with examples a conjugate acid and a conjugate base.
4. Bring out a conjugate acid-base pair out of the following reactions:

   (*i*) $HI + H_2O$ (*ii*) $MH_3 + HCl$ (*iii*) $NH_4^+ + Cl^-$
5. Arrange the following in the increasing order of acidity with explanation:

   (*i*) CH_4, NH_3, H_2O and HF

   (*ii*) HClO, $HClO_2$, $HClO_3$, $HClO_4$.
6. Define Lewis acids and Lewis base.
7. Compare Lewis and Bronsted-Lowry concepts of acids and base.
8. Which of the following are Lewis acids and Lewis bases. Justify by giving reasons.

   Ag^+, NH_3, F^-, BF_3
9. Give utility and limitations of Lewis acids and Lewis base.
10. Explain with examples the advantages of Lewis of acids and bases over others.
11. Which of the following is Lewis acid and a Lewis base:

    $TiCl_4$, SO_3^{2-}, BF_3 and SO_3

    Give reasons. All boron halides are Lewis acids. Explain.
12. Explain the following on the basis of hard and soft acid-base principle:

    (*i*) AgI_2^- is more stable than AgF_2^- ion

    (*ii*) BH_3F^- and BF_3H^- react to form BH_4^- and BF_4^-.

13. Explain in the light of SHAB principle 'It is not possible to detect Ag^+ ion in $[Ag(CN)_2]^-$ solution by halide ion?
14. Explain:
    (*i*) HCl is an acid in H_2O but not in C_6H_6.
    (*ii*) NH_3 is a base according to Bronsted Lowry concept as well as Lewis theory.
    (*iii*) Solution of $CuSO_4$ gives acidic reactions.
    (*iv*) A solution of NaCl is neutral.
    (*v*) The hydrogen ion in aqueous solution is represented by H_3O.
    (*vi*) F^- ion is a strong base.
    (*vii*) Metal cations are good Lewis acids.
    (*viii*) HNO_3, H_2SO_4 and HCl appear equally strong acids in H_2O but their strength differs in CH_3COOH.
    (*ix*) HCl is a covalent compound but gives H^+ ion in aqueous solution.
    (*x*) BCl_3 is a weak Lewis acid than BBr_3.
    (*xi*) $TiCl_4$ is a strong Lewis acid.
    (*xii*) Identify acids and bases in the following reactions. Indicate which acid-base definition applies to the each case:
    (*a*) $PCl_5 + ICl \longrightarrow |PCl_4|^+ + |ICl_2|^-$
    (*b*) $BF_3 + 2ClF \longrightarrow |Cl_2F|^+ + |BF_4|^-$
    (*c*) $NOF + ClF_3 \longrightarrow |NO|^+ + |ClF_4|^-$
    (*d*) $2ClO_3^- + SO_2 \longrightarrow 2ClO_2 + SO_4^{2-}$
    (*e*) $Pt + XeF_4 \longrightarrow PtF_4 + Xe$
    (*f*) $PtF_5 + ClF_3 \longrightarrow |ClF_2|^+ + |PtF_6|^-$
    (*g*) $2HF + SbF_5 \longrightarrow |H_2F|^+ + |SbF_6|^-$
15. Arrange in the increasing acid strength in aqueous solution:
    (*a*) $HMuO_4$ H_2AsO_4 H_2SO_3 H_2SO_4
    (*b*) HClO $HClO_4$ $HClO_2$ $HClO_3$

6

Aqueous and Non-aqueous Solvents

Most of the chemical reactions require a medium for the reactions to take place. The medium may lie aqueous or non-aqueous. Water is taken as a universal solvent. The properties of the solvents should be:

1. The range of temperature over which the solvent remains a liquid.
2. Solvent should be in liquid state at room temperature.
3. The dielectric constant of the solvent should be high (> 80.0).
4. It should dissolve a large number of inorganic and organic compounds.
5. It should have a wide range of physical and chemical properties.
6. It should be easy to handle.

Characteristics of Water (H_2O)

Freezing point = 4°C

Density = 1 g/cc

Boiling point = 100°C

Dielectric constant = 81.7

Specific conductance = 4×10^{-8} ohm^{-1} cm^{-1}

Viscosity = 1.01 g/cc

Ionic product = 1.008×10^{-14}

pH = 7.0

Chemical reaction do occur in solvents other than water (non-aqueous) solvents was highlighted by Cady, Frankline, Kraus and Welden in 1897. Some non-aqueous solvents are NH_3, SO_2, NO_2 (N_2O_4), HF and CH_3CHOOH etc.

For example NH_4Cl and $LiNO_3$ do not react in aqueous solution but give precipitate of LiCl in liquid NH_3,

$$NH_4Cl + LiNO_3 \xrightarrow[NH_3]{liq.} LiCl \downarrow + NH_4NO_3$$

Similarly $AgNO_3$ and $BaCl_2$ give precipitate of AgCl in aqueous solution.

$$2AgNO_3 + BaCl_2 \xrightarrow{H_2O} 2AgCl \downarrow + Ba(NO_3)_2$$

The above reaction in liquid NH_3 forms a precipitate of $BaCl_2$

$$2AgCl + Ba(NO_3)_2 \xrightarrow[NH_3]{liq.} BaCl_2 \downarrow + 2AgNO_3$$

i.e. the reaction takes place in reverse direction, therefore, the chemical reaction is dependent on the nature of the solvent.

Types of Solvents

1. **Protonic and non-protonic solvents:**
   - **(*a*) Protonic solvents or protic solvents:** Such solvents have hydrogen atom in the chemical formula. They are of two types:
     - **(*i*) Acidic or Protogenic solvents:** These solvents have a strong tendency to donate protons *e.g.* H_2SO_4, HF, CH_3COOH, HCN etc.
     - **(*ii*) Basic or protophilic solvents:** Such solvents have a strong tendency to accept protons *e.g.* NH_3, C_5H_5, RNH_2 or $ArNH_2$ amines and N_2H_4.
2. **Aprotic or non-protonic solvents:** These solvents may or may not contain hydrogen atom in their chemical formula. They neither donate or accept protons *e.g.* C_6H_6, $CHCl_3$, SO_2, BrF_3.
3. **Amphi-protonic or Amphoteric solvents:** In this category of solvents they donate or accept proton depending upon the nature of chemical reactions. They act both acids and bases in nature. They ionise into protons and anions *e.g.*

$$CH_3COOH \rightleftharpoons CH_3COO^- + H^+$$
$$H_2O \rightleftharpoons H^+ + OH^-$$

4. **Like dissolves like:** This includes both (*i*) ionising solvents and (*ii*) non-ionising solvents.
   - **(*i*) Ionising solvents** under go autoionisation

   e.g.
$$H_2O + H_2O \longrightarrow H_3O^+ + OH^-$$
$$H_3N + NH_3 \longrightarrow NH_4^+ + NH_2^-$$
$$SO_2 + SO_2 \longrightarrow SO^{2+} + SO_3^{2-}$$

   - **(*ii*) Non-ionising solvents:** They are non-polar and thus dissolve in non-polar solvents and so there is no question of ionic reactions. They have low dielectric constants. They have no associating or solvating tendency *e.g.* C_6H_6, CCl_4. These solvents do not undergo self-ionisation.
5. **Aqueous and non-aqueous solvents:** The solvents that do not contain water are called non-aqueous solvents. In H_2O the self ionisation between the water molecules give H_3O^+ and OH^- which is analogous to NH_4^+ and

NH_2^- ions obtained from self ionisation of liquid NH_3. This is called parent solvent concept. Some of the parent solvent and analogous groups are mentioned below:

$$H_2O + H_2O \rightleftharpoons H_3O^+ + OH^-$$

$$NH_3 + NH_3 \rightleftharpoons NH_4^+ + NH_2^-$$

$$NH_2OH + NH_2OH \rightleftharpoons NH_{3+}OH^+ + NH_4OH^-$$

$$H_2S + H_2S \rightleftharpoons H_3S + HS^-$$

Properties of the Self-ionising Solvents

Physical Characteristics

In Table 5.1 Are given some of the physical properties

Solvent	Viscosity Millipoise	Dipole Moments in Debye Units	Dielectric Constant
H_2O	10.08	1.84	78.5 (at 25°C)
NH_3	2.65	1.46	22.0 (at –33.5°C)
HF	2.40	1.90	83.6 (at 0.0°C)
HCN	2.00	2.93	106.8 (at 25°C)
SO_2	2.28	1.61	17.27 (at –16.5°C)

(*i*) **Dipole moment:** It is the product of charge and the distance between the charges. Solvent with higher value of dipole moment is easily miscible in polar solvents. Greater the polarity of solvent molecule greater is the solvation energy release on dissolution of a solute. Dipole moment value gives an idea about the extent of association in liquid state.

(*ii*) **Dielectric constant:** Coulombic force F between a cation and an anion in an ionic crystal is given by:

$$F = \frac{q^+ q^-}{Dr^2}$$

where q^+ and q^- are the charges on the cation and anion, '*r*' is the distance between cation and an anion and D is called the dielectric constant which depends upon the nature of the solvent in which the ionic crystal is kept. If the dielectric constant of the solvent is greater than the lattice energy of the solute then the solute will dissolve in it otherwise not.

Water has a dielectric constant of 78.5 while liquid NH_3 had $D = 22.0$ therefore water is a better solvent than ammonia.

(*iii*) **Electrical conductance:** All self-ionising solvents show electrical conductance. The value on electrical conductance scale shows the extent of self-ionisation. Greater is the ionisation, easier for the acid-base reactions to occur in the solvent.

(*iv*) **Viscosity:** It is the resistance offered to the flow of liquid solvent. Some solvents with low molecular weight are highly fluid *e.g.* H_2O, NH_3 and alcohols while others are viscous with high molecular weight *e.g.*, anhydrous H_2SO_4 and higher alcohols. Reactions like precipitation, acid-base, solvolysis, redox and crystallisation are easy to perform in solvents with low viscosity than those solvents with high viscosity.

(*v*) **Proton Affinity:** It is possible in protonic solvents only. It affects the behavior of a solute in a solvent system. Solvent NH_3 has a greater proton affinity than H_2O. Acetamide acts as a weak base in aqueous solution becomes an acid in NH_3 solvent.

$$CH_3CONH_2 + H^+OH^- \xrightleftharpoons{\text{Weak base}} CH_3CONH_3^+ + OH^-$$

$$CH_3CONH_2 + NH_3 \xrightleftharpoons{\text{Acid}} CH_3CONH^- + NH_4^+$$

Common Reactions in Ionising Solvents

1. **Acid-base reactions:** Acid-base reactions result in salt formation and solvent. Ionising solvents are polar solvents and therefore some self-ionisation takes place to some extent *e.g.*

$Acid_1$		$Base_1$		$Acid_2$		$Base_2$
H_2O	+	H_2O	$\rightleftharpoons$	H_3O^+	+	OH^-
NH_3	+	NH_3	$\rightleftharpoons$	NH_4^+	+	NH_2^-
SO_2	+	SO_2	$\rightleftharpoons$	SO^{2+}	+	SO_3^{2-}
HF	+	H_2F_2	$\rightleftharpoons$	H_2F^+	+	HF_2^-
NO_2	+	NO_2	$\rightleftharpoons$	NO^+	+	NO_3^-

According to Cady and Elsey (1928) acid is a substance which increases the amount of positive ions and base increases the negative ions concentration. For example in the reaction below:

$$\underset{\text{acid}}{NH_4^+Cl^-} + \underset{\text{base}}{K^+NH_2^-} \rightleftharpoons \underset{\text{salt}}{KCl} + \underset{\text{solvent}}{2NH_3}$$

2. **Solvation or solvate formation:** When one or more molecules of the solvent are attached to a solute cation, anion or neutral compound by a chemical bond the reaction is called solvation reaction. The product is

called solvate. The solvates obtained in case of H_2O are called hydrates, for ammonia ammoniates, for alcohol they are alocholates and ethertrates. The attachment is through ion-dipole, hydrogen bonding or a coordinate bond *e.g.*

$$\underset{\text{Lewis acid}}{F_3B} + \underset{\text{solvent}}{NH_3} \longrightarrow \underset{\text{Ammoniate}}{F_3B \leftarrow NH_3}$$

$$\underset{\text{acid}}{CuCl_2} + \underset{\text{base}}{4H_2O} \longrightarrow \underset{\text{hydrate}}{[Cu(H_2O)_4]^{2+}} + 2Cl-$$

$$\underset{\text{base}}{KI} + \underset{\text{acid}}{4SO_2} \longrightarrow \underset{\text{solvate}}{K^+[(SO_2)_4I]^-}$$

They are called addition compounds or adducts.

3. **Sowolysis or solvolytic reactions:** Solvolysis is that in which solvent molecule splits into two parts and one or both parts are attached to a solute molecule or ion. Due to self ionisation of solvent there is considerable increase in the concentration of cations and anions. Solvolytic reactions occur in H_2O, liquid NH_3 and alcohol which are called hydrolysis, ammolysis and alcoholysis respectively. *e.g.*

Hydrolysis:

(*i*) $H_2O + H_2O \rightleftharpoons H_3O^+ + OH^-$

(*ii*) $CuSO_4 + H_2O \rightleftharpoons Cu(OH)_2 + SO_4^{2-} + 2H_3O^+$

(*iii*) $CN^- + H_2O \rightleftharpoons HCN + OH^-$

In equation (*i*) and (*ii*) concentration of H_3O^+ increases while in equation (*iii*) OH^- concentration increases.

Ammonolysis: In this concentration of NH_4^+ and NH_2^- ions are increased.

$$NH_3 + NH_3 \rightleftharpoons NH_4^+ + NH_2^-$$

$$SiCl_4 + 8NH_3 \rightleftharpoons Si(NH_2)_4 + 4NH_4Cl$$

$$H^- + NH_3 \rightleftharpoons NH_2^- + H_2$$

In liquid SO_2 concentration of SO^{2+} or $SO^{2-}{}_3$ ion is increased.

$$SO_2 + SO_2 \rightleftharpoons SO^{2+} + SO_3^{2-}$$

$$Zn(C_2H_5)_2 + 2SO_2 \rightleftharpoons Zn^{2+} + SO_3^{2-} + (C_2H_5)_2SO$$

4. **Precipitation Reactions:** The precipitation of salt in a solution of solute and solvent depends upon the solubility product of the solute and the solvent. It largely depends upon the solvent used.

e.g. Aqueous solution of $BaCl_2$ and $AgNO_3$ when mixed together result in the precipitate of AgCl.

$$BaCl_2 + 2AgNO_3 \longrightarrow AgCl \downarrow + Ba(NO_3)_2$$

In liquid NH_3 on mixing the solution of $Ba(NO_3)_2$ and AgCl it is the $BaCl_2$ that is precipitated.

$$Ba(NO_3)_2 + 2AgCl \xrightarrow[NH_3]{liq} BaCl_2 \downarrow + 2AgNO_3$$

Many halides are precipitated as complex ammines.

$$2NaCl + Ca(NO_3)_2 + 8NH_3 \longrightarrow [Ca(NH_3)_8]Cl_2$$

5. **Redox Reactions:** Same type of redox reactions occur both in aqueous and non-aqueous media. Sometimes solvent-itself may be prone to redox reactions *e.g.* water is susceptible to reduction.

   Its electrode potential is above hydrogen. Therefore oxidation reactions can be carried out easily in aqueous solution. Reverse is applicable to NH_3. It is susceptible to oxidation. Hence water is suitable for oxidation and liquid NH_3 for reduction. Liquid SO_2 is inert towards redox reactions. Therefore it serves as a suitable medium for redox reactions.

Important Non-aqueous Solvents

Liquid ammonia solvent

Water-like properties of liquid ammonia have made it a useful solvent and as a reaction medium to carry out some organic and inorganic reactions.

It has low dielectric constant, therefore it is poor solvent for ionic compounds and a good solvent for non-polar compounds. Liquid NH_3 shows hydrogen bonding like H_2O.

It is associated with sp^3 hybridisation. It acquires a pyramidal structure owing to the presence of lone pair on nitrogen. The bond angle $\angle$ H—N—H is 107° which is less than a regular tetrahedron, Polarity, high basicity and low degree of auto photo lysis ammonia becomes a solvent for variety of purposes It has lesser tendency to react with dissolved solute, Liquid NH_3 possesses high thermal coefficient of expansion therefore its density decreases with rise in temperature.

Solubility in liquid NH_3

(*i*) In general inorganic compounds have a poor solubility in NH_3 due to low dielectric constant *e.g.* oxides, hydroxides, sulphates, carbonates, phosphate, sulphites and sulphides are insoluble while nitrates thiocyanates, perchlorates and cyanides are soluble. Compounds having many electrons for example iodine compounds and non-polar hydrocarbons are soluble in liquid ammonia. Order of solubility of halides in liquid NH_3 is $I^- > Br^- > Cl^- > F^-$. Most of the ammonium salts such as NH_4NO_3, NH_4SCN, CH_3COONH_4 are soluble. All metal amines except those of alkali metals are insoluble.

(*ii*) **Organic compound:** Alcohols, halogen compounds ($CHCl_3$, CCl_4), ketones, esters, ethers, phenol and phenol derivatives are soluble. Aromatic hydrocarbons are sparingly soluble.

(*iii*) **Non-metals:** Non-metals P, S and I dissolve in liquid NH_3 reacting with it.

$$\frac{5}{8}S_8 + 16NH_3 \longrightarrow N_4S_4 + 6(NH_4)_2S$$

Advantages of liquid NH_3

(*i*) All alkali metals dissolve in liquid NH_3 without any chemical reaction. The dissolved alkali metal can be recovered by evaporating alkali metal-ammonia solution.

(*ii*) For a stronger reducing agent alkali metal-ammonia solution is used which provides mobile ammoniated electron to reduce the substance soluble in liquid NH_3.

(*iii*) NH_3 poses a lesser tendency to undergo solvolysis with dissolved solutes than in case of H_2O.

Disadvantages of liquid NH_3

(*i*) Boiling point of liquid NH_3 is –33.5° therefor low temperature is recommended to work with this solvent to avoid high pressure.

(*ii*) White working with liquid NH_3 as a solvent or reaction medium special techniques are recommended.

(*iii*) Reactions in presence of liquid NH_3 are performed in sealed tubes out of contact of moisture.

Auto or self ionisation of liquid NH_3

Self ionisation of liquid ammonia is similar to H_2O owing to the lower conductance value in comparison to H_2O

$$NH_3 + NH_3 \rightleftharpoons NH_4^+ + NH_2^-$$

$$H_2O + H_2O \rightleftharpoons H_3O^+ + OH^-$$

The ionic product value of liquid NH_3 is low in comparison to water.

$$[NH_4^+][NH_2^-] = 1.9 \times 10^{-23} \text{ at } 50°C$$

$$[H_3O^+]\,[OH^-] = 1 \times 10^{-14} \text{ at } 25°C$$

Auto-ionisation shows that any substance which give NH_4^+ ions in liq, NH_3 is an ammono acid and that which gives NH_2^- ions is an ammono base (NH_2^-, NH^{2-} or N^{3-}). The neutralisation of ammono acid and ammono base result in ammono salt and liquid NH_3 solvent.

Chemical reactions in liquid ammonia

Some chemical reactions which can take place in liquid ammonia are given below:

1. **Precipitation reactions:** There is big difference in the solubility of some substances in liquid ammonia and water. Some precipitation reactions which do not occur in water may take place in liquid ammonia.

   (*i*) Except Na^+, Be^{2+} and NH_4^+ chlorides all other alkali chlorides are insoluble in liquid ammonia *e.g.*

   $$KI + NH_4Cl \longrightarrow \underset{\text{White ppt}}{KCl\downarrow} + NH_4I$$

   (*ii*) When nitrates of Sr^{2+} and Zn^{2+} are mixed withe NH_4Br the corresponding metallic bromides and iodides get precipitated.

   $$Sr(NO_3)_2 + 2NH_4Br \longrightarrow SrBr_2\downarrow + 2NH_4NO_3$$
   $$Zn(NO_3)_2 + 2NH_4I \longrightarrow ZnI_2\downarrow + 2NH_4NO_3$$

   (*iii*) Sodium carbonate forms a precipitate in liquid ammonia.

   $$NH_4[NH_2CO.O] + NaNO_3 \xrightarrow[NH_3]{\text{Liq.}} \underset{\text{sod.carbonate}}{Na[NH_2CO.O]\downarrow} + NH_4NO_3$$

   (*iv*) $(NH)_2S$ in liquid ammonia can precipitate metal sulphides such as MgS, ZnS, CdS, MnS, CoS, Ag_2S, PbS, HgS, Bi_2S_3 and BaS from the nitrate solution of these metals.

   (*v*) **Metal alcoholates:** Metal salt and K-alcoholate in liquid ammonia form the precipitate of metal alcoholates.

   $$Ba(NO_3)_2 + 2K(OC_2H_5) \longrightarrow Ba(OC_2H_5) + 2KNO_3$$

   (*vi*) Insoluble halides in liquid ammonia can be precipitated as complex halides.

   $$2NaCl + Ca(NO_3)_2 + 8NH_3 \longrightarrow [Ca(NH_3)_8]Cl_2\downarrow + 2NaNO_3$$

2. **Proton Donor-acceptor reactions:** Liquid NH_3 has a strong tendency to accept proton from H_2O molecule.

   $$H_2O \rightleftharpoons H^+ + OH^-$$
   $$NH_3 + H^+ \rightleftharpoons NH_4^+$$
   $$NH_3 + H_2O \rightleftharpoons NH_4^+ + OH^-$$

   At the same time liquid NH_3 has a high solubility in H_2O because of the above reaction.

   (*a*) (*i*) $\underset{\text{acetic acid}}{CH_3COOH} + NH_3 \rightleftharpoons CH_3COO^- + NH_4^+$

   (*ii*) $\underset{\text{Urea}}{NH_2CONH_2} + NH_3 \rightleftharpoons NH_2CONH^- + NH_4^+$

(*iii*) $CH_3CONH_2 + NH_3 \rightleftharpoons CH_3CONH^- + NH_4^+$

(*b*) Similarly $AgNH_2$ acts as an acid in the presence of liquid NH_3

$$AgNH_2 + 2NH_3 \rightleftharpoons [Ag(NH_2)_2]^- + NH_4^+$$

In all these reactions NH_4^+ ion concentration is increased in liquid NH_3 therefore these substances behave as acids in liquid NH_3.

3. **Ammoniates formation:** In ammoniate formation molecules of liquid NH_3 get attached to the solute substances that contain an acceptor atom by a chemical linkage to give the products.

e.g.

$$SiF_4 + 2NH_3 \longrightarrow SiF_4{\cdot}2NH_3$$
$$SO_3 + 2NH_3 \longrightarrow SO_3{\cdot}2NH_3$$
$$BF_3 + NH_3 \longrightarrow F_3B \leftarrow NH_3$$
$$Ca(H_2O)_4 + 4NH_3 \longrightarrow [Cu(NH_3)_4]^{2+} + 4H_2O$$
$$[Ag(H_2O)_2]^+ + 2NH_3 \longrightarrow [Ag(NH_3)_2]^+ + 2H_2O$$

The two types of ammoniates formed in above reaction are 1:2 or 1:1 adducts, Down the grouping alkaline earth metals Mg^{2+}, Ca^{2+}, Sr^{2+} and Ba^{2+} because of the increase in size NH_3 molecules are weakly attached. Overall NH_3 is a better Lewis base than H_2O.

4. **Reactions of Ammono-Acids**: Ammono-acids are substance which form NH_4^+ ions in liquid NH_3. Various reactions of metals and metals oxides with liquid NH_3 are listed below:

(*i*) Reaction with active metals:

$$Na + NH_4^+ \longrightarrow Na^+ + \frac{1}{2}H_2 + NH_3$$

The blue solution of alkali metals in liquid NH_3 react with ammonium salts and get decolourised.

Metal oxides and hydroxides dissolve in ammono acids.

$$Na_2O + 2NH_4Cl \xrightarrow{(l)\,NH_3} 2Nacl + 2NH_3 + H_2O$$

$$2NaOH + 2NH_4NO_3 \xrightarrow{(l)\,NH_3} NaNO_3 + 2NH_3 + 2H_2O$$

(*ii*) Magnesium silicide and germinides react with liquid NH_3 solution of NH_4Br to form lower silanes and germanes.

$$Mg_2Si \xrightarrow[(l)\,NH_3]{NH_4Br} SiH_4$$

$$Mg_2Ge \xrightarrow[(l)\,NH_3]{NH_4Br} GeH_4$$

(*iii*) Compounds which do not denote protons to H_2O do donate protons to liquid NH_3.

$$\underset{\text{urea}}{H_2N-\underset{\underset{O}{\|}}{C}-NH_2} + NH_3 \longrightarrow H_2N-\underset{\underset{O}{\|}}{C}-NH^- + NH_4^+$$

$$\underset{\text{acetamide}}{H_3C-\underset{\underset{O}{\|}}{C}-NH_2} + NH_3 \longrightarrow H_3C-\overset{\overset{O}{\|}}{C}-NH^- + NH_4^+$$

$$\underset{\text{sulphonamide}}{H_2N-SO_2-NH_2} + NH_3 \longrightarrow H_2N-SO_2-NH^- + NH_4^+$$

5. **Reactions of Ammono-Bases:** Some metal salts react with ammono-bases to give amides, imides and nitrides which give preceptates in liquid NH_3.

$$AgNO_3 + kNH_2 \xrightarrow[NH_3]{(l)} \underset{\text{amide}}{AgNH_2} \downarrow + kNO_3$$

$$PbI_2 + 2kNH_2 \xrightarrow[NH_3]{(l)} \underset{\text{imide}}{PbNH} \downarrow + 2KI + NH_3$$

$$BI_3 + 3kNH_3 \xrightarrow[NH_3]{(l)} \underset{\text{nitride}}{BiN} + 3kI + 2NH_3$$

6. **Formation of Ammono-Salt:** These are neutralisation reactions. In these ammono acids react with ammono-bases to form ammono salts in liquid NH_3.

e.g $$NH_4NO_3 + KNH_2 \xrightarrow[NH_3]{(l)} KNO_3 + 2NH_3$$

$$H_3C-\underset{\underset{O}{\|}}{C}-NH_2 + KNH_2 \xrightarrow[NH_3]{(l)} K^+[H_3C-\overset{\overset{O}{\|}}{C}-NH^-] + NH_3$$

$$H_2N-\overset{\overset{O}{\|}}{C}-NH_2 + kNH_2 \xrightarrow[NH_3]{(l)} [H_2N-\overset{\overset{O}{\|}}{C}-NH^-]K^+ + NH_3$$

$$2NH_4I + PbNH \xrightarrow[NH_3]{(l)} PbI_2 + 3NH_3$$

$$3NH_4I + BiN \xrightarrow[NH_3]{(l)} BiI_3 + 4NH_3$$

7. **Ammonolysis:** In such reactions there occurs increase in the concentration of NH_4^+ or NH_2^- ion.

$$NH_3 + NH_3 \rightleftharpoons NH_4^+ + NH_2^-$$
$$SiCl_4 + 8NH_3 \rightleftharpoons Si(NH_2)_4 + 4NH_4Cl$$
$$NH_3 + H^+ \rightleftharpoons NH_2^- + H_2$$

In these reactions ion, atom or radical is replaced by $—NH_2$, $=NH$ or $\equiv N$

(*i*) **Inorganic halides**

$$BCl_3 + 6NH_3\ (l) \longrightarrow B(NH_2)_3 + 3NH_4Cl$$
$$SO_2Cl_3 + 6NH_3\ (l) \longrightarrow SO_2(NH_2)_3 + 3NH_4Cl$$

$TiCl_4$ is ammonlysed in excess of liquid NH_3

$$TiCl4 \xrightarrow[\text{excess}]{NH_3\ (l)} Ti(NH_2)_4$$

Hg_2Cl_2 and $SnCl_4$ are partially ammonlysed

$$Hg_2Cl_2 + NH_3(l) \longrightarrow Hg + Hg(NH_2)Cl + NH_4Cl$$

$$SnCl_4 \xrightarrow[NH_3]{(l)} Sn(NH_2)_3Cl$$

(*ii*) **Organic compounds:**

Organic halides get ammonlysed at the boiling point of liquid NH_3.

$$RX + 2NH_3 \longrightarrow RNH_2 + NH_4Cl$$

$$RX + RNH_3 \xrightarrow[NH_3]{(l)} R_2NH + NH_4Cl$$

$$RX + R_2NH \xrightarrow[NH_3]{(l)} R_3N + NH_4Cl$$

(*iii*) **Ammonlysis of alkali metal hydrides and oxides**

$$NaH \xrightarrow[NH_3]{(l)} NaNH_2 + H_2$$

$$M_2O \xrightarrow[NH_3]{(l)} 2M(NH_2) + H_2O$$

8. Complex formation tendency

Insoluble hydroxide of some metal dissolved in excess of alkali hydroxides.

$$Zn(OH)_2 + 2NaOH \longrightarrow \underset{\text{soluble complex}}{Na_2[Zn(OH)_4]}$$

In the same way metal amides, imides and nitrides form soluble amido complex in excess of metal amides.

$$Zn(NH_2)_2 + 2NaNH_2 \longrightarrow \underset{\text{soluble amido complex}}{Na_2[Zn(NH_2)_4]}$$

Similarly soluble alumino complex is formed with amides and hydroxides of alkali metals.

$$Al^{3+} + 3NH_2^- \longrightarrow Al(NH_2)_3 \xrightarrow[NH_3]{(l)} \underset{\text{Soluble complex}}{[Al(NH_2)_4]}$$

9. Redox reactions: Some important oxidation and reduction reactions are carried in liquid NH_3.

(*a*) **Oxidation reactions:**

(*i*) Oxidation of NH_3 to NO is the presence of a catalyst (Ostwald's oxidation)

$$4NH_3 + 5O_2 \xrightarrow{Pt} 4NO + 6H_2O$$

(*ii*) Oxidation of active metals

$$3Mg + 2NH_3 \longrightarrow Mg_3N_2 + 3H_2$$

(*iii*) I_2 in liquid NH_3 acts as a mild oxidising agent

$$\underset{\text{Potassium aminono stanite}}{K_4[Sn(NH_2)_6]} \xrightarrow[\text{liq}\cdot NH_3]{I_2} \underset{\text{Potassium ammono stannate}}{K_2[Sn(NH_2)_6]}$$

(*b*) **Reduction reactions**

(*i*) Metal oxide is reduced by liquid NH_3 at elevated temperature

$$3CuO + 2NH_3 \longrightarrow 3Cu + 3H_2O + N_2$$

(*ii*) $$[Ni(Cr)_4]^{2-} + 3e \longrightarrow [\overset{1}{Ni}(CN)_3]^{2-} + CN^-$$

$$[Ni(CN)_3]^{2-} + CN^- \longrightarrow [\overset{0}{Ni}(CN)_4]^{4-}$$

(*iii*) $$CuI + Na \xrightarrow[NH_3]{(l)} Cu + NaI$$

(*iv*) $$4ZnI_2 + 9Na \xrightarrow[NH_3]{(l)} NaZn_4 + 8NaI$$

(*v*) $KMnO_4$ is reduced to K_2MnO_4 by liquid NH_3

$$6KMnO_4 + 6kNH_2 \xrightleftharpoons[NH_3]{(l)} 6k_2MnO_4 + 4NH_3 + N_2$$

(*vi*) Organo-metallic compounds are reduced in presence of Na in liquid NH_3

$$(C_6H_5)_2\,BiI + 2Na \xrightarrow[NH_3]{(l)} [(C_6H_5)_2Bi]Na + NaI$$

10. Metal-ammonia reaction:

(*a*) Some metals like alkali, alkaline earth, Al, Eu and Y readily dissolve in liquid. NH_3 and give metal ammonia solution. These solutions are blue in dilute and bronze in concentrated solutions. The blue solutions are very good electrolytes. The bronze (concentrated) solutions have high thermal coefficient of electrical conductance approaching free metal electrical conductance. Blue-coloured solutions is of the measure of electrolytic solutions.

Alkali metal can be recovered from the alkali-metal ammonia solution by evaporation where as alkaline-earth ammonia solution gives $[M(NH_3)_6]^{2+}$ which indicates solvent action on the metal forming a chemical bond, All alkali metal-ammonia solution on long standing or in the presence of a catalyst yield H_2 gas and metal oxide

$$M + nNH_3 \longrightarrow M(NH_3)_n + n/2\ H_2\uparrow$$

The blue metal solution is paramagnetic (attracted by the magnetic field) because of the presence of unpaired electrons. With the increase in the concentration of solution the magnetic susceptibility becomes less .

It is now considered that in concentrated solution the electrons are largely free *e.g*

$$M \rightleftharpoons M^+ + e$$

In dilute solutions the electrons and the M^+ ions get solvated reversibly:

$$M^+ + xNH_3 \rightleftharpoons \underset{\text{Solvated or ammoniated}}{M(NH_3)_x^-}$$

$$e^- + y(NH_3) \rightleftharpoons e(NH_3)_y^-$$

The overall reaction remains:

$$M^+ + (x - y)\ NH_3 \rightleftharpoons [M(NH_3)x]^+ + [e(NH_3)_y]^-$$

$$\text{or } M \xrightleftharpoons[NH_3]{(l)} M^+_{\text{ammonia}} + e_{\text{ammonia}}$$

The electrons occupy the cavities present in solvent NH_3 surrounded by other ammonia molecules whose protons are oriented toward these free electrons. With increase in concentration M^+ ammonia cationic bond together by e^- ammonia into aggregates such as M_2, M_3 and so on.

In the concentrated bronze-coloured solution, M^+ ammonia and e^- ammonia are bonded in a way similar to the metal ions in a molten metal.

(*b*) Reactions of metal-ammonia solutions

(*i*) Blue metal-ammonia solution contains mobile ammoniated electrons which can act as good reducing agent specially for substances soluble in liquid NH_3 same examples of reducing action are:

Non-metallic elements

$$S + \text{Metal-ammonia} \xrightarrow[NH_3]{Na} H_2S$$

(*ii*) Metal-ammonia solutions are decolorised by NH_4^+ salt or other acids

$$NH_4^+ + e^-_{(ammonia)} \longrightarrow NH_3 + \frac{1}{2} H_2$$

This conforms that the blue colour is due to ammoniated electrons.

$$SO_2\!\begin{cases} NH_2 \\ NH_2 \end{cases} + 2e_{(ammonia)} \longrightarrow SO_2\!\begin{cases} NH^- \\ NH^- \end{cases} + H_2$$

(*iii*) Organic compound undergo reduction in Na/NH_3 solution:

$$R\text{—}C{\equiv}CH + e^-_{(ammonia)} \xrightarrow{Na/NH_3} \underset{\text{acetylide ion}}{R\text{—}C{\equiv}C} + \frac{1}{2} H_2$$

$$R\text{—}C{\equiv}CH + \frac{1}{2} H_2 \longrightarrow \underset{\text{ethylene}}{R\text{—}C{=}CH_2}$$

(*iv*) Organo metallic compounds also react with metal-ammonia solution and form a number of compounds containing metal bonds.

Liquid Sulphur Dioxide

It is a water like solvent used for carrying out large number of chemical reactions. It has wide range of temperature with boiling point 10.0°C and freezing point – 75.5°C. Many organic and inorganic compounds are soluble in liquid SO_2 therefore it can be used as a solvent.

The autoionisation of liquid SO_2 is similar to water and ammonia

$$SO_2 + SO_2 \rightleftharpoons \underset{\text{Thionyl ion}}{SO^{2+}} + \underset{\text{Sulphite ion}}{SO_3^{2-}}$$

$$H_2O + H_2O \rightleftharpoons H_3O^+ + OH^-$$
$$NH_3 + NH_3 \rightleftharpoons NH_4^+ + NH_2^-$$

Dipole moment of SO_2 is 1.61 D and its dielectric constant is 17.27 at –16.5°C. The structure of SO_2 is a resonance hybrid of following structure:

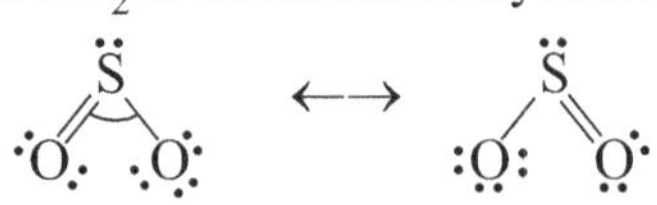

electron diffraction measurements are

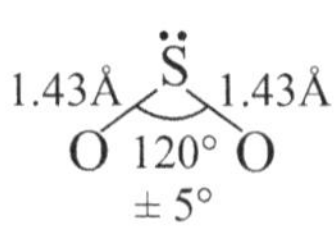

5-atom has a vacant orbital and therefore it can act as an electron acceptor to form complexes

Solubility of Compounds in liquid SO_2

(*a*) **Inorganic materials:** Inorganic substances like I_2 and SCN^- are highly soluble.

Solubilities of alkali halides is in the order MI > MBr > MCl > MF. Alkali metal sulphites and acetates are also soluble. Alkali metal sulphates, thiocyanates and cyanides are soluble whereas other metal sulphates, sulphides, oxides and hydroxides are insoluble. Ammonium, thalium and Hg^{2+} salts are soluble. Substances like IBr, PBr_3, CCl_4, $SiCl_4$, $GeCl_4$, $SnCl_4$, Br_2, ICl, BCl_3, PCl_3, $AsCl_3$, CS_2, $POCl_3$, $SOCl_2$, $SO(CH_3COO)_2$, SO_2X_2 are miscible with liquid SO_2.

(*b*) **Organic compounds:** Covalent organic compounds are more soluble in liquid SO_2 than ionic and semi-ionic compounds.

Aromatic hydrocarbons and alkenes are more soluble than alkanes. Friedel Croft's and bromination reactions take place in liquid SO_2.

Chemical reactions in liquid SO_2

1. **Precipitations reactions:** Chemical reactions given below indicate several insoluble materials formed by mixing the soluble compounds in liquid SO_2.

(*i*) $BaI_2 + Zn(SCN)_2 \longrightarrow Ba(SCN)_2\downarrow + ZnI_2$

(*ii*) $SbCl_3 + 3LI \longrightarrow SbI_3\downarrow + 3LiCl$

(*iii*) $AlCl_3 + 3NaI \longrightarrow 3NaCl\downarrow + AlI_3$

(*iv*) $PbF_2 + Li_2SO_4 \longrightarrow PbSO_4\downarrow + 2LiF$

Other thionyl halides can be prepared from miscible solution of $SOCl_2$ in liquid SO_2.

$$2Ag(CH_3COO) + SOCl_2 \xrightarrow[SO_2]{(l)} 2AgCl\downarrow + SO(CH_3COO)_2$$

$$2KI + SOCl_2 \xrightarrow[SO_2]{(l)} 2KCl\downarrow + SOI_2$$

$$2KBr + SOCl_2 \xrightarrow[SO_2]{(l)} 2KCl\downarrow + SOBr_2$$

$$2NH_4SCN + SOCl_2 \xrightarrow[SO_2]{(l)} 2NH_4Cl + SO(SCN)_2$$

Organic derivates of NH_3 with liquid SO_2 can be represented as:

$$2[\geqslant N] + 2SO_2 \rightleftharpoons [(\geqslant N)_2\,SO]\,SO_3$$

$$\Downarrow$$

$$[(\geqslant N)_2\,SO]^{2+} + SO_3^{2-}$$

2. **Acid-base reactions:** Auto ionisation of liquid SO_2 was proposed by cody and Elsey and supported by Jander tand Wikert

$$\underset{\text{Acid}_1}{SO_2} + \underset{\text{Base}_2}{SO_2} \rightleftharpoons \underset{\text{Acid}_2}{SO^{2+}} + \underset{\text{Base}_1}{SO_3^{2-}}$$

Some neutralisation reactions are summarised below:

Acid		Base		Salt		Solvent
$SOCl_2$	+	Cs_2SO_3	$\longrightarrow$	2CsCl	+	$2SO_2$
$SO(SCN)_2$	+	K_2SO_3	$\longrightarrow$	2K(SCN)	+	$2SO_2$

Reactions between SO_3^{2-} and thionyl compounds in liquid SO_2 occur as:

$$SO_3^{2-} + SO^{2+} \longrightarrow 2SO_2$$

3. **Solvolysis:** Certain halides undergo solvolysis in liquid SO_2.

(*i*) $PCl_5 + SO_2\ (l) \longrightarrow POCl_3 + SOCl_2$

(*ii*) $NbCl_5 + SO_2\ (l) \xrightarrow{70°C} NbOCl_3 + SOCl_2$

(*iii*) $WCl_6 + SO_2\ (l) \xrightarrow{70°C} WOCl_4 + SOCl_2$

While other metal holides like $SbCl_3$, $SiCl_4$, $SnCl_4$ undergo hydrolysis to give a clear solution but no solvolysis.

4. **Solvation:** Some salts form addition compounds withe the liquid SO_2. These are called solvates similar to hydrates and ammoniates in case of water and liquid ammonia.

For example NaI, $4SO_2$; SrI_2 $4SO_2$; Cs(SCN).$5SO_2$; K(SCN).$2SO_2$; $[(CH_3)_4N]$ $3SO_2$; Cs(SCN).$5SO_2$.

5. **Complex farmation tendency:** Complex compounds are formed by $SbCl_3$ and $SbCl_5$ in liquid SO_2 *e.g.*

$$SbCl_3 + 3KCl \xrightarrow[SO_2]{(l)} K_3[SbCl_6]$$

$$SbCl_5 + KCl \xrightarrow[SO_2]{(l)} K[SbCl_6]$$

$$SbCl_5 + [(CH_3)_4N]Cl \xrightarrow[SO_2]{(l)} [N(CH_3)_4]\ [SbCl_4]$$

liquid SO_2 is also used as a medium for the preparation of some complexes:

$$2SbCl_3 + 3SOCl_2 \xrightarrow[SO_2]{(l)} [SO]_3\ [SbCl_6]_2$$

$$SbCl_5 + NOCl \xrightarrow[SO_2]{(l)} [NO]\ [SbCl_6]$$

Hexachloro antimonate

6. **Amphoteric Behaviour:** Salts of metals whose hydroxides are amphoteric in aqueous medium behave in an anologous way in liquid SO_2 *e.g.*

$$AlCl_3 + 3OH^- \xrightarrow{H_2O} Al(OH)_3 \downarrow + 3Cl^-$$

$$2AlCl_3 + 3SO_3^{2-} \xrightarrow[SO_2]{liq} Al_2(SO_3)_3 \downarrow + 6Cl^-$$

The above precipitate dinolve in excess of the solvent to give soluble compex.

$$Al(OH)_3 \downarrow + \underset{\text{excess}}{OH^-} \xrightarrow{H_2O} \underset{\text{Soluble complex}}{[Al(OH)_4]^-}$$

$$Al(SO_3)_3 + 3SO_3^{2-} \xrightarrow[SO_2]{(l)} \underset{\text{Soluble complex}}{2[Al(SO_3)_3]^{3-}}$$

7. **Redox reactions:** Gaseous SO_2 has an oxidation number of +4 which can change to +6 or +2, zero, –2 therefore it can act both an oxidising as well as a reducing agent. On the other hand liquid SO_2 may serve as a medium for oxidation-reduction reactions.

$$I_2 + 2RSO_3 \xrightarrow[SO_2]{(l)} RSO_4 + RI_2 + SO_2$$

In this reaction I_2 is reduced by sulphite in liquid SO_2.

$$6KI + 3SbCl_4 \xrightarrow[SO_2]{(l)} 3I_2 + SbCl_3 + 2K_3[SbCl_4]$$

8. **Organic compounds soluble in liquid SO_2:** Liquid SO_2 serves as a medium for a number of organic reactions listed below:

(*i*) **Sulphonation:** Anomatic hydrocarbons undergo sulphonation in liquid SO_2

$$ArH + ClSO_3H \xrightarrow[SO_2]{(l)} ArSO_3H + HCl$$

(*ii*) **FCR** (Freidel Craft's reaction)

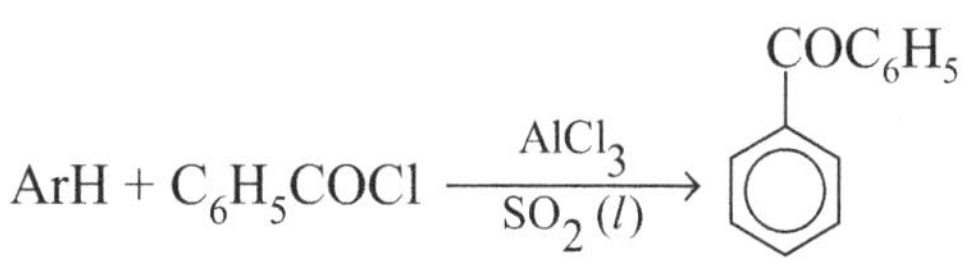

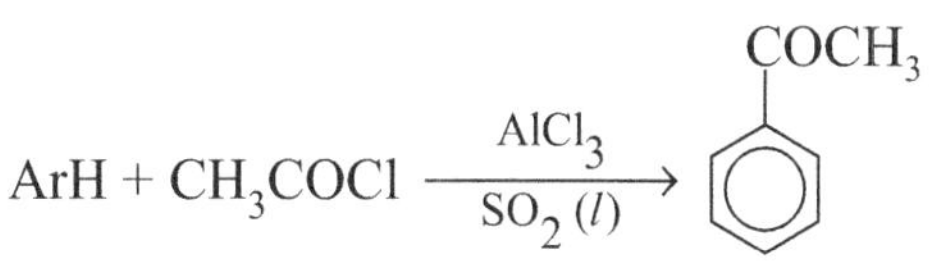

The above reactions are named as Freidel Craft's acylation and benzoylation

(*iii*) **Bromination:** Liquid SO_2 is used as a solvent in bromination reactions.

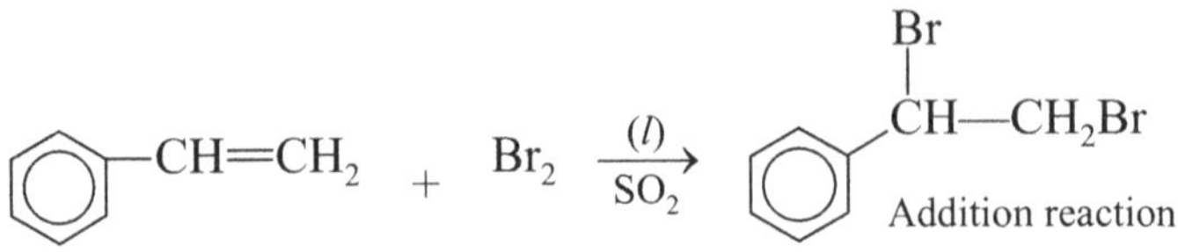

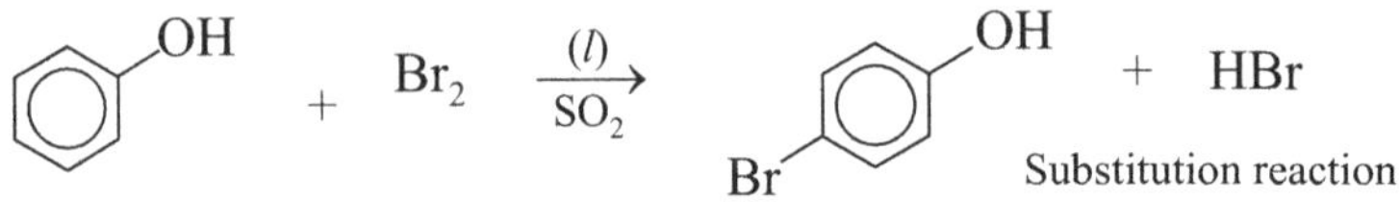

Liquid Dinitrogen Tetraoxide (N_2O_4)

On Cooling nitrogen dioxide gas turns into colourless liquid which is a dimer of NO_2 brown turmes

$$\underset{\text{brown gas}}{2NO_2} \xrightleftharpoons{\text{cooling}} \underset{\text{colourless liquid}}{N_2O_4}$$

Liquid N_2O_4 is a good oxidising agent. It oxidises organic compounds of reducing nature explosively. It has wide liquid range therefore can be used as a good solvent.

Auto ionisation of liquid N_2O_4 takes place

$$N_2O_4 + N_2O_4 \rightleftharpoons \underset{\text{Nitrosyl ion}}{2NO^+} + \underset{\text{Nitrate ion}}{2NO_3^-}$$

Chemical Reactions

1. **Acid-Base reactions or salt formation:** Substances giving NO^+ ion N_2O_4 solvent will behave as acids and those furnishing NO_3^- ions behave.

   Reaction of NOCl and $AgNO_3$ in liquid N_2O_4 form AgCl salt and N_2O_4 solvent.

$$\underset{\text{Acid}}{NOCl} + \underset{\text{Base}}{AgNO_3\,(s)} \longrightarrow \underset{\text{Salt}}{AgCl} + \underset{\text{Solvent}}{N_2O_4}$$

2. **Reaction with metals:** This reaction is similar to that in case of Water *e.g*

$$M + H_2O \longrightarrow MOH + \frac{1}{2}H_2$$

$$M + N_2O_4 \longrightarrow MNO_3 + \underset{\text{nitric oxide gas}}{NO}$$

In water solvent acidulated water increases the reactivity of metals, in the same way NOCl addition increases the reactivity of metals towards liquid N_2O_4.

$$Zn + 2NOCl \xrightarrow[N_2O_4]{(l)} ZnCl_2 + 2NO\uparrow$$

$$M + 2HCl \xrightarrow{H_2O} MCl_2 + H_2\uparrow$$

3. **Complex formation tendency:** $Zn(NO_3)_2$ dissolves $[Et_2\ NH_2]\ NO_3$ to yield zinc nitrato complex in liquid N_2O_4.

$$Zn(NO_3)_2 + x[Et_2NH_2]\ NO3 \xrightarrow[N_2O_4]{(l)} \underset{\text{Nitrato zincate}}{[Et_2NH_2]_x\ [Zn(NO_3)_{x+2}]}$$

This reaction shows amphoteric nature of the reaction.

4. **Solvate reactions:** Nitrates of Zn, Fe and U react with liquid N_2O_4 and form solvate adducts *e.g* $Fe(NO_3)_3 \cdot N_2O_4$; $Zn(NO_3)_2 \cdot 2N_2O_4$

   The solvates can also be written in the complex salt formation: $(NO)_2\ [Zn(NO_3)_4]$ and $NO\ [Fe(NO_3)_4]$

5. **Solvolysis:** Some examples of solvolysis reactions in liquid N_2O_4 are listed below:

   (*i*) $[Mg(H_2O)_6]Cl_2 + N_2O_4\ (l) \longrightarrow [Mg(H_2O)_6]\ (NO_3)_2 + 2NOCl$

   (*ii*) $Mg(ClO_4)_2 + 2N_2O_4\ (l) \longrightarrow Mg(NO_3)_2 + 2(NO)ClO_4$

   (*iii*) $Li_2CO_3 + 2N_2O4\ (l) \xrightarrow[H_2O]{} 2LiNO_3 + N_2O_3 + CO_2$

Liquid Hydrogen Fluoride (HF)

Its liquid range is –83°C to 19.4°C therefore it is a good solvent. Its disadvantages lies in the fact that (*i*) few substances dissolve in it and it has a poisnous nature. It has high dielectric constant hence it acts as a good ionising solvent like water. It dissolves inorganic substances more than organic substances to give a conducting solution.

1. **Solute in liquid HF**

   (*i*) Solutes like pentahalides of P, As and Sb dissolve in liquid HF as:

$$SbF_5 + HF \longrightarrow SbF_6^- + H^+$$

   (*ii*) Addition of HF to substances like acetic acid and KNO_3 give F^- ions

$$CH_3COOH + HF \longrightarrow CH_3COOH_2^+ + F^-$$

$$KNO_3 + 2HF \longrightarrow K^+ + H_2NO_3^+ + F^-$$

(*iii*) Compounds in which F^- replaces its anion.

$$KCN + HF \longrightarrow KF + HCN$$

$$KBr + HF \longrightarrow KF + HBr$$

(*iv*) Reactions involving more than simple replacement

$$H_2SO_4 + 2HF \longrightarrow HSO_3F + H_3O^+ + F^-$$

2. **Autoionisation:** It may be represented by the following

$$HF + HF \longrightarrow \underset{\text{Fluoronium ion}}{H_2F^+} + \underset{\text{fluoride ion}}{F^-}$$

Where H_2F^+ behaves as an acid and F^- ions as base.

(*i*) Strong acids like H_2SO_4 and HNO_3 behave as strong acids in aqueous solution become bases in liquid HF

$$HNO_3 + HF \longrightarrow H_2NO_3^+ + F^-$$

$$H_2SO_4 + HF \longrightarrow H_2SO_4^+ + F^-$$

(*ii*) $HClO_4$ is the strongest acid in aqueous solution behaves amphoteric in liquid HF.

(*a*) $HClO_4 + HF \longrightarrow H_2ClO_4^+ + F^-$

(*b*) $HClO_4 + HF \longrightarrow ClO_4^- + H_2F^+$

In reaction (*a*) $HClO_4$ behaves as a base and in (*b*) it acts as an acid.

(*iii*) Lewis acids like BF_3 and SbF_5 act as acids in liquid HF

$$F_3B + 2HF \longrightarrow H_2B^+ + BF_4^-$$

$$SbF_5 + 2HF \longrightarrow H_2F^+ + SbF_6^-$$

Weak acids behave as bases in liquid HF.

$$\underset{\text{weak acid}}{CH_3COOH} + HF \longrightarrow CH_3COOH_2^+ + F^-$$

3. **Redox reactions:** As an acid HF reactions with metals are not different than that of metals in HNO_3 in aqueous medium

$$3HAsF_6 + 2Ag \longrightarrow 2AgAsF_6 + AsF_3 + 3HF$$

$$4HNO_3 + 3Ag \longrightarrow 3AgNO_3 + NO + 2H_2O$$

4. **Precipitation reactions:**

$$Na_2SO_4 + 2AgF \xrightarrow[HF]{(l)} Ag_2SO_4 \downarrow + 2NaF$$

$$NaClO_4 + AgF \xrightarrow[HF]{(l)} AgClO_4 + NaF$$

$AgBF_4$ precipitate is obtained when $AgNO_3$ and BF_3 react in liquid HF.

$$AgNO_3 + BF_3 + 2HF \longrightarrow AgBF_4 + H_2NO_3F$$

Adduct formation

Liquid HF forms adducts with metal fluorides.

$$KF + HF \longrightarrow KF{\cdot}HF$$
$$KF + 2HF \longrightarrow KF{\cdot}2HF$$
$$NH_4F + HF \longrightarrow NH_4F{\cdot}HF$$
$$KF + 3HF \longrightarrow KF{\cdot}3HF$$

Acetic Acid as a Solvent

Acetic acid exists as a dimer in the liquid state. It has wide liquid range 16.6° to 118°C. It is non-ionic and stable at normal conditions . Preparation of anhydrous acetic is quite difficult. It has zero dipole moment and low dielectric constant *i.e* 9.7. It is expected to behave as a poor solvent but surprisingly large number of compounds are soluble in it.

1. **Self-ionisation and acid-base reaction:** Self-ionisation of acetic acid occurs as:

$$\underset{Acid_1}{CH_3COOH} + \underset{Base_2}{CH_3COOH} \rightleftharpoons \underset{Acid_2}{CH_3COOH_2^+} + \underset{Base_1}{CH_3COO^-}$$

Typical acid-base reactions are given below:

$$\underset{base}{CH_3COONa} + \underset{acid}{HCl} \longrightarrow \underset{solt}{NaCl} + \underset{solvent}{CH_3COOH}$$

Weak bases like acetamide and acetanilide can be titrated in CH_3COOH medium. $HClO_4$ is highly dissociated strong acid in CH_3COOH. Weak organic bases behave as strong bases in acetic acid.

$$\underset{\text{weak organic base}}{B} + CH_3COOH \longrightarrow BH^+ + CH_3COO^-$$

or $CH_3CONH_2 + CH_3COOH \longrightarrow CH_3CONH_3^+ + CH_3COO^-$

2. **Precipitation reaction:** Precipitation reactions taking place in liquid CH_3COOH are:

$$Pb(CH_3COO)_2 + 2NaCl \xrightarrow[CH_3COOH]{(l)} PbCl_2\downarrow + 2CH_3COONa$$

$$Cd(CH_3COO)_2 + H_2S \xrightarrow[CH_3COOH]{(l)} CdS\downarrow + 2CH_3COOH$$

3. **Amphoteric reaction:** $Zn(CH_3COO)_2$ exhibits amphoteric behaviour in liquid CH_3COOH.

$$Zn^{2+} + 2CH_3COO^- \longrightarrow \underset{\text{Insoluble ppt.}}{Zn(CH_3COO)_2}$$

$$\xrightarrow{2CH_3COO^-} \underset{\text{soluble complex}}{[Zn(CH_3COO)_4]^{2-}}$$

4. **Complex formation tendency:** This tendency exists in liquid CH_3COOH just like, aqueous solution *e.g.* Fe^{3+} and SCN^- ions react in liquid CH_3COOH to give red complex $[Fe(SCN)_6]^{3-}$. The same insoluble red complex complex is also obtained in aqueous medium.

$$Fe^{3+} + 6SCN^- \xrightarrow[CH_3COOH]{(l)} \underset{\text{Red insoluble Complex}}{[Fe(SCN)_6]^{3-}}$$

5. **Sovolysis:** The following reaction exemplifies the reaction:

$$SO_2Cl_2 + 4CH_3COOH \longrightarrow SO(CH_3COO)_2 + 2CHCOOH_2^+ + 2Cl^-$$
$$CN^- + CH_3COOH \longrightarrow CH_3COO^- + HCN$$

Solvate Reaction

The typical example of solvate formation is CH_3COOK, $2CH_3COOH$ in liquid acetic acid.

In the same way other non-aqueous solvents such as BrF_3, H_2SO_4 and Phosphoryl chloride $POCl_3$ behave and give similar reactions.

EXERCISES

1. Name at least three aprotic and aprotic solvents.
2. Give some advantages of liquid SO_2 solvent in spite on its toxic nature.
3. Explain solvolytic reaction in:
   (*i*) Liquid NH_3 and (*ii*) Liquid SO_2
4. What is the use of SO_2 as a solvent in organic reactions?
5. Discuss amphoterism in liquid SO_2 with that of water.
6. Give reactions of complex formation in liquid SO_2 and liquid NH_3.
7. Complete the following equations:
   (*a*) $PbI_2 + KNH_2 \xrightarrow{NH_3}$
   (*b*) $HgI_2 + KNH_2 \xrightarrow{NH_3}$
   (*c*) $SO_2Cl_2 + NH_3 \longrightarrow$
   (*d*) $WCl_6 + SO_2 \xrightarrow{SO_2}$
   (*e*) $C_6H_6 + SO_3 \xrightarrow{SO_2}$
   (*f*) $SbCl_5 + NOCl \xrightarrow{SO_2}$
8. Give two reactions of showing sulphuric acid as a
   (*i*) Dehydrating agent (*ii*) Protic acid
9. Discuss self ionization reaction of H_2SO_4 solvent.
10. Account for 'Solutes in H_2SO_4 are highly conducting inspite of high viscosity of H_2SO_4'.

11. Discuss:
    (*i*) Neutralization reaction in SO_2 solvent
    (*ii*) Solutions of alkali metals in liquid NH3 are good conductors and good reducing agents.
12. Discuss two examples of each reaction in liquid HF
    (*i*) acid-base reaction (*ii*) precipitation reactions
13. A strong acid HNO_3 behaves as a base in HF. Explain.
14. Discuss the behavior of CH_3COOH and SbF_5 in HF.
15. Why NH_3 is a better solvent than H_2O.
16. Liquid SO_2 acts as oxidizing and reducing solvent. Discuss.
17. Compare auto-ionization of
    (*i*) Liquid NH_3 (*ii*) H_2O and (*iii*) Liquid SO_2
18. What are the important factors to dissolve the substances?
19. Water is universal solvent. Discuss.
20. Complete the equations:

    $$SiCl_4 + NH_3 \xrightarrow[NH_3]{\text{liquid}}$$

    $$Zr(NO_3)_2 + KNH_2 \xrightarrow[NH_3]{\text{liquid}}$$

21. Discuss the oxidation reduction complex formation and ammonlysis is liquid NH_3 medium.

7

Oxidation and Reduction

Oxidation and reduction play a very important role in out daily life. The processes like digestion of food, fading colour of the clothes, torch light from the cell are all because of oxidation reactions. Industrial processes like extraction of metals, removal of stains from clothes involve reduction. Therefore, it is very useful to understand how the process of oxidation and reduction is brought about. The redox reactions require the understanding of the concepts of valency and oxidation number.

Valency: Valency is the combining capacity of an element. It refers to the number of hydrogen atoms or double the number of oxygen atoms which combine with one atom of the element. Covalency of an element relates to the number of covalent bonds by which the element is linked to the surrounding elements.

It is a pure number and has no plus or minus associated with it *e.g.* in water valency of oxygen is two (H_2O). It always remains a whole number. In $O{=}C{=}O$ covalency of C is 4 and O atom is 2. Some elements like C, N and sulphur exhibit fixed valency. Valency is real and not imaginary. Valency of N in all its compounds is three.

Oxidation Number: Oxidation number is a measure of the extent of oxidation of an element in its compounds. Oxidation number is defined as residual charge left on the atom when all other atoms from the molecule are removed as ions by counting the shared pair of electrons with more electronegative atoms. It can be determined by following a set of simple rules:

1. An element in any of its allotrophic form is assigned an oxidation number of zero, *e.g.* S in monoclinic, rhombic or colloidal form has an oxidation number 'zero'.
2. The oxidation number of monoatmoic ion is equal to the charge it contains *e.g.* Na^+ ion has an oxidation number +1, Mg^{2+} is assigned oxidation number +2, O^{2-} oxide has an oxidation number –2 etc.
3. The sum of oxidation number in a molecule or compound is zero *i.e.*, the positive oxidation number and negative oxidation number must be

equal, *e.g.* in NaCl oxidation number of Na^+ is +1 and Cl^- ion is –1 so that their sum, is zero, Fe_2O_3 iron has an oxidation number of +3 or Fe_2 is 6^+ like wise O^{2-} ion is –2 and $3O^{2-}$ is 6^- so that their sum is zero.

4. Elements like C, N and S show variable oxidation numbers in its compounds.

Compounds	Oxidation number of nitrogen
N_2O	+1
NO	+2
N_2O_3	+3
NO_2	+4
N_2O_5	+5
N_2	zero
NH_3	–3

Oxidation number is not real but imaginary. It is a hypothetical charge that is supposed to be present on the atom in its compound. Oxidation number may not always be a whole number.

Example: Assign the oxidation number to:

(*i*) Cr in $Cr_2O_7^{2-}$ ion

(*ii*) Mn in KMn O_4

(*iii*) S in $Na_2S_4O_6$

(*iv*) Mn in Mn_3O_4

(*v*) O in KO_3

(*vi*) O in RbO_2

Easiest way to find the oxidation number in above compounds and ions is noted below:

The fixed oxidation in O is oxide ion with O^{2-} and –2 oxidation number. In the case of molecule or compound the sum of the charges is zero while in the case of ions the sum of the charges is equal to the charge on the ions (oxocations or oxoanions)

(*i*) $Cr_2O_7^{2-}$ — O is taken as oxide O^{2-} with –2 oxidation number

Cr_2	O_7	$^{2-}$
$+6 \times 2$	14^-	

O.N. Cr is +6 sum of the charges $12^+ + 14^- = 2^-$ compound

(*ii*)

K	Mn	O_4
+1	+7	8^-

Mn is +7 sum of the charges $1^+ + 7^+ + 8^- = 0$

(*iii*)

Na_2	S_4	O_6
+2	+2.5 × 4	12^-

O.N. of S is 2.5 Sum of the charges
$2^+ + 10^+ + 12^- = 0$

(*iv*)

Mn_3	O_4
+8	8^-

i.e. 3 Mn = +8
or Mn = 8/3 = 2.67
Sum of the charges = $8^+ + 8^-$

(*v*)

K	O_3
+1	−1

i.e. $O_3 = -1$
or $O = -\frac{1}{3}$
Therefore KO_3 is an ozonide.

(*vi*) RbO_2

Rb	O_2
+1	−1

i.e. $O_2 = -1$
or $O = -\frac{1}{2}$
Therefore RbO_2 is a superoxide.

Classical Definition of Oxidation

Oxidation is a process in which oxygen or any other electronegative element is added while hydrogen or electropositive element is removed.

(*i*) *Addition of oxygen*

$$2Mg + O_2 \longrightarrow 2MgO$$
$$S + O_2 \longrightarrow SO_2$$

(*ii*) *Addition of electronegative element*

$$Na + \frac{1}{2}Cl_2 \longrightarrow NaCl$$
$$Hg_2Cl_2 + Cl_2 \longrightarrow 2HgCl_2$$

(*iii*) *Removal of hydrogen*

$$3Fe + 4H_2O \text{ (vapours)} \longrightarrow Fe_3O_4 + 4H_2$$
$$2Na + 2H_2O \longrightarrow 2NaOH + H_2$$

(*iv*) *Removal of electropositive element*

$$H_2S + Cl_2 \longrightarrow 2HCl + S$$

Classical Definition of Reduction

Reduction is a reverse process of oxidation *i.e.* it is process of addition of hydrogen or electropositive element or removal of oxygen or any electropositive element.

(*i*) *Additional of hydrogen*

$$2C + H_2 \longrightarrow C_2H_2$$
$$N_2 + 3H_2 \longrightarrow 2NH_3$$

(*ii*) *Addition of electropositive element*

$$HgCl_2 + Hg \longrightarrow Hg_2Cl_2$$
$$CuCl_2 + Cu \longrightarrow 2Cu\,Cl$$

(*iii*) *Removal of oxygen*

$$ZnO + C \longrightarrow Zn + CO$$
$$CuO + H_2 \longrightarrow Cu + H_2O$$

(*iv*) *Removal of electronegative element*

$$2HgCl_2 + SnCl_2 \longrightarrow Hg_2Cl_2 + SnCl_4$$
$$2FeCl_3 + H_2S \longrightarrow 2FeCl_2 + 2HCl + S$$

Electronic Concept of Oxidation and Reduction

Oxidation is a process in which an atom, ion or molecule loses one or more electrons. It is also called De-electronation.

Reduction is a process in which an atom, ion or molecule gains one or more electrons. It is also termed as electronation.

Examples of Oxidation: Loss of electrons

(*i*) $Zn^0 \longrightarrow Zn^{2+} + 2e$

(*ii*) $Fe^{2+} \longrightarrow Fe^{3+} + e$

(*iii*) *Loss of electron from and anion*

$$2I^- \longrightarrow I_2 + 2e$$
$$[Fe(CN)_6]^{4-} \longrightarrow [Fe(CN)_6]^{3-} + e$$

(*iv*) *Molecules*

$$H_2 \xrightarrow{-2e} 2H^+ \quad \text{or} \quad H_2 \longrightarrow 2H^+ + 2e$$
$$2H_2O \longrightarrow O_2 + 4H^+ + e$$

Examples of Reduction: Gain of electrons

(*i*) *Atoms*

$$F + e \longrightarrow F^-$$
$$Ce + e \longrightarrow Cl^- \qquad \left(\frac{1}{2}Cl_2 + e \rightarrow Cl^-\right)$$

(*ii*) *Cations*

$$2H^+ + 2e \longrightarrow H_2$$
$$Fe^3 + e \longrightarrow Fe^{2+}$$
$$Cu^2 + 2e \longrightarrow Cu$$

(*iii*) *Anions*

$$MnO_4^- + 8H^+ + 5e \longrightarrow Mn^{2+} + 4H_2O$$
$$Cr_2O_72^- + 14H^+ + 6e \longrightarrow 2Cr^{3+} + 7H_2O$$
$$2IO_3^- + 12H^+ + 10e \longrightarrow I_2 + 6H_2O$$

(*iv*) *Molecules*

$$O_2 + 4H^+ + 4e \longrightarrow 2H_2O$$
$$Cl_2 + 2e \longrightarrow 2Cl^-$$
$$2H_2O + 2e \longrightarrow H_2 + 2OH^-$$

Oxidation and Reduction on the basis of oxidation number: On this basis oxidation is defined as a process in which the oxidation number of the particular element in the molecule or specy increases where as reduction is the process in which the oxidation number of the concerned element decreases.

Some examples of oxidation:

(*i*) $Zn^0 \longrightarrow Zn^{2+} + 2e$ (0 to +2)

(*ii*) $Fe^{2+} \longrightarrow Fe^{3+} + e$ (+2 to +3)

(*iii*) $2I^- \longrightarrow I_2 + 2e$ (–1 to 0)

(*iv*) $2S_2O_3^{2-} \longrightarrow S_4O_6^{2-} + 2e$ (+2 to +2.5)

(*v*) $[Fe(CN)_6]^{4-} \longrightarrow [Fe(CN)_6]^{3-} + e$ (+2 to +3)

Some examples of reduction:

(*i*) $Fe^0 + e \longrightarrow F^-$ (0 to –1)

(*ii*) $Cu^{2+} + 2e \longrightarrow Cu^0$ (+2 to 0)

(*iii*) $MnO_4^- + 8H^+ + 5e \longrightarrow Mn^{2+} + 4H_2O$ (+7 to + 2)

(*iv*) $ClO_2 + 2H_2O + 5e \longrightarrow Cl^- + 4OH^-$ (+4 to –1)

Redox Reactions and Half Reactions

In a complete reaction oxidation and reduction go side by side. If one specy is oxidised, the other has to be reduced. This means that oxidation and reduction take place simultaneously. Oxidation is loss of electrons while reduction is gain of electrons. This implies that gain and loss of electrons within the overall reaction is called redox reaction. The part of the reaction that represents oxidation (loss of electrons) is called oxidation half reaction and that which represents reduction is called reduction half reaction.

At the same time oxidation number shows an increase in the process of oxidation and the oxidation number shows a decrease in the process of reduction.

e.g.

(1) $$Cl_2^0 + 2KI^{-1} \longrightarrow 2KCl + I_2$$

Oxidation Half Reaction

$$2I^- \longrightarrow I_2^0 + 2e \qquad (-2 \text{ to } 0)$$

Reduction Half Reaction

$$Cl_2^0 + 2e \longrightarrow 2Cl^- \qquad (0 \text{ to } -2)$$

(2) $$4Zn + 10HNO_3 \longrightarrow 4Zn(NO_3)_2 + NH_4NO_3 + 3H_2O$$

Oxidation

$$4Zn^0 \longrightarrow 4Zn^{+2}(NO_3)_2 \qquad (0 \text{ to } +2)$$

$$\begin{array}{c|c|c} H & N^{+5} & O_3 \\ \hline 1^+ & +5 & 6^- \end{array} \longrightarrow N^{-3}H_4NO_3 \qquad (+5 \text{ to } -3)$$

$$\left(NH_4^+ \longrightarrow \begin{array}{c|c} N & H_4 \\ \hline -3 & 4^+ \end{array}\right)$$

(3) $$2\overset{(+2)}{Hg}Cl_2 + \overset{+2}{Sn}Cl_2 \longrightarrow \overset{(+1)}{Hg_2}Cl_2 + \overset{+4}{Sn}Cl_4$$

Reduction: $Hg^{+2} \rightarrow Hg^{+1}$; Oxidation: $Sn^{+2} \rightarrow Sn^{+4}$

(4) $$\overset{+7}{Mn}O_4^- + 8H^+ + 5Fe^{2+} \longrightarrow Mn^{+2} + 5Fe^{3+} + 4H_2O$$

Reduction: $Mn^{+7} \rightarrow Mn^{+2}$; Oxidation: $Fe^{2+} \rightarrow Fe^{3+}$

Balancing Redox Reactions

Balancking the redox reactions involves the following steps:

1. Assign oxidation numbers to atoms undergoing oxidation and reduction and write separate equations for such atoms.
2. Add electrons on the right side of oxidation reaction and on the left side for reduction reaction.
3. The number of electrons added must be equal to the change in oxidation number.
4. Balance the number of electrons in oxidation and reduction equations and add the two equations to cancel out electrons.
5. Write the complete or full equation in the original form taking the coefficients of oxidation and reduction in step 3. Balance other atoms.
6. In acid medium, if oxygen remains unbalanced, add H_2O on the side deficient in O. If hydrogen remains unbalanced then add H^+ on the required side. Make sure that all the atoms and charges are balanced in the final equation.

7. In basic medium, if oxygen remains unbalanced, then add same number of H_2O as the unbalanced oxygen atoms on the side with excess oxygen and add double the number of OH^- on the other side so as to balance oxygen.

In basic medium, if H remains unbalanced then add same number of OH^- as the difference on the side with excess H and add same number of H_2O on the other side.

In the final equation all the participants and the charges must be balanced, and satisfied.

e.g. (*a*) Balanced the given equation taking place in acid medium:

$$Cr_2O_7^{2-} + H_2SO_3 \longrightarrow Cr^{3+} + HSO_4^-$$

Invoking

Step I. $\quad {}^{+6}Cr_2O_7^{2-} + H_2{}^{+4}SO_3 \longrightarrow Cr^{3+} + H^{+6}SO_4$

$$2Cr^{6+} \longrightarrow 2Cr^{3+} \qquad \text{... reduction}$$

$$S^{4+} \longrightarrow S^{6+} \qquad \text{... oxidation}$$

Step II.

$$2Cr^{6+} + 6e \longrightarrow 2Cr^{3+} \qquad ...(i)$$

$$S^{4+} \longrightarrow S^{+6} + 2e \qquad ...(ii)$$

To cancel electrons multiply (*ii*) by 3.

Step III.

$$2Cr^{6+} + 6e \longrightarrow 2Cr^{3+}$$

$$3S^{4+} \longrightarrow 3S^{6+} + 6e$$

Step IV. $\quad Cr_2O_7^{2-} + 3H_2SO_3 \longrightarrow 2Cr^{3+} + 3HSO_4^-$

In the above equation oxygen and hydrogen atoms are not balanced. To balance oxygen, add H_2O on the side deficient in oxygen (RHS).

$$Cr_2O_7^{2-} + 3H_2SO_3 \longrightarrow 2Cr^{3+} + 3HSO_4^- + 4H_2O$$

In acid medium add H^+ on LHS to balance hydrogen atom.

$$Cr_2O_7^{2-} + 3H_2SO_3 + 5H^+ \longrightarrow 2Cr^{3+} + 3HSO_4^- + 4H_2O$$

This balances all atoms and charges in the equation.

(*b*) Balance the ionic equation:

$$ClO_2 + SbO_2^- \longrightarrow ClO_2^- + Sb(OH)_6^-$$

Step I. $\quad Cl^{4+}O_2 + Sb^{3+}O_2^- \longrightarrow Cl^{3+}O_2^- + Sb^{5+}(OH)_6^-$

$$Cl^{4+} \longrightarrow Cl^{3+} \qquad \text{... reduction}$$

$$Sb^{3+} \longrightarrow Sb^{5+} \qquad \text{... oxidation}$$

Step II.

$$Cl^{4+} + e \longrightarrow Cl^{3+} \qquad ...(i)$$

$$Sb^{3+} \longrightarrow Sb^{5+} + 2e \qquad ...(ii)$$

Step III. Multiply (*i*) by 2 and add

$$2Cl^{4+} + Sb^{3+} \longrightarrow 2Cl^{3+} + Sb^{5+}$$

Step IV. $\quad 2ClO_2 + SbO_2^- \longrightarrow 2ClO_2^- + Sb(OH)_6^-$

Step V. As per rule in basic medium, add $4H_2O$ on the R.H.S. and OH^- on L.H.S.

$$2ClO_2 + SbO_2^- + 80H^- \longrightarrow 2ClO_2^- + Sb(OH)_6^- + 4H_2O$$

To balance the number of H atoms add 6 OH^- on R.H.S.

Now it has six excess H atoms. To balance this, add $4H_2O$ on L.H.S.

$$2ClO_2^- + SbO_2^- + 80H^- + 6H_2O \longrightarrow 2ClO_2^- + Sb(OH)_6^- + 4H_2O + 60H^-$$

By rearrangement we get:

$$2ClO_2 + SbO_2^- + 20H^- + 2H_2O \longrightarrow 2ClO_2^- + Sb(OH)_6^-$$

This equation has all atoms and charges completely balanced.

Some Oxidising Agents (Oxidants)

It is a substance which oxidises some other specy but itself gets reduced.

Conventionally oxidising agent:

(*a*) gives up oxygen or electronegative element

(*b*) accepts hydrogen or electropositive element

(*c*) gains one or more electrons

(*d*) shows decrease in the oxidation number of effective element.

Oxidising Agents and their Half Reactions

Reduction half-reacion

Reduction

(*i*) $$Cr_2O_7^2 + 14H^+ + 6I^- \longrightarrow 2Cr^{3+} + 7H_2O + 3I_2$$

Oxidation

Reduction half-reaction

$$Cr_2O_7^{2-} \longrightarrow 2Cr^{3+}$$

$$Cr^{(+6)} \times 2 \longrightarrow 2\,Cr^{3+}$$

i.e., $$12^+ + 6e \longrightarrow 6^+$$

Oxidation half-reaction:

$$6I^- - 6e \longrightarrow 3I_2^0$$

$$(-1) \longrightarrow (0)$$

(*ii*) $$Br_2 + 2KI \longrightarrow 2KBr + I_2$$

Reduction half-reaction

$$\underset{(0)}{Br_2} + 2e \longrightarrow \underset{(1)}{2Br^-}$$

Oxidation half-reaction

$$\underset{(-1)}{2I^-} - 2e \longrightarrow \underset{(0)}{I_2}$$

Examples of Oxidising Agents

In oxidation half-reactions oxidising agents are reduced therefore elecrons are added on the L.H.S. of the equation.

(*i*) **Ferric salts (Fe^{3+}):** It behaves as an oxidising agent only if it gets reduced to fersoces salt (Fe^{2+})

$$2Fe^{3+} + H_2S \longrightarrow 2Fe^{2+} \text{ Reduction half-reaction}$$

or $$Fe^{3+} + e^2 \longrightarrow Fe^{2+} + 2H^+ + S$$

(*ii*) **Hydrogen peroxide (H_2O_2):** It oxidises many substances depending when the medium used.

(*a*) **Acid Medium**

$$\underset{(0=-1)}{H_2O_2} + 2H^+ + 2e \longrightarrow \underset{(0=-2)}{2H_2O}$$

(*b*) **Basic Medium**

$$\underset{(0=-1)}{H_2O_2} + 2e \longrightarrow \underset{(0=-2)}{2OH^-}$$

(*iii*) **Halogens:** All halogens (x_2) act as oxidising agent as they get reduced to halides ($2x^-$)

$$\underset{(x=0)}{x_2} + 2e \longrightarrow \underset{(x=-1)}{2x^-}$$

(*iv*) **Acidified potassium dichromate:**

$$K_2Cr_2O_7 + 4H_2SO_4 \longrightarrow K_2SO_4 + Cr_2(SO_4)_3 + 4H_2O + 3|O|$$

or $$Cr_2O_7^{2-} + 14H^+ + 6e \longrightarrow 2Cr^{3+} + 7H_2O$$

(*v*) **Potassium chlorate ($KClO_3$):** $KClO_3$ is reduced to KCl

$$KClO_3 \longrightarrow KCl + 3O$$

$$\underset{(Cl=+5)}{ClO_3^-} + 6H^+ + 6e \longrightarrow \underset{(Cl=-1)}{Cl^-} + 3H_2O$$

(*vi*) **Copper sulphate ($CuSO_4$):** Aqucous solution of $CuSO_4$ oxidisc KI solution to I_2 by itself getting reduced to cupurous ion.

$$2CuSO_4 + 2KI \longrightarrow K_2SO_4 + Cu_2SO_4 + I_2$$

$$\underset{(Cu=+2)}{2Cu^{2+}} + 2I^- \longrightarrow \underset{(Cu=+1)}{2Cu^+} + I_2$$

(*vii*) **Potassium permanganate ($KMnO_4$):** Potassium permanganate acts as a strong oxidising agent in acid, neutral and alkaline medium.

Acid Medium: In acid medium $KMnO_4$ gets reduced to Mn SO_4 (Mn^{2+})

$$2KMnO_4 + 3H_2SO_4 \longrightarrow K_2SO_4 + 2MnSO_4 + 3H_2O + 5O$$

$$\underset{(Mn=+7)}{MnO_4^-} + 8H^+ + 5e \longrightarrow \underset{(Mn=+2)}{Mn^{2+}} + 4H_2O$$

In **Neutral Medium** it gets reduced to MnO_2

$$2KMnO_4 + H_2O \longrightarrow 2MnO_2 + 2KOH + 3O$$

$$\underset{(Mn = +7)}{MnO_4^-} + 2H_2O + 3e \longrightarrow \underset{\substack{\text{(insoluble)} \\ (Mn = +6)}}{MnO_2\downarrow} + 4OH^-$$

In **Basic Medium** it is reduced to manganate ion (MnO_4^{2-})

$$2KMnO_4 + 2KOH \longrightarrow 2K_2MnO_4 + H_2O + O$$

$$\underset{(Mn = +7)}{2MnO_4} + 2OH \longrightarrow \underset{(Mn = +4)}{2MnO_4^{2-}} + H_2O + \frac{1}{2}O_2$$

or $$MnO_4^- + e \longrightarrow Mn\,O_4^{2-}$$

Finally it forms MnO_2.

(*viii*) Nitric Acid: It acts as a strong oxidising agent both in dilute and concentrated form.

(*a*) Hot and conc. HNO_3

$$2HNO_3 \longrightarrow 2NO_2 + H_2O + O$$

$$\underset{(N = +5)}{NO_3^-} + 2H^+ + e \longrightarrow \underset{(N = +4)}{NO_2} + H_2O$$

(*b*) Dilute (50%) HNO_3

$$2HNO_3 \longrightarrow 2NO + H_2O + 3O$$

$$\underset{(N = +5)}{NO_3^-} + 4H^+ + 3e \longrightarrow \underset{(N = +2)}{NO} + 2H_2O$$

(*c*) Dilute (20%) HNO_3

$$2HNO_3 \longrightarrow N_2O + H_2O + 4O$$

$$\underset{(N = +5)}{2NO_3} + 10H^+ + 8e \longrightarrow N_2O + 5H_2O$$

(*d*) Very Dilute (6%) HNO_3

$$HNO_3 + 8H \longrightarrow NH_3 + 3H_2O$$

$$\underset{(N = +5)}{NO_3^-} + 10H^+ + 8e \longrightarrow \underset{(N = -3)}{NH_4^+} + 3H_2O$$

(*e*) V. Dilute

$$\underset{(H = +1)}{2HNO_3} + 2e \longrightarrow \underset{(H = 0)}{H_2} + 2NO_3^-$$

(*ix*) Sodium hypochlorite (NaOCl)

$$NaOCl\,(aq) \longrightarrow NaCl(aq) + O$$

$$\underset{(Cl = +1)}{OCl^-} + 2H^+ + 2e \longrightarrow \underset{(Cl = -1)}{Cl^-} + H_2O$$

(*x*) **Mercuric salts (Hg^{2+})** are reduced to mercurous salts (Hg_2^{2+}) and finally Hg^0.

$$2Hg^{2+} + 2e \longrightarrow Hg_2^{2+}$$
$$Hg_2^{2+} + 2e \longrightarrow 2Hg^0$$

Reducing Agents and their Half-Reactions

Oxidation half-reactions form the products. These oxidation products are produced when the reducing agent reduces the oxidising agent and itself gets oxidised. In a way in oxidation half-reactions elections are released by the reducing agent or get added to R.H.S. of the oxidation reaction.

e.g. $Br_2 + H_2O + HNO_2 \longrightarrow HNO_3 + 2HBr$

In the equation Br_2 acts as an oxidising agent and HNO_2 oxidised to HNO_3 therefore HNO_2 behaves as a reducing agent.

Complete reaction represented by the molecular equation is:

$$\underset{(Br = 0)}{Br_2} + H_2O + \underset{(N = +3)}{HNO_2} \longrightarrow \underset{(N = +5)}{HNO_3} + \underset{(Br = -1)}{2HBr}$$

According to ionic equation

$$Br_2 + H_2O + NO_2^- \longrightarrow NO_3^- + 2H^+ + 2Br^-$$

This redox reaction is split into two half-reactions.

Oxidation half-reaction

$$\underset{(N = +3)}{NO_2^-} + H_2O \longrightarrow \underset{(N = +5)}{NO_3^-} + 2H^+ + 2e$$

Reduction half-reaction

$$\underset{(Br = 0)}{Br_2} + 2e \longrightarrow \underset{(Br = -1)}{2Br^-}$$

Combining the two half-reactions the overall reaction becomes

$$Br_2 + N_2O + NO_2^- \longrightarrow NO_3^- + 2H^+ + 2Br^-$$

Some Reducing Agents: Electrons released by the reducing agent added on the R.H.S. of the equation.

1. Oxalic acid ($H_2C_2O_4$)

Its reducing half-reaction

$$H_2C_2O_4 + \frac{1}{2}O_2 \longrightarrow 2CO_2 + H_2O$$

or

$$\underset{(C = +1)}{C_2O_4^{2-}} \longrightarrow \underset{(C = +4)}{2CO_2} + 2e$$

2. Acidified $FeSO_4$

$$FeSO_4 + H_2SO_4 + \frac{1}{2}O_2 \longrightarrow Fe_2(SO_4)_3 + H_2O$$

or $$Fe^{2+} \longrightarrow Fe^{3+} + e$$

3. Sodium thiosulphate ($Na_2S_2O_3$)

(*i*) $$Na_2S_2O_3 + \frac{1}{2}O_2 \longrightarrow Na_2SO_4 + S$$

or $$\underset{(S=+2)}{S_2O_3^{2-}} \longrightarrow \underset{(S=+6)}{SO_4^{2-}} + S + 2H^+ + 2e$$

(*ii*) $S_2O_3^{2-}$ is oxidised to dithionate $S_4^{2-}O_6$ ion

$$\underset{(S=+2)}{2S_2O_3^{2-}} \longrightarrow \underset{(S=+2.5)}{S_4O_6^{2-}} + 2e$$

4. Stannous Chloride ($SnCl_2$)

It gets oxidised to $SnCl_4$

$$Sn^{2+} \longrightarrow Sn^{4+} + 2e$$

5. Hydrogen sulphide (H_2S)

$$H_2S + \frac{1}{2}O_2 \longrightarrow H_2O + S$$

Therefore it gets oxidised to S

or $$S^{2-} \longrightarrow S^0 + 2e$$

6. Aqueous (SO_2) or (H_2SO_3)

$$\underset{(S=+4)}{SO_2} + H_2O + \frac{1}{2}O_2 \longrightarrow \underset{(S=+6)}{H_2SO_4}$$

or $$\underset{(S=+4)}{SO_3^{2-}} + H_2O \longrightarrow \underset{(S=+6)}{SO_4^{2-}} + 2H^+ + 2e$$

7. Potassium iodide (KF)

(*a*) Acidified KI is oxidised to I_2

$$2KI + H_2SO_4 + \frac{1}{2}O_2 \longrightarrow K_2SO_4 + H_2O + I_2$$

(*b*) **Alkaline KI**

$$KI + \frac{3}{2}O_2 \longrightarrow KIO_3$$

$$\underset{(I=-1)}{I^-} + 6OH^- \longrightarrow \underset{(I=+5)}{IO_3^-} + 3H_2O + 6e$$

8. Hydrogen peroxide (H_2O_2)

(*a*) Acidified H_2O_2

(*i*) $$\underset{(O_2=1)}{H_2O_2} + \frac{1}{2}O_2 \longrightarrow H_2O + \underset{(O=0)}{O_2}$$

(*ii*) $$H_2O_2 \longrightarrow O_2 + 2H^+ + 2e$$

(*b*) Basic medium

$$\underset{(O=-1)}{H_2O_2} + 2OH^- \longrightarrow \underset{(O=0)}{O_2} + 2H_2O + 2e$$

Redox Reactions

Oxidation and reduction reactions take place side by side. In redox reaction, in the oxidation reaction proceeds by loss of electrons and reduction reaction takes place by gain of electrons. In a complete reaction one of species has to supply electrons and the other specy has to gain or accept these electrons *i.e.* the reaction is possible only when there is an oxidising agent and a reducing agent.

There are two types of redox reactions (*a*) Direct redox reactions and (*b*) Indirect redox reactions.

Direct redox reactions: In such redox reactions oxidation and reduction take place in the same vessel simultaneously. The decrease in energy in these reactions appears as heat energy *i.e.* the chemical energy is converted into heat energy.

Indirect Redox Reactions

A redox reaction in which oxidation and reduction take place in separate vessels is called indirect redox reaction. In this the transfer of electrons from reducing agent to the oxidising agent take place over conducting wires. The decrease in energy in all these reactions appears as electrical energy *i.e.* the chemical energy is converted into electrical energy. Indirect redox reactions happen in electro chemical cells, calvanic cells or Voltaic cells.

Importance in Life

In our daily life we observe these reactions in the case of (*i*) Rusting of iron articles (*ii*) fading colour of clothes due to air and sunlight (oxidation), (*iii*) burning of coal, cooking gas and wood (combustion).

Large number of industrial processes like (*i*) extraction of metals, (*ii*) electroplating, (*iii*) bleaching of wood pulp and manufacture of NaOH.

Redox reactions also take place in electrochemical or electrolytic cells.

EXERCISES

1. Differentiate between oxidation state and oxidation number.
2. Calculate the oxidation number of the underlined.

   (*i*) $\underline{U}O_2(NO_3)_2$ (*ii*) $\underline{N}H_2OH$ (*iii*) $H_2\underline{P}O_3$

   (*iv*) $\underline{N_2}H_4$ (*v*) $\underline{N}H_4\underline{V}O_3$
3. Oxidation and reduction are the process of electron transfer. Comment.
4. Find the equivalent weight of $KMnO_4$ in acid, neutral and alkaline media.

5. What are redox reactions? Explain the various step involved to balance the following equation by ion-electron method in acid medium:

   $Cr_2O_7^{2-} + NO_2^- \longrightarrow Cr_3^+ + NO_3^-$

   Find out the equivalent weight of $K_2Cr_2O_7$ in the reaction.

6. Complete and balance the following equations:

   (*i*) $Fe_2^+ + NO_3^- + H^+ \longrightarrow Fe_3^+ + ... + H_2O$

   (*ii*) $NH_3 + Cl_2$ (excess) $\longrightarrow ..... + HCl$

   (*iii*) $KMnO_4 + H_2SO_4 + H_2O_2 \longrightarrow ..... + K_2SO_4 + H_2O + ...$

   (*iv*) $Ag + CN^- + O_2 + ... \longrightarrow OH^- + ...$

7. Find the oxidation number of the element mentioned in the bracket against each molecule or ion:

   (*i*) H_3PO_4 (P) (*ii*) CrO_2Cl_2 (Cr) (*iii*) H_2PtCl_6 (Pt)

   (*iv*) NH_4VO_3 (N) (*v*) NH_3 (N)

8. Write the oxidation number of Cr in:

   (*i*) $[Cr(H_2O)_6]Cl_3$ (*ii*) $[Cr(H_2O)_5 Cl]Cl_2$

   (*iii*) $[Cr(H_2O)_4Cl_2]Cl$ (*iv*) $[Cr(H_2O)_3 Cl_3]$

9. Complete and balance the following equations:

   (*a*) $Cd^{2+} + CN^- \longrightarrow$ ....

   (*b*) $MnO_4^- + H_2C_2O_4 \longrightarrow$ ....

   (*c*) $Cr_2O_7^{2-} + H^+ + Fe^{2+} \longrightarrow$ ....

   (*d*) $S_2O_3^{2-} + I^- + SO_4^{2-} \longrightarrow$ ....

   (*e*) $Mn^{2+} + PbO_2 + H^+ + SO_4^{2-} \longrightarrow$ ....

   (*f*) $IO_3^- + I^- \longrightarrow I^2 +$ ....

   (*g*) $Cu^{2+} + I^- \longrightarrow I_2 +$ ....

   (*h*) $Cr_2O_7^{2+} + H^+ + Fe^{2+} \longrightarrow Cr^{3+} + Fe^{3+} + H_2O$

   (*i*) $SO_2 +$ .... $\longrightarrow SOCl_2 +$ ....

   (*j*) $H_2O + Cl_2 \longrightarrow HCl +$ ....

   (*k*) $Na_2SO_3 + I_2 \longrightarrow NaI +$ ....

   (*l*) $Na_2SiO_3 + HF \longrightarrow$ .... $+ H_2O$

   (*m*) $MnO_4^- + H^+ + Fe^{2+} \longrightarrow Mn^{2+} + H_2O +$ ....

   (*n*) $BiO_3^- + H^+ \longrightarrow Bi^{3+} + 3H_2O$

   (*o*) $Cr(OH)_3 + IO_3^- \longrightarrow I^- + CrO_4^{2-}$ (alkaline medium)

Unit II
Hydrogen-Unique Elements

8

Hydrogen and Hydrides

INTRODUCTION

Hydrogen is the uniquest element in the periodic table. It was discovered by Henry Cavendish in 1766 and named by Antoine Lavosier in 1783. It is the lightest element discovered. The Greek meaning of hydrogen stands; Hydro-means water and ginner-means producer. Hydrogen has the simplest atomic structure with one proton and one electron. The electronic structure is given by $1s^1$. It is very active and forms more compounds than any other element. Many aspects of hydrogen such as protonic acids and aqueous hydrogen ion are discussed in other chapters in the book.

Hydrogen atom may acquire stability in three possible ways:

1. It can share its electron with another atom by forming an electron pair *e.g.*, H_2, H_2O, HCl(g) or CH_4.
2. Hydrogen can lose its electron to form H^+. It is the element or atom where the loss of electron reduces the element to a base proton, therefore, it is a unique element. Proton is extremely small ($r = 1.5 \times -10^{-5}$ Å) compared to hydrogen atom 0.7414 Å and 1 to 2 Å for most other atoms. It has a high polarising power and distorts the electron cloud of other atoms. Free protons do not exist under normal conditions but can be detected by mass spectrophotometer at low pressure.

   Protons are attached to other atoms or molecules *e.g.*, in H_2O it exists as H_3^+O; in aqueous solutions of HCl and H_2SO_4 also it exist as $H(H_2O)^+_n$.
3. It stands to gain electron to form H^- ions and formed by highly electro-positive elements of alkali and alkaline earth metals.

Hydrogen atom has an electro-negativity of 2.1 therefore, it can act both electro-positive and electronegative. However, H^- ions are uncommon. Although it can form bonds by any of the above methods but covalent bonds are most common.

Unique Position of Hydrogen in the Periodic Table

Hydrogen and Helium are the only two elements which exist in period I. In actual, the properties of hydrogen cannot be related to any other element in the periodic table. Hydrogen is best discussed on its own. In a way hydrogen resembles in parts with (*i*) alkali metals (*ii*) halogens and the group 14 elements such as carbon, silicon and germanium.

Alkali Metals: Hydrogen and alkali metals resemble in their electronic structures. Both have one electron in their outermost shell. Alkali metals have a tendency to lose an electron and become M^+. Though hydrogen forms H^+ but it has a greater tendency to form covalent bond.

In a way structure of hydrogen resembles that of halogens as both of them are short of one electron from a noble gas configuration. Halogens have a tendency to gain an electron to form a negative ion X^-. Hydrogen has no tendency to form a negative ion but a few highly electropositive elements like alkali metals and calcium form ionic hydrides *e.g.*, Na^+H^-, CaH_2 in which H^- does form.

There is some resemblance of hydrogen with carbon family of elements as both have half filled shell of electrons. There is similarity with organo metallic compounds *e.g.*, group CH_3^- and H^- can have one more valency. Thus, hydride is a part of organo metallic compounds *e.g.*, LiH, LiMe, NH_3, NMe_3, CH_3SiH_3, $(CH_3)_2SiCl_2$, $(CH_3)_3SiCl$ and $(CH_3)_4Si$.

The best representation of the position of hydrogen in the periodic table is:

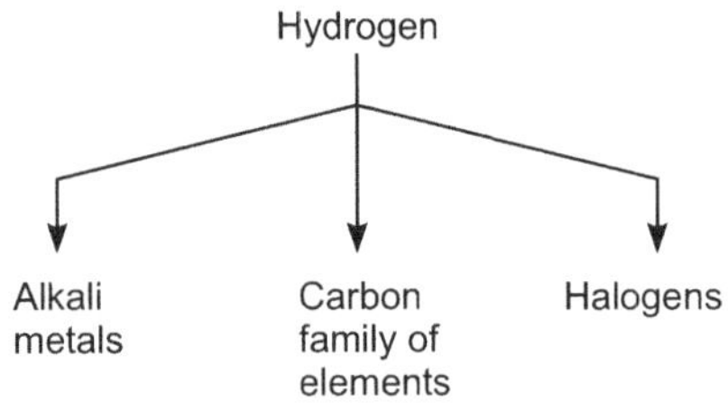

Inspite of showing similarities to different group of elements it is best represented as a group in itself.

Occurrence: It is one of the most abundant element in the universe. It is estimated that the universe comprises of 92% hydrogen, 7% helium and 1% are all other elements. The element is very light therefore, its presence on earth's atmosphere is very small. It is found in volcanic gases. It occurs in vast quantities in the form of ocean water. Large number of compounds contain hydrogen in living matter *e.g.*, carbohydrates, organic compounds such as fossil feed, coal, petroleum natural gas, ammonia and acids. Its compounds are more abundant than any other element. Being a very light element, it escapes earth's atmosphere and is most likely to be present round the sun.

Manufacture of Hydrogen

Various methods to manufacture hydrogen on a large scale are:

1. Passing steam on red hot coke:

(i) $C + H_2O \xrightarrow{1000^o} \underset{\text{water gas}}{CO + H_2}$

(ii) $CO + H_2 + O_2 \longrightarrow CO_2 + H_2O + \text{heat}$

The product water gas ($CO + H_2$) is used as an important industrial fuel since it burns evolving lot of heat. CO and H_2 are difficult to be separated.

Hence, it is difficult to obtain pure H_2.

CO is liquified at low pressure and thus H_2 gets removed. Alternatively water gas may be passed over iron oxide to convert to CO_2 and H_2.

$$CO + H_2 \xrightarrow[450^oC \ Fe_2O_3]{H_2O \text{ steam}} 2H_2 + CO_2$$

CO_2 formed as a product is separated by dissolving in water or reacting with K_2CO_3 giving $KHCO_3$ setting H_2 gas free.

2. Steam is mixed with light hydrocarbons over Ni catalyst at 800 to 900°C.

$$CH_4 + H_2O \xrightarrow[800-900^oC]{Ni} CO + 3H_2$$

$$CH_4 + 2H_2O \xrightarrow[800-900^oC]{Ni} CO_2 + 4H_2$$

The gas emerging from the furnace contains CO, CO_2, H_2 and excess steam. The gas mixture in further mixed with steam cooled to 400°C over Fe/Cu catalyst and CO gets converted to CO_2.

$$CO + H_2O \longrightarrow CO_2 + H_2$$

CO_2 formed is absorbed by K_2CO_3 or ethanolamine which is then regenerated by heating.

$$K_2CO_3 + CO_2 + H_2O \longrightarrow 2KHCO_3$$

$$2HO{-}CH_2{-}CH_2{-}NH_2 + CO_2 + H_2O \longrightarrow (HO{-}CH_2{-}CH_2NH_2)CO_3$$

3. Cracking of high molecular weight hydrocarbons. Hydrocarbon in oil refineries are cracked to get low molecular weight hydrocarbons alongwith H_2 obtained as a by product.

$$C_6H_{14} \longrightarrow \underset{\text{cyclohexane}}{C_6H_{12}} + H_2$$

4. Electrolysis of H_2O or solutions of NaOH or KOH. This method is costlier to electrolyse aqueous solutions of NaOH or KOH Nickel anode and Fe cathodes are used. The gases at anode and cathode are not allowed to combine.

Anode $2OH^- \longrightarrow H_2O + ½O_2 + 2C$

Cathode $2H_2O + 2e \longrightarrow 2OH^- + H_2$

Overall reaction $H_2O \longrightarrow H_2 + ½O_2$

5. Pure H_2 can also be obtained by electrolyzing aqueous NaCl *e.g.*,

$$NaCl \rightleftharpoons Na^+ + Cl^-$$

$$H_2O \rightleftharpoons H^+ + OH^-$$

Anode: $2OH^- \longrightarrow H_2O + 2e + \frac{1}{2}O_2$

$$2Cl^- \longrightarrow Cl_2$$

Cathode: $Na^+ + e \longrightarrow Na^o$

$$Na + 2H_2O \longrightarrow NaOH + H_2$$

Laboratory Preparation: More active metals like Zn, Al etc. react with dilute acids or alkali to produce H_2.

$$Zn + H_2SO_4 \longrightarrow 2nSO_4 + H_2\uparrow$$

$$2Al + 2NaOH + 6H_2O \longrightarrow 2Na[Al(OH)_4] + 3H_2\uparrow$$

7. Ionic hydrides react with H_2O to form H_2.

$$LiH + H_2O \longrightarrow LiOH + H_2\uparrow$$

Properties of Hydrogen: Hydrogen is the lightest gas known, has a low density and is used to fill balloons. It forms diatonic molecule H_2. It is odourless, colourless and insoluble in H_2O. It is very reactive under standard conditions. Reactivity is attributed to kinetics while lack of reactivity is because of strong H—H bond which requires 435:9 kJ/mol therefore its reactions are slow. Due to high activation energy most of the reactions take place at high temperature or in the presence of a heterogeneous catalyst. catalyst helps to lower the activation energy or reacts with H_2 to weaken the bond H—H some example of such reactions are:

(*i*) **Habers process for the manufacture of NH_3**

$$N_2 + 3H_2 \underset{\substack{380-450^\circ C\\ 200\text{ at}}}{\overset{Fe/Mo}{\rightleftharpoons}} 2NH_3$$

(*ii*) **Hydrogenation:** Unsaturated oils are hydrogenated using finely divided Ni, Pt or Pd as catalysts.

$$\underset{\text{unsaturated oil}}{>C{=}C<} + H_2 \xrightarrow[450-500^\circ C]{Ni} \underset{\text{saturated compound}}{-\overset{|}{\underset{|}{C}}-\overset{|}{\underset{|}{C}}-}$$

(*iii*) **Production of methanol**

$$CO + 2H_2 \xrightarrow[300^\circ C]{Cu/Zn} CH_3OH$$

Hydrogen burns in air or oxygen giving out a large amount of heat. This concept is used in producing oxy-hydrogen flame for welding which gives the temperature of 3000°C.

(*iv*) Hydrogen reacts with halogens in the presence of sunlight (Photolysis)

$$H_2 + F_2 \xrightarrow{hv} 2HF \text{ violent reaction}$$

$$H_2 + Cl_2 \rightarrow 2HCl \quad \text{slow in the dark but faster in the daylight.}$$

(*v*) **Metals:** Metals react with H_2 to form hydrides. The reaction in non-explosive but all the same high temperature is required.

Besides, H_2 is used for hydrogenation and to reduce nitrobenzene to aniline. This is further used in the dye industry. Hydrogen can be dissociated at high temperature produced in an electric arc or ultra violet light. The atomic hydrogen exists for less than half a second it further recombines to give large amount of heat.

Uses: H_2 can be made to replace as a major source of energy. In future times it can be made to replace oil and coal. Liquid hydrogen is used as a fuel in space rockets and space shuttle. Presently, car engines are being modified to run on hydrogen fuel. The only draw back that it involves the risk of explosion.

Electronic structure: Hydrogen has $1s^1$ configuration quite in resemblance with alkali metals.

$$Li — 1s^2, 2s^1$$

$$Na — 1s^2, 2s^2, 2p^6, 3s^1$$

$$K — 1s^2, 2s^2 2p^6, 3s^2 3p^6, 4s^1$$

All these elements have one electron in the outermost *s*-orbital.

Affinity for non-metals

Hydrogen combines with non-metals such as Cl, S and O to form: HCl, H_2S and H_2O respectively.

In all these compounds it prefers to form bond by sharing of electrons to form hydrides, oxides and sulphides *e.g.*,

$$H_2 \longrightarrow Cl_2 \longrightarrow 2HCl$$

$$H_2 + \tfrac{1}{2}O_2 \longrightarrow H_2O$$

$$H_2 + S \longrightarrow H_2S$$

Non-metals hydrides are covalent in nature.

Aflinity for metals: Hydrogen is moderately electronegativity with an electronegative value of 2.1 therefore it combines with alkali, alkaline earth and transition metals like La, Ti, Pd, Zr, V, Yb, and Nb etc. Metal hydrides are ionic or interstitial hydrides. Transition metal hydrides are non-stoichiometric *e.g.*,

$$Na + \tfrac{1}{2}H_2 \longrightarrow Na^+H^- \text{ (s)}$$

$$Ca + H_2 \longrightarrow CaH_2 \text{ (s)}$$

The bond is electrovalent therefore they are also called electrovalent hydrides.

Reducing character of Hydrogen: Like alkali metals, hydrogen is good reducing agent.

$$CuO + H_2 \longrightarrow Cu + H_2O$$

$$B_2O_3 + 6K \longrightarrow 2B + 3K_2O$$

Molecular state: Like most of the gases H_2 is diatomic under normal temperature and pressure conditions. Similarly, alkali metals like Li exists as Li_2 at its boiling point.

Isotopes of Hydrogen: Isotopes are atoms of the same element having different atomic masses. The difference in atomic masses arises out of different number of neutrons present in the nucleus. Hydrogen occurring in nature shows three isotopes 'H protium, ^2H Deuterium or D and ^3H Tritium or T on the basis of 0, 1 and 2 neutrons respectively in the nucleus. Protium is most abundant out of the lightest of the three (99.986%). It is the lightest of the three isotopes. These isotopes have the same electronic configuration therefore they show the same chemical properties but owing to large difference in their masses, there occurs difference in their rates of reaction and equilibrium constants.

Some differences in properties are attributed to their different masses.

(*i*) Molecular hydrogen, H_2 is rapidly absorbed on the surface of D_2.

(*ii*) Molecular hydrogen reacts 13 times faster with Cl_2 than with D_2 because H_2 has a low energy of activation. Differences in properties that are connected to their masses are called isotope effect. Some of the physical constants of hydrogen, deuterium and tritium are listed in table given below:

Table 8.1 Physical Constants for Hydrogen, Deuterium and Tritium

Physical countant	H_2	D_2	T_2
Mass of atom (amu)	1.0078	2.0141	3.0160
Freezing point (°C)	–259.0	–254.3	–252.4
Boiling point (°C)	–252.6	–249.3	–248.0
Bond length (Å)	0.7414	0.7414	(0.7414)
Heat of dissociation⁺ (kJ mol^{-1})	435.9	443.4	446.9
Latent heat of fusion (kJ mol^{-1})	0.117	0.197	0.250
Latent heat of vaporization (kJ mol^{-1})	0.904	1.226	1.393
Vapour pressure* (mm Hg)	54	5.8	–

* Measured at – 259.1 °C.

⁺ Measured at 25°C.

Thus the isotopes of hydrogen show much greater differences in physical properties than are found between the isotopes of other elements. Some physical constants for H_2, D_2 and T_2 are given in Table 8.1.

The percentage difference between masses of hydrogen deuterium and tritium is greater than in the case of isotopes of other elements. Thus, the physical constants of H_2, D_2 and T_2 show big difference as shown in the table.

The equilibrium constant for dissociation of H_2O is 1.0×10^{-14} while D_2O is 3.0×10^{-15}.

$$H_2O \rightleftharpoons H^+ + OH^-$$
$$D_2O \rightleftharpoons D^+ + OD^-$$

Protium bond is broken more easily than deuterium. On electrolysis H_2O forms H_2 faster than D_2 from D_2O. The remaining H_2O goes on getting richer in D_2O. In this way almost pure D_2O can be obtained. This is also a process of separating D_2. Heavy water gives all the reactions given by protium water.

Some physical properties of H_2O and D_2O are compared in the Table 8.2 Deuterium is exchanged by hydrogen in the compounds. D_2 and H_2 react at high temperature to form HD. It exchanges with hydrogen in compounds like NH_3 and CH_4 in stages to give NH_2D, NHD_2 and ND_3 and similarly CH_3D, CH_2D_2, CHD_3 and CD_4 respectively. D_2O is better source of getting deuterium compounds, than D_2.

Table 8.2 Physical Constants for Water and Heavy Water

Physical constant	**H_2O**	**D_2O**
Physical point (°C)	0	3.82
Boiling point (°C)	100	101.42
Density at 20°C (g cm^{-3})	0.917	1.017
Temperature of maximum density (°C)	4	11.6
Ionic product K_w at 25°C	1.0×10^{-14}	3.0×10^{-15}
Dielectric constant at 20°C	82	80.5
Solubility g NaCl/100 g water at 25°C	35.9	30.5
Solubility g $BaCl_2$/100 g water at 25°C	35.7	28.9

Some exchange reactions using D_2O are like this:

$$NaOH + D_2O \longrightarrow NaOD + HDO$$
$$NH_4Cl + D_2O \longrightarrow NH_3DCl + HDO$$

Tritium is used in fusion reactions to produce energy. It is widely used as a radioactive tracer. It is easy to handle, emits low energy β-radiations alongwith some β-radiation. It is non-toxic unless swallowed.

Tritium compounds are prepared from T_2 gas. T_2O is obtained by

$$T_2 + CuO \longrightarrow T_2O + Cu$$
$$2T_2\ O_2 \xrightarrow[\text{Cat}]{\text{Pd}} 2T_2O$$

Tritium compound can be prepared by catalytic exchange using T_2 gas dissolved in H_2O or T_2O.

$$NH_4Cl + \underset{\text{or}\\ \text{HTO}}{T_2O} \rightleftharpoons NH_3TCl$$

Ortho and Para Hydrogen: Molecular hydrogen exists in two different forms called Ortho and Para hydrogen. Each atom in molecular hydrogen has spin of the nucleus and the spin of the electron. In hydrogen molecule electron always spins in the same direction in both the atoms in a hydrogen molecule whereas the two nuclie may spin in the same direction or the opposite direction. This may give rise to spin isomerism. Hence, hydrogen (H_2) molecule may exist in two different forms which are called Ortho and Para hydrogen as illustrated in the diagram below.

$$Mg_3N_2 + 3D_2O \longrightarrow 2ND_3 + 2MgO$$

Some direct reactions of D_2O are:

$$SO_3 + D_2O \longrightarrow D_2SO_4$$

$$P_4O_{10} + 6D_2O \longrightarrow 4D_3PO_4$$

(^3_1T) Tritium is radioactive and decays to give β-emission.

$$^3_1T \longrightarrow {}^3_2He + {}^{0}_{-1}e \quad (\beta\text{-emission})$$

It has a half life of 12.26 years. Therefore, T present when the earth was formed has almost decayed and small amount that is present on earth has been formed as a result of cosmic rays reaction in the upper atmosphere.

$$^{14}_7N + {}^1_0n \longrightarrow {}^{12}_6C + {}^3_1T$$

$$^{14}_7N + {}^1_1H \longrightarrow {}^3_1T + \text{other fragments}$$

$$^2_1D + {}^2_1D \longrightarrow {}^3_1T + {}^1_1H$$

The amount of T present is one part in 7×10^{17} parts of H_2, Tritium was first prepared by D_3PO_4 and $(ND_4)_2SO_4$ with deuterons.

$$^2_1D + {}^3_1D \longrightarrow {}^3_1T + {}^1_1H$$

It is now prepared by irradiating Li with slow neutrons.

$$^6_3Li + {}^1_0n \longrightarrow {}^4_2He + {}^3_1T$$

Parallel spins (Ortho form) Opposite spins (Para form)

Such isomerism is also found in other symmetrical molecules *e.g.*, D_2, N_2, F_2 and Cl_2. These isomers show wide differences in their physical properties namely boiling points, specific heats and thermal properties due to difference in their internal energy.

The percentage of para form increases at low temperature so much so that at absolute zero the gas contains 100% para form. At high temperature para gets converted to other form. At high temperature only 75% of ortho form can be isolated. Ortho para conversion is slow in the absence of a catalyst *e.g.*, Fe, Ni, Pt and W.

Hydrides

Hydrogen combined with elements to form hydrides. The type of hydrides depend on the type of elements. There are three main types of hydrides:

(*i*) Ionic hydrides,

(*ii*) Molecular or covalent hydrides,

(*iii*) Metallic hydrides.

Ionic Hydrides: Alkali metals and Alkaline earth metals form ionic or salt-like hydrides at high temperature *e.g.*, NaH, CaH_2. These hydrides are solids and have high melting points. Their characteristic properties are discussed below.

(*i*) LiH melts at 691° and conducts electricity where H_2 is obtained at anode confirming H^- ion.

(*ii*) CaH_2 dissolves in LiCl/ KCl mixture. On electrolysis of the mixture H_2 is evolved at anode.

(*iii*) Crystal structures of these hydrides are well known which points to ionic bonding.

Lithium has a very small size therefore, it is prone to forming covalent compounds although LiH is largely ionic and contains H^- hydride ion. Lithium metal hydride has greater density than other metal hydrides. Here H^- ions occupy the holes created by the metal lattice. The ionic hydrides have greater heat of formations and are stochiometric. These types of hydrides are only formed by metals having electro-negatively lower than that of hydrogen (2.1) thus, forming M^+ and H^- ions alkali metal hydrides are more reactive than the alkaline earth metal hydrides. The reactivity increases down the group.

Alkali metals hydrides decompose into their constituents on heating at 400–500°C except LiH (Covalent). These hydrides react with H_2O to liberate H_2.

$$LiH + H_2O \longrightarrow LiOH + H_2$$

$$CaH_2 + 2H_2O \longrightarrow Ca(OH)_2 + 2H_2$$

These are strong reducing agents at high temperatures. Due to reaction towards H_2O their use gets limited.

$$2CO + NaH \longrightarrow HCOONa + C$$

$$SiCl_4 + 4NaH \longrightarrow SiH_4 + 4NaCl$$

$$PbSO_4 + 2CaH_2 \longrightarrow PbS + 2Ca(OH)$$

NaH is a very good reducing agent. It helps to form other important hydrides which are important reducing agents both in inorganic and organic synthesis.

Lithium Aluminium

$$4LiH + AlCl_3 \longrightarrow \underset{\text{Lithium Aluminium hydride}}{Li[AlH_4]} + 3LiCl$$

$$4NaH + B(OCH_3)_3 \longrightarrow \underset{\text{Sodium borohydride}}{Na[BH_4]} + 3NaOCH_3$$

Covalent Hydrides: The hydrides belonging to *p*-block elements are all covalent hydrides of the two atoms have a small difference in their electro-negativity value.

The molecules of these hydrides are held by weak van der Waals forces. These hydrides are given by a definite formula. They are usually volatile, have low melting and boiling points and do not conduct electricity. These hydrides are given by the formula XH_n or XH_{g-n} where g is the group number of X. They are obtained by the following methods.

1. Direct combination of elements and hydrogen. This can occur in elements belonging to boron, carbon, nitrogen and oxygen family of elements plus the halogens.

For examples

(*a*) $3H_2 + N_2 \xrightarrow[\text{+press catalyst (Fe/Mo)}]{\text{H.T}} 2NH_3$ (Haber's Process)

(*b*) $2H_2 + O_2 \xrightarrow{\text{Explosive}} 2H_2O$

(*c*) $H_2 + Cl_2 \xrightarrow{\text{Bum}} 2HCl$

2. Reaction of halide with $LiAlH_4$

$$4BCl_3 + 3LiAlH_4 \xrightarrow[\text{Ether}]{\text{Dry}} 2B_2H_6 + 3AlCl_3 + 3LiCl$$

$$SiCl_4 + Li[AlH_4] \xrightarrow{\text{D/ether}} SiH_4 + AlCl_3 + LiCl$$

3. Action of borides, carbides, sulphides and phosphides with acids

(*i*) $2Mg_3B_2 + 4H_3PO_4 \longrightarrow B_4H_{10} + 2Mg_3(PO_4)_2 + H_2$

(*ii*) $Al_4C_3 + 12HCl \longrightarrow 3CH_4 + 4AlCl_3$

(*iii*) $FeS + H_2SO_4 \longrightarrow FeSO_4 + H_2S$

(*iv*) $Ca_3P_2 + 3H_2SO_4 \longrightarrow 2PH_3 + 3CaSO_4$

4. **Action of oxo acids with $NaBH_4$ in aqueous solution**

$$4H_3AsO_3 + 3Na[BH_4] \longrightarrow 4AsH_3 + 3H_3BO_3 + 3NaOH$$

5. **Converting one hydride into another by cracking or pyrolysis**

$$B_4H_{10} \xrightarrow{\text{pyrolysis}} B_2H_6 + \text{other products}$$

6. Silent electric discharge can give long chain hydrides from simple hydrides:

$$GeH_4 \longrightarrow Ge_2H_6 \longrightarrow Ge_3H_8 \longrightarrow Ce_9H_2O$$

Hydrides of Boron family are electron deficient and polymeric. The starting hydride of boron is B_2H_6 *i.e.,* diborane. Other polymeric boranes are B_4H_{10}, B_5H_9, B_5H_{11}, B_6H_{10} and $B_{10}H_{14}$. Aluminium hydride is also polymeric $(AlH_3)n$. In polymeric hydrides hydrogen is bonded to two or more atoms and this is known as multi-centre bonding. The tendency to form poly nuclear hydrides is strongest with C, N and O. Several hundreds of atoms of carbon an hydrogen form C-Chain compounds known as carbanes. These are aliphatic and aromatic hydrocarbons, *i.e.,*

$$\left.\begin{array}{l} CH_4, C_2H_6, C_3H_8, C_4H_{10}.... \\ C_2H_4, C_3H_6, C_4H_8.............. \\ C_2H_2, C_3H_4, C_4H_6.............. \end{array}\right] \text{Aliphatic}$$

$$C_6H_6......................\text{Aromatic}$$

These hydrocarbons are saturated and unsaturated. Si and Ge form only saturated hydrocarbons *e.g.,* $Si_{10}H_{22}$. Other Covalent hydrides that are formed are Sn_2H_6, N_2H_4, P_3H_5, As_3H_5, H_2O_2, H_2O, $H_2S_2H_2S_3$, H_2S_4, H_2S_5 and H_2S_6. The melting and boiling points show higher values for NH_3 an HF. This is due to hydrogen bonding.

Metallic or Interstitial Hydrides

Most of the *d*- and *f*-block elements form metallic hydrides on reacting with hydrogen. The middle elements in the *d*-block elements do not form hydrides. This gap in the formation of hydrides is called Hydrogen Gap. The metallic hydrides are prepared by heating metal and hydrogen under pressure. These hydrides decompose when heated to higher temperature. This serves as a very good method for getting pure hydrogen. These metal hydrides have properties similar to the parent metal. The hydrides are hard, have a metallic brilliance, conduct electricity and have magnetic properties. They have an extended crystal lattice to accommodate hydrogen. Due to distortion of the crystal lattice they become brittle. The solid metal hydride can be turned into fine powder. Finely divided powder of hydrides decomposes at high temperature. The finely divided metallic hydrides can be used as a catalyst zirconium hydride is used as a moderator in nuclear reactors.

Many of these metal hydrides are non-stoichiometric *e.g.,* $LaH_{2.87}$, $YbH_{2.55}$, $TiH_{1.8}$, $ZnH_{1.9}$, $VH_{1.6}$, $NbH_{0.7}$ and $PdH_{0.7}$.

The hydrogen occupies the interstitial sites and therefore called interstitial hydrides shown in figure.

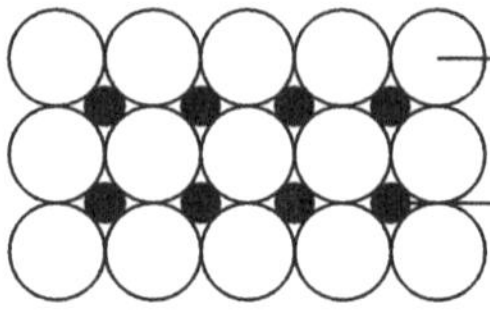

The interstices may be occupied by H and O. The metallic hydrides can be effectively used as catalysts in the process of hydrogenation as they make available hydrogen atom rather than H_2 molecules. Titanium dissolves H_2 and makes it brittle. Titanium is used in the body of supersonic aircraft. The non-stoichiometric compounds are regarded as solid solutions varying amounts of hydrogen can dissolve in metals which can act as catalysts for hydrogenation reactions. The catalysts can provide hydrogen atoms used in the reactions. Hydrogen dissolved in the metals make it brittle.

The bonding in metal hydrides is quite complex.

1. Many metal hydrides have hydrogen present in tetrahedral holes in cubic close packed metal atoms. If all the tetrahedral voids are occupied then the formula should be MH_2 with a fluorite structure. The fact remains that some of the sites are unoccupied therefore they represent non-stoichiometry and hence lose hydrogen. The formula then becomes $MH_{1.5 \text{ to } 2.0}$. The hydrides are formed by Sc and Ti and most of the lanthanides and actinides.
2. Two of the lanthanide elements like Eu and Yb form ionic hydrides with the formula EuH_2 and YbH_2 which resemble CaH_2.
   The Eu^{2+} and Yb^{2+} form stable configuration such as Eu $\rightarrow 4f^7$ (half filled) and $4f^{14}$ (Fully filled).
3. YH_2 and LaH_2 plus many lanthanides and actinides form hydrides with the formula MH_2 which can absorb more hydrogen to form MH_3. $LaH_{2.76}$ and $CeH_{2.69}$ are found. Their structure may cubic or sometimes hexagonal. The third hydrogen is loose and may occupy octahedral hole.
4. Uranium is different from other and forms two crystalline forms with the formula UH_3 that is stoichiometric.
5. V, Nb, Ta, Cr, Ni and Pd hydrides have the formula MH. They are less stable and non-stoichiometric with a wide ranging composition *e.g.*, $NbH_{0.7}$ and $PdH_{0.6}$.
   Pd/H_2 is extraordinary, H_2 occludes 935 times more than its own volume.

It is in interesting method to separate H_2 from D_2 and He. As H_2 is absorbed the metallic character of Pd decreases and finally becomes semi-conductor.

Intermediate Hydrides: These hydrides are in-between ionic and covalent hydrides. $(BeH_2)_n$ is polymeric with chain consisting of hydrogen bridge.

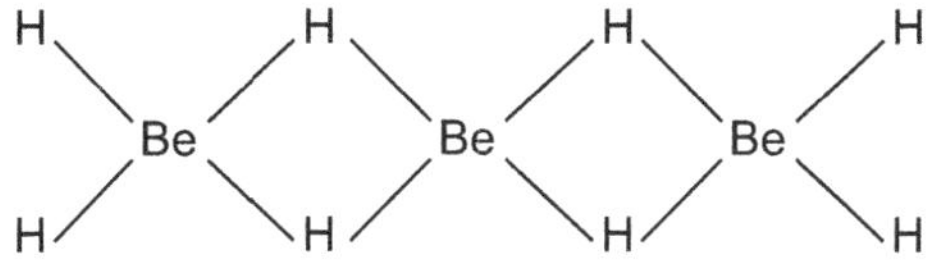

CuH, ZnH_2, CdH_2 and HgH_2 hare properties intermediate to metallic and covalent hydrides. They are electron deficient like $(AlH_3)_n$ CuH is endothermic and is formed by reducing Cu^{2+} with H_3PO_2. Hydrides of Zn, Cd and Hg are made from their chlorides by reducing with $LiAlH_4$.

Hydrogen Ion: The energy required to remove an electron from the shell in a neutral gaseous hydrogen atom *i.e.*, the ionization energy is 1311 kJ mol^{-1}. The ionization energy of hydrogen atom is quite high. The hydrogen in the gas phase is purely covalent. In hydrogen fluoride the ionic hydrogen (H^+) exists because due to large electro-negativity difference the bond is 45% ionic. H^+ cannot exist independently in gas phase since it is two small in size, it gets dissolved in water and gives out a large amount of hydration energy $\sim 10^{91}$ kJ mol^{-1} which compensates the high ionization energy to some extent. The remaining difference in energy is provided by electron affinity (energy is released) and the solvation energy of the negative ion.

$$H^+ + H_2O \longrightarrow H_3O^+$$

Compounds which give solvated hydrogen ions are called acids.

The structure of H_3O^+ is given by:

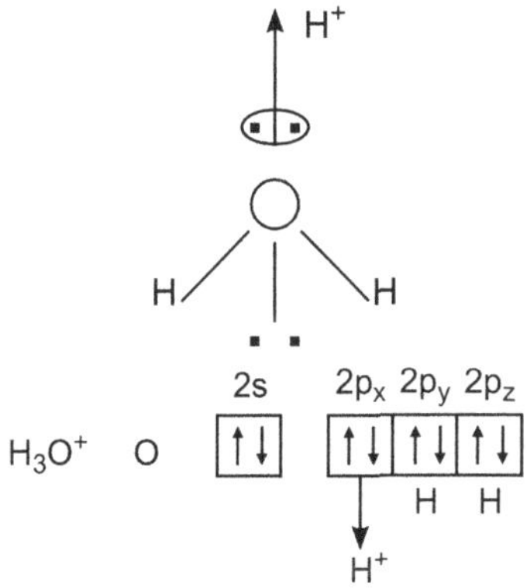

Therefore, hybridization is sp^3 and the structure is distorted tetrahedral with a bond angle ∠HOH of 104.5° due to the presence of two lone pairs present on 2*s* and 2*pz*.

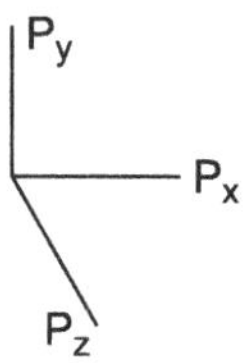

Although H^+ does not exist independently thus H_3O^+ is generally written as H^+ for the sake of convenience.

The detailed discussion of the other important terms in the chapter namely (*i*) Hydrogen bonding (*ii*) Acids and Bases has been taken up in the chapter on chemical bonding in this book.

Important Compounds of Hydrogen

Hydrogen Peroxide (H_2O_2)

Hydrogen peroxide was discovered by the French chemist Thenard. It was first prepared from BaO_2.

It is present in trace amount in water and decay of some plants.

Preparation

(*i*) It is obtained by the treatment of metal peroxide and acids.

$$Na_2O_2 + 2HCl \longrightarrow 2NaCl + H_2O_2$$
$$BaO_2 + H_2SO_4 \longrightarrow BaSO_4 + H_2O_2$$

(*ii*) Electrolysis of conc. H_2SO_4

$$H_2SO_4 \rightleftharpoons H^+ + HSOS_4^-$$

At anode $2HSO_4^- \longrightarrow H_2S_2O_8 + 2e$

At cathode $2H^+ + 2e \longrightarrow H_2$

$H_2S_2O_8$ on distillation provides H_2O_2

Distillation of $H_2S_2O_8$

$$H_2S_2O_8 \xrightarrow{\text{Dist.}} H_2SO_5 + H_2SO_4$$
$$H_2SO_5 \xrightarrow{H_2O} H_2SO_4 + H_2O_2$$

Or $$H_2S_2O_8 + 2H_2O \xrightarrow[\Delta]{\text{heating}} 2H_2SO_4 + H_2O_2$$

Physical Properties: It exists as a blue liquid soluble in H_2O giving acid. It forms blisters when in contact with skin. It boils at 85°C. Its density is 1.64 at 0°C. It decomposes at its boiling point.

Chemical properties

(*i*) It is weekly acidic

$$H_2O_2 \rightleftharpoons H^+ + HO_2^-$$
$$HO_2^- \rightleftharpoons H^+ + O_2^-$$

(*ii*) It reacts with alkali to form salt

$$H_2O_2 + NaOH \longrightarrow Na(OH)_2 + H_2O$$
$$H_2O + Na(OH)_2 \longrightarrow Na_2O_2 + 2H_2O$$

(*iii*) At room temperature it decomposes due to sunlight.

$$H_2O_2 \longrightarrow H_2O + \tfrac{1}{2}O_2$$

(*iv*) It acts as a reducing agent

$$H_2 \,|\, O_2 \longrightarrow O_2 + 2H^+ + 2e$$
$$(-1) \quad O$$

It decolorise acidified $KMnO_4$ solution

$$2kMnO_4 + 3H_2SO_4 + 5H_2O_2 \longrightarrow K_2SO_4 + 2MnSO_4 + 5O_2 + 8H_2O$$

(*v*) It also acts as on oxidizing agent

$$H_2O_2 + 2H^+ + 2e \longrightarrow 2H_2O$$
$$O = (-1)$$

In acidic KI solution

$$H_2O_2 + 2KI + H_2SO_4 \longrightarrow K_2SO_4 + 2H_2O + I_2$$

According to ionic equation

$$H_2O_2 + 2H^+ + 2e \longrightarrow 2H_2O$$
$$2I^- \longrightarrow I_2 + 2e$$

$$H_2O_2 + 2I^- + 2H^+ \longrightarrow I_2 + 2H_2O$$

(*vi*) It disproportionate to show auto oxidation and reduction:

$$H_2O_2 \longrightarrow H_2O + O$$
$$H_2O_2 \longrightarrow O_2 + 2H$$
$$2H + O \longrightarrow H_2O$$

$$2H_2O_2 \longrightarrow 2H_2O + O_2$$

(*a*) Oxidising property

$$H_2O_2 \longrightarrow H_2O + O] \times 4$$
$$PbS + 4O \longrightarrow PbSO_4$$

$$PbS + 4H_2O_2 \longrightarrow PbSO_4 + 4H_2O$$

(*b*) Oxidises Arsenite to arsenate

$$H_2O_2 \longrightarrow H_2O + O$$
$$Na_3AsO_3 + O \longrightarrow Na_3AsSO_4$$

$$Na_3AsO_3 + H_2O_2 \longrightarrow Na_3AsO_4 + H_2O$$

(*c*) Oxidises Ferrocyanide to ferricyanide

$$2K_4[Fe(CN)_6] + H_2O_2 \longrightarrow 2K_3[Fe(CN)_6] + 2KOH$$

(*d*) Oxidises Hg

$$Hg + H_2O_2 \longrightarrow HgO = H_2O$$

Reducing Property

(*a*) alkaline Ag_2O is reduced to Ag

$$Ag_2O + H_2O_2 \rightarrow 2Ag + H_2O + O_2$$

(*b*) O_3 is reduced to O_2

$$H_2O_2 + O_3 \longrightarrow H_2O + 2O_2$$

(*c*) Fe^{3+} is reduced to Fe^{2+}

$$2K_3[Fe(CN)_6] + 2KOH + H_2O_2 \longrightarrow 2K_4[Fe(CN)_6]$$

(*d*) Hypochlorite reduces to chloride

$$NaOCl + H_2O_2 \longrightarrow NaCl + H_2O + O_2$$

$$NaOI + H_2O_2 \longrightarrow NaI + H_2O + O_2$$

Bleaching actions of H_2O_2 is due to its oxidizing actions. It bleaches silk, hair and wool.

Uses

(*i*) It is used both an oxidizing as well as a reducing agent.

(*ii*) It is used as an analytical reagent.

Tests of H_2O_2

(*i*) It liberates brown iodine from KI solution

(*ii*) It forms blue solution with K_2Cr_2O, solution

(*iii*) It given orange colour with titanium salts and forms per titanic acid.

Strength of H_2O_2: H_2O_2 container is labelled 10 V, 20 V. The strength of H_2O_2 is measured is terms of O_2 obtained from H_2O_2 at STP. 20 V H_2O_2 indicates that at STP H_2O_2 gives 20 ml of O_2. Thus, 10 V H_2O_2 gives 10 ml of O_2 at STP. H_2O_2 is a dibasic acid and normality of 20 V solution is 3.6 N.

Structure: X-ray studies of the crystalline H_2O_2 shows non-planer and non-linear with a dipole moment of 2.1. The O—H bond is asymmetric with a bond angle of 97°. The O—O bond length is 1.48 Å.

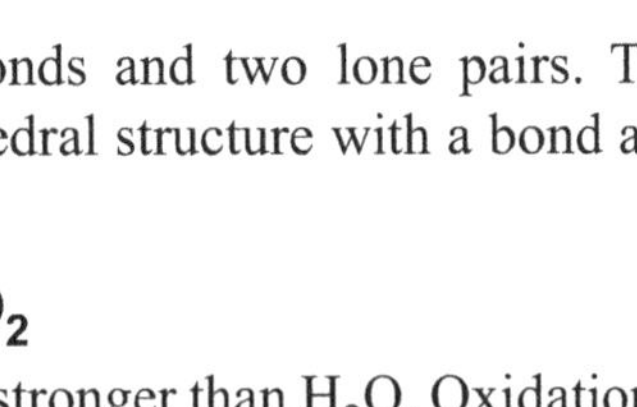

Each O has two bonds and two lone pairs. Therefore, each O is sp^3 hybridized with a tetrahedral structure with a bond angle 97° due *lp lp* and *lp* bp repulsions.

Acid Strength of H_2O_2

H_2O_2 is a weak acid but stronger than H_2O. Oxidation state of oxygen in H_2O_2 is (–1) and in H_2O it is (–2). Dissociation of H_2O_2 is greater than H_2O *e.g.*,

$$H_2O_2 \rightleftharpoons H^+ + HO_2^- \text{ and } HO_2^- \rightleftharpoons H^+ + O_2^{2-}$$

$$H_2O \rightleftharpoons H^+ + OH^-$$

This indicates the H^+ concentration in H_2O_2 is greater than in case of H_2O.

Utility

(*i*) H_2O_2 is used as fuel in jet and rocket engine.
(*ii*) It is used as a bleaching agent in silk, paper and hair.
(*iii*) It serves as an antiseptic in surgical treatments.
(*iv*) It is used as an oxidizing reagent.
(*v*) It restores the colour of Pb paintings.

Heavy Water (D_2O)

Introduction: The oxide of heavy hydrogen (Deuterium) is known as heavy water. Ordinary water contains trace amount of D_2O. Its presence can be observed in minuscule amount in rain water and banyan trees. It is also present in snow melting by sun in the mountains.

Preparation of D_2O

It can be obtained from natural water by subsequent electrolysis in which O_2 and H_2 are liberated. Due to high density D_2 remains in water making the water richer in D_2O. Amount of D_2O in water increases after every stage and after 8 successive stages residual water becomes rich in D_2O. Residual water is then electrolysed further to get pure O_2 which burns to give 100% D_2O.

The isotopic exchange can be observed in the chemical reactions.

$$H_2O + D_2O \rightleftharpoons 2HDO$$

$$H_2 + D_2 \rightleftharpoons 2HD$$

$$H_2O + DH \rightleftharpoons HDO + H_2$$

According to law of mass actions in equation (*iii*)

$$K = \frac{[HDO]}{[H_2O]} = \frac{[H_2]}{H_2O} = \frac{[H_2]}{[H_2O]} = \frac{[HDO]}{DH}$$

$$K = 3.7 \text{ At } 2.5°C$$

According to the law [HDO] [HD] is 3.7 in the abundance of H_2 and H_2O. Hence, abundance of D in water between 3 to 4 times greater at equilibrium. This is ably represented by the flow diagram which indicate enrichment of D_2 in water.

Flow diagram of Electrolysis of H_2O

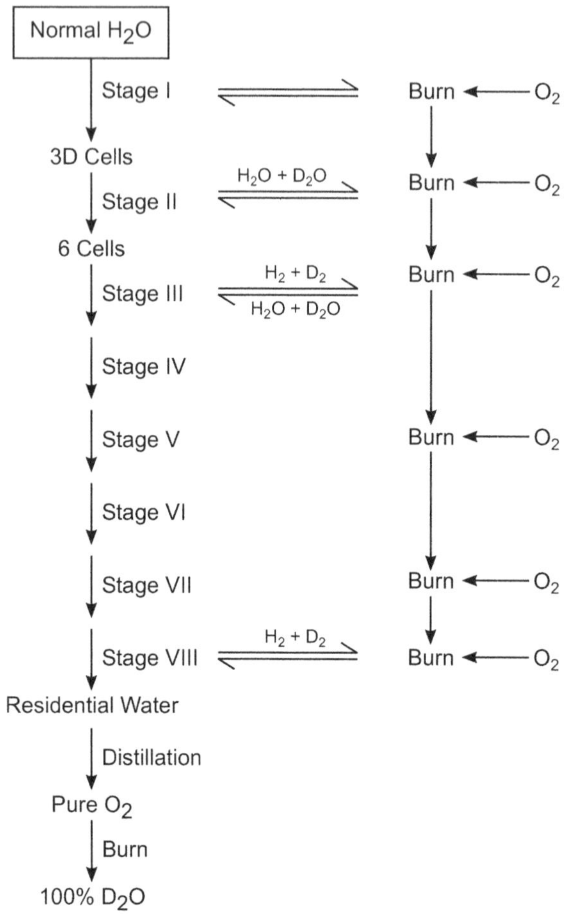

Taylor Eyring and Frost found out that electrolysis of 2310 litres of water gave 82 ml of D_2O as shown in the table.

Pure D_2O can be obtained by the action of D_2 gas on O_2 gas

$$2D_2 + O_2 \longrightarrow 2D_2O$$

Physical Properties

Heavy water is tasteless, colourless and odourless like ordinary water. Density of D_2O is greater than H_2O due to high molecular weight. It also has a higher viscosity.

Chemical Properties: High density of D_2O makes it slow in reactions compared to ordinary H_2O.

(*i*) **Electrolysis:** D_2O decomposes to D_2 and O_2 in the presence P_2O_5 or Na_2CO_3.

$$2D_2O \xrightarrow[\text{or } Na_2CO_3]{P_2O_5} 2D_2 + O_2$$

(*ii*) **Action with metals is slow**

$$2Na + 2D_2O \longrightarrow 2NaOD + D_2$$
$$Ca + 2D_2O \longrightarrow Ca(OD)_2 + D_2$$

(*iii*) **Metallic Carbides:** This reaction gives hydrocarbons

$$Al_4C_3 + 12D_2O \longrightarrow 4Al(OD)_3 + 3CD_4$$
$$CaC_2 + 2D_2O \longrightarrow Ca(OD)_2 + C_2D_2$$

(*iv*) **Action with metallic phosphides and nitrides**

$$2AlN + 6D_2O \longrightarrow 2Al(OD)_3 + 2ND_3$$
$$Mg_3N_2 + 6D_2O \longrightarrow 3Mg(OD)_2 + ND_3$$
$$Ca_3P_2 + 6D_2O \longrightarrow 3Ca(OD)_2 + 2PD_3$$

(*v*) **Exchange Reactions:** Compounds containing hydrogen are treated with D_2O give these reactions.

$$NaOH + D_2O \longrightarrow NaOD + HDO$$
$$HCl + D_2O \longrightarrow DCl + HDO$$
$$NH_3 + DCl \longrightarrow NH_2D + HCl$$
$$NH_2D + DCl \longrightarrow NHD_2 + HCl$$
$$NHD_2 + DCl \longrightarrow ND_3 + HCl$$

(*vi*) **Deutrolysis:** Heavy water also deutorlyses with inorganic salts.

$$AlCl_3 + 3D_2O \longrightarrow Al(OD)_3 + 3DCl$$
$$BaS + 2D_2O \longrightarrow Ba(OD)_2 + D_2S$$
$$SiCl_4 + 4D_2O \longrightarrow D_4SiO_4 + 4DCl$$
$$SOCl_3 = 3D_2O \longrightarrow D_4SO_3 + 2DCl$$

(*vii*) **Biological Properties:** Heavy water retards the growth of many plants and animals. Pure D_2O kills aquatic life but some moulds develop better. Tobacco seeds do not grow in D_2O.

Uses

(*i*) D_2O is used to slow down the speed of moving neutrons which are required in the fission of uranium atoms

(*ii*) D_2O is used as a tracer to study the mechanism of many chemical reactions.

(*iii*) Used for preparing Pure D_2.

EXERCISES

1. Explain why the element hydrogen does not find place in the main group of the periodic table.
2. What is present day position of hydrogen in the periodic table. Discuss
3. Describe the four ways in which hydrogen is produced on a commercial scale.

4. Describe the convenient method of preparing hydrogen in the laboratory.
5. Give important used of hydrogen.
6. Represent the reaction of hydrogen with:

   (*a*) Na (*b*) Ca (*c*) Co (*d*) N

   (*e*) S (*f*) Cl (*g*) CuO.

   Write the appropriate chemical equation.
7. Explain the various types of hydrides on the basis of pH value.
8. Describe the division of hydrides on the basis of bond formation.
9. Discuss the different types of isotopes of hydrogen and their properties.
10. Explain the formation of ortho and Para hydrogen. Elaborate
11. Describe hydrogen ion formation in aqueous medium and its structure.
12. What is hydrogen bonding? Explain its effect on the properties of the compound. Give examples
13. Explain the variation in the boiling points of:

    (*a*) HF = 20°C (*b*) HCl = –85°C (*c*) HBr = –67°C and HI = –36°C
14. Describe self ionization of H_2O
15. How the properties of H_2O, NH_3 and HF affected by the hydrogen bonding.

Unit III
Main Group of Elements

9

Group I or Alkali Metals

The main group elements encompass the chemistry of *s*-block and *p*-block elements. The chemistry of *s*-block elements has in its domain (*i*) The Alkali Metals and Alkaline earth metals. The study of alkali metals includes metals like lithium, sodium, potassium, Rubidium, caesium and Francium. These elements have closely related properties because of their similar electronic structures.

Electronic Structures Table

All these elements have one electron in the outer most valency shell (n) in the *s*-orbital. Their valency shell configuration is represented by ns, therefore, they are assigned group I (vertical column). The hydroxides of these elements are strong alkalis. Thus, they are given the name alkali metals. Some of the alkali metal salts are known form very early times namely sodium chloride. Sodium chloride in used as an antiseptic (gargles) food preservative and for flavouring of food. The alkali metals are difficult to be reduced. Sodium and potassium levels are well regulated in the human body.

Table 9.1 Electronic Structures

Element	Symbol	Electronic structure	
Lithium	Li	$1s^2 2s^1$	or [He] $2s^1$
Sodium	Na	$1s^2 2s^2 2p^6 3s^1$	or [Ne] $3s^1$
Potassium	K	$1s^2 2s^2 2p^6 3s^2 3p^6 4s^1$	or [Ar] $4s^1$
Rubidium	Rb	$1s^2 2s^2 2p^6 3s^2 3p^6 3d^{10} 4s^2 4p^6 5s^1$	or [Kr] $5s^1$
Caesium	Cs	$1s^2 2s^2 2p^6 3s^2 3p^6 3d^{10} 4s^2 4p^6 4d^{10} 5s^2 5p^6 6s^1$	or [Xe] $6s^1$
Francium	Fr		[Rn] $7s^1$

The elements lithium (Li), sodium (Na), potassium (K), rubidium (Rb), caesium (Cs) and francium (Fr) fall in its domain. These elements are given the name alkali metals because their hydroxides are found to be strong bases (alkalis). Sodium is derived from the Arabic word 'soda', which implies shift apart as it was considered cure for spilling headache potassium is derived from 'potash', which stands for ash of plants or the ash contains potassium carbonate. Rubidium name is derived from its ruby coloured spectral lines. Caesium means sky blue after its blue spectral lines. Francium was discovered by madam curie in 1939. It was named after its radioactive nature. Out of all these elements caesium is liquid and francium is radioactive. Not much is known about the element francium because of short half life.

General Characteristics

1. **Occurrence:** All these elements are very reactive therefore, they are not available in the free state. They exist as cations and combine with electronegative elements or anions. Sodium chloride is present in sea water and as deposits on earth (rock salt) Potassium is present in carnallite KCl. $MgCl_2$, $6H_2O$. Lithium, rubidium and caesium are present in alumino silicates in small quantities. Sodium and potassium are present in abundance on earth's crust.
2. **Extraction of Alkali Metal:** Extraction by normal reduction of metal oxides cannot be carried out in the case of alkali metals owing to following reasons:
   (*i*) The bond between the strongly electropositive and strongly electronegative elements is very strong.
   (*ii*) The alkali metal themselves are very strong reducing agents.
   (*iii*) The elements react violently with water therefore, displacing from their solutions by the elements higher in electrochemical series is not possible.
   (*iv*) These elements can be prepared from their aqueous solutions by electrolysis using Hg cathode where they form amalgams. Recovery of the metals from amalgams is difficult. The metal liberated at cathode reacts with water to form metal hydroxide.

   The easiest option to obtain the metal is by the electrolysis of fused metal halides. These metal halides are mixed with other metal halides to lower the melting point.
3. **Electronic Configuration:** Atoms of the alkali metals in their ground state contain one electron in the outermost *s*-orbital outside the noble gas core. Their general configuration is represented by ns^1 (Table). Due to the similarity in their electronic configuration, these elements exhibit similarities in their physical and chemical characteristics.

Physical Characteristics: For all practical purposes the characteristics of the first five elements will be discussed as francium is a radioactive element.

Table 9.2 Physical Properties of Alkali Metals

Property	Lithium Li	Sodium No	Potassium K	Rubidium Rb	Cesium Cs
Atomic number	3	11	19	37	55
Electronic configuration	[He] $2s^1$	[Ne] $3s^1$	[Ar] $3s^1$	[Kr] $4s^1$	[Xe] $6s^1$
Atomic radius, pm	123	157	203	216	235
Ionic radius, M^+, pm	68	97	133	148	169
Ionisation energies kJ/g atom 1st	527	502	427	410	381
Ionisation energies kJ/g atom 2nd	7303	4570	3076	2662	2427
E°, M^+/M(V)	–3.04	–2.71	–2.92	–2.99	–3.02
Electronegative	1.0	0.9	0.8	0.8	0.7
Heat of atomization (at 25°C, 1 atm), kcal mol^{-1}	38	26	21	19.5	18.7
Atomic volume	12.97	23.68	45.39	55.80	69.95
Atomic weight	6.94	22.99	39.10	85.47	132.91
Density, g/cc	0.53	0.97	0.86	1.53	1.90
Melting point, °C	180	98	63.5	39	28.5
Boiling point, °C	1336	883	759	700	670
Colour of the flame	carmine red	yellow	violet	reddish violet	blue
Ionic conductances of M ions Ω^{-1}	33.5	43.5	64.6	67.5	68.0

Most of the physical characteristics of the elements are related to the facts.

(*a*) Loose binding of *s*-electrons

(*b*) Size or atomic radii of the alkali metals.

Noble gas core shields the outer *s*-electrons from the nuclear charge. As a consequence of loose binding the following properties are affected.

(*i*) **Ionization Energy:** The energy required to remove the outer *s*-electron from a neutral gaseous atom is called the ionization energy or in other words it is the ease of loosely bound *s*-electron.

First ionization energy of all alkali metals is low. Therefore all alkali metals behave as electropositive elements as the metals form the positive ions readily.

$$M(g) \longrightarrow M(g)^+ + e^-$$

Second Ionization energy is quite high as it requires more energy to remove the second electron from the noble gas core. Therefore, the chemistry of alkali metals gets limited to unipositive ions. Trend in the alkali metal group (I_A).

Ionization energy down the group shows a decrease because *s*-electron is getting farther away from the nucleus due to the addition of more shells. Hence attraction of the nucleus for the outer most electron decrease thus increasing the electropositive character (remove of *s*-electron becomes more and more easy).

(*ii*) **Reducing Action:** The reducing property of these elements becomes obvious from the following reaction.

(*a*) **Liberate H_2 from water and acids**

$$2M + 2H_2O \longrightarrow 2MOH + H_2$$

$$M + 2HCl \longrightarrow MCl_2 + H_2$$

(*b*) **Strong tendency to lose electron:** The tendency to lose electron is measured in terms of reduction potential.

$$M^+ + e \longrightarrow M^\circ$$

They are powerful reducing agents. Lithium has the highest value E° = –3.04 V. Therefore, it is the strongest reducing agent among alkali metals. Due to small size of lithium ns^1 electron is tightly held by the nucleus but the ion formed cannot remain in gas form as the reducing action takes place in aqueous solution.

$$M(s) \rightarrow M^+ \text{ aq} + e^- \text{ (aq)}$$

The steps involved are

(*i*) M(s) + sublimation energy $\longrightarrow$ M(g) sublimation energy for alkali metals do not vary much.

(*ii*) Mg + I. E $\longrightarrow$ M^+ (g) + *e*

I,E energy value for Li is maximum

(*iii*) M^+ (g) + H_2O $\longrightarrow$ M^+ (aq) + H.E.

Li^+ (g) ion is the smallest gets hydrated to the maximum extent. Therefore, *e*/*r* value is maximum. Hence Li^+ has the maximum hydration energy. The H.E. decreases gradually down the group (Cs).

The E° reduction potential is the sum of

$$E^\circ = (i) + (ii) + (iii)$$

On comparing with alkali metals E° value is maximum –ve value for lithium.

(*iii*) Photons (low energy) can eject from the loosely held *s*-electron from there alkali metals. Cs is particularly used in photo-electric cells which are sensitive to blue light.

(*iv*) Electronegativity is the tendency of gain of electron of the neutral gaseous atom. These elements have a tendency to lose electron thus gain of electron is a remote possibility. The value decrease from Li to Cs.

(*v*) **Flame Colour:** The alkali metal chlorides are used for the flame test as they are more volatile and give coloured vapours on a platinum wire. The *s*-electron gets excited to higher energy states and this extra energy is emitted as coloured light when the electron reverts back to its original level. The colour of the flame is because it conforms to the visible part of the spectrum each metal atom gives a characteristic colour to the flame.

(*vi*) **Electrical Conductivity:** Metals are held by the metal atoms in the crystal. Metals have shielded nuclei in a 'sea' of *s*-electrons. In alkali metal the 'sea' is diffused and the binding in the solid state is weak. The valence electrons are relatively free to move from one metal ion to another with the result metals exhibit high electrical conductivity.

(*vii*) **Ionic Compound Formation:** The highly electropositive alkali metals compare with highly electronegative elements by transfer of *s*-electron. This forms ionic bonds. On the other hand these elements can form covalent bonds in some cases. However, vapours of alkali metals are covalently bonded such is Na_2 and Cs_2. The strength of covalent bend formation decreases down the group. Another example of covalent bond formation is in the form of organic metallic compounds *e.g.*, CH_3Li, $C_2H_5C_6H_5$, CH_2Na.

(*b*) **Size of Atoms or Ions:** Atomic volume atomic radii and ionic radii have been listed in Table 9.1. Alkali metals have relatively large values with one *s*-electron for per atom alkali metals results in loosely packed metal lattices.

(*i*) Alkali metals are soft malleable and ductile.

(*ii*) They have high luster.

(*iii*) They have low melting and boiling points.

(*iv*) They have low heats of atomization due to weak metal-metal bonds.

(*v*) They exhibit low density because of large atomic volume.

Alkali metals crystallize with a body centered cubic lattice with lattice points occupied by the metal ions (M^+). Valence electrons are spread over the whole lattice. These electrons more freely over whole of the metal and this is responsible for the electrical conductivity when a beam of light falls on these electrons they are set into vibrations and that accounts for their lustre. The softness, malkeability and ductility are explained in terms of binding forces between the metals.

Melting and Boiling Points: Alkali metals have weak cohesive forces therefore, they are very soft and can be easily cut with knife. They exhibit low values of melting and boiling points. The melting points decrease down the group. Lithium melts at 181°C while caesium at 28.5°C. Most of the hard transition metals melt above 1000°C. Melting point of Li is twice as high as that of sodium. Other alkali metals besides Li have melting points close together.

It is observed that in all the main group elements the first element differs in properties than the rest of the elements in the group. Melting and boiling points of alkali metals are given in the Table 9.3.

Table 9.3 Melting and boiling Points

	Melting Point (°C)	Boiling pint (°C)
Li	181	1347
Na	98	881
K	63	766
Rb	39	688
Cs	28.5	705

Flame Test for Alkali Metals: When the alkali metals are irradiated by light, the light energy absorbed may make them lose an electron (low ionization energy). The emitted electrons are called photo-electrons. K and Cs act as cathodes in photo electric cells.

To perform their flame test the metal chloride or metal salt dipped in concentrated HCl is heated on a platinum were in an oxidizing flame (chloride salts are most volatile). On heating the orbital electron gets excited to higher energy level. The excited electron falls back to its original level and gives out the extra energy. Energy is related to the wave number Ω as

given: $E = h\,\Omega$ ('h' is Planck's constant)

The energy appeas in the range of visible light which has seven colours (VIBGYOR) *e.g.,*

$$Na^{+} + e \longrightarrow Na$$

In sodium atoms the electronic transition $3s^1 \rightarrow 3p^1$ forms the colours of the flame. The colours for different elements do not arise from the same transition. The electronic transition may be of still higher level and when it drops back to the lower level the energy may be different and different colour may be emitted *e.g.,* Li shows red line. Melting and boiling points give colour of flame.

Trend in The Group: On moving down atomic volume, atomic and ionic radii increase. Large alkali metal atoms exert weaker attractions because of electrostatic forces decrease with increase in inter-nuclear distance, therefore these metals become soft and their melting and boiling points and the heats of atomization decrease.

Chemical Properties: The chemical reactivity of the alkali metals is related to some of the physical characteristics namely.

(*i*) First Ionization energy

(*ii*) Electro negativity

(*iii*) Heat of atomization.

The alkali metals have low value of these properties hence they show high chemical reactivity. The chemical properties of the alkali metals are discussed: Preterably (i) to (viii) on one page

Hydrides: Alkali metals react with hydrogen to form ionic hydrides (M^+H^-). These hydrides contain H^- ion as H_2 gets liberated at anode although hydrogen usually forms H^+ ions. These hydrides react with H_2O liberating hydrogen. It is a source of hydrogen which is used to fill balloons for weather studies.

$$LiH + H_2O \longrightarrow LiOH + H_2\uparrow$$

Lithium forms complex hydride with AlH_3 which is a very good reducing agent

$$LiH + AlH_3 \xrightarrow[\text{ether}]{\text{Dry}} Li[AlH_4]$$

$$4LiH + AlCl_3 \longrightarrow Li[AlH_4] + 3LiCl_3$$

The complex is ionic and in which AlH_4^- is tetrahedral. $LiAlH_4$ is mostly used in organic reaction as a powerful reducing agent *e.g.*,

$$\underset{\text{Carboxylic acid}}{R{-}\overset{\overset{\displaystyle O}{\|}}{C}{-}OH} \quad + 4H \xrightarrow[\text{Dry Ether}]{LiAlH_4} \underset{\text{alcohols}}{>CH_2\,OH}$$

The reaction is carried out in dry ether as Li undergoes violent reaction with water. $LiAlH_4$ is also used in many inorganic reactions.

$$BCl_3 + Li[AlH_4] \xrightarrow[\text{Ether}]{\text{D/}} B_2H_6 \qquad \text{Diborane}$$

$$PCl_3 + Li[AlH_4] \longrightarrow PH_3 \qquad \text{Phosphine}$$

$$SiCl_4 + Li[AlH_4] \longrightarrow SiH_4 \qquad \text{Silane}$$

Sodium forms $Na(BH_4)$ which is another complex hydride that is used as powerful reducing agent. It is prepared from sodium hydride and trimethylborate.

$$4NaH + B(OCH_3)_3 \xrightarrow{250°C} Na(BH_4) + 3NaOCH_3$$

$Na(BH_4)$ and $K(BH_4)$ are less reactive to water than $Li(AlH_4)$.

Reaction with water: All alkali metals react with water forming hydroxides and liberating hydrogen. The reaction becomes more violent down the group. Only Lithium reacts gently, sodium melts on the water surface and may catch fire while potassium invariably catches fire.

$$2Li + H_2O \longrightarrow 2LiOH + H_2$$

$$2Na + 2H_2O \longrightarrow 2NaOH + H_2$$

$$2K + 2H_2O \longrightarrow 2KOH + H_2$$

Lithium has the most negative value on the standard electrode potential scale out of all elements in the periodic table $E° = -3.05$ volts. This is because of high hydration energy. $E°$ and ΔG are related by the equation

$$\Delta G = -n\,FE°$$

Where 'n' is the number of electrons removed and F is the Faraday constant and ΔG is Gibbs free energy.

$Li^+ + e \longrightarrow Li$ has the largest E° negative value and highest ΔG positive value. According to this Li should react violently than potassium. This is explained by the reaction kinetics.

Reaction with air (oxygen and nitrogen) alkali metals are very reactive and get transited by dry air. This is due to the formation of oxides peroxides, superoxide and nitrides Li_3 in ruby red. Nitrides have a tendency to hydrolyses to form hydroxides.

$$Li_3N + 3H_2O \longrightarrow 3LiOH + NH_3$$

Alkali metals burn in air to form oxides. Lithium forms monoxides Li_2O plus some Li_2O_2 peroxide. Sodium forms peroxide Na_2O_2 plus some Na_2O while other alkali metals form superoxide of the formula MO_2. All these metals dissolve in liquid ammonia and by controlled reaction with dioxygen are known to form normal oxides, peroxides and superoxide. Monoxides are ionic $2Li^+O^{2-}$ Li_2O and Na_2O are white K_2O is yellow, Rb_2O is bright yellow and Cs_2O is orange. These oxides are basic, they dissolve in water to give strong bases (Alkalis)

$$Li_2O + H_2O \longrightarrow 2LiOH$$
$$Na_2O + H_2O \longrightarrow 2NaOH$$
$$K_2O + H_2O \longrightarrow 2KOH$$

NaOH is called caustic soda and KOH caustic potash. They have corrosive action an skin. These hydroxides are water soluble but LiOH is less soluble.

These strong bases reaction with acid to form salts and water. This process is called neutralization.

$$NaOH + HCl \longrightarrow NaCl + H_2O$$

With CO_2 being acidic they form carbonates.

$$2NaOH + CO_2 \longrightarrow Na_2CO_3 + H_2O$$

With Al_2O_3 they form aluminates SiO_2 gives silicates SnO_2 stagnates PbO_2 plumbates and ZnO gives zincates

$$2NaOH + Al_2O_3 \longrightarrow \underset{\text{Sodium aluminates}}{2NaAlO_2}$$

With NH_4 salts NaOH liberates NH_3 gas

$$NH_4Cl + NaOH \longrightarrow NaCl + NH_3 + H_2O$$

Similarly, NaOH reacts with H_2S and alcohols to form sulphides and alkoxides.

$$NaOH + H_2S \longrightarrow NaSH \longrightarrow Na_2S$$
$$NaOH + C_2H_5OH \longrightarrow \underset{\text{alkoxides}}{C_2H_2NHa} + H_2O$$

Peroxides and Superoxide: Peroxides and superoxides react with water to form H_2O_2

$$Na_2O_2 + 2H_2O \longrightarrow 2NaOH + H_2O_2$$
$$KO_2 + 2H_2O \longrightarrow 2KOH + H_2O_2 + \frac{1}{2}O_2$$

Both are strong oxidizing agents but superoxide are stronger. The $|\text{-O-O-}|^{2-}$ ion has 18 electrons while $|O_2|^-$ has 17 electrons. Superoxide can be prepared from peroxides by heating at 450°C in dioxygen.

$$Na_2O_2 + O_2 \longrightarrow 2NaO_2$$

Sulphides: Alkali metals react with sulphur forming sulphides (Na_2S) and polysulphides (Na_2S_x) where x may vary from 2 to 6. The polysuphides form zig-zag chains.

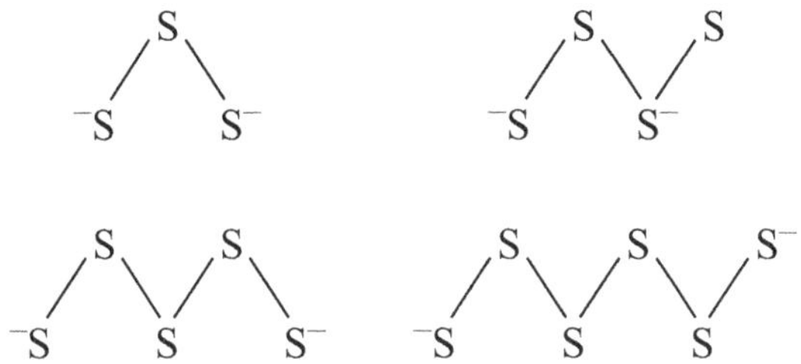

Sodium sulphide can be obtained by heating Na_2SO_4 with carbon or by passing H_2S in NaOH solution.

$$Na_2SO_4 + 4C \longrightarrow Na_2S + 4CO$$
$$NaOH + H_2S \longrightarrow NaHS + H_2O$$
$$NaOH + NaHS \longrightarrow Na_2S + H_2O$$

Alkali metal suliphides are appreciably hygroscopic giving strong alkali solutions.

$$Na_2S + H_2O \longrightarrow NaSH + NaOH$$

Na_2S is used to make organic dyestuffs and in leather industry. It can be easily oxidized to form $Na_2S_2O_3$ which is further used in photography. Sodium thiosulphate is known as hypo solution. It can dissolve silver halides and can be used as a reagent in iodine titrations.

$$Na_2S + 2O_2 + H_2O \longrightarrow Na_2S_2O_3 + 2NaOH$$
$$2Na_2S_2O_3 + I_2 \longrightarrow 2NaI + Na_2S_4O_6$$

Sodium tetrathionate

Oxo. Salts

Alkali metals form very stable oxo salts such as carbonates, bicarbonates, nitrite nitrates and carbonates decompose into oxides and CO_2 at 1000°C Li_2CO_3 decomposes more readily.

Bicarbonates of ammonium and potassium exist as solids (NH_4HCO_3 and $KHCO_3$). They evolve CO_2 on global warming.

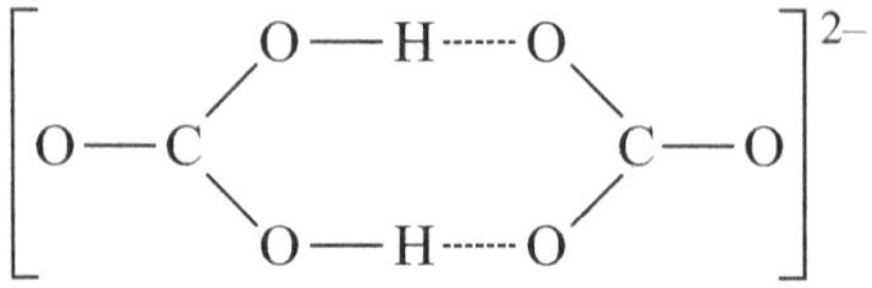

Lithium does not form solid bicarbonate. All carbonates and bicarbonates are water soluble. Na_2CO_3 is a called washing soda and $NaHCO_3$ as baking powder.

Nitrates can be purchased from the corresponding carbonate or hydroxide by the action of NHO_3.

$$2HNO_3 + R_2CO_3 \longrightarrow 2KNO_3 + CO_2 + H_2O$$
$$NaNO_3 + KCl \longrightarrow KNO_3 + NaCl$$

Nitrates are deliquescent. Gunpowder is a mixture of KNO_3 + S and charcoal. They are low melting solids. On strong heating they decompose into nitrites and oxygen.

$$2NaNO_3 \xrightleftharpoons{500°C} 2NaNO_2 + O_2$$

At 800°C

$$4NaNO_3 \rightleftharpoons 2Na_2O + 5O_2 + 2N_2$$

Nitrites: Nitrite produce brown fumes of NO_2 with dilute acids

$$2NaNO_2 + 2HCl \longrightarrow 2NaCl + H_2O + NO + NO_2$$
$$2NO + \underset{\text{(air)}}{O_2} \longrightarrow 2NO_2$$

Nitrites can also be formed by thermal decomposition of nitrates and reduction of nitrates.

$$2NaNO_3 + C \longrightarrow 2NaNO_2 + CO_2$$
$$KNO_3 + Zn \longrightarrow KNO_2 + NO$$

Reaction of NO and KOH also produces nitrite.

$$4KOH + 6NO \longrightarrow 4KNO_2 + N_2 + 2H_2O$$

Halides of Polyhalides: Li^+ ion has the smallest size in the group therefore it forms hydrated salts $Li\,X.3H_2O$ while other alkali halides are anhydrous. They show NaCl type structure with a co-ordination number of 6 except Cs salts which show body centered cubic structure with a co-ordination of 8.

They are prepared by the direct combination of metals and halogens at elevated temperature. These metal halides react with halogens and inter halogens Li form polyholides.

$$KI + I_2 \longrightarrow KI_3$$
$$KBr + ICl \longrightarrow K(Br)Cl$$
$$KF + BrF_3 \longrightarrow K(BrF_4)$$

Metals in liquid ammonia: Alkali metals react with liquid NH_3 in the presence of a catalyst or purities

$$M + NH_3 \xrightarrow{Fe} MNH_2 + H_2$$

In the absence of catalysts and impurities alkali metals; Ca Sr and Ba; and Eu and yb dissolve in liquid ammonia at high temperature. The metals can be recovered by boiling of NH_3.

Dilute solutions of alkali metals is liquid ammonia give dark blue colour because of solvated metal ions and solvated electrons.

$$M \xrightarrow[NH_3]{\text{dissolve}} M^+ \text{(amm.)} + [e(NH_3)_x]^-$$

Both metal ions and electron are solvated by ammonia, the dissociation into cations and electrons accounts for electrical conductivity.

Blue solution exhibits para magnetism because of large number of unpaired electrons. Blue colour is owing to the absorption of photons by the electrons.

The electrons are excited to a higher energy level and at absorption band 5000A° blue colour appears. The dilute solution consists of lot of free electrons hence they show reducing behaviour.

Blue Solution Exhibits

1. Blue colour is independent of the metal.
2. Density of the solution is similar to that of pure ammonia.
3. Conductivity is same as that of other electrolytes in ammonia.
4. Unpaired electrons indicate paramagnetic.
5. Strongly reducing nature.
6. Solutions are meta stable and reversible.

$$[e(NH_3)_x]^- \xrightarrow{Fe_2O_3} \underset{\text{amide}}{NH_2^-} + H_2 + (x - 1)NH_3$$

The bronze solution have these characteristics.

1. Bronze colour has a metallic luster
2. Low densities
3. Conductivities are in the range of metals
4. Magnetic properties similar to pure metals.

The electrons behave in a manner as they would do in the metals. The metals solutions behave as strong reducing agents much stronger than could exist in water because the oxidizing action of water is greater than that of ammonia some examples of the reducing power are:

$$Be + Na/NH_3 \longrightarrow Na_3Bi \text{(Bi 0 to − 111)}$$

$$S + Na/NH_3 \longrightarrow Na_2S \text{ (S 0 to − 11)}$$

$$[Ni(CN)_4]^{2-} + 2e \longrightarrow [Ni(CN)_4]^{4-} \text{ (Ni from + 2 to 0)}$$

Unique behaviour of Lithium: Lithium differs from its congeners. The main reason for its different properties as compared to other alkali metals are

1. Extremely small size of lithium atom or ion.
2. Polarizing rouser of lithium is the greatest of all the alkali metals which results in its tendency towards solution and covalent bond formation.
3. Lithium has the highest ionization energy and electro negativity value in caparison to other alkali metals.
4. Lithium is harder and lighter than other alkali metals.
5. It hah high melting and boiling point.
6. High ionization energy does not allow it to react with oxygen below 373 K, reacts slowly with water or liquid bromine.
7. It combines with small amines such as hydrides, carbides and nitrides to from stable compounds. They show high lattice energy in the crystalline form lithium gives exothermic reactions. Li H is more stable to heat.
8. Only lithium forms monoxide Li_2O while other alkali metals form peroxides and superoxide.
9. Lithium dissolves in water quietly while other react violently.
10. Li OH is a weak base while NaOH and KOH are strong alkalis.
11. Li OH decomposes at red heat to form Li_2O where as other hydroxide sublime undecomposed.
12. Li Cl is deliquescent and soluble is alcohol and pyridine.
13. Li_2CO_3 decomposes to Li_2O and CO_2Li_2O lattice is more stable than Li_2CO_3. Other alkali metal oxides give less stable lattice (M_2O).
14. Lithium nitrate decomposes to give Li_2O, NO_2 and O_2 while other give nitrites and oxygen.

$$4LiNO_3 \longrightarrow 2Li_2O + 4NO_2 + O_2$$

$$2KNO_3 \longrightarrow 2KNO_2 + O_2$$

15. Lithium does not form lithium acetylide while other metals do form metal acetylides.
16. LiSH is unstable while NaSH and KSH fuse without decomposition.
17. Lithium salts like carbonates, phosphates and fluorides are insoluble in water whereas other alkali metal salts are soluble.
18. Lithium combines with silicon to give lithium silicide Li_6Si_2 dark violet hydroscopic crystals.
19. Lithium NH_3 form Li_2NH imides while other alkali metals form MNH_2 amide.
20. $LiClO_4$ is soluble in alcohol, acetone and ethyl acetate white other alkali metal per chlorates are insoluble.

21. Lithium does not form double salts while other alkali metals do.
22. Li forms complexes with NH_3 and H_2O to form organo lithium compounds.

Diagonal Relationship between Lithium and Magnesium

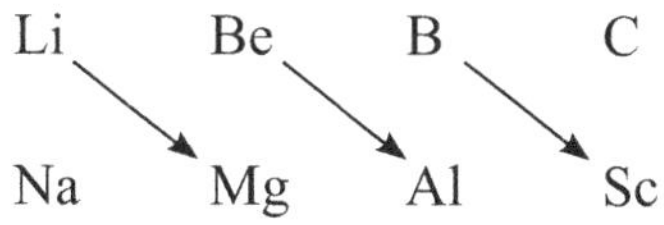

Diagonally opposite elements pair are indicated by the away in the diagram.

It has been observed that diagonally opposite elements resemble more than their own congeners. The reasons for similarities between such elements are:

1. The elements becomes more electropositive down the group and less electropositive from left to high. Therefore it results is the similarity in the electropositive character of diagonally opposite elements.
2. On moving from left to right charge on the ions increases but their size decrease which causes an increase in the polarizing power of the ions. Down the group, the size increase and polarizing power decrease. In the diagonally opposite elements these two effects cancel each other so the behavoiur of these elements is quite similar. Similarly ionization energy and electro negativity values become similar in such elements. Hence their properties and compounds are very similar.

Lithium resembles more with its diagonally opposite elements magnesium than with the other alkali metals. This is accentuated with following observations.

(*i*) The ionic radii of Li^+ (60 pm) and Mg^{2+} (65) pm are almost equal.

(*ii*) The atomic radii are also comparable.

Li = 122 pm Mg = 136 pm

(*iii*) The atomic volumes are quite similar.

Li = 12.93 ml Mg = 14.00 ml

(*iv*) Electro negativity values are comparable.

Li = 1.00 Mg = 1.20

(*v*) Polarizing power of both elements is similar.

(*vi*) H (enthalpies of vaporization) are comparable.

(*vii*) Boiling points are comparable

Li = 1609 K Mg = 1373 K.

(*viii*) Lithium and magnesium are equally hard.

(*ix*) Lithium and magnesium burn in air or oxygen to form monoxides of the formula Li_2O respectively.

(*x*) Their carbonates and nitrates give Li_2O and MgO.

(*xi*) Both form ionic nitrides when heated in the atmosphere of nitrogen Li_3N and Mg_3N_2.

(*xiii*) Lithium ion Li^+ and magnesium compounds are heavily hydrated $MgCl_2 . 6H_2O$, $MgSO_4 . 7H_2O$.

(*xiv*) Li_2CO_3, LiF and Li_3PO_4 are insoluble in water like magnesium counter parts.

(*xv*) Both combine directly with carbon forming carbides.

(*xvi*) LiCl and $MgCl_2$ are soluble in alcohol and pyridine.

(*xvii*) $LiClO_4$ and $Mg(ClO_4)_2$ are extremely soluble in alcohol.

(*xiii*) Complex hydrides of Li and Mg are soluble in organic solvents *e.g.*, $LiAlH_4$, RMgX are soluble in ether.

Complexes of Alkali Metal Ions

Coordination complexes fall mostly in the domain of transition metals. Conventional ligands *e.g.*, NH_3 do not form complexes with alkali and alkaline earth metals due to their weak complexing ability. The metal ion have small charge to radius ratio.

X-ray crystallography showed that some of the molecules (legends) show strong complication ability for alkali and alkaline earth metal ions (Na^+, K^+, Ca^{2+} and Mg^{2+} ions).

The alkali metals replace hydrogen from the organic acids forming salts such as sodium acetate and potassium benzoate. Soap is made up of mixture of sodium salts of palmitic, oleic and stearic acids.

$$\begin{array}{l} CH_2-O-OC-C_{17}H_{31} \\ | \\ CH-O-OC-C_{17}H_{31} \\ | \\ CH2-O-OC-C_{17}H_{31} \end{array} + 3NaOH \xrightarrow{\text{ether}} \begin{array}{l} CH_2-O-OC-C_{17}H_{31} \\ | \\ CH-O-OC-C_{17}H_{31} \\ | \\ CH2-O-OC-C_{17}H_{31} \end{array} + 3NaOH$$

Lithium stearate is a soap made of LiOH and natural fat. Lithium shows stronger tendency to form covalent bonds. Lithium shows diagonal relationship to magnesium therefore both lithium and magnesium form a number of alkyl and aryl compounds called Grignard reagents.

$$RCl + Li \longrightarrow LiR + LiCl$$

Lithium ethyl is tetrameric $[Li(C_2H_5)]_4$ but is hexameric when dissolved in hydrocarbons $[LiC_2H_5]_6$.

N-Butyl lithium is a tetrameric solid $[LiB_4]_4$. It is a very good reagent to synthesize aromatic and unsaturated derivatives such as vinyl and alkyl lithium

$$LiB_4 + ArI \longrightarrow LiAr + B_4I$$

$$4LiAr + Sn(CH = CH) \longrightarrow 4LiCH = CH_2 + Sn(Ar)_4$$

This gives a route to form large number of organ metallic and organic compounds

$$3LiR + BCl_3 \longrightarrow BR_3 + 3LiCl$$
$$4LiR + SnCl_4 \longrightarrow SnR_4 + 4LiCl$$
$$LiR + CO_2 \longrightarrow RCOOH + LiOH$$
$$3LiR + P(OCl^-)_3 \longrightarrow PR_3 + 3LiOCl^-$$
$$LiR + R'I \longrightarrow R—R' + LiI$$
$$LI'R + Cl_2 \longrightarrow R—Cl + LiCl$$

Alkyls of Na, K, Rb and Cs are prepared from organic mercury compound

$$2K + HgR_2 \longrightarrow Hg + 2KR$$

These compounds are ionic and catch fire in air.

Biological Use: Both Na^+ and K^+ ions in the animal body maintain the electrolytic balance of the body. The Na^+ and K^+ ratio inside and outside cells produce electrical potential across the cell membrane which is necessary for the functioning of nerve and muscle cells. The Na^+ ions enter into cells along with glucose. This is favoured by high concentration gradient. The Na^+ ions entering the cells must be expelled, K^+ ion inside the cell are essential for the metabolism of glucose, the synthesis of proteins and the activation of enzymes, potassium ions control the nervous system. Decrease of potassium ions in the cell membrane can prove fatal. In care of deficiency of K^+ ions potassium chloride (KCl) is given to the patient so as to maintain the level of K^+ ions.

EXERCISES

1. Why are the alkali metals soft and volatile?
2. Why the alkali metals have low ionization energies?
3. LiH, LiF and Li_3N show high thermal stabilities. Discuss.
4. Why the salts of Li^+ with large anions CO_3^{2-}, ClO_4^- and NO_3^- are less stable than its salts with small anions?
5. Show by examples that lithium is least reactive out of all the alkali metals.
6. $Mg(ClO_4)_2$ is an excellent drying agent. Comment.
7. Compare the chemistry of extraction of the following metals:
   (*i*) Lithium
   (*ii*) Sodium?
8. Discuss the following properties of the alkali metals:
   (*i*) Electropositive character
   (*ii*) Ionization energy
   (*iii*) Hydration energy
   (*iv*) Strength of their covalent bond in M_2 molecules.
9. Discuss the following properties of the alkali metals with increasing atomic number:
   (*i*) Ionization energy (*ii*) Reactivity towards air and water.

10. K_2O, K_2O_2, KO_2, K_3N which of the above is paramagnetic.
11. Give reasons:
    (*i*) Low ionization energies of alkali metal.
    (*ii*) Strong reducing action of alkali metals.
    (*iii*) Poor complexing agents
    (*iv*) Alkali metals lose electrons when exposed to sunlight.
    (*v*) Alkali metals have low melting and boiling points.
    (*vi*) Alkali metals use soft, volatile and electro positive.
12. Lithium resembles more with magnesium than with congeners. Comment.
13. Why does Li form the normal oxide, Na the peroxide and K_2, Rb and Cs the superoxide?
14. Why do alkali metals in liquid ammonia show high conductivity?
15. Electrolysis of an aqueous solution of NaCl does not yield metallic sodium. Comment.
16. Li halides are covalent while other alkali metal halides ionic. Discuss
17. Give reasons:
    (*i*) Why Lithium has maximum hydration energy?
    (*ii*) Why chemical reactivity increase from Li to CS.
    (*iii*) Alkali metals in liquid ammonia are strong reducing agents than alkali metals in water.
18. How do you extinguish a fire by sodium?
19. Discuss electrolysis of fused NaCl and aqueous NaCl.
20. What are the main uses of chlorine, sodium and caustic soda?
21. What is Na_2CO_3 used for? Why its use has declined? How NaOH is converted to Na_2CO_3 and vice-versa?

10

Group II or Alkaline Earth Metals

The elements the alkaline earth metals included in this group are listed in the Table 10.1 along with their electronic configuration.

Table 10.1 Electronic Structures

Element	Symbol	Electronic structure	
Beryllium	Be	$1s^2 2s^2$	or [He] $2s^2$
Magnesium	Mg	$1s^2 2s^2 2p^6 3s^2$	or [Ne] $3s^2$
Calcium	Ca	$1s^2 2s^2 2p^6 3s^2 3p^6 4s^2$	or [Ar] $4s^2$
Strontium	Sr	$1s^2 2s^2 2p^6 3s^2 3p^6 3d^{10} 4s^2 4p^6 5s^1$	or [Kr] $5s^2$
Barium	Ba	$1s^2 2s^2 2p^6 3s^2 3p^6 3d^{10} 4s^2 4p^6 4d^{10} 5s^2 5p^6 6s^2$	or [Xe] $6s^2$
Radium	Ra		[Rn] $7s^2$

They are called alkaline earth because some of them are abundantly available on the earth's surface *e.g.,* calcium and barium oxides. There oxides react with water to form hydroxide which are called alkalis.

Their electronic configuration show two electrons in their outermost shell. The electronic structure may be written as $2s^2$, $3s^2$, $4s^2$, $5s^2$, $6s^2$ and $7s^2$.

Element Be stands unique from the other elements in the group. The main reason for this uniqueness is due to exceptionally small size and high charge/radius ration as compared to the other elements in the group. Beryllium shows semilarities with diagonally opposite aluminium in group 11. Be and Ba compounds are toxic. They are all bivalent and highly reactive elements. Their compounds are colourless and ionic.

Occurrence: All these elements do not occur in the free state due to their highly reactive nature, Be does not occur in abundance and its extraction is difficult. Mg and Ca occur in abundances on earth's surface, Sr and Ba occur in small amounts in the ores but their extraction is very easy. Radium is radioactive and scarce and not much is known about the chemistry of radium.

Extraction of the Metals: Both alkali and alkaline earth metals have highly reactive nature. Therefore, similar problems are encountered in their extraction. The metals are usually extracted by the electrolysis of fused metal halides containing NaCl.

By mixing with NaCl the melting point of the metal halides is lowered and conductivity of alkaline earth metal increases.

Extraction Beryllium

Beryllium chloride is mixed with sodium chloride in the ratio 2:1. The mixture is fused in a nickel crucible which acts as a cathode and carbon anode is also used in it.

On electrolysis the metal is obtained at cathode, since at high temperature Be–Ni alloy is formed therefore nickel crucible is kept in cool water. The metal is further purified by distillation in vacuum in an atmosphere of Argon.

Physical Properties: Important physical properties of alkaline earth metals have been given in Table 10.2.

Table 10.2 Physical Properties of Alkaline Earth Metals

Property	Beryllium Be	Magnesium Mg	Calcium Ca	Strontiu Ca	Barium Ba	Radium Ra
Atomic number	4	12	20	38	56	88
Electronic Configuration	$[He]2s^2$	$[Ne]3s^2$	$[Ar]4s^2$	$[Kr]5s^2$	$[Xe]6s^2$	$[Rn]7s^2$
Atomic radii pm	105	162	197	213	217	—
******	31	65	99	113	135	150
*******	894	734	587	547	500	508
******	1748	1443	1140	1059	960	962
	14784	7691	4909	4130	3423	—
E, M^{2+} M(V)	–1.85	–2.34	–2.87	–1.89	–2.90	–2.92
Electronegativity	1.50	1.2	1.1	1.0	0.9	0.9
Heat of atomisation (at 298 K and I atm) kcal/mol	78.2	35.1	42.3	39.0	42.0	—
Atomic weight	9.0	24.3	40.0	87.6	137.4	226
Density of solid at 293 K, g/c.c	1.86	1.75	1.55	2.6	3.39	5
Melting point, K	3043	1380	1710	1639	1790	1413
Colour of the flame	—	—	Brick red	Crimson	Apple green	—
Enthalpy of hydration ****	570	460	395	355	305	—

Binding of *s*-electrons in the alkaline earth metals is loose as they are shielded by the noble gas core. As a result they form dipositive metal ions (M^{2+}). Further loss of electron is almost ruled out due to very high third ionization energy.

Metal atoms and Ions: Alkaline earth metal atoms are large but smaller than the alkali metal atoms. This is attributed to the extra charge on the nucleus which pulls the orbital electrons inwards. Same trend holds for the ions. Removal of two *s*-electrons increases the effective nuclear charge still further. The elements have higher densities than the alkali metals.

Group 2 metals are silvery white in appearance. These metals have two valency electrons which participate in metallic bonding compared to one in the alkali metals. Therefore these metals are harder due to higher cohesive force and have high melting and boiling points than the alkli metals which are soft. The melting and boiling points show no regular trend because of different crystal structures Table 10.2.

Table 10.3 Melting and Boiling Points of Group 1 and 2 Elements

	Melting Point (°C)	Boiling Point (°C)		Melting Point (°C)	Boiling point (°C)
Be	1287	(2500)	Li	181	1347
Mg	649	1105	Na	98	881
Ca	839	1494	K	63	766
Sr	768	1381	Rb	39	688
Ba	727	(1850)	Cs	28.5	705
Ra	(700)	(1700)			

Ionization Energy

Alkaline earth metals form M^{2+} ions because the 2*s*-electrons are loosely bound with the nucleus. M^{3+} ions are not possible as the third ionization energy is considerably high. The ionization energy of Be^{2+} is also high and its compounds formed are covalent. Mg also forms some covalent compounds. Compounds of Mg, Ca, Sr and Ba are divalent and ionic. After removing one electron the second electron is more tightly bound and the *e*/*r* ratio increases. This make the energy for the second electron double that required for the first, The ionic compounds formed show that the energy given out by a crystal lattice more than compensation the energy required to form ions.

Table 10.4 Ionization Energies and Electronegativity

	Ionization energy (kJ mol^{-1})			Pauling's Electronegativity
	1st	2nd	3rd	
Be	899	1757	14847	1.5
Mg	737	1450	7731	1.2
Ca	590	1145	4910	1.0
Sr	549	1064		1.0
Ba	503	965		0.9
Ra	509	979	(3281)	

Electronegativity

Electronegativity values for group 2 elements are higher than the values for group 1 elements. From Be to Ra the electronegativity values decrease which shows greater tendency to form ionic compounds down the group. The higher electronegativity value of Be suggests the formation of covalent compounds of Be. BeF_2 has the biggest electronegativity difference and the compound is ionic. When fused BeF_2 behaves as a covalent compound.

Hydration

Reducing Action: The high standard reduction potentials $E°$ (Table 10.1) indicate that their aqueous solutions are good reducing agents like alkali metals. The hydration energies of Group 2 ions is 4 to 5 times greater than Group 1 ions. ΔH hydration decreases down the group. NaCl and KCl are anlydrous but $MgCl_2 6H_2O$; $CaCl_2 6H_2O$ and $BaCl_2 2H_2O$ have water of crystallisation. All these divalent ions have a noble gas core and two paired electrons therefore their compounds are diamagnetic and colourless unless the negative radical is coloured.

$$M^0 \longrightarrow M^{2+} + 2e$$

$$M^{2+} + 6H_2O \longrightarrow [M(H_2O)_6]^{2+} \longrightarrow \Delta H = \text{high}$$

Those ions with high negative values react vigorously with water.

Colour to the Flame: Except Be the chlorides of all these elements are volatile their electrons are very easily excited to higher energy levels. When the excited electrons fall back to their original level then they emit a coloured flame. Be and Mg require high excitation energy therefore they do not impart colour to the flame.

Metallic Bonding: Two electrons for metal atom are involved in the metallic bonding in alkaline earth metals. Due to this metallic bonding is stronger which results in close packed structures. Because of smaller size than the alkali metals their crystals are (*i*) harder, (*ii*) denser, (*iii*) higher melting and boiling, points and (*iv*) higher heats of atomization. The bond strength decreases with increasing size of the atoms.

Conducts heat and electricity: In the metallic form they have 2*s* electrons which can resonate between the nearest neighbours in the crystalline form. (6 neighbours will mean 12 electrons). These electrons can move through the lattice under electrical potential. As a result they act as good conductors of heat and electricity. This is also the reason for reflecting light and hence a silvery shine.

Metals in liquid ammonia: The alkaline earth metals dissolve in liquid ammonia like group 1 metals. The dilute solutions of these metals appear bright blue due to the solvated electron. Their solutions decompose slowly forming amides and evolving hydrogen. This reaction is speeded up by the presence of transition metal and their compounds.

$$2NH_3 + 2e \longrightarrow \underset{\text{amide}}{2NH_2^-} + H_2$$

Evaporation of ammonia yields hexamine complexes which decompose to give amides.

$$M(NH_3)_6 \longrightarrow M(NH_2)_2 + 4NH_3 + H_2$$

The concentrated solutions of these metals are bronze owing to the formation of metal clusters.

Chemical Properties: The smaller size and increased charge make alkaline earth alkali metal ions better polariser than the alkali metal ions. Their oxides are more covalent or less ionic and hydroxides are less basic than the Group 1 metals. Be and Mg oxides are insoluble in water. Due to extremely small size Be has a unique behaviour. Be compounds are more covalent Mg is partly covalent and partly ionic while the rest of the compounds of Ca, Sr, Ba and Ra are ionic.

As a consequence of small size and high charge alkaline earth metals are less soluble in water.

Reaction with Water: The standard electrode potential value $E°$ of Be is much less than the other members of alkaline earth metals. Therefore it is less electropositive than others. Therefore Be does not react with water.

Ca, Sr and Ba have $E°$ values similar to Group 1 metals. They react with cold water readily liberating hydrogen and forming corresponding metal hydroxides

$$Ca + 2H_2O \longrightarrow Ca(OH)_2 + H_2$$

The trend of basic properties increase with increasing atomic number, $Ca(OH)_2$ is called lime water and $Ba(OH)_2$ as baryta. Both hydroxides become turbid on passing CO_2 through them. More of CO_2 makes it clear due to the formation of bicarbonates.

(*i*) $Ca(OH)_2 + CO_2 \longrightarrow \underset{\text{Turbid}}{CaCO_3\downarrow}$

(*ii*) $CaCO_3 + H_2O \xrightarrow[CO_2]{\text{excess}} \underset{\text{Clear soluble}}{Ca(HCO_3)_2}$

Table 10.2 Physical Properties of Alkaline Earth Metals

Property	Beryllium Be	Magnesium Mg	Calcium Ca	Strontium Sr	Barium Ba	Radii Ra
Atomic number	4	12	20	38	56	88
Electronic configuration	[He]2s^2	[Ne]3s^2	[Ar]4s^2	[Kr]5s^2	[Xe]6s^2	[Rn]7s^2

(Contd...)

Atomic radii, pm	105	162	197	213	215	—
Ionic radii, M^{++}, pm	31	65	99	113	135	150
Ionizaton energies, kJ 1st	894	734	587	547	500	508
2nd	1748	1443	1140	1059	960	962
3rd	14784	7691	4909	4130	3423	—
E°, M^{2+}/M(V)	–1.85	–2.34	–2.87	–1.89	–2.90	–2.92
Electronegativity	1.50	1.2	1.1	1.0	0.9	0.9
Heat of atomization (at 298 K and 1 atm) kcal/mol	78.2	35.1	42.3	39.0	42.0	—
Atomic weight	9.0	24.3	40.0	87.6	137.4	226
Density of solid at 293 K, g/c.c.	1.86.	1.75	1.55	2.6	3.39	5
Melting point, K	3.43	1380	1710	1639	1790	1413
Colour of the flame	—	—	Brick	Crimson	Apple green	—
Enthalpy of hydration, $kcal/mol^{-1}$	570	460	395	355	305	—

Hard water: Hard Water contains carbonates, chlorides and sulphates as dissolved salts with soap. With sodium stearate they form calcium and magnesium stearate. Calcium and magnesium stearates form scum (insoluble).

Temporary hardness can be removed by boiling

$$2HOC_3^- \rightleftharpoons CO_3^{2-}\downarrow + CO + H_2O$$

It can also be removed by lime softening.

$$Ca(HCO_3)_2 + Ca(OH)_2 \rightleftharpoons 2CaCO_3 + 2H_2O$$

Permanent hardness cannot be removed by boiling because of the presence of $CaCl_2$ $CaSO_4$ and $MgCl_2$ and $MgSO_4$ in solution, Premanent hardness can be removed by ion-exchange methods.

Chemically permanent hardness can be removed by trisodium phosphate Na_3PO_4, sodium pyrophosphate $Na_4P_2O_7$, sodium tripolyphosphate $Na_5P_3O_{10}$ or Graham's salt $(NaPO_3)_n$ $(NaPO_3)_n$ (Calgon). It can also be removed by using large quanities of lime soda.

$$CaSO_4 + Na_2CO_3 \longrightarrow CaCO_3\downarrow + Na_2SO_4$$

Hydrides: Mg, Ca, Sr and Ba hydrides can be obtained by the direct combination with hydrogen (MH_2). BeH_2 is prepared indirectly from $BeCl_2$ and $LiAlH_4$. Pure BeH_2 can be obtained from $BeCl_2$ and $Li[BH_4]$ to give BeB_2H_8. Be BeH_8 on heating with triphenyl phosphoric PPh_3 give pure Be H_2.

$$2BeCl_2 + LiAlH_4 \longrightarrow 2BeH_2 + LiCl + AlCl_3$$

$$BeB_2H_8 + 2PPh_3 \longrightarrow BeH_2 + 2Ph_3\ PBH_2$$

$$BeCl_2 + 2Li[BH_4] \longrightarrow BeB_2H_8 + 2LiCl$$

All these hydrides are reducing agents. They hydrolyse by water and evolve H_2.

$$CaH_2 + 2H_2O \longrightarrow Ca(OH)_2 + H_2$$

CaH_2, SrH_2 and BaH_2 are ionic and contain H^- ion. BeH_2 are covalent and polymeric $(BeH_2)_n$

They from three centre bonds Be...H...Be and contain two electrons. This is called three centre two electron bond BeH_2 is a cluster compound. BeH_2 is formed with normal bonds which would result in 4 electrons in the outer shell of Be atom. By clustering each atom shares its electrons with several neighbours.

Be...H...Be. Three atoms three centre 2 electron bond. This is an example of a cluster compounds, where the monomeric molecule BeH_2 would result in only four electrons in the outer shell of Be atom. This is called electron deficient situation.

Oxides and oxo-salts: All the alkaline earth metals burn in O_2 above 600°C to form and oxides. In air they form mixture of oxides and nitrides. Mg burns with a dazzling flame in air evolving heat.

$$Mg + air \longrightarrow MgO + Mg_3N_2$$

BeO is obtained by ignition of metal while the other metal oxides are obtained by the decomposition of carbonates MCO_3. Oxosalts such as $M(NO_3)_2$, $M(OH)_2$ and MSO_4 also decompose to form oxides MO because they are less stable than Group 1 metals *e.g.*,

$$CaCO_3 \xrightarrow{\Delta} CaO + CO_2$$

These oxides are used for furnace lining because they have high melting points. CaO, SrO and BaO react with water with the evolution of heat.

$$CaO + H_2O \longrightarrow Ca(OH)_2$$

$Mg(OH)_2$ is insoluble while $Ca(OH)_2$, $Sr(OH)_2$ and $Ba(OH)_2$ are soluble in excess of OH^- ions. $Mg(OH)_2$ is weakly basic whereas other hydroxides are strong bases so much so that $Ca(OH)_2$ is called slaked lime. The predicted structure for BeO, MgO, CaO has 4 or 6 coordination number but SrO and BaO has a coordination number of 8.

BeO reacts with acids and alkalis to show amphoteric behaviour MgO dissolves in water to form weakly basic solution.

CaO and H_2O make moderately basic solution SrO and BaO make strong bases with water $CaCO_3$ has two crystalline forms (*i*) calcite and (*ii*) aragonite. Aragonite is usually prepared in the laboratory by precipitation in hot solution. It is stable and changes to calcite.

Lime is of industrial importance. In can be used

(*i*) to remove slag ($CaSiO_3$) in steel industry

(*ii*) Mixed with SiO_2 and alumina to make cement

(*iii*) for making glass

(*iv*) for lime-soda process for removing hardness.

(*v*) to prepare CaC_2

(*vi*) to prepare bleaching powder

$$Ca(OH)_2 + Cl_2 \longrightarrow CaOCl_2 + Ca(OH)_2 \cdot CaCl_2 2H_2O$$

Peroxides: Barium peroxide BaO_2 is obtained by passing air at 500°C

$$2BaO + O_2 \longrightarrow 2BaO_2$$

The peroxides are white ionic solids containing $|O — O|^{2-}$. They are salts of very weak acid, BaO_2 on treatment with acids H_2O_2 is liberated.

$$BaO_2 + 2HCl \longrightarrow BaCl_2 + H_2O_2$$

Oxy salts

Sulphates: $BeSO_4$ and $MgSO_4$ are water soluble while Ca, Sr and Ba sulphates are insoluble. Be SO_4 and $MgSO_4$ are soluble due to high enthalpy of solvation of smaller Be^{2+} and Mg^{2+} ions. $MgSO_4 \cdot 7H_2O$ is known as epsom salt is used as a laxative, $CaSO_4 . \frac{1}{2} H_2O$ is used in construction as Plaster of Paris. This is made from gyprum

$$CaSO_4.2H_2O \xrightarrow{150°C} CaSO_4.\frac{1}{2}H_2O \xrightarrow{200°C} \underset{\text{anhydrous}}{CaSO_4} \xrightarrow{1100°C} CaO + SO_3$$

Powdered plaster of Pasis $CaSO_4 . \frac{1}{2} H_2O$ when mixed with equal amount of water changes into a solid mass used for plastering, making moulds and in hospitals for broken bones. $BaSO_4$ is used for barium meal test for stomach. All these sulphates decompose on heating

$$MgSO_4 \longrightarrow MgO + SO_3$$

Heating the sulphates with C reduces them to sulphides.

$$BaSO_4 + 4C \longrightarrow 4CO + BaS$$

Alkaline earth metals form $M(ClO_4)_2$ perschlorates which are used as strong oxidising agents. Anhydrous $CaSO_4$ is used as a dehydrating agent.

Nitrates: Nitrates are prepared by the reaction of HNO_3 on metal carbonates oxides or hydroxides. Heating the nitrates give solid oxides and not anhydrous nitrates. Anhydrous nitrates can be obtained by using liquid N_2O_4 and ethyl acetate.

$$BeCl_2 + N_2O_4 \text{ (liq.)} \longrightarrow Be(NO_3)_2.2N_2O_4$$

$$\Big\downarrow \text{Warm to 50°C in vacuum}$$

$$\underset{\text{Basic beryllium nitrate}}{[BeO(NO_3)_6]} \xleftarrow{125°C} Be(NO_3)_2$$

Halides: Metal halides MX_2 can be prepoared by:

(*i*) Direct combination of metal and halogen at high temperature

$$M + Cl_2 \longrightarrow MCl_2$$

$$Ca + Cl_2 \longrightarrow CaCl_2$$

(*ii*) Metal and halogen acid or metal carbonate and halogen acid

$$M + 2HCl \longrightarrow MCl_2 + H_2$$

$$MCO_3 + 2HCl \longrightarrow MCl_2 + H_2O + CO_2$$

$BeCl_2$ is covalent, hygroscopic and fumes in air due to hydrolysis, Anhydrous $BeCl_2$ cannot be obtained when made in aqueous solution as hydrated ion is formed $[Be(H_2O_4)]^{2+}$ erg $[Be(H_2O)_4]Cl_2$ or $[Be(H_2O)_4]F_2$ $BeCl_2$ does not conduct electricity.

The anhydrous halides can be prepared by the following reactions:

(*i*) $BeO + 2NH_3 + 4HF \longrightarrow (NH_4)_2[BeF_4] \xrightarrow{\Delta} BeF_2 + 2NH_4F$

(*ii*) $BeO + C + Cl_2 \xrightarrow{700°C} BeCl_2 + CO$

(*iii*) $2BeO + CCl_4 \xrightarrow{800°C} 2BeCl_2 + CO_2$

Anhydrous $BeCl_2$ are polymeric Like BeH_2, $BeCl_2$ show cluster involving three centre bonding. Instead of H-bridging they show chlorine bridging $BeCl_2$.

(*i*) Cl—Be—Cl
Monomer

(*ii*) 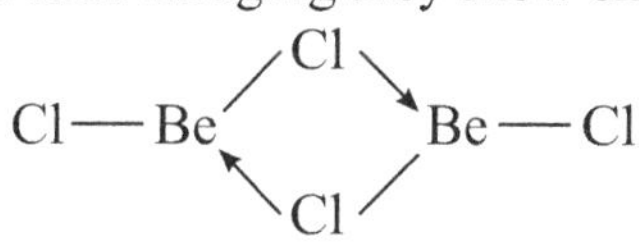

Dimer

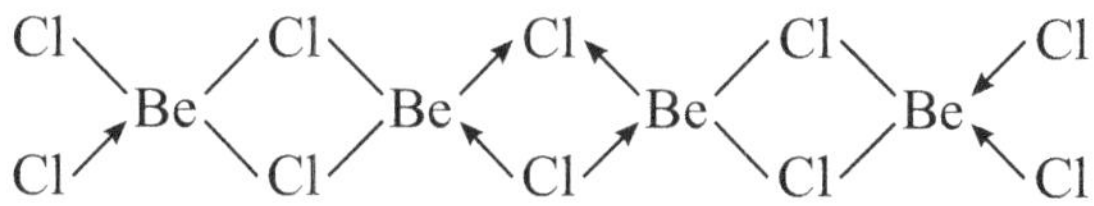

Polymeric

Fluorides of Ca and other higher alkaline earth metals are ionic, insoluble in H_2O having high melting points.

$$CaF_2 + H_2SO_4 \longrightarrow CaSO_4 + 2HF$$

$$HF + KF \rightarrow KHF_2 \xrightarrow{\text{electrolysis}} F_2$$

Chlorides, bromides and iodides have much lower melting points and are readily soluble in water. Halides also form hydrates and therefore they are hygroscopic, $MgCl_2$ is electrolyzed to obtain Mg.

Nitrides and Carbides

Nitrides: Alkaline earth metals burn in dinitrogen to form ionic nitrides M_3N_2.

$$3Ca + N_2 \longrightarrow Ca_3N_2$$

N_2 is very stable therefore lot of energy is required to convert N_2 to N^{3-} ion. This energy comes from lattice energy of the crystalline solids. The nitrides react with water to liberate NH_3.

$$Ca_3N_2 + 6H_2O \longrightarrow 3Ca(OH)_2 + 2NH_3$$

Carbides: The carbide of beryllium is prepared from BeO heated with carbon at 1500°C to 2000°C

$$2BeO + 3C \xrightarrow{1500-2000°\,C} Be_2C + 2CO$$

Be_2C is an ionic red coloured carbide. It easily reacts with H_2O to form methane.

$$Be_2C + 4H_2O \longrightarrow 2Be(OH) + CH_4$$

other group 2 metals form carbides with the formula MC_2 *e.g.,*

$$Ca + 2C \xrightarrow{1100°C} CaC_2$$

or

$$CaO + 3C \xrightarrow{1200°C} CaC_2 + CO$$

MC_2 formula carbides have NaCl type structure.

calcium carbide reacts with water liberating acetylene (ethyne)

$$CaC_2 + 2HO \longrightarrow Ca(OH)_2 + C_2H_2$$

CaC_2 when heated in air at 100°C forms calcium cyanamide

$$CaC_2 + N_2 \longrightarrow \underset{\text{calcium cyanamide}}{CaNCN} + \underset{\text{graphite}}{C}$$

Calcium cynamide reats with water to form NH_4OH

$$CaNCN + 5H_2O \longrightarrow CaCO_3 + 2NH_4OH$$

Organometallic Compounds: Victor Grignard a french chemist was awarded the noble prize for chemistry in 1912 to prepare organomagnesium compound presently called Grignard reagents, Grignard reagent is a very useful compound to manufacture a variety of alochols, aldehydes, ketones, carboxylic acids, esters, amides and alkenes. Grignard reagents are prepared by a slow addition of alkyl or aryl halides (Cl, Br or I) to a stirred mixture of magnesium turnings in dry ether ($C_2H_5OC_2H_5$) *i.e.,* ether with no moisture or water. A trace of iodine addition hastens the start of the reaction.

$$Mg + RBr \xrightarrow[\text{ether}]{\text{Dry}} \underset{\text{Grignard reagent}}{RMgBr}$$

Alkyl Grignard reagent is more reactive than aryl Grignard reagent. All Grignard reagents are quickly hydrolysed by water. Grignard reagents are not stored. They are prepared when required. Some of their common reactions are mentioned below:

$$RMgBr + CO_2^{+} \xrightarrow{\text{acid}} RCOOH$$

$$RMgBr + R_2CO \xrightarrow{+H_2O} R_3COH \text{ (Tertiary alcohol)}$$

$$RMgBr + RCHO \xrightarrow{+H_2O} R_2CHOH \text{ (Secondary alcohol)}$$

$$RMgBr + O_2 \xrightarrow{+\text{acid}} ROH \text{ (Primary alcohol)}$$

$$RMgBr + S_2 \longrightarrow RSH + R_2S$$

$$RMgBr + HCHO \xrightarrow{+\text{acid}} RCH_2OH$$

$$RMgBr + I_2 \longrightarrow RI$$

$$RMgBr + H^- \longrightarrow RH$$

$$RMgBr + BeCl_2 \longrightarrow BeR_2$$

$$RMgBr + Li\,R \longrightarrow MgBr_2$$

$$RMgBr + BCl_3 \longrightarrow BR_3$$

$$RMgBr + SiCl_4 \longrightarrow RSiCl_3, R_2SiCl_2R_3SiCl + R_4Si$$

Complexes: Complexes of group 2 metals are well known. Magnesium forms very stable complexes with oxygen-donor ligands eg $[Be(C_2O_4)_2]^{2-}$, $[Be_4O(CH_3COO)_6]$, chlorophyll and [Mg edta]

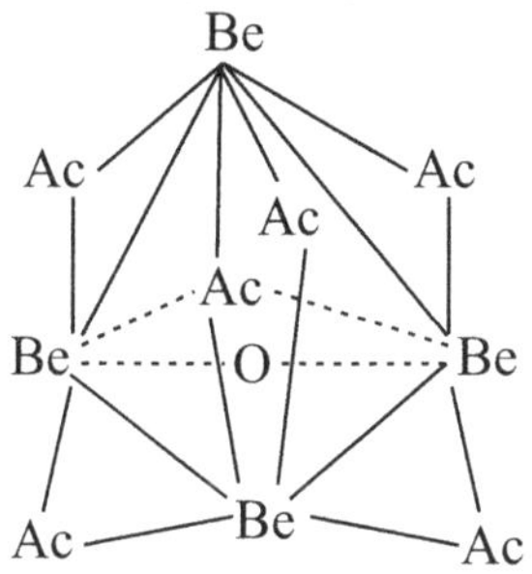

$[Be_4O\,(CH_3COO)_6]$

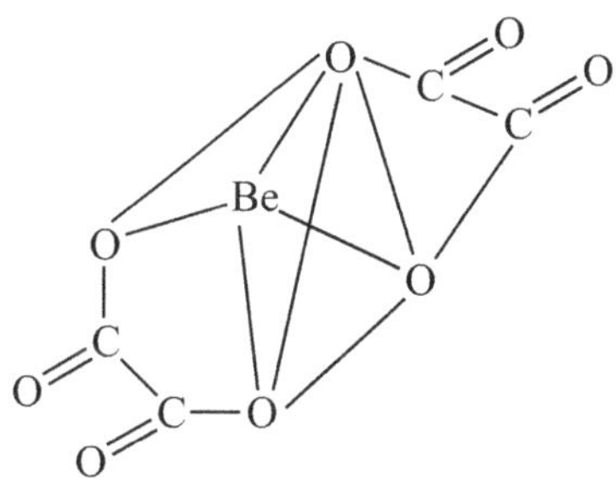

$[Be_4(OX)_2]^{2-}$

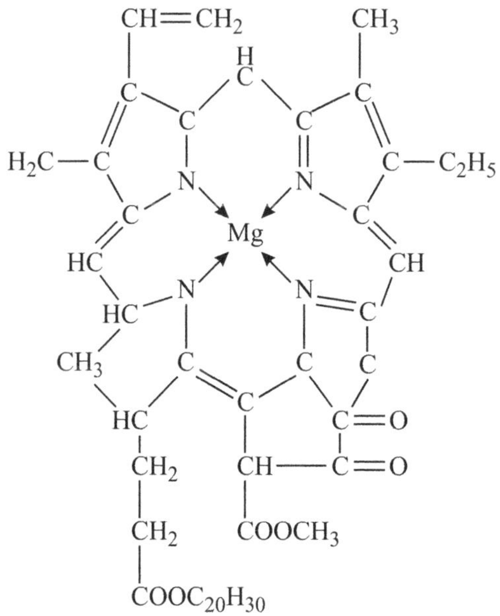

Chlorophyll

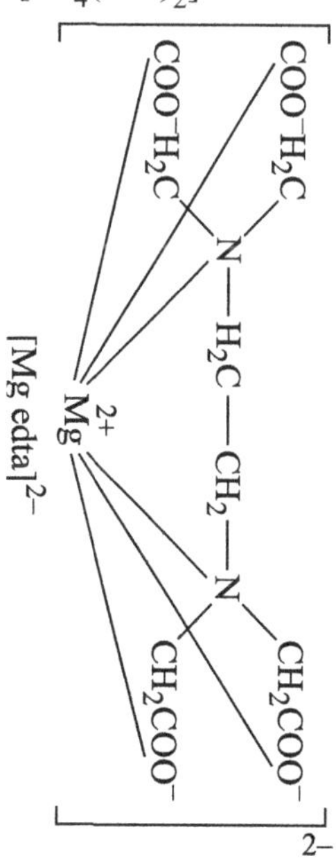

Differences of Be with other group 2 elements: Due to exceptionally small size and high charge density it shows unique properties compared to other elements in the group and its similarity in properties with diagonal opposite aluminium.

Uniqueness and Diagonal relationship to Al

1. Be has a strong tendency to covalency. Its compounds have low melting points *e.g.*,

$$BeF_2 — 800°C \text{ m.pt.}$$

$$\text{other group 2 elements} \longrightarrow 1300°C \text{ m.pt.}$$

2. Be halides are covalent, soluble in organic solvents, hydrolyse in water like Al halides.
3. Be hydride is electron deficient and polymeric like Al hydride.
4. Halides of Be are electron deficient polymeric with halogen bridges, $BeCl_2$ forms chains like $AlCl_3$ being dimeric.
5. Be forms large number of comlexes while groups 1 and 2 form fewer complexes.
6. $Be(OH)_2$ and $Al(OH)_3$ are amphoteric.
7. Be and Al are made passive by nitric acid
8. E° value of Be = – 1.85 V

   Al = – 1.66 V

   are closer than E° value for

   Ca = – 2.87 V

   Sr = – 2.89 V

   Ba = – 2.90 V
9. Be_2C and Al_4C_3 are unusual. Both form CH_4 on hydrolysis.
10. Be salts have high solubility in water and most rapidly hydrolysed.

    Diagonal relation of Li and Mg, and Be and Al are based on similarity in atomic sizes and charge densities.

EXERCISES

1. Why the size of alkaline earth elements smaller than alkali metals.
2. What is the chemistry of extraction of the following metals:

   (*a*) Beryllium (*b*) Magnesium (*c*) Calcium
3. Compare the solubilities of the following salts of the alkaline earth salts

   (*a*) Sulphates (*b*) Hydroxides (*c*) Iodates

   Give reasons.

4. Compare the following properties of the alkaline earth metals and the alkali metals.
   (*a*) Ionic radii (*b*) Density (*c*) Reducing property
   (*d*) Melting and boiling points (*e*) Flame spectra
   (*f*) Ionization energies.
5. What is the structure of $BeCl_2$ in the gaseous state and solid state why is $BeCl_2$ acidic when dissolved in water.
6. Explain the difference in the structure between BeH_2 and CaH_2.
7. Be shows more covalency than the other elements in the group. Comment.
8. Why are halides and hydrides of Be polymerize.
9. What are the common coordination number of Be^{2+} and Mg^{2+}? Why do they differ?
10. Be salts seldom contain more than four molecules of water of crystallization. Explain.
11. Give equation of reaction between Ca and:
    (*a*) H_2O (*b*) H_2 (*c*) C (*d*) N_2 (*e*) O_2 (*f*) Cl_2 (*g*) NH_3
12. Give reasons why:
    (*i*) NaOH s stronger base than $Ba(OH)_2$?
    (*ii*) K is more reactive than Ca?
13. What causes hardness in water? What is temporary or permanent hardness?
14. How do we remove hardness in water? Explain the chemical and instrumental methods used to remove hardness.
15. Ca and Mg form stable complexes with edta. What indicator is used? Why is the titration performed by using alkaline buffer?
16. Describe the preparation, properties and structure of basic beryllium acetate.
17. On treatment with cold water an element (A) reacted quietly liberating a colourless, odorless gas (B) and a solution (C). Lithium reacted with (B) yielding solid (D) which effervesced with water to give a strongly basic solution (F) when CO_2 was bubbled through solution (C) an initial white precipitate (G) was formed but this resolved forming solution (H) when more CO_2 was bubbled precipitate (G) effervesced when HCl was added and gave a deep red colour to the Bunsen burner, when (G) was heated with C at 1000°C a caustic white compound (I) was formed which when heated carbon at 1000°C gave a solid (J) of commercial importance. Name the substances A to (J) and give balanced equations for each reaction.

11

Group 13 Elements or Boron Family of Elements

General Discussion

The elements that refer to the outermost *s* and *p*-orbitals are called main group elements. These elements are considered both along horizontal row called a 'period' and along vertical column called a 'group'. Along the period we come across their graded properties and along a group we see the similarities in their properties. In these elements we shall review the periodic variations in their properties both along a period and along a group. In the chapter of *p*-block elements we shall discuss both physical and chemical properties of these elements and their periodic variations.

Boron (B), aluminium (Al), gallium (Ga), Indium (In), and thallium (Tl) are the members of group 13 elements, their electronic configuration and oxidation states are Table 11.1 below:

Table 11.1 Electronic Structures and Oxidation States

Element	Symbol	Electronic configuration		Oxidation states*	
Boron	B	[He]	$2s^2\ 2p^1$		III
Aluminium	Al	[Ne]	$3s^2\ 3p^1$	(I)	III
Gallium	Ga	[Ar]	$3d^{10}\ 4s^2\ p^1$	I	III
Indium	In	[Kr]	$4d^{10}\ 5s^2\ 5p^1$	I	III
Thallium	Tl	[Xe]	$4f^{14}\ 5d^{10}\ 6s^2\ 6p^1$	I	III

* The most stable and common oxidation state of these elements is shown in the bold numerals while other oxidation states are shown in the normal type.

The unstable oxidation state are given in the bracket ( ). Their general configuration is given by $ns^2\ np^1$. The first member boron is a non-metal and forms covalent compounds with sp^2 hybridization. All BX_3 compounds are electron deficient therefore they act as lewis acids.

Al, Ga, In, Tl form trivalent compounds, element Tl shows inert pair effect and forms univalent compounds. These compounds are borderline between ionic and covalent.

Occurrence

Boron is a less abundant element. It is present in large deposits as borax Na_2 $[B_4O_5(OH_4)].8H_2O$ and kernite Na_2 $[B_4O_5(OH_4)]$ $2H_2O$. The deposits are found in california in the desert region. The largest producers. Russia 53% and USA and Turkey 19% each. In the group Al is the most abundant element. Aluminium metal is produced on a vast scale. It is a commercially important metal. The ore bauxite $Al_2O.H_2O$ and $Al_2O_3.3H_2O$. It also occurs in rocks fluidizer in the form of alumina silicate. Gallium is twice as abundant as boron. Indium and Thallium are not common. Ga, In and Tl occur as sulphides.

Extraction: Boron

Boron of low purity can be obtained from B_2O_3 by reduction with Mg or Na at high temperature, it generally contains impurities of metal bodies and is black in colour.

$$Na_2[B_4O_5(OH)_4]\cdot 8H_2O \xrightarrow{\text{acid}} \underset{\text{orthoboric acid}}{H_3BO_3} \xrightarrow{\text{heat}} B_2O_3$$

$$B_2O_3 + 3Mg \xrightarrow{\Delta} MgO + 2B$$

In is difficult to obtain boron in crystalline form. It has very high melting point of 2180°C.

Methods of preparing crystalline boron are:

(*i*) Reducing BCl_3 with H_2

$$2BCl_3 + 3H_2 \xrightarrow[\text{Ta}]{\text{W}} 6HCl + 2B$$

(*ii*) Cracking of BI_3.

$$2BI_3 \xrightarrow[\text{Ta}]{\text{W}} 2B + 3I_2$$

(*iii*) Thermal decomposition of diborone

$$B_2H_6 \xrightarrow{\text{heat}} 2B + 3H_2$$

Aluminium

Baeyer's Process: Aluminium is obtained from the ore bauxite $Al_2O_3 \,.\, 3H_2O$, NaOH is added to the ore. Al dissolves to form sodium aluminate $NaAlO_2$. Next aluminium hydroxide is obtained from alkaline aluminate. This can be done by bubbling CO_2 that lowers the pH or seeding with Al_2O_3 solution. $Al(OH)_3$ is heated strongly which converts it to Al_2O_3. Finally Al_2O_3 is reduced by carbon or some reducing metals.

$$Al_2O_3 + 3Mg \longrightarrow 3MgO + 2Al$$

$$Al(OH)_3 + 3NaOH + 6HF \longrightarrow Na_3[AlF_6] + 6H_2O$$

Cryolite improves the conductivity of the cell because Al_2O_3 is poor conductor.

Electrolysis

$$Na_3(AlF_6) \longrightarrow 3Na^+ + [AlF_6]^{3-}$$

$$2Al_2O_3 \rightleftharpoons 4Al^{3+} + 6O^{2-}$$

Anode:

$$6O^{2-} - 12e \longrightarrow 3O_2$$

Cathode:

$$4Al^{3+} + 12e \longrightarrow 4Al^0$$

Therefore aluminium metal gets collected at cathode.

Extraction of Ga, In and Tl

Small amount of Ga is found in bauxite in the ratio of 1 part Ga in 5000 parts Al. First of all Al_2O_3 is purified in the process. The concentration of Ga in the alkaline solution increase to $\underset{Ga}{1} : \underset{Al}{250}$, Ga is finally extracted by electrolysis of the solution. In and Tl are present in minute quantities in ZnS and PbS ores.

In and Tl are extracted from these ores by the usual method of extraction from sulphide ores (roasting and smelting).

These metals are soft, silver coloured and reactive. They are used in transistors. There are no large scale use of these metals.

Uses of Boron and Aluminium Boron

1. It is used to make steel.
2. Boron carbide is used in nuclear reaction to capture neutrons.
3. Borax and orthoboric acid are used in making fibre glass.
4. Borax is used as a flame retarder in textiles.
5. Borax is used as a flux, reacts with metal oxides and restores the shine of the metal.
6. Borax is used in making enamel and in leather tanning.
7. Boric acid is a constituent of washing powders and detergents.
8. B_2O_3 is used in heat resistant borosilicate.
9. Pgrex glass contains B_2O_3.

10. Peroxo borates absorb UV light and emit visible light (brighteners).

Uses of Aluminium

1. Aluminium being light weight is used in making air crafts, cars and ships.
2. In buildings it is used in doors, windows and panels.
3. Al is used in making cans and tooth paste tubes.
4. Al is used in cooking utensits.
5. Used in electric power cables. They conduct twice as strong as copper.
6. Fine powder of aluminium is used in paint.

Physical Properties

1. Oxidation States

Group 13 elements contain three outer electrons $ns^2\ np^1$ in their configuration. They usually form three bonds exhibiting an oxidation state of (+ 111). The bonds may be ionic or covalent, bonds suggested are because of:

(*i*) Small size of ions and high charge of +3 suggests covalent bond (Fajan's rules)

(*ii*) Third ionization energy is quite high

(*iii*) Electornegativity values are quite high. The electro negativity difference between the elements in compounds is not large,

Boron, due to smaller size in comparison to other elements is almost always covalent, $AlCl_3$ and $GaCl_3$ are covalent in anhydrous state. The type of bonds formed depends upon the energy considerations. If the hydration energy exceeds the ionization energy *e.g.*,

In $AlCl_3$, 5137 kJ mol^{-1} energy is required to convert Al to Al^{3+}

ΔH hydration for Al^{3+} is –4665 kJ mol^{-1}

ΔH hydration for Cl^- is –381 kJ mol^{-1}

Total hydration energy is

$$-4665 + (3x - 381) = -5805 \text{ kJ mol}^{-1}$$

Therefore hydration energy is more than the ionization energy so $AlCl_3$ ionizes in solution.

(+1) Oxidation State

The electronic configuration of group 13 elements is represented by $ns^2\ np^1$. As we descend the group for example in Ga, In or Tl (+1) state becomes more stable. In Ga and In (+1) state is less stable than the (+111) state whereas in Tl (+1) compounds are more stable than the (+111) compounds. In all such compounds the paired *s*-electrons do not partcipate in bonding. This is called inert fair effect. Tl(+1) are thallous compounds and Tl + 111) are thallic compounds. This happens owing to the bond energy *i.e.,* the energy to form bonds with *s*-electrons is not available (+1) state ions are larger and form ionic compounds. This tendency is not limited to group 13 elements. It occurs in group 14, group 15 and other elements.

(+11) Compounds are very limited Gallium is (+11) divalent. It is derived from $Ga^+[GaCl_4]^-$ where Ga is (+1) and (–111).

Crystal structures and melting and boiling points: Melting and boiling points of group 13 elements do not show a regular trend. Boron and gallium have uncommon crystal structures. Boron has very high melting point, boron forms electron deficient molecules. Other elements exhibit metallic bonding to solve deficiency of electrons. Boron has small size and high ionization energy, therefore the problem of insufficient electrons remains unsolved.

Boron exists in four allotropic forms and each one forms icosahedral unit (12 corners and 20 faces). Al, In and Tl have close-packed metal structures. Ga has an unusual structure. Gallium in the form of liquid expands to form the less dense solid this is a unique property. Due to such a wide difference in the structures of Group 13 elements their melting and boiling points show irregular trend as illustrated in the Table 11.2.

Table 11.2 Melting and Boiling Points

	Melting point (°C)	Boiling point (°C)
B	2180	3650
Al	660	2467
Ga	30	2403
In	157	2080
Tl	303	1457

Size of Atoms and Ions

These is no steady increase in the metallic radii from boron to thallium.

Table 11.3 Ionic and Covalent Radii and Electronegativity Values

	Metallic radius (Å)	Ionic radius		Pauling's electronegativity
		M^{3+} (Å)	M^+ (Å)	
B	(0.885)	(0.27)	–	2.0
Al	1.43	0.535	–	1.5
Ga	(1.225)	0.620	1.20	1.6
In	1.67	0.800	1.40	1.7
Tl	1.70	0.885	1.50	1.8

Boron is not a metal and the radius is half of the closest-packed structure. Ga being a liquid has an unusual structure. Its value is half of the closest-packed. The remaining have the closert-packed metal structures. The ionic radii for M^{3+} increases on descending the group. The increase in radii is not as regular as in Group 1 and Group 2.

The possible explanation to this is:

1. These is no proof of existence of B^{3+}. It is estimated only from the compounds of boron which exist under normal condition.
2. Ga, In and Tl come after the row of ten transition elements *i.e.,* they have 10 *d*-electrons. The shielding effect *d*-electrons is poor in comparison to *s* and *p* electrons. Due to this outer electrons are pulled firmly by the nucleus. Therefore the expected increase in size on descending the group is not observed. In the case of Tl the effect is felt even more because of the presence of 14*f* electrons which shield the nuclear charge even to a lesser extent. Hence the contraction in size is noticed. This is called lanthanide contraction.

Table 11.4 Ionic and Covalent Radii and Electronegativity Values

	Metallic radius Å	Ionic radius		Pauling's electronegativity
		M^{3+} (Å)	M^{+} (Å)	
B	(0.885)	(0.27)	–	2.0
Al	1.43	0.535	–	1.5
Ga	(1.225)	0.620	1.20	1.6
In	1.67	0.800	1.40	1.7
Tl	1.70	0.885	1.50	1.8

Large difference in the size of B and Al is seen, hence the great difference in the properties of the two. Boron forms acidic oxide Al forms amphoteric oxide. Al is a metal while B is a non-metal.

Some standard electrode potential values have been tabulated ($E°$) which give an idea to the extent of reduction that can take place.

Electropositive Nature

The metallic nature increases from B to Al and then falls from Al to Tl. This is amply proved by the $E°$ value for the reaction

$$M^{3+} + 3e \longrightarrow M°$$

This trend is further related to the increase in size on going down the group Ga, In and Tl do not follow the trend. These elements are less prone to loose electrons due to poor shelding by *d* and *f*-electrons.

The $E°$ value becomes less negative from Al $\rightarrow$ Ga $\rightarrow$ In whereas it becomes positive in Tl.

In the reaction

$$Al^{3+} + 3e \longrightarrow Al^{\circ}$$

ΔG is positive ($\Delta G = -n\text{FE}^{\circ}$)

Therefore reaction is difficult.

$$Al \longrightarrow Al^{3+} + 3e$$

Table 11.5 Standard Reduction Potentials (volts)

Oxidation states: Acid solution (+III, +II, +I, 0)	Oxidation states: Basic solution (+III, +II, +I, 0)
$Al^{3+} \xrightarrow{-1.66} Al$	$Al(OH)_3 \xrightarrow{-2.31} Al$
$Ga^{3+} \xrightarrow{-0.44} Ga^{+} \xrightarrow{-0.79} Ga$ (overall $Ga^{3+} \to Ga$: −0.56)	$H_2GaO_3^{-} \xrightarrow{-1.22} Ga$
$In^{3+} \xrightarrow{-0.44} In^{+} \xrightarrow{-0.18} In$ (overall $In^{3+} \to In$: −0.34)	$In(OH)_3 \xrightarrow{-1.0} In$
$Tl^{3+} \xrightarrow{+2.06} Tl^{+} \xrightarrow{-0.34} Tl$ (overall $Tl^{3+} \to Tl$: +1.26)	$Tl(OH)_3 \xrightarrow{-0.05} Tl(OH)$

This reaction is spontaneous because E° value is less negative. Down the group E° becomes still less negative therefore on going down the group it becomes less difficult for

$$M^{3+} + 3e \longrightarrow M^{\circ}$$

This shows that (+111) oxidation state becomes less stable in aqueous solution on descending the group. In the case of Tl the (+1) state is more stable than the (+111) state.

$$M^{+} + e \longrightarrow M^{\circ}$$

Table 11.6 Standard Electrode Potentials E°

	M^{3+} | M (volts)	M^{+} | M (volts)
B	(–0.87*)	
Al	–1.66	+0.55
Ga	–0.56	–0.79**
In	–0.34	–0.18
Tl	+1.26	–0.34

* For $H_3BO_3 + 3H^{+} + 3e^{-} \longrightarrow B + 3H_2O$)

** Value in acidic solution.

Ionizations Energy

The ionization energies of group 13 elements are quite high. The sum of first, second and third ionisation energies as given in the table below are very high, Boron has very little tendency to form ions rather it forms covalent bonds, other elements in the group also form covalent compounds except in solution. Ionization energy values decrease from B to Al in the normal trend. On going down the group decrease in the values is related to the increase in size. The poor shielding effect of *d*-electrons affect the values of later elements.

Table 11.7 Ionization Energies

	Ionization energy (kJ mol^{-1})			
	1st	**2nd**	**3rd**	**Sum of three**
B	801	2427	3659	6887
Al	577	1816	2744	5137
Ga	579	1979	2962	5520
In	558	1820	2704	5082
Tl	589	1971	2877	5437

Chemical Properties

Pure crystalline boron is quite unreactive. Finely divided amorphous boron reacts with strong oxidizing agents at high temperature. It burns in dioxygen and dinitrogen at white heat. It reacts with halogens to form trihalides. It reacts with oxidizing agents like H_2SO_4 and HNO_3 liberating slow H_2. It also liberates H_2, from fused NaOH.

Some Reactions of Boron

1. $4B + 3O_2 \xrightarrow[\text{temp.}]{\text{high}} 2B_2O_3$

2. $2B + 3S \xrightarrow{1200°C} B_2S_3$

3. $2B + N_2 \xrightarrow{\Delta} 2BN$

4. $2B + 3Cl_2 \xrightarrow[\text{Temp.}]{\text{H}} 2BCl_3$

   with other halogens BF_3, BBr_3, BI_3

5. $2B + 6\,NaOH \xrightarrow{\text{Fused}} 2Na_3\,BO_3 + 3H_2$

6. $2B + 2NH_3 \xrightarrow[\text{H Temp.}]{\text{Very}} 2BN + 3H_2$

7. With metals it forms borides $xM + yB \longrightarrow Mx\,By$

Reactions of other elements in the group: Al, Ga, In and Tl are silvery white. Al is stable in both water and air.

1. $4M + 3O_2 \xrightarrow[\text{Temp.}]{\text{V. high}} 2M_2O_3$

   Tl forms Tl_2O as well.

2. $2Al + N_2 \xrightarrow[\text{Temp.}]{\text{V.H.}} 2AlN$

3. $2M + 3F_2 \xrightarrow[\text{Temp.}]{\text{H}} 2MF_3$

   $2M + 3Cl_2 \xrightarrow[\text{Temp.}]{\text{H}} 2MCl_3$

   $2M + 3Br_2 \xrightarrow[\text{Temp.}]{\text{H}} 2MBr_3$

   $2M + 3I_2 \xrightarrow[\text{Temp.}]{\text{H}} 2MI_3$

   Thallous forms $TlPI_3$

4. **With acids:** All the group metals react with dilute mineral acids. Al is passivated with conc. HNO_3.

   $2M + 6HCl \longrightarrow 2MCl_3 + 3H_2$

5. **With Alkali:** Only Al and Ga react.

   $Al + NaOH + H_2O \longrightarrow NaAlO_2 + H_2$

   or

   $2Na^+[Al(OH)_4]^- + H_2$

6. **With ammonia**

   $M + NH_3 \longrightarrow \underset{\text{metal amides}}{MNH_2} + \frac{1}{2}H_2$

Thermite Process

Aluminium may be used in alumina thermite process for the extraction of Mn.

$$8Al + 3Mn_3O_4 \longrightarrow 4Al_2O_3 + 9Mn$$

Alums

Aluminium ions may separate out from the aqueous solutions as double salts. They are called aluminium alums or alums. They are represented by the general formula $[M'(H_2O)_6]\,[Al(H_2O)_6]\,(SO_4)_2$. Some of the important alums are potash alum $[K(H_2O)_6]\,[Al(H_2O)_6]\,(SO_4)_2$, ferric alum and chrome alum.

Cement

Aluminium compounds such as tricalcium aluminate is an important constituent of portland cement. $Ca_9[Al_6O_{18}]$. Cement is manufactured by mixture of $CaCO_3$, SiO_2 (sand) and alumna silicate at 1450 – 1600°C in a rotary kiln. High alumina cement contains 40% Al_2O_3.

Ga, In and Tl are not attacked by air and water. Tl is more reactive and is oxidized by air. All these metals dissolve in dilute acids liberating hydrogen. Ga and Al are amphoteric Ga dissolves in NaOH forming gallates and liberating H_2. Oxides and hydroxides of Al and Ga are amphoteric. Oxides and hydroxides of In and Tl are basic. All of them react with halogens.

Properties of Tl(1)

Tl(1) or thallous compounds are colourless. They are highly poisonous. Tl^+ ion is stabler than Tl (111). Ionic radius of Tl^+ is 1.5Å. Tl^+ resembles with alkali metal ions. TlOH is yellow, on heating to 100°C it turns black (Tl_2O) TlF is water soluble while halides of Tl^+ are insoluble. TlCl is sensitive to sunlight. TlCl darkens on exposure to air and moisture like AgCl.

Compounds of Boron

Boron Hydrides

Not a single element in group 13 reacts directly with hydrogen but lot of boron hydrides, about 20 are known. These hydrides are volatile and inflammable. Stock and his collaborators prepared these volatile hydrides under conditions of vacuum between 1912 to 1936. The area of boron hydrides was developed in nineteen sixteen. Several of boron hydrides and isomers were established during this period. Boron hydrides are called boranes. They are similar to carbon hydrides or hydrocarbons. They quickly react with water and oxygen BH_3 is similar to BF_3 does not exist because hydrogen lacks electrons needed to form B ← H back bonding to account for incomplete octet of boron. In other words BH_3 dimerises immediately to from diborane B_2H_6.

The general formula for boron hydrides is $B_nH(n + 4)$ and $B_nH(n + 6)$. Out of two the later is less stable. The most important of boron hydrides is diborane B_2H_6.

Preparation

1. Diborane can be easily prepared on a laboratory scale by the reaction

$$2Na\,BH_4 + I_2 \xrightarrow{\text{diglyme}} B_2H_6 + 2NaI + H_2$$

2. By the action of ortho phosphoric acid on magnesium boride

$$Mg_3B_2 + H_3PO_4 \longrightarrow \text{Mixture of boranes} \xrightarrow{\Delta} B_2H_6$$

3. Reduction of boron trihalides:
   (*i*) $8BF_3 + 6LiH \longrightarrow 6LiBF_4 + B_2H_6$
   (*ii*) $4BCl_3 + 3LiAlH_4 \longrightarrow 3LiCl + 3AlCl_3 + 2B_2H_6$
   (*iii*) $2BCl_3 + 6H_2 \longrightarrow 6HCl + B_2H_6$
   (*iv*) $4BF_3 + 3NaBH_4 \longrightarrow 3NaBF_4 + 2B_2H_6$

Properties: Diborane is a colourless gas b.pt. –92.6° and m.p. –164.9°C with a foul smell.

1. **Thermal Stability:** On heating at 100°C and above it decomposes to give higher boranes

$$B_2H_6 \xrightarrow{\text{heat}} B_4H_{10}, B_5H_9, B_5H_{11} \text{ and } B_6H_{10} \text{ etc.}$$

2. **Action with oxygen:** It burns in air liberating heat or energy.

$$B_2H_6 + 3O_2 \longrightarrow B_2O_3 + 3H_2O$$
$$\Delta H = -2165 \text{ kJ mol}^{-1}$$

3. **Hydrolysis:** It is quickly hydrolysed by water to make available hydrogen. Therefore it is used as a reducing agent.

$$B_2H_6 + 6H_2O \longrightarrow 2B(OH)_3 + 6H_2\uparrow$$

   Similar reaction with CH_3OH.

4. **Reaction with halogens:** Different conditions yield different products.

$$B_2H_6 + Cl_2 \xrightarrow{< 25°C} B_2H_5Cl + HCl$$

$$B_2H_6 + Br_2 \xrightarrow[\text{at } 100°C]{\text{rapid}} B_2H_5Br + HBr$$

5. **With Halogen acids HX**

$$B_2H_6 + HCl \longrightarrow B_2H_5Cl + H_2$$
$$B_2H_6 + HI \longrightarrow B_2H_5I + H_2$$

6. **With Ammonia**

   (*a*) $B_2H_6 \xrightarrow[\text{low temp.}]{\text{excess } NH_3} B_2H_6.2NH_3$
   Animaniate

   or $[(NH_3)_2BH_2]^+[BH_4]^-$

   (*b*) $B_2H_6 \xrightarrow[\text{H. temp.}]{\text{excess } NH_3} (BN)_x$
   boron nitride

   Boron nitride is a white crystalline solid. It has a graphite like structure *i.e.*, boron and nitrogen alternate in the ring structure with B – N bond length of 145 pm. The bonds are sp^2 hybridized in case of boron and nitrogen atoms, the left out electrons are used to form *pi*-bonds. It gives layer structure, boron atoms in one layer are over nitrogen atoms in the adjacent layer.

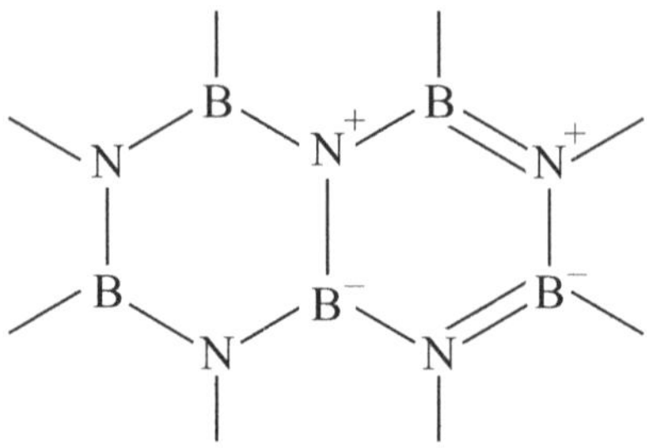

(c) $3B_2H_6 + 6NH_3 \longrightarrow B_3N_3H_6 + 12H_2$

Borazole
or
Borazine

Borozine is called inorganic benzene.

7. With Alkalie

$$B_2H_6 + 2KOH + 2H_2O \longrightarrow 2KBO_2 + 6H_2$$

Potassium metaborate

8. Action with metals: Reacts with electropositive Na, K or Ca to give amalgams

$$2B_2H_6 + 2Na \longrightarrow NaBH_4 + NaB_3H_8$$

B_3H_8 is a polyborate anion. It is also obtained by:

$$B_2H_6 + NaBH_4 \xrightarrow[100°C]{\text{diglyme}} NaB_3H_8 + 2H_2$$

9. Action with Amines: Tertiary amines give amine-borane complex

$$B_2H_6(g) + 2R_3N(g) \longrightarrow 2BH_3—NR_3(l)$$

10. With alkali hydrides: It forms a complex hydride.

$$2NaH + B_2H_6 \longrightarrow 2Na^+[BH_4]^-$$

sodium borohydride

11. Reaction with CO

$$B_2H_6 + 2CO \xrightarrow[100°C]{\text{20 atm}} 2BH_3CO$$

12. Action with boron halides

$$BCl_3 + B_2H_6 \longrightarrow B_2H_5Cl + BHCl$$

$$B\,Br_3 + B_2H_6 \longrightarrow B_2H_5\,Br + BH\,Br_2$$

13. With ether and thio ethers

$$B_2H_6 + 2(CH_3)_2O \longrightarrow 2BH_3 \longleftarrow O(CH_3)_2$$

$$B_2H_6 + 2(CH_3)_2S \longrightarrow 2BH_3 \longleftarrow S(C_2H_5)_2$$

14. Pyridine

$$2B_2H_6 + 2C_5H_5N \longrightarrow 2H_3B^- \longleftarrow N^+C_5H_5$$

15. Action with trimethyl aluminium

$$2B_2H_6 + (CH_3)_3Al \longrightarrow B(CH_3)_3 + \underset{\text{aluminium borohydride}}{Al(BH_4)_3}$$

16. Reaction with alcohols

$$B_2H_6 + 6CH_3OH \longrightarrow 2B(OCH_3)_3 + 6H_2$$

$$B_2H_6 + 4CH_3OH \longrightarrow 2BH(OCH_3)_2 + 4H_2$$

17. Tetraalkyl lead

$$B_2H_6 + (CH_3)_4Pb \longrightarrow (CH_3)_2PbH + B_2H_5\,(CH_3)$$

18. With dialkyl phosphine

$$2B_2H_6 + 4(CH_3)_2PH \longrightarrow \underset{\text{Cyclo Tetramer}}{[(CH_3)_2PBH_2]_4} + 4H_2$$

$$3/2\ B_2H_6 + 3(CH_3)_2PH \longrightarrow \underset{\text{Cyclo trimer}}{[(CH_3)_2PBH_2]_3} + 3H_2$$

19. Addition of alkenes (Hydroboration) diboranes adds on to alkenes and alkynes in ether solvent at room temperature. This is called hydroboration this forms organ boranes.

$$3RCH{=}CH_2 + \frac{1}{2}B_2H_6 \xrightarrow{RT} B(CH_2{-}CH_2{-}R)_2$$

Here boron is attached to the least substituted carbon which is anti-markownikov

(*i*) Refluxing with anhydrous carboxylic acid give alkane product

$$B(CH_2 - CH_2 - R)_3 \xrightarrow[RCOOH]{an} \underset{\text{alkane}}{3RCH_2 - CH_3}$$

(*ii*) With alkaline hydrogen peroxide ($NaOH/H_2O_2$) it forms primary alcohol.

$$B(CH_2{-}CH_2R)_3 \xrightarrow[H_2O_2]{NaOH} \underset{\text{primary alcohol}}{3R{-}CH_2{-}CH_2{-}OH}$$

(*iii*) Diborane is an electrophile reducing agent it attacks BH_4^-. Here reduction can occur without bond breaking. Alkenes and alkynes are reduced to alkanes and $>C{=}O$ group is reduced to $> CH_2$

(*a*) $RCHO \xrightarrow{Red} RCH_2OH$

Similarly,

$$(b)\ R{-}C{\equiv}N \xrightarrow{\text{Red}} \underset{\text{Amines}}{RCH_2NH_2}$$

Above reactions take place due to electron deficient nature of diborane.

Structure of B_2H_6

Boron hydrides exhibit interesting structures. William tips comb in 1976 was awarded Noble prize for unfolding structures of boranes.

Diborane

The simplest boron hydride B_2H_6 poses an interesting bonding feature. In this boron these are insufficient electrons to cater to the expected number of covalent bonds B_2H_6 has 12 valence electrons where B = 6 valence electrons and 6H atoms have 6 valence electrons B_2H_6 has an ethane C_2H_6 type structure which should need 14 electrons to form the required covalent bonds. Therefore B_2H_6 molecule is electron deficient

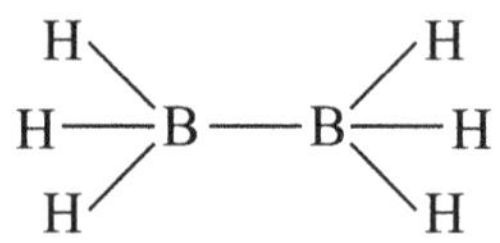

To make up for the deficiency hydrogen bridge structure was proposed, this was confirmed by electron diffraction studies. Due to the evidence following structure was given.

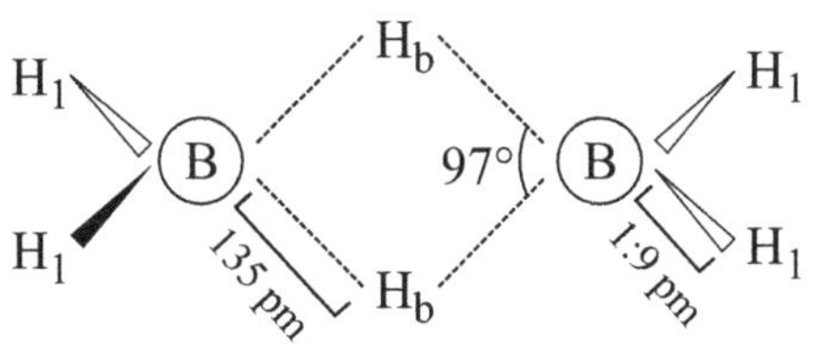

Structure of Diborane

The structure depicted shows two types of atoms. H_1 forms normal covalent bonds and H_b forms bridge between two B atoms. Two bridged hydrogen atoms are nearly at right angle to the rest of the molecule. This restricts rotation between 2B atoms.

The above structure is further supported by the following studies:

(*i*) Specific heat measurements indicate that no free rotation is possible between two B atoms.

(*ii*) Raman spectra and NMR spectra confirm that 4H atoms are different from the two hydrogen atoms.

(*iii*) Methylation of B_2H_6 further confirm that four H atoms are different from the two H atoms as suggested by the product $(CH_3)_4 B_2H_2$.

The hydrogen bridge atoms form bonds which are different from the conventional two electron covalent bonds. The bridgebonds involve one electron from each B atom and one electron from each H atom. This means that the three atoms are sharing only two electrons.

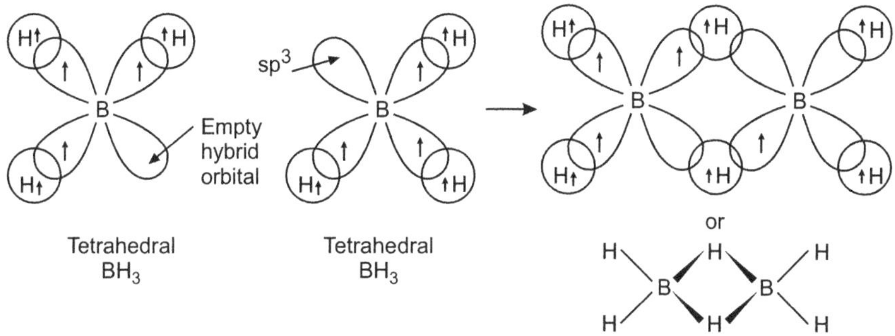

Each boron contains $4sp^3$ hybrid orbitals. Two of there are used to form bonds with terminal H atoms and the remaining two overlap with is orbital of bridged hydrogen atoms forming a delocalized cloud which cover the three nulei.

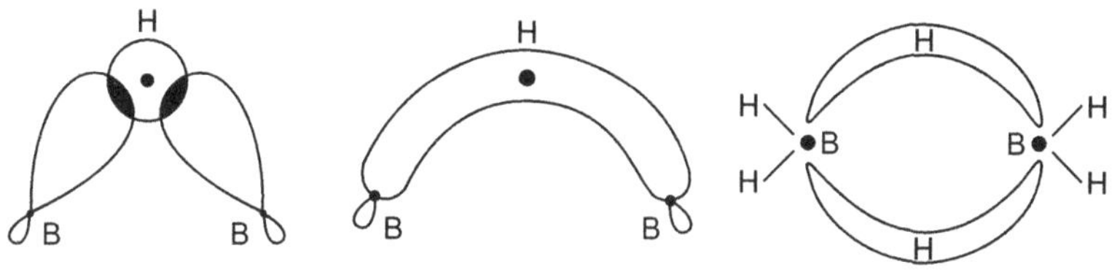

Fig. 11.1 Three centre bond in diborane

Diborane is thus defined as having four two centre bond (B – H) and two three-centre B.....H_b.....B bonds. All these bonds are σ covalent bonds sharing two electrons. This accounts for the overall 12 electron involved in the bond formation. Hence terminal B–H bonds are 2*e* bonds. The delocalization covering three nuclei is known as three centre electron pair bonds or banana bonds. The overall molecule is thus electron deficient. In B_2H_6 there are four B – H bonds (2*e* – 2*e*) and two 3*e* – 2*e* B....H....B bonds.

M.O. theory: According to the M.O. theory a three centre B....H....B linkage may be represeuted as a combinations of three orbitals (two sp^3 orbital belonging to B and *s*-orbital of H atom) resulting in three molecular orbitals.

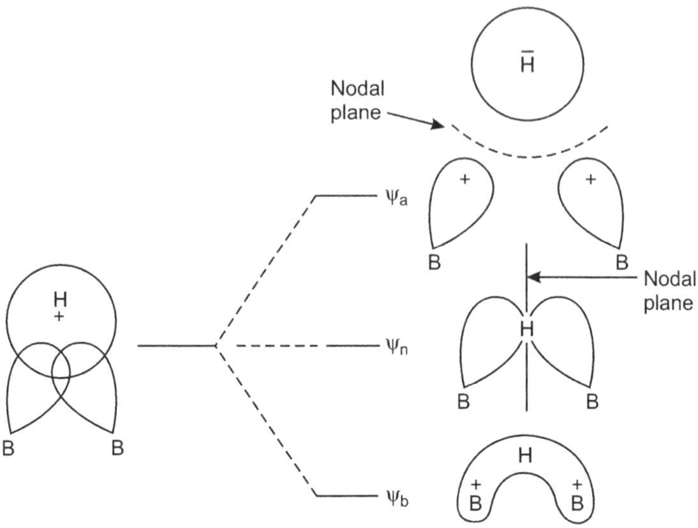

Note each abridging BHB bond contains two electrons in the ψ_b bonding M.O.

Other Boranes

Other boron hydrides are higher boranes. These are quite large in number *e.g.*, B_4H_{10}, B_5H_9, B_5H_{11}, B_6H_{10} and $B_{10}H_{14}$.

They are divided into two categories based on their stoichiometry and structure.

1. Nido-Boranes B_nH_{n+4}

B_2H_6, B_5H_9, B_6H_{10} and $B_{10}H_{14}$ come in this category. The word into is a Latin word meaning 'nest' this suggests that they have a nest like structure.

2. Arachno-boranes

B_4H_{10} and B_5H_{10}, the word arachno means spider's web the structure is open of both ends. Boron atoms occupy a contiguous corners of an $n + 2$ polyhedron.

Nomenclature: Their nomenclature is similar to hydrocarbons or carbanes. The number of boron atoms is written as prefixes.

B_2H_6	Diborane (6)
B_4H_{10}	Tetra borane (10)
B_5H_9	Penta borane (9)
B_5H_{11}	Penta borane (11)
B_6H_{10}	Hexa borane (10)
$B_{10}H_{14}$	Decaborane (14)

Arabic numbers in parenthess stand for H atoms.

Preparation

1. **B_4H_{10}** is best prepared by keeping B_2H_6 under pressure at 25°C for 10 days.

$$2B_2H_6 \longrightarrow B_4H_{10} + H_2$$

2. **B_5H_9** can be prepared by the pyrolysis of B_2H_6 at 180°C for 2.5 days in a reactor.
3. **Penta borane** B_5H_{11} can be prepared by the reaction of B_2H_6 and B_4H_{10} in a reactor at 120–130°C.
4. **Decaborane-14** $B_{10}H_{14}$ is prepared by the pyrolysis of B_2H_6 at 100–200°C in the presence of a lewis base (dimethyl ether).
5. **Preparation boron hydrides via hydride ion**

(*i*) B_2H_6 from BH_4^-

$$M[BH_4] + BX_3 \xrightarrow{CH_2Cl_2} \frac{1}{2} B_2H_6 + M(HBX_3)$$

where M may be $(n\text{-}C_4H_9)_4$, $BX_3 = BCl_3$ or BBr_3

(*ii*) B_4H_{10} from $B_3H_8^-$

$$M[B_3H_8] + BX \longrightarrow B_4H_{10} + M(HBX_3) - \text{BH residue}$$

$M = (n\text{-}C_4H_9)_4^+$, $BX_3 = BF_3, BCl_3, BBr_3$

(*iii*) B_5H_{11} from $C_4H_9^-$

$$K[B_4H_9] + BX_3 \longrightarrow B_5H_{11} + K(HBX_3) + \text{BH residue}$$

Physical Properties

Boranes are colourless, diamagnetic compounds with low thermal stability. They are very reactive and burn in air. All of them are highly toxic. The lower boranes are gases at room temperature whereas higher hydrides are volatile liquids and solids. The higher boranes are less reactive. Decaborane is stable in air.

Chemical Properties

1. **Acidic behaviour:** B_5H_9 is a bronsted acid. Higher boranes are more acidic. Arachno-boranes are more acidic than nido boranes.

   B_5H_9 can be deprotonated at low temperatures due to the loss of bridging H atom.

$$C_5H_9 + NaH \longrightarrow NaB_5H_8 + H_2$$

2. **Adduct formation:** Tetra borane undergoes cleavage both symmetric and unsymmetric

(*i*) $B_4H_{10} + \underset{\text{Lewis base}}{2(CH_3)N} \longrightarrow (CH_3)_3NB_3H_7 + (CH_3)_3NBH_3$

(*ii*) Hard Lewis bases like OH^- or NH_3

$B_4H_{10} + 2NH_3 \longrightarrow [H_2B(NH_3)_2]^+ + B_3H_8^-$

$2B_4H_{10} + 4OH^- \longrightarrow [B(OH)_4]^- + BH_4^- + 2[B_3H_8]^-$

(*iii*) B_5H_9 forms adducts with NH_3 and $P(CH_3)_3$

$B_5H_9 + 2P(CH_3)_3 \longrightarrow B_5H_9\,[P(CH_3)_3]_2$

$B_5H_9 + 2NH_3 \xrightarrow{-78°C} [(NH_3)_2\,BH_2]^+\,[B_4H_7]^-$

(*iv*) $B_{10}H_{14}$ reacts with sodium metal in liquid NH_3

$B_{10}H_{14} + 2Na \longrightarrow Na_2B_{10}H_{14}$

Adducts can be prepared directly with methyl cyanide

$B_{10}H_{14} + 2CH_3CN \longrightarrow B_{10}H_{12}\,(CH_3CN)_2 + H_2$

3. **Substitution reactions:** The substitution of B_5H_9 occurs by

(*i*) Halogenation (*ii*) Freidel Craft's alkylation with RX.

Halogenation (or Iodination)

$C_5H_9 \xrightarrow[AlCl_3]{I_2} C_5H_8—I$

$C_5H_9 \xrightarrow[AlC_3]{R-X} C_5H_8—R$

4. **Cluster expansion or degradation reaction**

(*i*) Cluster expansion

$LiB_5H_8 + B_2H_6 \xrightarrow[-78°C]{Me_2O} LiB_7H_{14} \longrightarrow B_6H_{10} + LiBH_4$

This method is used to prepare B_6H_{10}

(*ii*) Degradation reaction

$[Me_4N][B_5H_8] + 6PrOH \longrightarrow [MeN]B_3H_8—2B(OPr)_3 + 3H_2$

Structure of B_4H_{10}

It contains 22 valence electrons, at the end there are 12 electron (4 × 3) from B atoms and 10 from H atoms. It consists of six $2e - 2e$ B – H bonds, four $3e - 2e$ B – H bonds and one $2e - 2e$ B – B bonds.

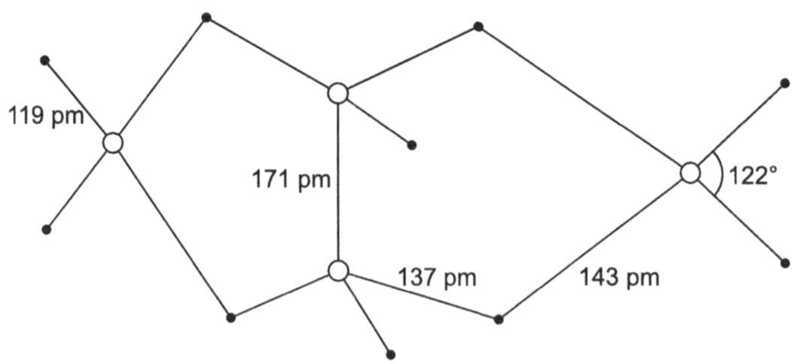

• Stands for B atom

• Stands for hydrogen H atom other structures of boranes.

B_5–H_9

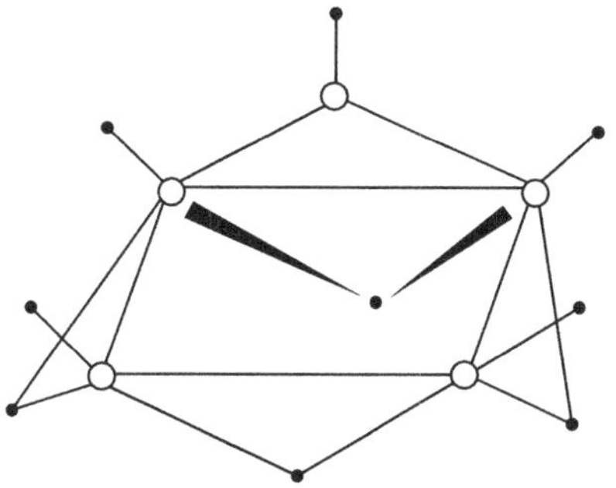

or

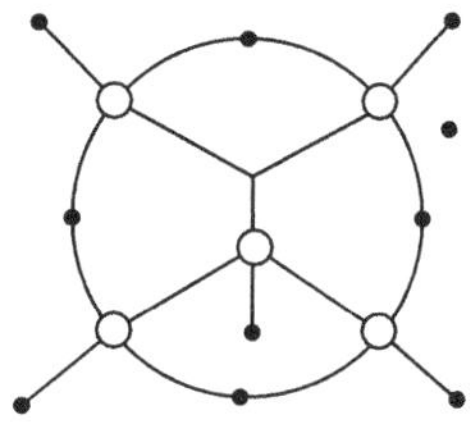

B_5H_{11}

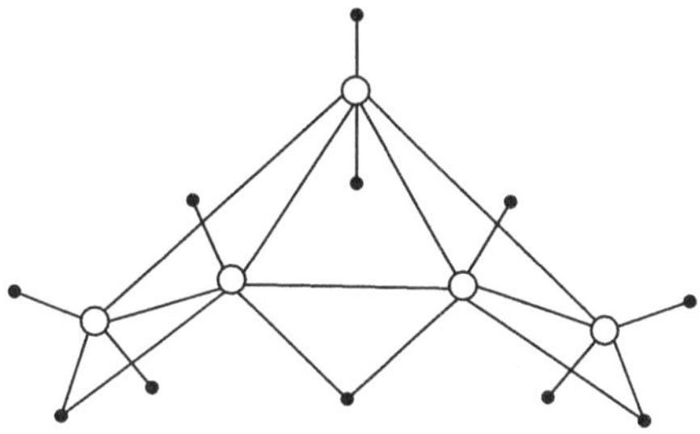

or

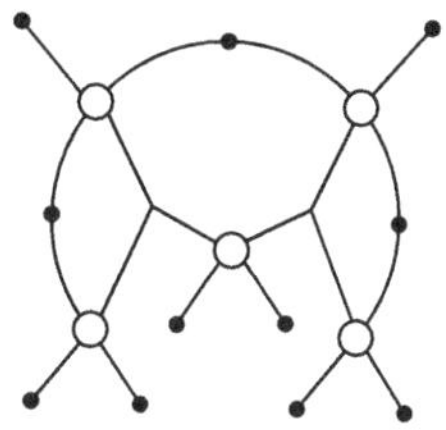

B_6H_{10}

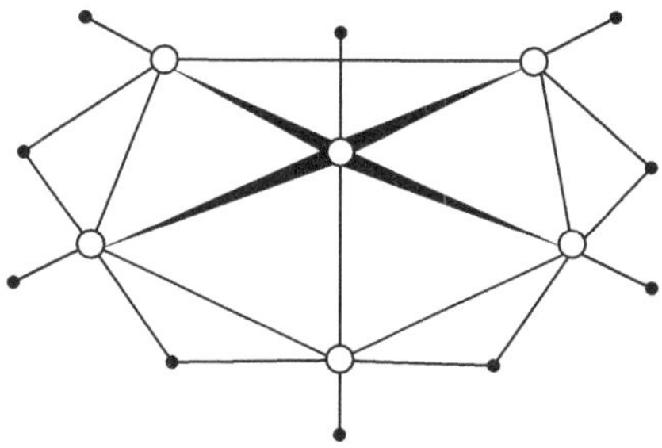

or

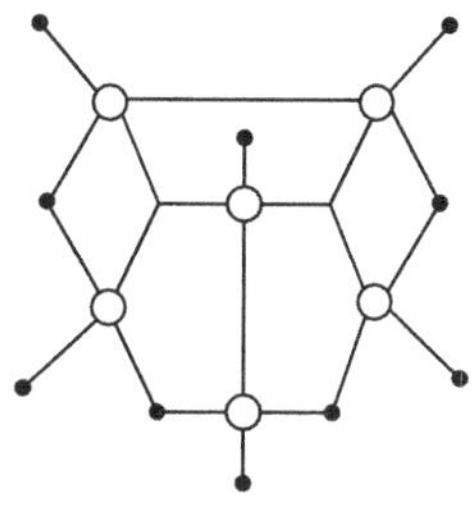

$B_{10}H_{14}$

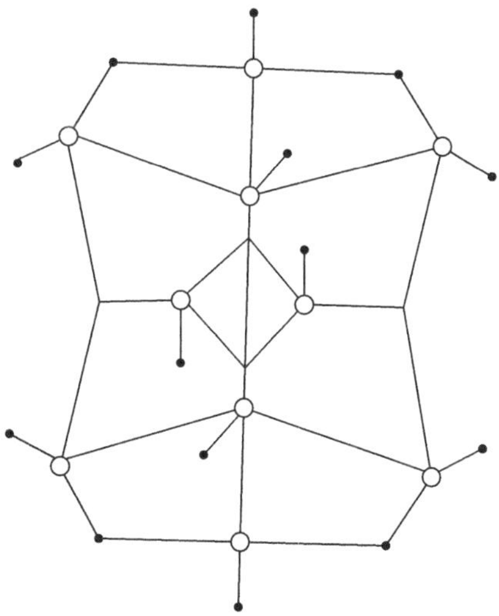

Boric acid and Borax

Boric Acid $B(OH)_3$ or H_3BO_3.

It is also known as orthoboric acid.

Preparation: It is obtained by adding a fixed amount of HCl on a concentrated solution of borax. The resulting solution is allowed to cool when crystals of boric acid separate out.

$$Na_2B_4O_7 + 2HCl + 5H_2O \longrightarrow 2NaCl + 4H_3BO_3$$

Properties: It is a white crystalline substance. It consists of unsymmetrical H bonds as drawn in the figure.

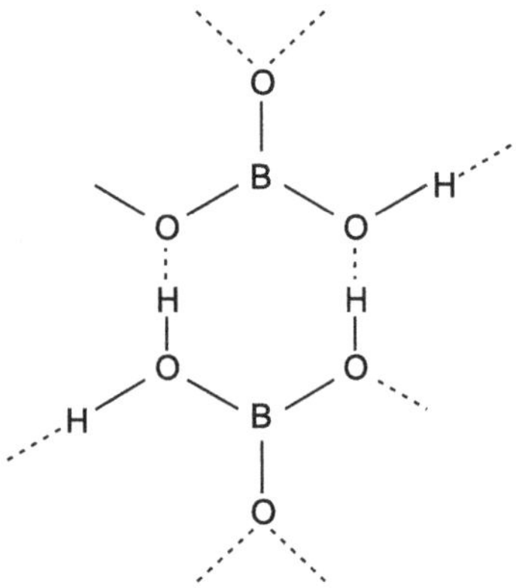

Fig. 11.2 Layer structure of $B(OH)_3$

It entails weak bonds between the layers therefore it has week density of 1.48g cm^{-3}. It is a weak monobasic acid its acidic property is shown by OH^- acceptance.

1. $B(OH)_3 + 2H_2O \rightleftharpoons H_3O^+ + B(OH)^-{}_4$ pKa = 9.25
2. Dehydration of boric acid yields mataboric acid which has different crystalline forms.

$$\underset{\text{cubic}}{HBO_2} \xrightarrow{170°C} B(OH)_3 \xrightarrow{140°C} \underset{\text{monoclinic}}{HBO_2} \xrightarrow{\text{Rapid}} \underset{\text{ortho rhombic}}{HBO_2}$$

3. Boric acid when heated to red hot gives B_2O_3

$$2H_3BO_3 \xrightarrow[\text{hot}]{\text{red}} \underset{\text{boric oxide or boric anhydride}}{B_2O_3} + 3H_2O$$

4. Boric acid is a weak acid therefore it is difficult to titrate against NaOH as the end point cannot be located correctly although it can be easily

titrated against polyhydroxy alcohols *e.g.*, glycerol, catechol, mannitol etc.

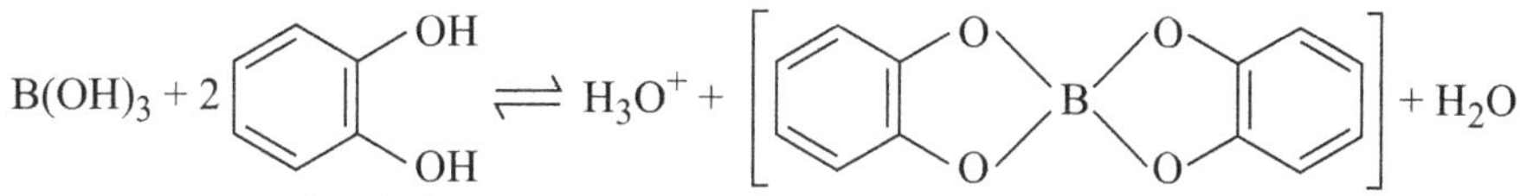

Catechol

The complex ion shows that boron has acquired the maximum covalency of 4 and therefore does not interact with H^+. This facilitates the titration of boric acid and strong base like NaOH and detection of end point.

5. Boric acid when heated with C_2H_5OH and conc. H_2SO_4 gives triethyl borate which is a sure test for borate as it burns with a green edged flame.

$$H_3BO_3 + 3C_2H_5OH \xrightarrow[H_2SO_4]{\text{Conc.}} (C_2H_5)_3BO_3 + 3H_2O$$

6. Boric acid reacts with KHF_2 to product KBF_4 salt.

$$2KHF_2 + H_3BO_3 \longrightarrow \underset{\substack{\text{Potassium} \\ \text{fluoroborate}}}{K[BF_4]} + KOH + 2H_2O$$

7. H_3BO_3 when heated with CaF_2 and conc. H_2SO_4, BF_3 results

$$3CaF_2 + 3H_2SO + 2H_3BO_3 \longrightarrow 3CaSO_4 + 2BF_3 + 6H_2O$$

Uses: Boric acid finds its use in enamels and for glazing the pottery. Boric acid solution is used for eye-wash. Its is used as a preservative for milk, food articles, leather industry and for wicks of the candle. It is also used for the manufacture of borax and pigments.

Borax or Sodium Tetraborate Decayhydrate

It occurs naturally in the dried lakes of India, Tibet, Sri Lanka and California, USA.

Preparation

The salt borax is dissolved in the boiling water and later filtered to remove insoluble clay and sand. The filterate is evaporated to obtain the crystals of borax.

Properties

The is a white orystalline solid, sparingly soluble in cold water and readily in hot water. The solution in water is alkaline.

1. On heating it swells due to the loss of water of crystallization as steam. The anhydrous products melts at a temperature > 1000 K.

2. Borax and NH_4Cl are heated together to get $(BN)_x$.

Uses

1. Used in soldering and welding.
2. Used in making enamels, glazes and candle wicks.
3. It is used in making heat resistance glasses boro silicate (pyrex and corning)
4. Famous borax-bead test for the coloured salts is performed by mixing borax and coloured salt, taking it in a platinum wire loop and exposing to the oxidation flame. It forms the transparent glassy sodium metaborate and B_2O_3.

$$Na_2B_4O_7.10H_2O \xrightarrow{\text{heat}} \underset{\substack{\text{sodium}\\ \text{metaborate}\\ \text{(glassy)}}}{2NaBO_2} + B_2O_3 + 10H_2O$$

With coloured salts this transparent bead forms the coloured beads when introduced on the flame.

$$CoO + B_2O_3 \xrightarrow[\text{on flame}]{\text{heat}} Co(BO_2)_2$$

$$NiO + B_2O_3 \xrightarrow{\text{heat}} Ni(BO_2)_2$$

$$Cr_2O_3 + 3B_2O_3 \xrightarrow{\text{heat}} 2Cr(BO_2)_3$$

$$CuO + B_2O_3 \xrightarrow{\text{heat}} Cu(BO_2)_2$$

These are transparent glassy coloured metaborates.

Borates

Anhydrous borates may be prepared by the action of burning boric acid and metal oxides. The hydrated borates may be crystallized from its aqueous solution. The composition of borates $KB_5O_8 . 4H_2O$, CaB_2O_4, $Na_2B_4O_7 . 10H_2O$ provides very little knowledge about the structures of borate anions. This is attributed to their linear and cyclic polymer formation. BO_3 and BO_4 units are linked together through oxygen atoms. Borates with 5B atoms are known to exist. Borates mainly consist of:

(*i*) Boron atom linked to three oxygen atoms (BO_3 unit) or four oxygen atoms (BO_4 tetrahedron).

(*ii*) In the event of polynuclear anions corner sharing of boron oxygen triangles or tetrahedrons.

(*iii*) These units can be protonated to different limits.

Ring Borates

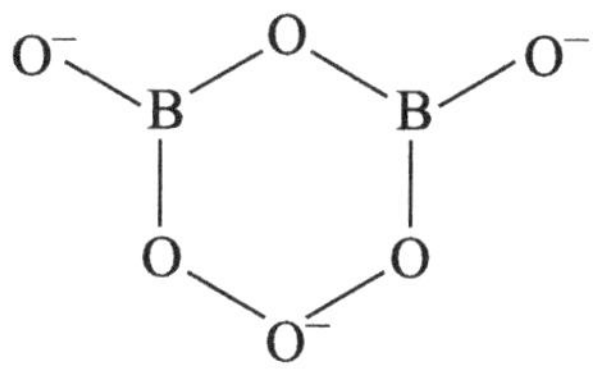

$K_3B_3O_4$

Chain Borates

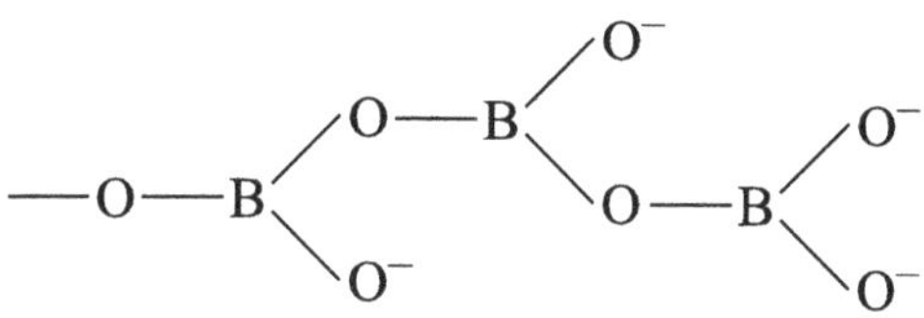

CaB_3O_6

In contrast to borates carbonate ion does not form polymeric structures owing to the presence of strong C–O π bonds

Borax $Na_2B_4O_7.10H_2O$ contains tetranuclear units of BO_4 and BO_3. Hence it is formulated as $Na_2[B_4O_5(OH)_4]5H_2O$

Structure of $[B_4O_5(OH)_4]^{2-}$

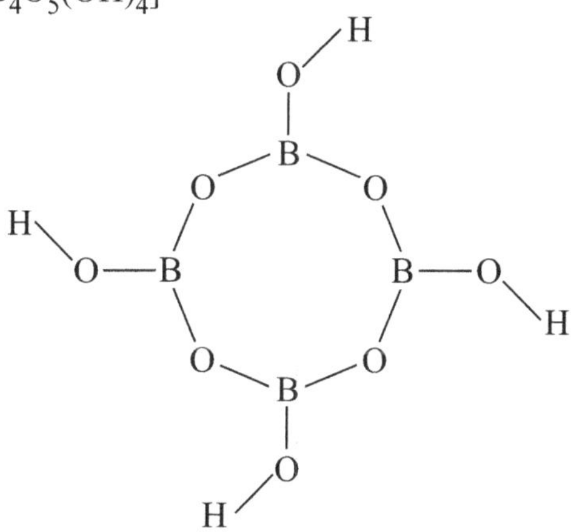

Halides of Boron (BX_3)

1. BF_3 is prepared by the action of HF on B_2O_3

$$B_2O_3 + 6HF \longrightarrow 2BF_3 + 3H_2O$$

2. Other boron halides are obtained by the action of halogens with B_2O_3 in the presence of carbon at high temperatures.

$$B_2O_3 + 3C + 3Cl_2 \longrightarrow 2BCl_3 + 3CO$$

$$B_2O_3 + 3C + 3Br_2 \longrightarrow 2BBr_3 + 3CO$$
$$B_2O_3 + 3C + 3I_2 \longrightarrow 2BI_3 + 3CO$$

Properties

1. BF_3 and BCl_3 are gases, BBr_3 is a liquid and BI_3 is a solid at room temperature.
2. The order of melting and boiling points is:
   $BI_3 > BBr_3 > BCl_3 > BF_3$
3. Boron trihalides are covalent due to small size and high ionization energy of boron.
4. They do not conduct electric current in the liquid state.
5. They exhibit low boiling points in comparison to Group I and Group II elements.
6. They show discrete molecules in all the three states.

Structure of BX_3 and AlX_3 Compounds

Boron trihalides exist as monomers but aluminium trihalides are dimers. This is attributed to the difference in the size of boron and aluminium atoms. Boron atom is extremely small and therefore unable to form four bonds with large halide ions. Boron is small therefore it is difficult to carry $4F^-$ ions. This suggests that addition bond to boron atom is not enough to compensate for the loss in energy to the system in overcoming the B–F $p\pi$–$p\pi$ interactions (or π bonding) in the monomers unit.

Boron Halides as Lewis Acids

Boron trihalides have sp^2 hybridised plane triangular structure where boron overlaps with $2px$ orbital of F atom to form BF_3 molecule. This gives one $2p$ orbital vacant on boron. BX_3 contains six valence electrons (*i.e.,* it is two electrons short of the octet). This makes BX_3 as an electron pair acceptors or Lewis acids.

Valence shell 2*s*

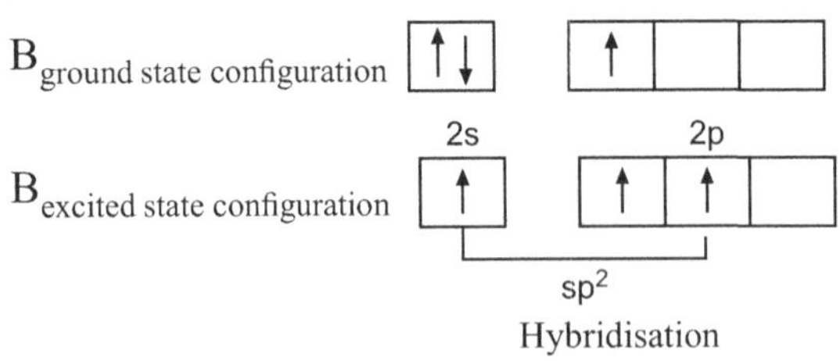

Structure of BCl_3

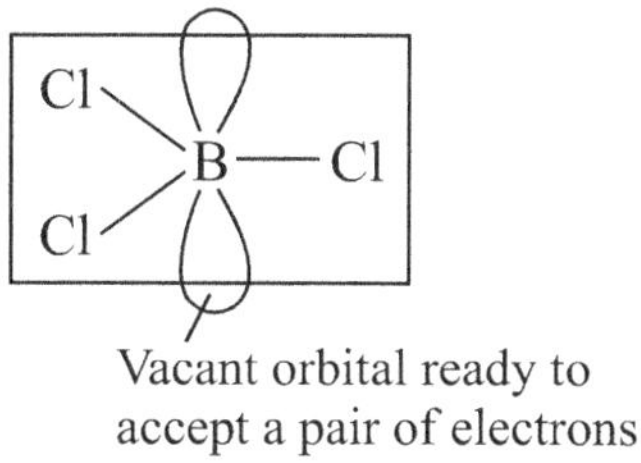

Relative Strength of BX_3 Acids

All boron trihalides are electron pair acceptors therefore they act as Lewis acids. The relative strength of lewis acids depends upon the acceptor strengths of BF_3, BCl_3 and BBr_3. This can be gauged by measuring heats of formations and dipde moments of their adducts. The results indicate that the acceptor power decreases in the order:

$$BBr_3 \geq BCl_3 > BF_3$$

The order is opposite to the expectation on the basis of electronegativity of X and on steric considerations. The anomaly is explained by the boron hydrogen π-bonding.

Concept of Back-bonding in BX_3 Compounds

In BF_3, fluorine has a filled 2*p* and boron has a vacant 2*p* orbital. Since the two orbitals are of the same energy they can overlep effectively and as a consequence fluorine electrons can be transferred to the vacant 2*p* orbitals of boron giving rise to *p*π–*p*π bond. Due to the two way bonding the B–F bond acquires some double bond character.

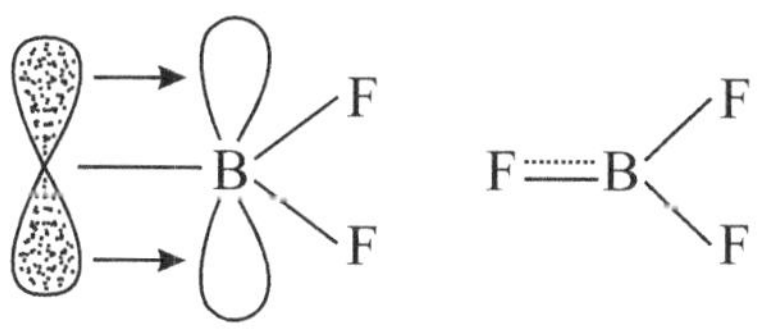

*p*π - *p*π overlap

The type of bond formed is a nature bond or back bonding. The same back bonding may take place in the other to B°–F bonds.

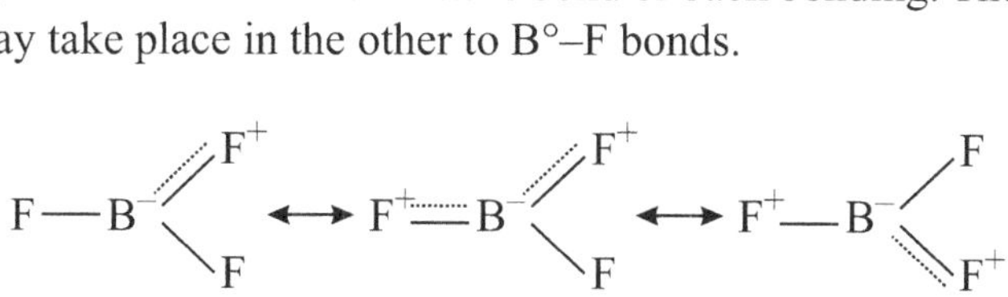

As a consequence of donation of electrons from fluorine to boron compensates electron deficiency of boron. This tendency decreases in the case of BCl_3 and BBr_3 because the overlap of $2p$ orbital of boron and higher energy filled $3p$ and $4p$ orbitals of chlorine and bromine cannot be that effective as in the case of BF_3.

Uses: BF_3 can be used as a catalyst, in FCR reaction and polymerization reactions compared to other boron halides.

Boron Nitride $(BN)_x$

Boron nitride is a polymeric white solid isoelectronic with graphite. It is available in group allotropic forms, hexagonal form being more common. It consists of layers of fused six membered $(BN)_3$ rings like graphite.

Preparation

1. Pure hexagonal form can be prepared by the action of NH_3 or BCl_3 or BF_3

$$2BCl_3 + 6NH_3 \xrightarrow{-6HCl} 2B(NH_2)_3 \xrightarrow[750°C]{\text{Pyrolysis}} 2(BN)_3$$

2. Boron nitride ion also be prepared by the action of B_2O_3 and NH_3.

$$B_2O_3 + 2NH_3 \longrightarrow 2(BN)_x + 3H_2O$$

3. In the laboratory it can be prepared by fusion of borax with NH_4Cl or by fusing area and boric acid.

Structure

The polymeric boron nitride has a graphite like layer structure with alternate B and N atoms rings, the B–N bond is 145 pm and sp^2 hybridised. The remaining two electrons on nitrogen are used in forming *pi*-bonds by donation form N to B. They form alternate single and double bonds between B and N atoms.

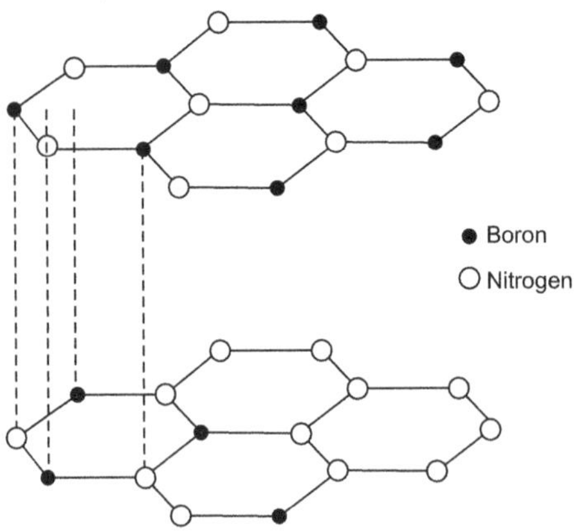

Layer structure of boron nitride

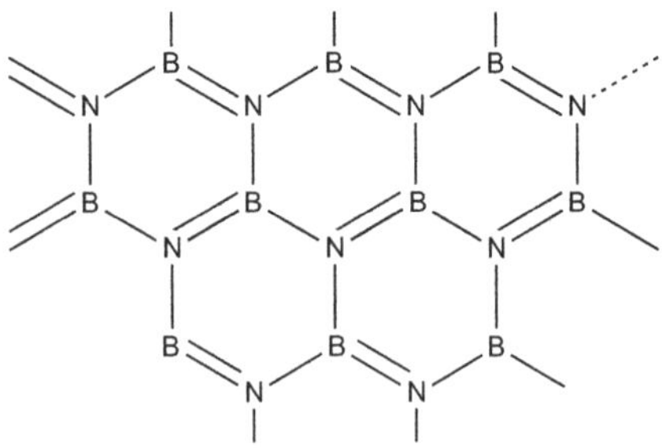

One layer structure of boron nitride

The sheets are so arranged that boron atoms are directly over nitrogen atoms.

This structure gives sliding of planes over each other layer. Hence boron nitride is used as a lubricant. Its properties are similar to graphite.

It is colourless electrical insulator because there is large energy gap between filled and vacant π-bonds.

Layer structure of Boron nitride.

Borazine or Borazole ($B_3N_3H_6$)

It was first prepared by a stock in 1926 by reacting B_2H_6 and NH_3 in the ratio of 1:2 between 250–300°C.

(*i*) $3B_2H_6 + 6NH_3 \longrightarrow 2B_3N_3H_6 + 12N_2$

Its structure is similar to benzene so it was called inorganic benzene. It is isoelectronic with benzene and shows similar properties.

(*ii*) Reduction of BCl_3 which gives trichlaro derivative of borazine and then reduced by $NaBH_4$

$$3BCl_3 + 3NH_4Cl \xrightarrow{150°C} B_3N_3H_3Cl_3$$

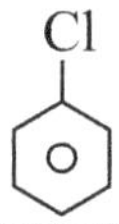

–9HCl

$$B_3N_3H_3Cl_3 + 3NaBH_4 \xrightarrow{\text{Triglyme}} B_3N_3H_6 + 3/2B_2H_6$$

(*iii*) It can also be prepared by the direct action of $NaBH_4$ with NH_4Cl

$$3NaBH_4 + 3NH_4Cl \xrightarrow{\text{Triglyme}} B_3N_3H_6 + 3NaCl + 9H_2$$

Borazine is a planar hexagonal ring in which B and N are planar and sp^2 hybridized. The difference between benzene and borazine is that the π-bond in borazine is a dative bond formed by filled $2p$ orbitals of nitrogen and vacant $2p$ orbitals of nitrogen.

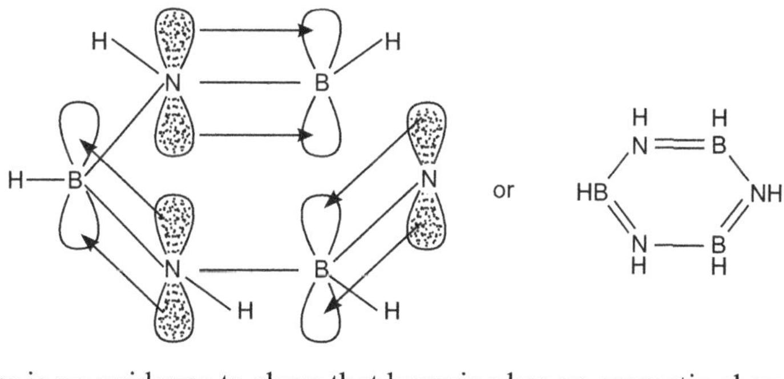

There is no evidence to show that borazine has an aromatic character.

Addition Reactions

It can add H_2O, alcohol and HX to yield 1:3 adducts which on heating to 100°C lose hydrogen.

$$B_3N_3H_6 + 3H_2O \xrightarrow{0°C} [BH(OH)(NH_2)]_3 \xrightarrow{100°C} [B(OH)NH]_3 + 3H_2$$

$$B_3N_3H_6 + 6ROH \xrightarrow{0°C} [BH(OR)NH_2]_3 \xrightarrow{100°C} [BH(OR)NH]_3 + 3H_2$$

Action of Heat

On heating borazines give polymeric substances similar to nepthalene and diphenyl etc.

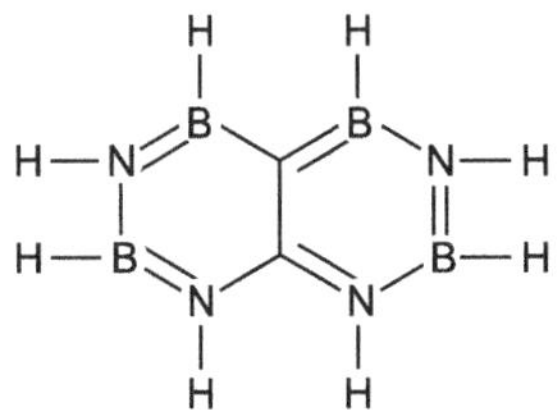

(Inorganic naphthalene)

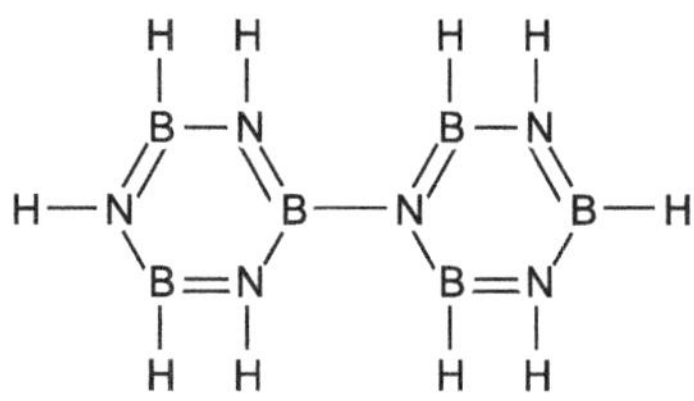

(Inorganic diphenyl)

Inorganic cyclohexane can also be obtained:

$$B_3N_3H_6 + 3HCl \longrightarrow B_3N_3H_9Cl \xrightarrow{NaBH_4} B_3N_3H_{12}$$

Inorganic cyclo hexane

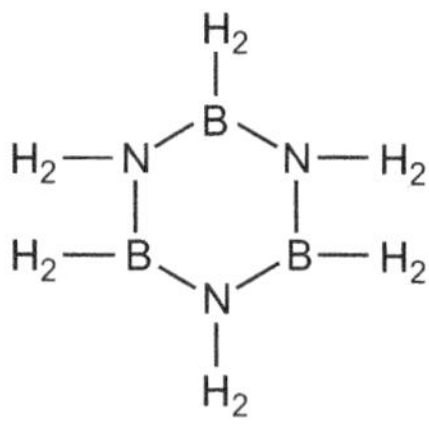

Aluminium

Occurrence

It is the most abundant metal on earth's crust, Al metal is widely used on a very vast scale. The most important ore of Al is bauxite which has a varying composition between $Al_2O_3 . H_2O$ and $Al_2O_3 . 3H_2O$. It also occurs as aluminosilicate and feldspar. The extraction of Al is most profitable from bauxite.

Excraction

There are two main processes for the extraction of Aluminium:

Baeyer's Process: The ore bauxite is dissolved in NaOH to form sodium meta aluminate. Then $Al(OH)_3$ is precipitated from the strongly alkaline aluminate solution. This may be done either by bubbling CO_2 or by seeding with Al_2O_3, $Al(OH)_3$ is heated strongly to convent it to Al_2O_3.

Hall-Heroult's Process: Al_2O_3 is melted with cryolite $Na_3[AlF_6]$ and electrolysed in graphite-lined steel cathode. The anode is made of graphite. The cell runs continuously and at intervals molten Al is taken out from the bottom of the cell.

Cryolite improves the electrical conductivity as Al_2O_3 is a non-conductor. Cryolite serves as an impurity to lower the melting point of the mixture to 950°C. At anode we get O_2,CO_2 and F_2 and at cathode molten Al is obtained.

$$2Al_2O_3 \longrightarrow 4Al^{3+} + 6O^{2-}$$

At cathode: $4Al^{3+} + 12e \longrightarrow 4Al^{\circ}$

At Anode: $6O^{2-} - 12e \longrightarrow 3O_2$

Qualitative test for Aluminium

Aluminium is precipitated as a white gelatinous precipite $Al(OH)_3$ when NH_4OH is added to a solution containing aluminium salt. $Al(OH)_3$ dissolves in NaOH giving sodium aluminate ($NaAlO_2$). $Al(OH)_3$ can be reprecipitated by adding a solution of CO_2 showing that $Al(OH)_3$ has a weak acidic property.

$$Al(OH)_3 \xrightarrow[\text{NaOH}]{\text{excess}} NaAl(OH)_4$$

$NaAlO_2.2H_2O$ sodium aluminate when a drop of blue litmus solution is added then the colour of blue litmus is laken up by $Al(OH)_3$ in preference to water. This we get a blue take floating in the colourless solution (famous blue lake test). The aluminate formation shows amphoteric nature of $Al(OH)_3$.

Alums: Aluminium has a tendency to crystallise out from its aqueous solution as double salts and alums. Alums may contain singly charged cations *e.g,* Na^+, K^+ or NH_4^+. Solutions containing NH^+_4 and Fe^{2+} prefer double salt formation with six water molecules of crystallization *e.g.,* $FeSO_4$ $(NH_4)_2SO_4.6H_2O$ called Mohr's salt.

When the solution contains singly charged cations and triply charged cations then its crystals contain 24 water molecules of crystallization *e.g.,* potash Alum $K_2SO_4 . Al_2(SO_4)_3 . 24H_2O$. Alums may or may not contain aluminium. There are examples of ferric alum and chrome alum. Their general formula is shown by $KAl(SO_4)_4.24H_2O$. Double salts break up in solution into its constituent ions. Potash alum is used as a mordant in dyeing. Crystals of potash alum break up as $[M(H_2O)_6]^+$, $[Al(H_2O)_6]^{3+}$ and SO_4^{2-} ions.

Trihalides of Aluminium

Aluminium (*iii*) Fluoride (AlF_3)

It is prepared by heating Al_2O_3 with HF at 700°C.

$$Al_2O_3 + 6HF \xrightarrow{700^\circ C} 2AlF_3 + 3H_2O$$

It is different from other trihalides of Al. It is insoluble and volatile. Al in AlF_3 has coordination number of 6 whereas in $AlCl_3$ coordination number changes from 6 to 4 at its melting point while it stays 4 for $AlBr_3$ and AlI_3, In AlF_3 each Al is surrounded by a distorted octahedron of 6F atom with 1:3 (Al:F) stoichiometry. This is achieved by sharing of corners (F atoms) between two octahedral.

Aluminium Chloride Preparation ($AlCl_3$)

(*i*) $AlCl_3$ (anhydrous) is obtained by heating Al in a current of Cl_2 gas or HCl gas

$$2Al + 3Cl_2 \longrightarrow 2AlCl_3 + \Delta H$$

$$2Al + 6HCl \longrightarrow 2AlCl_3 + 3H_2 + \Delta H$$

(*ii*) $AlCl_3$ is manufactured by passing Cl_2 gas on a mixture of Al_2O_3 and coke at 900°C

$$Al_2O_3 + 3C + 3Cl_2 \longrightarrow 2AlCl_3 + 3CO$$

(*iii*) It is also prepared by passing a mixture of CO and Cl_2($COCl_2$ or phosgene gas) over heated alumina at 600°C

$$Al_2O_3 + 3CO + 3Cl \longrightarrow 2AlCl_3 + 3CO_2$$

$$Al_2O_3 + 3COCl_2 \longrightarrow 2AlCl_3 + 3CO_2$$

Properties: Pure anhydrous $AlCl_3$ is a white solid while commercial sample is yellowish due to the impurity of $FeCl_3$. It is strongly hygroscopic and fumes in moist air. If sublimes at 183°C and its melting point is 193°C under pressure.

It exists as a dimer Al_2Cl_6 between 200–400°C and beyond 400°C it forms a monomer that exists up to 800°C.

It is a strong lewis acid and therefore forms addition compounds and complexes with bases.

(*i*) With ammonia $AlCl_3$ forms $AlCl_3$, $6NH_3$ while with trimethylamine Me_3N it forms complexes of the composition $AlCl_3.NMe_3$ and $AlCl_3.2NMe_3$. The former is tetrahedral, Al while the later is trigonal bipyramidal structure.

(*ii*) Phosphoryl chloride ($POCl_3$) forms compounds of three different compositions *e.g.*, $AlCl_3$, $POCl_3$, $AlCl_3$ $2POCl_3$ and $AlCl_3$ $6POCl_3$.

(*iii*) It gives non-chelate complex ion with alkali metal chlorides

$$AlCl_3 + NaCl \longrightarrow [AlCl_4]^- + Na^+$$

$$AlCl_3 + 3NaCl \longrightarrow Na_3[AlCl_6]^{3-}$$

(*iv*) It can form non-metal chlorides from their fluorides.

$$AlCl_3 + BF_3 \longrightarrow BCl_3 + AlF_3$$

The transhalogenation takes place due to higher electronegativity of fluorine.

(*v*) An $AlCl_3$ acts as a catalyst in FCR reactions

$$C_6H_6 + CH_3AlCl_4 \longrightarrow C_6H_5CH_3 + HCl + AlCl_3$$

(*vi*) Anhydrous $AlCl_3$ has a very strong affinity for water therefore H_2O as a ligand goes inside the square bracket and chloride ion goes out of the coordination sphere *e.g.*,

$$[Al(H_2O)_6]Cl_3$$

There is a strong Al–O linkage therefore hydrate cannot be dehydrated to anhydrous $AlCl_3$ by simple heating.

$$2Al(H_2O)_6\,Cl_3 \xrightarrow{\Delta} Al_2O_3 + 6HCl + 9H_2O$$

(*vii*) With NaOR it forms $Al(OR)_3$ and $NaAl(OR)_4$, R may be H or alkyl group

$$AlCl_3 + 3NaOR \xrightarrow{-3\,NaCl} Al(OR)_3 \xrightarrow{NaOR} Na[Al(OR)_4]$$

Structure

Anhydrous $AlCl_3$ possesses a closely packed layer structure with 6 coordinated Al with octahedral arrangement. The rectangles exist, are shown by the dotted line representing layer structures indicate the release of Al_2Cl_6 molecules on heating.

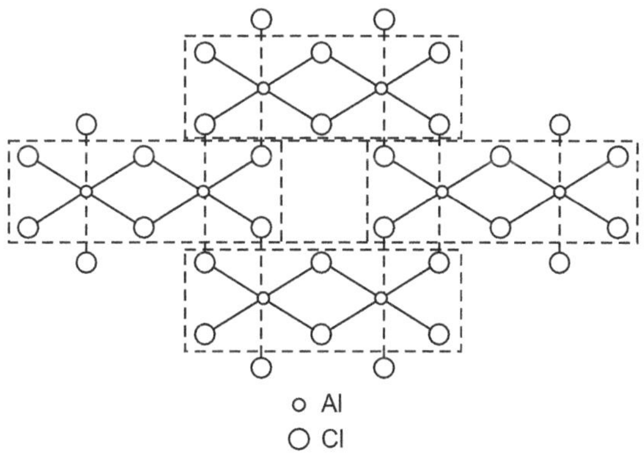

Fig. 11.3 Structural arrangement of Al Cl_3 in crystalline aluminium chloride.

The structure of molten $AlCl_3$ is dimeric Al_2Cl_6 in which each Al atom is sp^3 hybridised with 3 covalent bonds and one coordinate bond. Overall it is covalent and does not allow the passage of electric current.

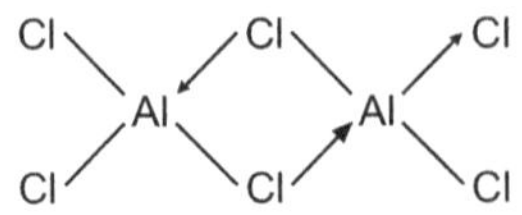

Fig. 11.4 Structure of $2Al_2Cl_6$ (Molten state)

At higher temperature the dimeric molecular form Al_2Cl_6 changes to plane triangular molecule $AlCl_3$ iso structural with BCl_3.

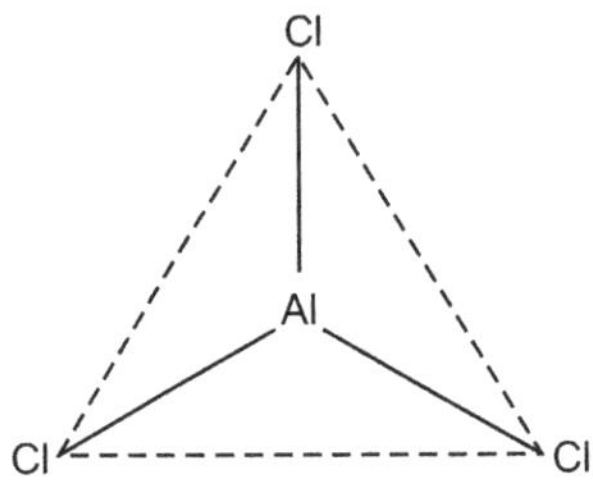

Fig. 11.5 Plane triangular 3-coordinated Al (vapour state)

$AlBr_3$ and AlI_3 form dimeric molecule Al_2Br_6 and Al_2I_6 in the solid as well as liquid state.

The dimeric formation tendency is attributed to the tendency of Al to complete the octet. The dimeric structure is dissociated by the donor molecules.

$$Al_2Cl_6 + 2R_3N \longrightarrow 2AlCl_3 + 2NR_3$$

Uses: Anhydrous $AlCl_3$ is used as a catalyst in the FCR reactions, cracking of mineral oils and manufacture of dyes and perfumes, AlF_3 is used in the metallurgy of Al.

Organometallic Compounds

Besides alkylboranes discussed before trihalides of group 13 elements undergo reaction with grignard reagents and organolithium reagents to form trialkyl or triaryl compounds.

$$F_3B + 3C_2H_5MgI \longrightarrow B(C_2H_5)_3$$
$$Cl_3Al + 3CH_3MgI \longrightarrow Al(CH_3)_3$$
$$GaCl_3 + 3C_2H_5Li \longrightarrow Ga(C_2H_5)_3$$
$$InBr_3 + 3C_6H_5Li \longrightarrow In(C_6H_5)_3$$

Out of these Al has a dimeric structure because of three centre bonds Al–C–Al

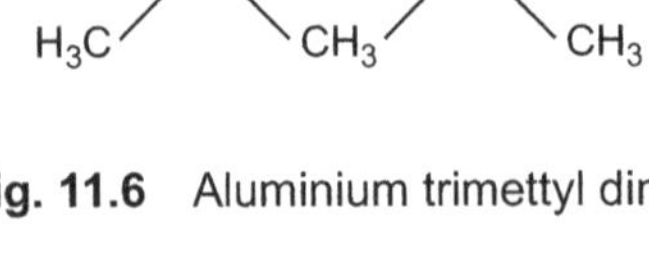

Fig. 11.6 Aluminium trimettyl dimer

Alkenes may be added to Al–H bonds

$$Et_2AlH + H_2C = CH_2 \longrightarrow Et_2—Al—CH_2—CH_2$$

Unique Behaviour of Boron Among Group 13 Elements

Boron atoms are very small among group 13 elements. It is non-metallic and forms covalent compounds. It is similar in properties to diagonally opposite silicon in group 14.

1. B_2O_3 is acidic like SiO_2. However, Al_2O_3 is amphoteric.
2. $B(OH)_3$ *i.e.,* boric acid is acidic while $Al(OH)_3$ is amphoteric.
3. Borates and silicate ions can polymerise giving isopolyacids.
4. Boron hydrides are gaseous, hydrolyse instantly and are inflammable $(AlH_3)_x$ are polymeric solid whereas SiH_4 is gaseous, readily hydrolyse and inflammable.
5. Boron and silicon halides hydrolyse vigourously excepts BF_3. Aluminium halides are partially hydrolysed in water.

EXERCISES

1. Why boron forms no compound in +1 state but thallium +1 state is quite stable?
2. Aluminum forms $[AlF_6]^{3-}$ complex ion but boron does not form $[BF_6]^{3-}$ ion.
3. List the *p*-block elements in which inert pair effect is predominant.
4. Why boron does not form B^{3+} ions?
5. Al is a good reducing agents. Give example.
6. Give three properties showing similarities of properties between boron and silicon.
7. Which is the weakest Lewis acid?

   (*i*) BF_3 (*ii*) BBr_3
8. Give a method of preparation of diborane. Describe its reaction with:

   (*a*) Water (*b*) Chlorine (*c*) Carbon monoxide

   (*d*) Oxygen on burning (*e*) ammonia (*f*) HCl

   (*g*) Tertiary amines.
9. Draw the structure of diborane. Explain the electron deficient nature of B_2H_6 and the three centre electron pair bond.
10. Explain:

    (*i*) Why Al halides are dimeric while boron halides are monomeric.

    (*ii*) $p\pi$-$p\pi$ back bonding occurs in boron halides but not in the case of Al.

    (*iii*) AlF_3 has a high melting point while $AlCl_3$ is a volatile, low melting solid.

(*iv*) Boric acid can be titrated against NaOH using phenolphthalein indicator in the presence of glycerin.

11. Give reasons:
    (*i*) Certain metal oxides are reduced by Al than carbon.
    (*ii*) Discuss the structure of BF_3
    (*iii*) Why BF_3 acts as a Lewis acid?
    (*iv*) Why BF_3 is a weaker Lewis acid than BCl_3?
12. Explain why
    (*i*) Al forms $[Al(OH)_6]^{3-}$ while $[B(OH)_6]^{3-}$ is not known.
    (*ii*) TlF_3 is known but TlI_3 is not known.
13. Explain why $B(OH)_3$ is acidic, $Al(OH)_3$ is amphoteric and $Tl(OH)_3$ is basic.
14. (*i*) Write a short note on borates.
    (*ii*) What is back bonding? Why does it occur in boron halides and not in Aluminum halides?
15. Explain the relative Lewis acid strength of boron tri-halides.
    $BI_3 > BBr_3 > BF_3$
16. Explain the methods of preparation of borazine. Discuss structure of borazine, what are the addition reactions and pyrolysis of borazine?
17. What happens when deterrence reacts with:
    (*i*) NH_3 at low and high temperature
    (*ii*) Chlorine at 298 K
18. (*i*) What is inorganic benzene?
    (*ii*) Why is $MgSO_4$ soluble in water whereas B_4SO_4 is insoluble.

12

Group 14 Elements or Carbon Family of Elements

Carbon has one of the most widespread occurrence in nature. It is one of the constituent element of living matter. It exists in nature as proteins, carbon hydrates and fats. Carbon dioxide is used in photosynthesis in plants and in their process of respiration. The property of catenation of carbon is responsible for very large of carbon compounds. Some of these compounds are classified in organic chemistry while others form a part of inorganic chemistry such as carbon black, coke, graphite, carbonates, sulphides, halides, carbon dioxide, carbon monoxide and carbides etc. These days lot of interest is shown in organometallic compounds, carbonyls and π-bonding complexes. Presently silicon is finding use in human technology namely cement, ceramics, clays, bricks, glass, and other silicon polymers. Its use is important in transistors and computer and mobile chips.

Germanium is a less known element but tin and lead are wall known. Tin and lead are in use from very early times.

Electronic Configuration and Oxidation States*

The elements included in group 14 are carbon, silicon, germanium, tin and lead. The electronic configurations of these elements has been given is the Table 12.1.

Table 12.1 Electronic Structures and Oxidation States

Element		Electronic structure		Oxidation states	
Carbon	C	[He]	$2s^2\ 2p^2$		IV
Silicon	Si	[Ne]	$3s^2\ 3p^2$	(II)	IV
Germanium	Ge	[Ar]	$3d^{10}\ 4s^2\ 4p^2$	II	IV
Tin	Sn	[Kr]	$4d^{10}\ 5s^2\ 5p^2$	II	IV
Lead	Pb	[Xe]	$4f^{14}\ 5d^{10}\ 6s^2\ 6p^2$	II	IV

* The most important oxidation states (generally the most abundant and stable) are shown in bold. Other well-characterized but less important states are shown in normal type. Oxidation states that are unstable, or in doubt, are given in parentheses.

Occurrence: Carbon is 17th and silicon is the second most aboundant element present on earth's crust by weight.

Table 12.2 Abundance of the Elements in the Earth's Crust by Weight

	ppm	Relative abundance
C	180	17
Si	272000	2
Ge	1.5	54
Sn	2.1	49 =
Pb	13	36

Carbon is found in its elemental form. Graphite is found is large quantities and extremely small quantity of diamond are mined. Germanium is very rare and is found in traces in the ores of other elements like carbon. Sn and Pb are present in their own ores. CO_2 and CO are industrially important compounds, CO_2 occurs in small amounts in the atmosphere, CO_2 has an important role in the carbon cycle, photosynthesis and respiration, CO forms carbonyl complexes, Silicon occurs as SiO_2 (sand) and silicates. Ge is present in Zinc and sliver ores. Sn is found in ore casseterite SnO_2 and lead as galena PbS.

General Group Discussion

Carbon block (soot) is produced by incomplete combustion of hydrocarbons from natural gas or oils. It is in the form of fine particles. It is used in rubber industry (tyres) and black ink.

Coke is obtained by burning or carbonisation of coal (in the absence of air) coke is important in metallurgical processes for the extraction of iron and other metals (reduction).

$$SiO_2 + 2C \xrightarrow{\Delta} SiC + 2CO \xrightarrow{2500°C} C\ \text{graphite (s)} + Si(g)$$

Graphite is used in making electrodes used in steel industry. It is also used in nuclear reactors to slow down the speed of neutrons.

Activated Charcoal

It is produced by oxidising sawdust or peat. It is used to purify and bleach sugar. It absorbs poisonous gases in gas masks (mines), sewage plants and as a catalyst in many reactions.

Silicon large quantities of silicon are manufactured annually. It is used in the production of corrosion resistant steel. Ferro silicon alloy is produced by reducing SiO_2 and with scrap iron and coke.

$$SiO_2 + Fe + 2C \longrightarrow FeSi + 2CO$$

Elemental Si is made by reducing SiO_2 with coke at high temperature. The excess SiO_2 is added to prevent the formation of SiC. High quality silicon is produced by converting Si to $SiCl_4$ and then reducing it with Mg or Zn.

$$SiO_2 + 2C \longrightarrow Si + 2CO$$

$$Si + 2Cl_2 \longrightarrow SiCl_4$$

$$SiCl_4 + 2Mg \longrightarrow Si + 2MgCl_2$$

Silicon is an insulator when pure. It is used to make *p*-type and *n*-type semi-conductors by doping. Very pure silicon is used in electronics industry and computer chips. Semi-conductor quality of Si can be made by sodium reduction of fluroapatite.

$$Na_2[SiF_6] + 4Na \longrightarrow Si + 6NaF$$

Germanium (Ge)

It can be obtained from coal ash but it is preferred to get it from flue dust recovered from melting Zn ores. This is a multistep process, GeO_2 recovered from flue dust is reduced by H_2 at 500°C.

$$GeO_2 + 2H_2 \xrightarrow{500^\circ C} Ge + 2H_2O$$

Purified Ge can be obtained by zone refining treatment. Very pure Ge is used in making transistors and semi-conductor devices. It is transparent to intra-red light therefore it is used in making prism and lenses.

Tin (Sn)

Sn occurs as cassiterite ore (SnO_2). It is reduced to the metal by reducing with carbon at 1200 – 1300°C, in an electric furnace. It is often mixed with Fe. This makes Sn very hard. To remove Fe, air is passed through the molten mixture to oxidise Fe to FeO. It is used in electroplating steel to make tin-plate. It is very useful in making number of alloys.

(*i*) Solder is Sn/Pb alloy

(*ii*) Bronze is Cu/Sn

(*iii*) Gun metal Cu/Sn/Pb/Zn

(*iv*) Pewter is Sn/Sb/Cu.

SnO_2 is used as a glaze in ceramics. SnO_2 can be mixed with other metal oxides to make a pigment for making pottery. It is used to toughen the glass and make the glass scratch proof.

Lead (Pb)

The most important ore of lead is galena PbS. Lead is black, shiny and weighty metal. Galena is purified and separated from other minerals by the process of froth flotation. The element is finally extracted by two methods.

(*i*) Roasting in air to form PbO and then reduce with coke in a blast furnace.

$$2PbS + 3O_2 \longrightarrow 2PbO + SO_2\uparrow$$
$$2PbO + 2C \longrightarrow 2Pb + 2CO$$

(*ii*) PbS is partially oxidised by heating by flowing air through it. Then after sometime air is turned off but heating is continued.

(a) $$\underset{\text{(air)}}{3PbS + O_2} \xrightarrow{\Delta} PbS + 2PbO$$

$$PbS + 2PbO \longrightarrow 3Pb + SO_2\uparrow$$

This process is called self-reduction. Pb so obtained contains metallic impurities like Cu, Ag, Au, Sn, As, Sb, Bi, and Zn. The impurities are removed by cooling near the freezing point of Pb, first Cu and Zn containing Ag and Au solidify. As, Sb and Sn get oxidised to their oxides As_2O_3, Sb_2O_3 and SnO_2.

Allotropic Forms

Carbon shows a large number of allotropic forms. Allotropy is a phenomenon in which an element shows different physical shapes and structures with different physical properties but the same chemical properties. The allotropes of carbon are diamond, α and β-graphite, rare hexagenal diamond and C_{60} which are clusters of carbon atoms collectively called fullerenes. Si, Ge and Sn have diamond like structure. Ge is a liquid but it expands on heating and forms a solid.

$$\underset{\substack{\text{diamond}\\\text{structure}}}{\alpha - Sn} \xrightleftharpoons{13.2°C} \underset{\text{metallic}}{\beta - Sn}$$

Diamond is unreactive while graphite is very reactive.

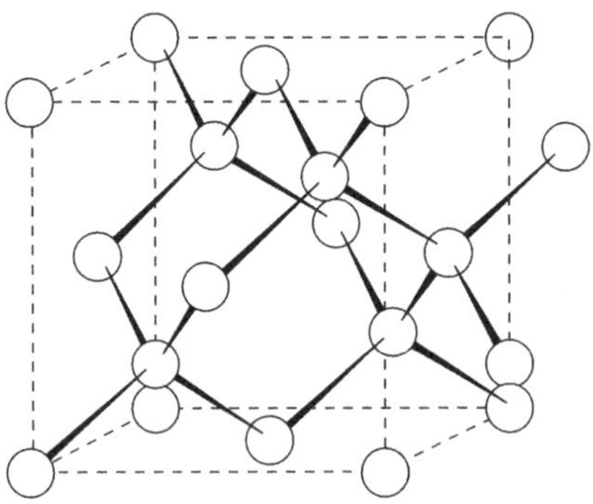

Fig. 12.1 The crystal structure of diamond.
(Wells, A.F., *Structural Inorganic Chemistry, Clarendon Press, Oxford.*)

Raw diamond is black and treated diamonds are transparent. Blue diamond contains a trace of Al. Diamond has each C-atoms tetrahedrally disposed and surrounded by 4 other C-atoms with bond length —C—C— = 1.54 Å. The tetraherals are linked together three dimensionally to form giant molecules. They form strong covalent bond and melt at 3930°C. The structure is very hard and it is utilised to cut glass. The tetrahedra are modified to give Wurtzite structure.

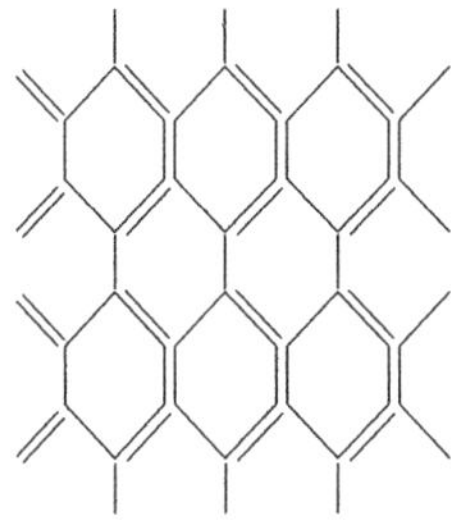

(a)

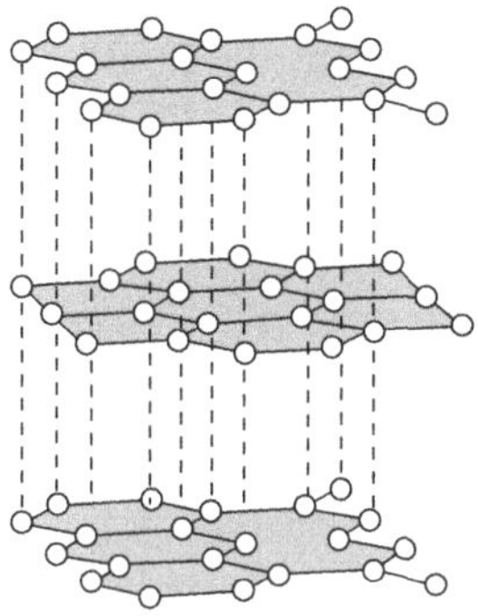

(b)

Fig. 12.2 (a) The structure of a graphite sheet. (b) Structure of α-graphite.

Graphite is a two-dimensional sheets of carbo atoms, graphite has a layer structure in which sheets are held together by weak van der. Waal's forces.

Graphite involves three electrons of each C–atom. The fourth electron forms π-bonds with sp^2 hybridisation. The π-electrons surround the whole sheet. As the fourth electron is mobile therefore it acts as a good conductor of electricity. Graphite can be converted to synthetic diamonds at 1600°C at a pressure of 50,000 atmospheres.

Carbon Dating

This technique can be used to measure the approximate age of trees, rocks and archaeological objects. Ordinarily carbon occurs as ^{12}C isotope but it is always mixed with a small amount of ^{14}C isotope. ^{14}C is a radioactive isotope. It is bombarded by cosmic neutrons.

$$^{14}_{7}N + {}^{1}_{0}n \longrightarrow {}^{14}_{6}C + {}^{1}_{1}H$$

Carbon reacts with dioxygen in air to form CO_2 which is used by the green plants in photosynthesis to get glucose and other carbohydrates. Glucose in turn is used by plants to make starch, proteins and cellulose. All plant tissues contain small amounts of ^{14}C. Animal eat plants and ^{14}C goes to animals. ^{14}C is weakly radioactive and undergoes β-decay. It has a half life of 5668 years. In plants and animals when they are alive there exists a balance between intake of radiocarbon and that of loss due to decay. When plants and animals die radiocarbon ceases but β-decay continues. Thus an old sample of wood, cloth, paper and leather is less radioactive than a fresh sample. By measuring the radioactive decay rate it can be estimated how long ago the plant or animal died. This technique is now used to know the age many other objects.

Physical Properties

Covalent Radii: Covalent radii increases down the group. Difference in the size of Si and Ge is less than the expected difference because Ge has a full 3*d* shell which shields the nuclear charge ineffectively. Sn and Pb has a small difference in size because Pb is intercepted by 4*f* shell. Due to poor shielding effect of 4*f* nuclear charge pulls the shells inwards and therefore the expected increase in size is not observed.

Table 12.3 Radii, Melting Points and Electronegativity Values

1	Covalent radius (Å)	Ionization energies ($kJ\ mol^{-1}$)				Melting point (°C)	Boiling point (°C)	Pauling's electro-negativity values
		1st	2nd	3rd	4th			
C	0.77	1086	2354	4622	6223	4100		2.5
Si	1.17	786	1573	3232	4351	1420	3280	1.8
Ge	1.22	760	1534	3300	4409	945	2850	1.8
Sn	1.40	707	1409	2943	3721	232	2623	1.8
Pb	1.46	715	1447	3087	4081	327	1751	1.8

Ionization Energy: The decrease in ionization energy is observed from C to Si but then onwards the effect is irregular due to filling of *d* and *f* shells. The energy required to form M^{4+} is very large therefore ionic compounds are seldom formed. The elements like F and O due to very high electronegativity can give ionic compounds. SnF_2, PbF_2, SnF_4, SnO_2 and PbO_2 are considerably ionic.

Covalent Compounds: Most of the covalent compounds are 4-covalent. Here, all the four valency electrons take part in bonding. In accordance with the VBT theory electrons are promoted from the ground state to the excited state. The energy required to unpair and promote to the higher level is compensated by the energy release by forming two extra covalent bonds. The four orbitals are tetrahedrally disposed with sp^3 hybridisation.

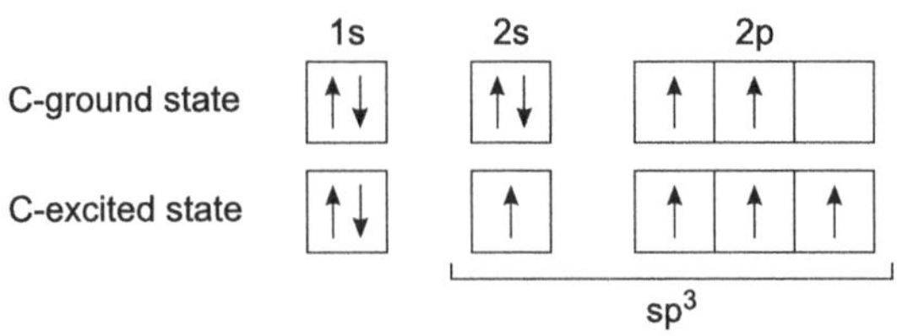

Metallic and Non-Metallic Behaviour

C and Si are non-metals, Ge has some metallic behaviour and Sn and Pb are metals. Due to stepwise increase in the metallic character the metals show malleability and electrical conductivity. Down the group the elements show increase in the tendency to form M^{2+} ions and likewise formation of acidic and basic oxides.

Melting Points: Melting points of C is extremely high Si has considerably low melting point in comparison to C. Si and Ge are still high melting. They all show diamond like lattice where the covalent bonds are quite strong and hence lot energy is required to break these bonds. In Sn and Pb we come across M–M bonds which give lower melting points due to increase in size. They do not use all the four valency electrons in metallic bonding.

Chemical Reactivity: The reactivity increases down the group because down the group M^{2+} oxidation state becomes stable. Pb often exhibits its unreactiveness due to the coating of oxide film on its surface and also due to high potential of reduction of H^+ to H_2.

C, Si and Ge do not react with water. Sn forms SnO_2 and H_2 with steam and Pb is unreactive to water due to protective oxide film.

C, Si and Ge are not affected by dilute acids. Pb dissolves in dilute HCl to form insoluble $PbCl_2$, Pb dissolves is dilute HNO_3 forming $Pb(NO_3)_2$ and oxides of nitrogen. Pb also dissolves in organic acid like oxali acid and acetic acid but it does not dissolve in dilute H_2SO_4 and forms insoluble $PbSO_4$. Diamond is not affected by the concentrated acids but graphite does dissolve.

Alkalis: C is unaffected by alkalis, Si reacts readily with hot alkalis to form $[SiO_4]^{4-}$ silicates. Sn and Pb attached quickly by hot alkali forming stamates and plumbates. Sn and Pb are amphoteric

With Halogens: Carbon and graphite react at 500°C forming fluorocarbons and fluorides, Si and Ge react with all halogens forming SiX_4 and GeX_4. Sn and Pb are less reactive, Sn reacts with Cl_2 and Br_2 in cold and F_2 and I_2 on warming giving SiX_4.

Inert Pair effect: On moving down the group there occurs decrease in the stability of +IV oxidation state and increase in the stability of +II oxidation state. Sn (+11) is ionic and Sn(+IV) is covalent. Sn(+II) is a strong reducing agent. Pb(+II) is stable, ionic and more common than Pb(+IV). According to Fajan's rule the M^{2+} radius is greater than M^{4+} ion. The smaller ion has greater tendency to form covalent compounds.

Standard Reduction Potential (Volts)

Acid solution	Basic solution
Oxidation state +IV +II 0 $Sn^{2+} \xrightarrow{+0.13} Sn^{2+} \xrightarrow{-0.14} Sn$ $PbO_2 \xrightarrow{+0.33} Pb^{2+} \xrightarrow{-0.13} Pb$	+IV +II 0 $[Sn(OH)]_6^{2-} \xrightarrow{-0.90} HSnO_2^- \xrightarrow{-0.91} Sn$ $PbO_2 \xrightarrow{+0.28} PbO \xrightarrow{-0.54} Pb$

Compounds of Carbon

Carbides: Carbon compounds are generally named as carbides although carbon is less electronegative except with N, P, O, S and halogens. There are three different types of carbides.

(*a*) Ionic or salt-Like

(*b*) Interstitial (metallic)

(*c*) Covalent

Ionic carbides: In these carbides, carbon behaves as an anion (C_2 or C_3 anions) Be_2C and Al_4C_3 belong to this category. Be_2C and Al_4C_3 react with H_2O to produce methane and therefore they may be called **Methanides**.

Group 1 and group 2 carbides are C_2 unit, coinage metals Cu, Ag, Au, Zn and Cd come in their ambit. Some of lanthanons also come in this category LnC_2 or Ln_4C_2 where carbon is 2^-. The most important of these is CaC_2 (calcium acetylide). It liberates acetylene gas on reaction with water.

Interstitial carbides: The transition metals and some of the lanthanides and actinides Cr, Mn, Ni, Fe, Co form a wide variety of carbides. They are typically infusible mass with high melting points TaC has a melting point of 3900°C. Due to their hardness they are used in cutting tools. They do show some metallic properties hence they are also called **metallic carbides**. In these compounds carbon occupies octahedral holes in the close packed arrangements. If all the octahedral holes are occupied then they exhibit the formula MC, Fe_3C cementite is an important constituent of steel. These carbides are hydrolysed by dilute acids and sometimes by water producing hydrocarbons and H_2.

Some carbides like TiC, ZnC, HfC, VC, NbC, TaC, CrC and MoC represent NaCl type structure.

Covalent Carbides

The most important of these are SiC and B_4C. SiC is very hard, infusibles and chemically inert. It is used as abrasive called Carborundum. It is produced by heating sand with coke at 2000–2500°C

$$SiO_2 + 2C \longrightarrow CO_2 + SiC$$

SiC is unreactive. It reacts with NaOH, air and Cl_2 at 100°C.

$$SiC + 2NaOH + 2O_2 \longrightarrow Na_2SiO_3 + CO_2 + H_2O$$

$$SiC + 2Cl_2 \longrightarrow SiCl_4$$

SiC is often dark purple, black or dark green. Pure SiC is pale yellow to colourless. SiC is three dimensional tetrahedral structure. Boron carbide is harder. It is used as abrasive and a shield for radiation.

Hydrides

Hydrides are formed by all the group 14 elements, carbon forms number of chain and ring compounds namely:

(*a*) Alkanes or Paraffins $C_nH_{2n} + 2$

(*b*) Alkenes of olefins C_nH_{2n}

(*c*) Alkynes or acetylenes C_nH_{2n-2}

(*d*) Aromatic compounds.

Due to the unique property of catenation, *i.e.*, —C—C— chain formation a very large number of carbon compounds are formed which forms the basis of organic chemistry a very important branch of chemistry.

Silicon forms a limited number of saturated hydrides Si_nH_{2n+2} or silanes, silicon-silicon chain is limited to 8 silicon atoms. Analogues of alkanes and alkynes are not reported. Monosilane SiH_4 is the only hydride that is important continuation.

It was first prepared by treating Al/Si alloy with dilute HCl. A mixture of silicon hydrides were obtained by treating Mg_2Si hydrolysis with dilute H_2SO_4 or phosphoric acid. Silanes are colourless gases or volatile liquids.

Silanes are very reactive and explode in air. Barring SiH_4 other silanes are thermally unstable. Their properties can only be studied under vacuum conditions.

$$\begin{aligned} Mg_2Si + H_2SO_4 &\longrightarrow SiH_4 \quad (40\%) \\ &\longrightarrow Si_2H_6 \quad (30\%) \\ &\longrightarrow Si_3H_8 \quad (15\%) \\ &\longrightarrow Si_4H_{10} \quad (10\%) \\ &\longrightarrow \left.\begin{matrix} Si_5H_{12} \\ Si_6H_{14} \end{matrix}\right] \quad (5\%) \end{aligned}$$

Currently SiH_4 is being produced by reducing $SiCl_4$ with LiH, NaH or $LiAlH_4$ in ether at low temperature. This method gives only SiH_4 and not a mixture of silanes. The yield of the product is quite appreciable.

$$SiCl_4 + Li[AlH_4] \longrightarrow SiH_4 + AlCl_3 + LiCl$$

$$Si_2Cl_6 + 6LiH \longrightarrow Si_2H_6 + 6LiCl$$

$$Si_3Cl_8 + 8NaH \longrightarrow Si_3H_8 + 8NaCl$$

Another method of preparation involves direct heating of Si or Fe/Si with anhydrous HX or RX in the presence of Cu catalyst.

$$Si + 2HCl \xrightarrow{\Delta} Si\,H_2\,Cl_2$$

$$Si + 3HCl \xrightarrow{\Delta} Si\,HCl_3 + H_2$$

$$Si + 2RCl \xrightarrow{\Delta} R\,Si\,HCl_2 + H_2 + C$$

Silanes are more reactive than alkanes. Silanes are strong reducing agents, ignite or explode in Cl_2. Pure silanes do not react with water or dilute acids in silica equipment. They readily hydrolyse in alkaline solution

$$Si_2H_6 + (4+n)\,H_2O \xrightarrow[\text{alkali}]{\text{minute}} 2SiO_2 + nH_2O + 7H_2$$

They undergo hydrosilation reaction with alkenes in the presence of Pt.

$$R-CH=CH_2 + SiHCl_3 \xrightarrow{\text{Pt}} RCH_2CH_2SiCl_3$$

Difference between alkanes and silanes

(*i*) On the electronegativity scale

$$C = 2.5$$
$$Si = 1.8$$
$$H = 2.1$$

Therefore bonding electrons between C and H or Si – H are not equally shared. Hence C and Si have δ^- charge and δ^+ charge. Thus Si will invite nucleophilic attack.

(*ii*) Si has a large size therefore the attack becomes easier.

(*iii*) Si entails lower energy due to *d* orbitals thus it forms an intermediate compounds so as to lower the activation energy.

Germanium hydrides or germanes are known up $n = 5$. They are straight chain, colourless gases or volatile liquids. They are less volatile, less inflammable and unaffected by water acids or alkalis in comparison to silanes.

GeH_4 is obtained by:

$$GeCl_4 + Li[AlH_4] \xrightarrow[\text{ether}]{\text{Dry}} GeH_4 + LiCe + AlCl_3$$

$$GeO_2 + Na[BH_4] \xrightarrow[\text{solution}]{\text{aqueous}} GeH_4 + NaBO_2$$

stannane SnH_4 is less stable. It can be obtained by reducing $SnCl_4$ with $Li[AlH_4]$. It does not react with dilute acids or alkalis. Si_2H_6 is still less stable. Plumbane PbH_4 is even less stable and difficult to prepare.

Oxygen Compounds

Carbon forms a large number of oxides CO, CO_2, C_3O_2, C_4O_2, $C_{12}O_9$, C_2O and C_2O_3. Out of these CO and CO_2 are most important. There oxides are formed as a result of pπ–pπ multiple bonds between C and O atoms.

Carbon Monoxide CO: The laboratory preparation of CO involves dehydration of formic and with concentrated H_2SO_4.

$$HCOOH + H_2SO_4 \longrightarrow CO + H_2O$$

CO is a colourless and odourless gas which burns with a blue flame. It reduces aqueous $PdCl_2$ to metallic Pd.

It liberates I_2 from I_2O_5,

(*i*) $$PdCl_2 + CO + H_2O \longrightarrow Pd + CO_2 + 2HCl$$

(*ii*) $$5CO + I_2O_5 \longrightarrow 5CO_2 + I_2$$

Reaction (*ii*) is used to estimate CO as the amount of I_2 is titrated against $Na_2S_2O_3$.

The toxic nature of CO is because of the formation of complex with haemoglobin in the blood $[Fe(CO)_5]$. This prevents haemoglobin to carry oxygen to the different parts of the body. Due to deficiency of oxygen it may cause unconsciousness and finally death. CO is sparingly soluble in water and is neutral oxide. CO keeps the body warm because it gives good amount of heat when it burns in air.

$$2CO + O_2 \longrightarrow 2CO_2 \qquad \Delta H = -565 \text{ kJ mol}^{-1}$$

It acts as an **industrial fuel**.

(*i*) Water gas ($CO + H_2$). It is an equimolecular mixture of CO and H_2.

(*ii*) Producer gas ($CO + N_2$)

(*iii*) Coal gas is a mixture of CO, H_2, CH_4 and CO_2 produced by distilling coal.

It is supplied to homes as cooking gas.

It is now being replaced by natural gas CH_4 but it is still in use.

The functioning of these fuel gases has been briefed below:

Water gas: It is made by blowing steam over red hot coke.

$$C + H_2O \xrightarrow[\text{hot}]{\text{red}} \underbrace{CO + H_2}_{\text{water gas}}$$

Amount of heat

$$\Delta H^\circ = +131 \text{ kJ mol}^{-1}$$
$$\Delta S^\circ = +134 \text{ kJ mol}^{-1}$$

The burning of water gas is strongly exothermic therefore coke cools down and steam is discontinued, water gas has a high calorific value so it acts as a good fuel. Both CO and H_2 burn and give out heat.

Producer gas ($CO + N_2$): It is produced by blowing air through red hot coke.

$$C + \underbrace{O_2 + 4N_2}_{\text{air}} \longrightarrow CO_2 + 4N_2$$
$$\downarrow + C$$
$$\underbrace{2CO + 4N_2}_{\text{Producer gas}}$$

The overall reaction is exothermic. The coke does not cool down as in water gas.

$$2C + O_2 \longrightarrow 2CO \Delta H^\circ = -221 \text{ kJ mol}^{-1}$$
$$\Delta S^\circ = +179 \text{ kJ mol}^{-1}$$

Producer gas has a low calorific value.

It is less efficient than water gas. In producer gas N_2 does not burn.

The contents of gases in the producer gas are:

$N_2 = 70\%$, $CO = 25\%$ and $CO_2 = 4\%$

plus traces of CH_4, H_2 and O_2.

CO is a good reducing agent. It is used in the metallurgy of Iron.

$$Fe_2O_3 + 3CO \xrightarrow[\text{furnace}]{\text{Blast}} 3CO_2 + 2Fe$$

$$ZnO + CO \longrightarrow CO_2 + Zn$$

Complexing agent: CO is an important ligand. It can donate electron pair to transition metals forming carbonyls. The bonding in carbonyls is more complicated than it appears.

Table 12.4 Binary metal carbonyls formed by the row transition elements.

Sc	Ti	V	Cr	Mn	Fe	Co	Ni	Cu	Zn
		$V(CO)_6$	$Cr(CO)_6$	$Mn_2(CO)_6$	$Fe(CO)_5$	$Co_2(CO)_8$	$Ni(CO)_4$		
					$Fe_2(CO)_5$	$Co_2(CO)_{12}$			
					$Fe_3(CO)_{12}$	$Co_6(CO)_{16}$			

Carbonyls may be prepared by the following reactions:

(*i*) $Ni + 4CO \xrightarrow{50°C} Ni\,(CO)_4$

(*ii*) $Fe + 5CO \xrightarrow[\text{under press.}]{200°C} Fe(CO)_5$

(*iii*) $2Fe\,(CO)_5 \xrightarrow{\text{Photolysis}} Fe_2(CO)_9 + CO$

(*iv*) $CrCl + 3Fe(CO)_5 \xrightarrow{\Delta} Cr(CO)_6 + 3FeCl_2 + 9CO$

The reaction is used for purifying Ni metal at 50°C it decomposes to give pure Ni metal.

$$Ni(CO)_4 \xrightarrow{230°C} Ni + 4CO$$

Water gas is mainly used as a sources of CO.

The bonding in CO may be shown as 3 electron pairs shared between C–O atoms.

$$\overset{\times}{\underset{\times}{}}C\vdots\vdots O: \quad \text{or} \quad C{\equiv}O$$

According to molecular orbital theory it is represented as:

$$\sigma 1S^2, \sigma^* 1S_2, \sigma 2S^2, \sigma^* 2S^2 \begin{bmatrix} \grave{A}^* p_y^2 + \\ \grave{A}^2 p_z^2 + \end{bmatrix} \sigma^2 p^2{}_z + \sigma^* 2p^0{}_x + \begin{bmatrix} \grave{A}^* 2p_y^0 \\ \grave{A}^* 2p_z^0 \end{bmatrix} \text{increasing}$$

energy

The metal-carbon bond is given by donation of electron pair from carbon to the metal atom $M \leftarrow : C \equiv O$. The original Metal-carbon σ bond is weak which is reinforced by back bonding called a dative π bond. This entails sideways overlap of a *dxy* orbital on the metal with empty anti-bonding π* 2*py* orbital of the carbon atom forming a π M → C bond. The overall band is thus M═C═O. The bond order between C—O triple bond is changed to C—O double bond. Therefore metal carbonyls show an increase in bond length from 1.128 Å in CO to 1.15Å in carbonyls. CO is a strong ligand and is called π acceptor ligand. The π electron drift due to Pauling's electro neutrality principle from metal to carbon makes the ligand more negative.

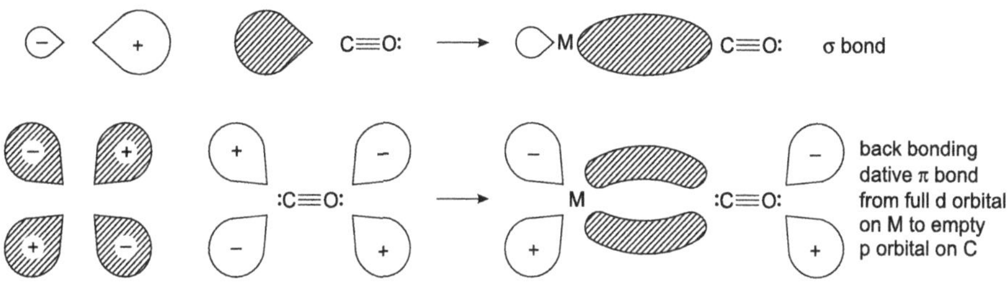

Fig. 12.3 Schematic of orbital overlaps in metal carbonyls. (After N.N. Green wood and A. Earnshaw, *Chemistry of the Elements*, Pergamon, 1984, p. 351.)

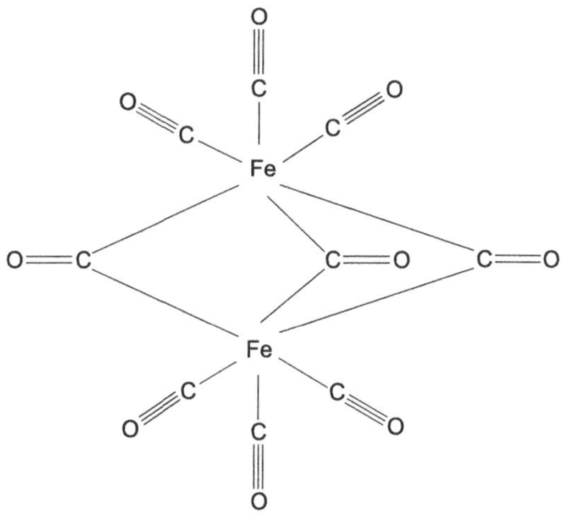

Fig. 12.4 Structure of $Fe_2(CO)_9$

This highlights the versatility of CO ligand. It may also form bridging groups between two metal atoms, *e.g.*, $[Fe_2(CO)_9]$

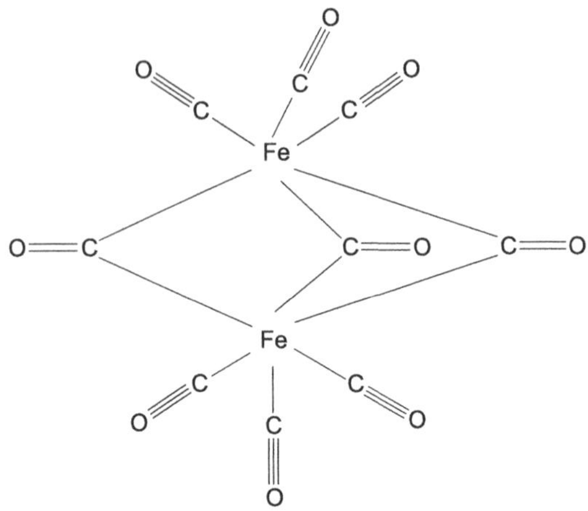

Fig. 12.5

CO readily combines with S, O and halogens

$$CO + \frac{1}{2}O_2 \longrightarrow CO_2$$

$$CO + S \longrightarrow COS \text{ carbonyl sulphide}$$

$$CO + Cl_2 \longrightarrow COCl_2 \text{ Carbonyl chloride.}$$

Carbonyl halides readily hydrolyse and react with ammonia to give urea.

$$COCl_2 + H_2O \longrightarrow 2HCl + CO_2$$

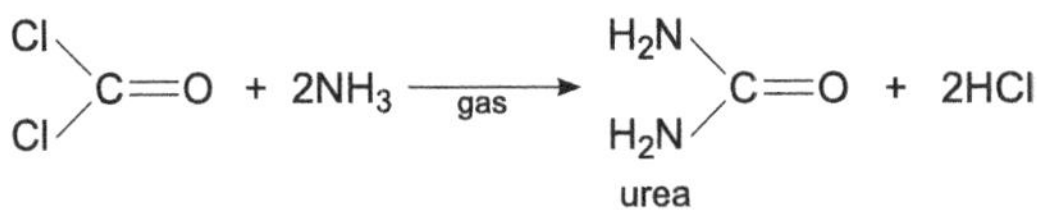

Fig. 12.6

Carbon Dioxide (CO_2)

It is a colourless, odourless gas. In small amounts it is used as a fire extinguisher. It is harmful when inhaled in large amounts.

It is a major by-product obtained in the manufacture of hydrogen which in turn is used for making ammonia.

$$CO + H_2O \rightleftharpoons CO_2 + H_2$$

$$CH_4 + 2H_2O \longrightarrow CO_2 + 4H_2$$

It is also produced in the fermentation process. CO_2, can be stored by absorbing in aqueous Na_2CO_3 from where it can be recovered time-to-time.

$$\underset{\text{glucose}}{C_6H_{12}O_6} \xrightarrow{\text{Yeast}} 2C_2H_5OH + 2CO_2\uparrow$$

It can be obtained from $CaCO_3$ by strong heating

$$CaCO_3 \xrightarrow{\Delta} CaO + CO_2$$

On smaller scale it can be produced by the action of dilute acids on carbonates or by burning carbon in excess of air or oxygen.

$$CaCO_3 + 2HCl \longrightarrow CaCl_2 + CO_2 + H_2O$$

$$C + O_2 \underset{\text{hot}}{\overset{\text{cool}}{\rightleftharpoons}} CO_2$$

CO_2 can also be recovered from $NaHCO_3$.

(*i*) $Na_2CO_3 + CO_2 + H_2O \underset{\text{hot}}{\overset{\text{cool}}{\rightleftharpoons}} 2NaHCO_3$

Gurbotol Process:

(*ii*) $\underset{\text{ethanolomine}}{2HOCH_2CH_2NH_2} + CO_2 + H_2O \underset{100-150°C}{\overset{30-60°C}{\rightleftharpoons}} (HOCH_2CH_2NH_3)_2CO_3$

Under pressure CO_2 gas can be liquefied between –57°C and + 31°C.

It is 80% liquid and 20% solid, solid CO_2 appears like snow and therefore it is called 'Dry ice'.

Solid CO_2 is expanded and filled in gas cylinders. It sublimates into the vapour state without changing into liquid state, it is used as a refrigerant to preserve frozen foods and ice creams. It finds immense use in aerated or carbonated drinks. Other uses include manufacture of urea fertilizer, formaldehyde resins and inflating life rafts.

$$CO_2 + 2NH_3 \xrightarrow[\text{press}]{180°} H_4NCO_2NH_2 \rightarrow \underset{\text{urea}}{H_2NCONH_2} + H_2O$$

Test for CO_2 gas: CO_2 gas when passed through lime water $Ca(OH)_2$ or $Ba(OH)_2$ give white precipitate of $CaCO_3$ and $BaCO_3$. If more of CO_2 is passed the mixture the haziness disappears, and it turns clear due to the formation of a bicarbonate.

$$Ca(OH)_2 + CO_2 \longrightarrow CaCO_3\downarrow + H_2O$$

$$Ba(OH)_2 + CO_2 \longrightarrow \underset{\text{white ppt.}}{BaCO_3\downarrow} + H_2O$$

More of CO_2

$$CaCO_3 + CO_2 + H_2O \longrightarrow \underset{\text{soluble}}{Ca(HCO_3)_2}$$

CO_2 is an acidic oxide like most other oxides of non-metals. It reacts with bases to form salts.

$$CO_2 + H_2O \longrightarrow \underset{\text{Carbonic Acid}}{H_2CO_3}$$

The solution may contain carbonates and bicarbonates. Carbonic acid is never isolated but forms salts such as carbonates and bicarbonates.

$$H_2CO_3 + NaOH \longrightarrow NaHCO_3 \text{ (sodium bicarbonate)}$$
$$H_2CO_3 + NaOH \longrightarrow Na_2CO_3 \text{ (sodium carbonate)}$$

CO_2 can also behaves as a ligand. CO_2 complexes are $[(RhCO_2)\,Cl\,(PR_3)_3]$ and $[Co(CO_2)\,(PPh_3)_3]$. In the first case CO_2 is bonded to the metal and in the second case CO_2 acts as a bidentate ligand with one C and one O atom bonded to the metal, *i.e.,* CO_2 is bent molecule.

Structure of CO_2: The structure of CO_2 is linear O—C—O. Both C—O bonds are of same length. In addition to 2σ bonds between C and O atoms there is 3 centre four electron π bond spread over all the three atoms. Therefore there exist 2 α bonds and 2 π bonds in CO_2.

Photosynthesis: CO_2 is involved biologically by the process of photosynthesis in plants, CO_2 and H_2O undergo photosynthesis in plants to form glucose sugar. Animal and plant life depends upon this reaction.

$$6CO_2 + 6H_2O \xrightarrow{h\nu} \underset{\text{glucose}}{C_6H_{12}O_6} + 6O_2$$

The reverse reaction takes place during respiration both in plants and animals and that releases energy.

$$C_6H_{12}O_6 + 6O_2 \longrightarrow 6CO_2 + 6H_2O + \text{energy}$$

Carbon suboxides C_3O_2: It is foul smelling gas boiling at 6°C. It is prepared by dehydrating malonic acid with P_4O_{10}.

$$CH_2\begin{matrix}\diagup COOH \\ \diagdown COOH\end{matrix} \xrightarrow[\text{with } P_4O_{10}]{150^\circ} C_3O_2 + 2H_2O$$

It is stable at –79°C and has a linear structures It is yellow solid at room temperature. At higher temperature it turns red and purple solid.

Properties

(*i*) Reacts with H_2O to give malonic acid

$$O{=}C{=}C{=}C{=}O + H_2O \longrightarrow CH_2\begin{matrix}\diagup COOH \\ \diagdown COOH\end{matrix}$$

(*ii*) Reaction with HCl

$$C_3O_2 + 2HCl \longrightarrow H_2CCO\begin{matrix}\diagup Cl \\ \diagdown Cl\end{matrix} \quad \text{acid chloride}$$

(*iii*) With NH_3

$$C_3O_2 + 2NH_3 \longrightarrow H_2C\text{–}(CONH_2)_2 \text{ acid amide}$$

The other important suboxide is $C_{12}O_9$ which is a white solid.

Carbonates

The important salts derived from carbonic acid are carbonates CO_3^{2-} and bicarbonates (hydrogen carbonate) HCO_3^-. The CO_4^{4-} does not exist because it is flat while SiO_4^{4-} does exist. This is attributed to the small size of carbon. The NO_3^- ion and PO_4^{3-} ion are having similar situation.

Structure of CO_3^{2-} Ion

π-bond is delocalised covering all the four atoms, H^{11} CO_3 have the same structure. Na_2CO_3, $NaHCO_3$ and Li_2CO_3 are produced in huge amounts.

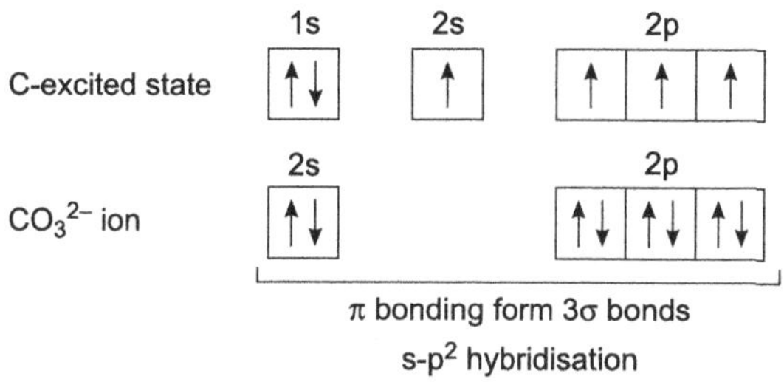

CO_3^{2-} ions are colourless hence alkali and alkaline earth metal carbonates are white. Ag_2CO_3 is yellow. $(NH_4)_2CO_3$ and alkali metal carbonates are water soluble except Li_2CO_3 which is only slightly soluble. Tl_2CO_3 is moderately soluble. Group 13 carbonates are insoluble. All the carbonates react with acids liberating CO_2.

$$Na_2CO_3 + 2HCl \longrightarrow 2NaCl + CO_2 + H_2O$$

Alkali metal carbonates are stable to heat and melt without decomposition. Alkaline earth metal carbonates decompose when heated strongly. The stability of alkaline earth metal carbonates increases with the increase in the size of metal ion.

$$CaCO_3 \xrightarrow{\Delta} CaO + CO_2$$

The decomposition temperature of group 2 carbonates is in the order.

$BeCO_3$	$MgCO_3$	$CaCO_3$	$SrCO_3$	$BaCO_3$
< 100°C	540°C	900°C	1290°C	1360°C

The solid bicarbonates known are those of alkali metals and NH_4^+ ion. They are less soluble than the carbonates. They decompose easily on heating.

Sulphides

The most important sulphide is that of carbon (CS_2). The boiling point of CS_2 is 46°C. It is highly inflammable with a flash point of 30°C. It ignites at 100°C. It affect brain and the nervous systems. Pure CS_2 smells like ether but impurities give it a very foul smell. It is produced by a gas phase reaction between CH_4 and sulphur in the presence of Al_2O_3 catalyst.

$$CH_4 + 4S \xrightarrow[Al_2O_3]{600°C} CS_2 + 2H_2S$$

It is an industrially important chemical.

1. It is used in the manufacture of viscose-rayon. CS_2 reacts with cellulose and NaOH.

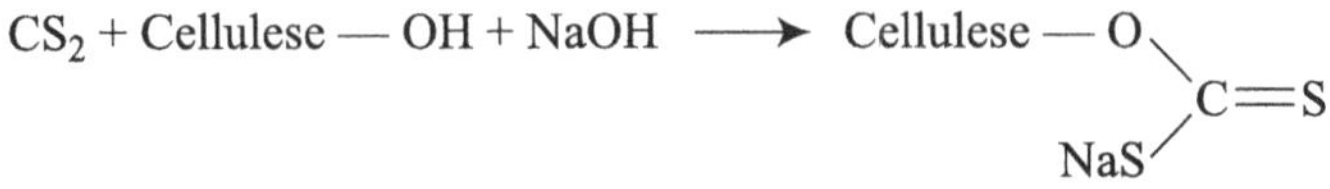

Sodium celluosic xanthate

   This is dissolved in aqueous alkali to give a viscous liquid called viscose. On acidification it reverts back to cellulose in the form of fibres or thin film.

2. It is used in the manufacture of CCl_4. ($CS_2 + 3Cl_2 \longrightarrow CCl_4 + S_2Cl_2$)
3. It is used as a solvent for the manufacture of rubber.

Properties

(*i*) **Reaction with aqueous NaOH:** With aqueous NaOH it gives a mixture of Na_2CO_3 and sodium trithio carbonates.

$$3CS_2 + 6NaOH \longrightarrow Na_2CO_3 + 2Na_2CS_3 + 3H_2O$$

(*ii*) With NH_3 it reacts to give ammonium dithio carbonate:

$$CS_2 + 2NH_3 \longrightarrow NH_4[H_2NCS_2]$$

CS_2 is a linear molecule quite akin to CO_2. It forms complexes very easily *e.g.*, $[Pt(CS_2)(PPh_3)]$. It acts as a bidenttate ligand with one carbon and one S bonded to the metal. CS_2 has a bent molecule. The S atoms may form bonds to other metal atoms. It is decomposed by sunlight therefore it is stored in colour bottles.

Oxides of Silicon

Silicon forms two oxides SiO and SiO_2. Silicon monoxide is prepared by reduction of SiO_2 with Si at high temperature.

$$SiO_2 + Si \longrightarrow 2SiO$$

The common term silica stands for SiO_2. It exists widely in sand and quartz. Silicon cannot form fouble bonds as it cannot utilise $p\pi$-$p\pi$ orbitals but instead it forms $p\pi$-$d\pi$ bonds because of the availability of *d* orbitals in silicon. It forms infinite three dimensional structure. SiO_2 has a high melting point.

In quartz, tridymite and cristobalite Si is tetrahedrally surrounded by four atoms. Each corner is linked to other tetrahedrons giving it an infinite array. The difference in these structures lies because of the way in which tetrahedral SiO_4 are arranged.

Silica in any form is unreactive. It is acidic in nature therefore it does not react with acids. It does react with HF to form silicon tetrafluoride which is hydrolysed to silicic acid.

$$4HF + SiO_2 \longrightarrow SiF_4 + 2H_2O$$
$$SiF_4 + H_2O \longrightarrow \underset{\substack{\text{white solid}\\\text{over drop}\\\text{of water}}}{Si(OH)_4} \text{ or } SiO_2 \cdot 2H_2O$$

SiO_2 being acidic dissolves in aqueous alkali or fused carbonate M_2CO_3 to give silicates.

$$SiO_2 + NaOH \longrightarrow (Na_2SiO_3)n \text{ and } Na_4SiO_4$$

This reaction is responsible for ground glass stoppers in reagent bottles. Out of halogens only fluorine attacks SiO_2.

$$SiO_2 + 2F_2 \longrightarrow \underset{\text{dense frames}}{SiF_4 + O_2}$$

Quartz is used in gramophone pick-ups as it is piezo-electric. Vitreous, silica is resistant to shock. It is transparent to visible and UV light and is used in spectrophotometers. Silica gel is amorphous and porous. It is used as a drying agent, Kieselguhr is a form of silica which is a white powder. It is used in filtration plants. All these forms are best written as $SiO_2 . nH_2O$.

Oxides of Germanium, Tin and Lead

All these oxides are dioxides MO_2. They adopt TiO_2 structure with 6 : 3 coordination. GeO_2 is less strongly acidic than SiO_2. SnO_2 and PbO_2 are amphoteric. With NaOH they form germinates, stannates and plumbates, Germinates resemble silicates. Stannates and plumbates are represented by $(Sn(OH)_4]^{2-}$ and $[Pb(OH)_4]^{2-}$ or $MO_2(H_2O)_n$. All these oxides are soluble in acids. $[GeF_6]^{2-}$ and $[SnCl_6]^{2-}$ complex ions are soluble in acids. GeO, SnO and PbO are more basic than the dioxides. Lead forms mixed oxide Pb_3O_4 red lead (सिन्दूर). It is made of $2PbO.PbO_2$ (Pb_3O_4). PbO_2 is an oxidising agent and is used in car battery. Black oxide is a mixture of PbO and Pb.

Silicates

Silicates are present in the earth's crust in huge amounts. Rocks and sands are all made up of silicates. Clay and soils are all silicates. Slates, granites, bricks, cement, ceramics and glass are all silicates.

Alkali metal silicates are soluble silicates. They are prepared by metal carbonates with sand in an electric furnace at 1400°C.

$$Na_2CO_3 \xrightarrow{1400^\circ C} Na_2O + CO_2$$
$$Na_2O + SiO_2 \longrightarrow Na_2SiO_3, (Na_2SiO_3)n$$

The product is a soluble glass of sodium and potassium silicates. They are used in liquid detergents. In case of hard water Ca^{2+} and Mg^{2+} are available which react to form insoluble $CaSiO_3$.

Structures of Silicates: Most of the silicates are insoluble, ionic with infinite structures. This is due to strong Si—O bonds. X-ray studies have revealed the principles in silicate structures.

(*i*) Difference in the electronegativity between Si and O is 1.7 which suggests that Si–O bonds are 50% ionic and 50% covalent.

(*ii*) The Si^{4+} and O^{2-} ratio is 0.29 which indicates that Si is four coordinate surrounded by 4 oxygen atoms at the corners of a tetrahedron. Thus silicates SiO_4^{4-} are tetrahedral units.

(*iii*) These SiO_4 tetrahedra are discrete units or may share corners to polymerise into large units.

(*iv*) O Atoms may be close-packed to give tetrahedral and octahedral holes and metal ions may occupy these holes. Al ions can fit into either. Al can replace a metal from one of the holes or Si atom in the lattice, *e.g.*, aluminosilicates.

Classification of Silicates

The classification is based on the way SiO_4^{4-} tetrahedral units are joined together.

Orthosilicates

Orthosilicates have the formula $M^{iv}(SiO_4)$ where M may be Be, Mg, Fe, Mn Zn, or Zr SiO_4. In Willemite Zn_2 (SiO_4) and phenacite Be_2 (SiO_4), Be and Zn have a coordination number 4 and they occupy tetrahedral holes. In forsterite $Mg_2(SiO_4)$ we have a coordination number 6 and octahedral holes are occupied. In forsterite one tenth of Mg^{2+} may be replaced by Fe^{2+} ions without changing the structure. These structures are related to hexagonal close-packing.

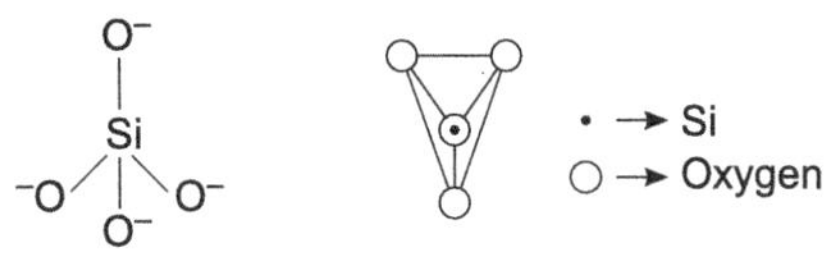

Fig. 12.7 Orthosilicates

$ZrSiO_4$ shines like diamond and is called zircon. Zircon has a coordination number 8 and it does not have close-packed structure. Garnet is another important Gemstone (red).

Pyrosilicates

In this two tetrahedral units share a common O atom at one corner giving $Si_2O_7^{6-}$ unit. It is a condensed silicate and has a similarity in structure with pyrophosphates $Na_4P_2O_7$. The word pyro is used because they can be formed by heating orthosilicates. Pyrosilicates are very few, *e.g.* $Sc_2[Si_2O_7]^-$ thortveitite.

In lanthanide disilicates $Ln_2 [Si_2O_7]$ the structure is same but Si-O-Si angle is not 180°. In Sc Compound it varies to 133°. The metal may have coordination number 6, 7 and then 8.

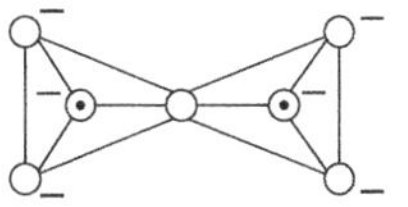

Pyro silicates $Si_2O_7^{6-}$

Cyclic Silicates

In this two oxygen atoms are shared for tetrahedral unit to give a ring structure of the formula $(SiO_3)_n^{2n-}$. Ring containing 3, 4, 6 and 8 tetrahedral units are known but those with 3 and 6 units are more important. The cyclic ion $Si_3O_9{}^{6-}$ exists in Wollastonite $Ca_3[Si_3O_9]$ and Ba Ti $[Si_3O_9]$ benitonite. $Si_6O_{18}{}^{12-}$ unit exists in Be_3Al_2 $[Si_6\ O_{18}]$ beryl. $Si_6O_{18}^{12-}$ unit exists in Be_3Al_2 $[Si_6O_{18}]$ beryl. Si_6O_{18} units are placed one above the other. Beryl is found in granite as pale green crystals. Emerald has the same formula and it contains 1-2% Cr which is cause of its intense green colour.

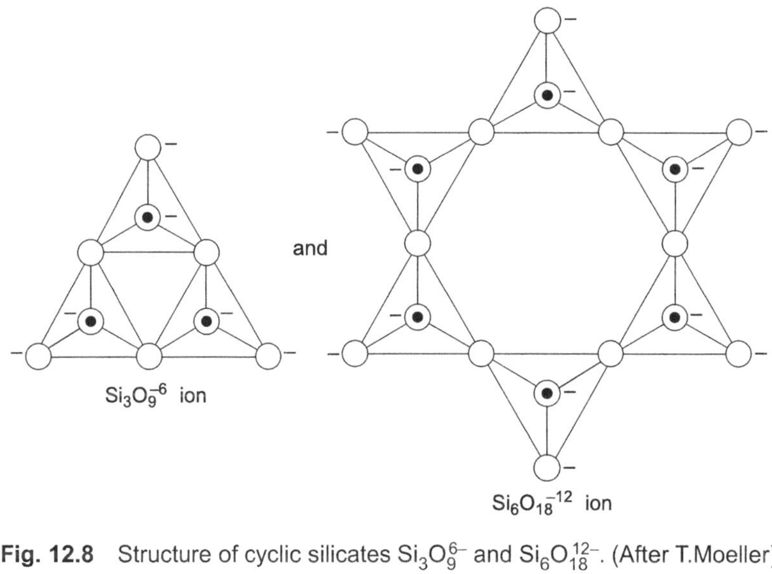

Fig. 12.8 Structure of cyclic silicates $Si_3O_9^{6-}$ and $Si_6O_{18}^{12-}$. (After T.Moeller)

Chain Silicates

In this O atoms on two corners of each tetrahedron are shared with other tetrahedra giving rise to the formula $(SiO_3)_n^{2n-}$. Many minerals may form such chains but variety of structures are formed because the arrangement of tetrahedra in space which may affect the repeat distance along the chain, *e.g.*,

LiAl $[(SiO_3)_2]$, $Mg_2[(SiO_3)_2]$, Ca $Mg[(SiO_3)_2$. In these examples the common arrangement repeats after every second tetrahedron. In $Ca[(SiO_3)_2]$ there is repeat unit of three tetrahedron. Others with repeat unit of 4, 5, 6, 7, 8, 9 and 12 are known.

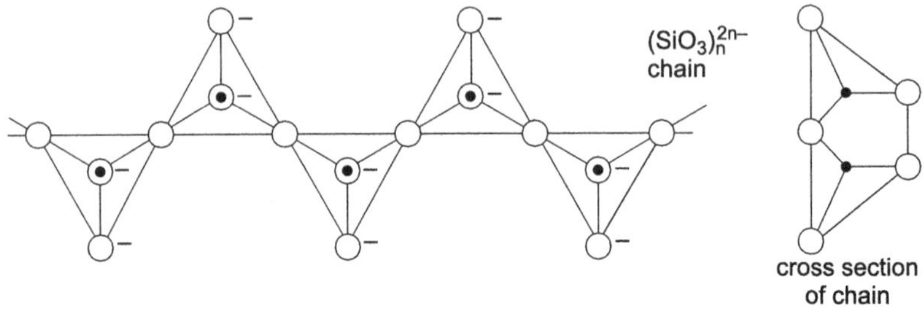

Fig. 12.9 Structure of $(SiO_3)_n^{2n-}$

Double chain structure are quite common in which simple chains are joined together by shared oxygens. These are called amphiboles, e.g., $(Si_2O_5)_n^{2n-}$, $(Si_4O_{11})_n{}^{2n-}$, $(Si_6\,O_{17})_n{}^{2n-}$ and more.

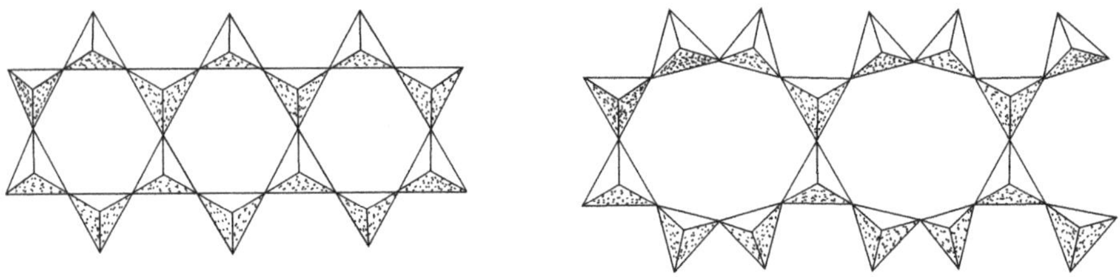

Fig. 12.10 Structure of various double chains. $(Si_4O_{11})_n^{6n-}$

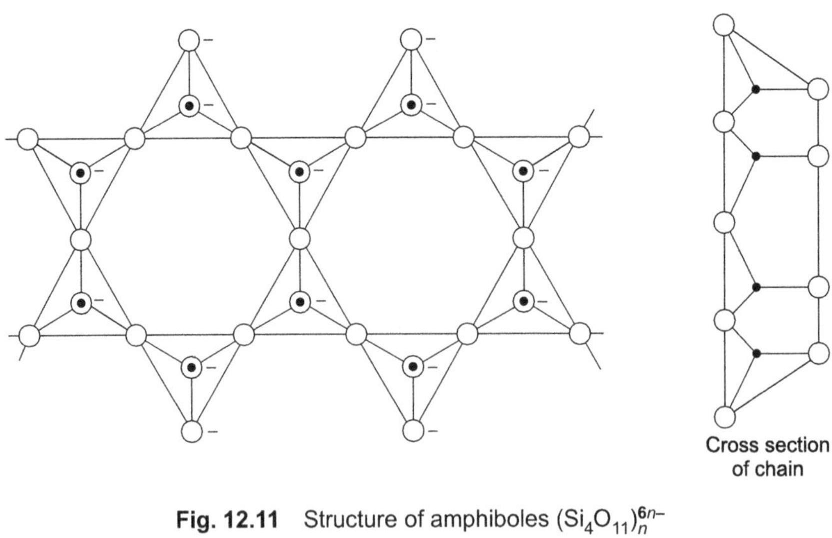

Fig. 12.11 Structure of amphiboles $(Si_4O_{11})_n^{6n-}$

Sheet Silicates

In case SiO_4 share three corners of the tetrahedra an infinite two dimensional sheet of formula $(Si_2O_5)_n^{2n-}$ are formed shown in the figure. These are strong bonds within the Si–O sheet but the forces holding one sheet from the next are weak. Therefore the minerals break into thin sheets. A large number of sheet silicates are known. They are made of two or three sheets joined together. The examples are:

(*i*) Clay minerals (talc)

(*ii*) White asbestos (chrysotile, biotite)

(*iii*) Micas (muscorite)

(*iv*) Montmorillonites (Fullers earth)

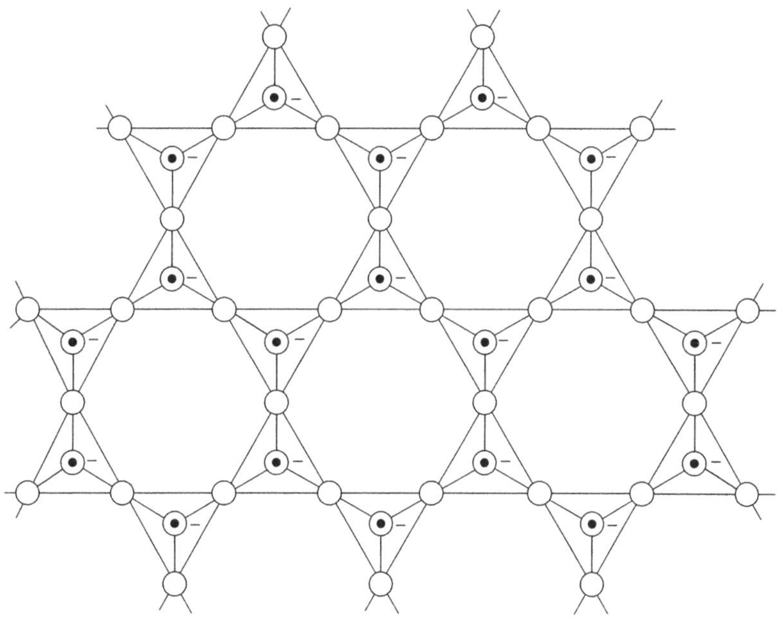

$(Si_2O_5)_n^{2n-}$ sheet

Fig. 12.12 Structure of sheet silicates $(Si_2O_5)_n^{2n-}$

Minerals Kaolinite having the formula $Al_2(OH)_4$ $[Si_2O_5]$ formed by the decomposition of granite is a white solid and has a two layer structure. Pyrophyllite Al_2 $(OH)_2$ $[Si_2O_5]_2$ has a three layer structure.

Three-dimension Silicates

Sharing of all the four corners of SiO_4 tetrahedron results in three-dimensional lattice of SiO_2 (quartz, tridymite and cristobalite). Replacement of Si^{4+} by Al^{3+} plus an additional metal ion forms the basis of three-dimensional structures. The extra metal ion can occupy the holes of the lattice. Si^{4+} replacement with

Al^{3+} gives $AlSi_3O_8$. The smaller cations K^+, Na^+, Ca^{2+}, Ba^{2+}, Fe^{3+}, Cr^{3+} and Mn^{2+} do not exist in three-dimensional silicates because the holes are larger, Replacement of $\frac{1}{4}$th or $\frac{1}{2}$ of Si atoms are quite common in these minerals, *e.g.*, (*i*) Feldspar (*ii*) Zeolites and (*iii*) Ultramarines. Granite is made of feldspar plus micas quartz, Feldspar is further divided into classes.

Orthoclase		Plagioclase	
(*i*) Orthoclase	$K[AlSi_3O_8]$	albite	$Na[AlSi_3O_8]$
(*ii*) Celsion	$Ba[Al_2\,Si_2O_8]$	anorthite	$Ca[Al_2Si_2O_8]$

Orthoclase are more symmetrical than plagioclase. Zeolites have an open structure than feldspar. Zeolites are used in ion-exchangers. Natrolite $Na_2[Al_2Si_2O_8]\,2H_2O$ is a natural ion exchanger. They take up Ca^{2+}/Mg^{2+} from hard water and replace them by Na^+ ions thus softening hard water. The zeolite after use becomes calcium or magnesium zeolite. It is regenerated by treatment with strong NaCl solution. Many synthetic zeolites hav been made in addition to natural zeolites.

Silicates in Industry

Clay silicates are used for absorbing chemicals. Micas are used as insulators in electrical appliances. Asbestos is used as thermal insulators. Silicates are also used as gemstones in jewellery.

Cement, ceramics and glass are made out of silicates, Metallurgical processes produce silicates as waste products or slag because the mineral contains silicate impurity, Alkali silicates are used as adhesives, Both Portland and high alumina cement contain silicates.

Ceramics

They are inorganic substances which can be made in the form of paste and shaped at normal temperatures. Ceramics are fired at higher temperature to give strength, melting the paste and sintering the crystals together.

Kaolinite $Al_2(OH)_4\,[Si_2O_5]$ loses water at 500-600°C resulting in Al_2O_3, $2SiO_2$. On further heating to 950°C it forms a solid solution mullite $3Al_2O_3$, $2SiO_2$ and SiO_2. It remains solid tell 1595°C and then softens slowly into a white product. Fe^{3+} in the clay gives red or purple colour, CaO imparts yellow or flue colour. Pigments are applied to give it a glaze.

Glass

Glass is a super-cooled liquid of high viscosity. A fraction of glass comprises of silica. Very high temperatures are required to produce it. It is an expensive variety of glass. The high temperature can be reduced by adding oxides such as Na_2O, H_2O, MgO, LaO, BaO, B_2O_3, Al_2O_3, ZnO and PbO. Glass is known as a solid solution. SiO_4 tetrahedra has an important role in its structure. The composition of glass may very within a limit. Borosilicate glass may contain boron and Al as well. Cheap laboratory glass using K_2CO_3 is known as potash glass. Glass with low coefficient of expansion and can be heated strongly without cracking is called pyrex, corning or borosil glass. Fe^{3+} and Fe^{2+} give green and light blue glass respectively. Co^{24} gives deep blue, copper gives ruby-red and CaF_2 gives opal glass.

Silicones

Silicones are organo-silicon compounds. Organo-silicon compounds have a strong Si—C bonds and are quite stable to heat. Silicon has poor catenation property with regard to carbon. This is due to weak Si—Si bonds.

Carbon forms $p\pi - p\pi$ double bonds whereas Si has no such bonding. Si = Si, Si = C and Si=N are transient reaction species.

Silicon forms $p\pi$–$d\pi$ double bonds in which silicon uses *d* orbitals.

Organosilicon Compounds

They can be prepared by using a grignard reagent

(*i*) (*a*) $SiCl_4 + CH_3MgCl \longrightarrow H_3CSiCl_3 + MgCl_2$

(*b*) $H_3CMgCl + H_3CMgCl \longrightarrow (H_3C)_2\,SiCl_2 + MgCl_2$

(*c*) $(CH_3)_2\,SiCl_2 + CH_3MgCl \longrightarrow (H_3C)_3\,SiCl + MgCl_2$

(*d*) $(H_3C)_3\,SiCl + CH_3MgCl \longrightarrow (H_3C)_4\,Si + MgCl_2$

This method is used as a laboratory preparation

(*ii*) On a smaller scale organs lithium compound may be used.

$$4Li\,R + SiCl_4 \longrightarrow SiR_4 + 4LiCl$$

(*iii*) ***Rochow Direct process:*** Acyl or alkyl halide is heated in the presence of fluid si using copper catalyst at 300°C.

$$Si + 2CH_3Cl \xrightarrow[300°C]{Cu} (H_3C)_2\,SiCl_2$$

This method is utilised for large scale preparation of organosilicon products.

Direct method yields a mixture of silicones. Their boiling points are close therefore they are fractionally distilled.

(*iv*) ***Addition of Si-H to an alkene:*** It is a general method of preparation and is not applicable to methyl or phenyl silanes required for silicones. The products are volatile, reactive and flammable.

Silicones

Silicones are organosilicon polymers. They are commercially useful as oils, elastomers and resins. They are now produced in large amounts than other organometallic compounds.

Hydrolysis of $SiCl_4$ produces SiO_2 which has a three dimensional structure.

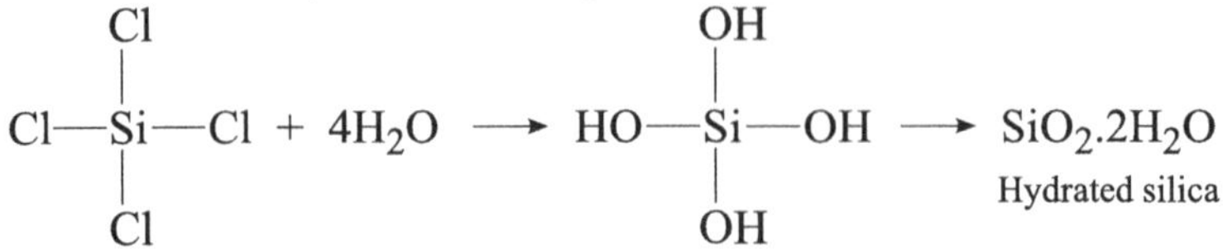

Alkyl substituted chlorosilane hydrolysis gave expected silicon compounds.

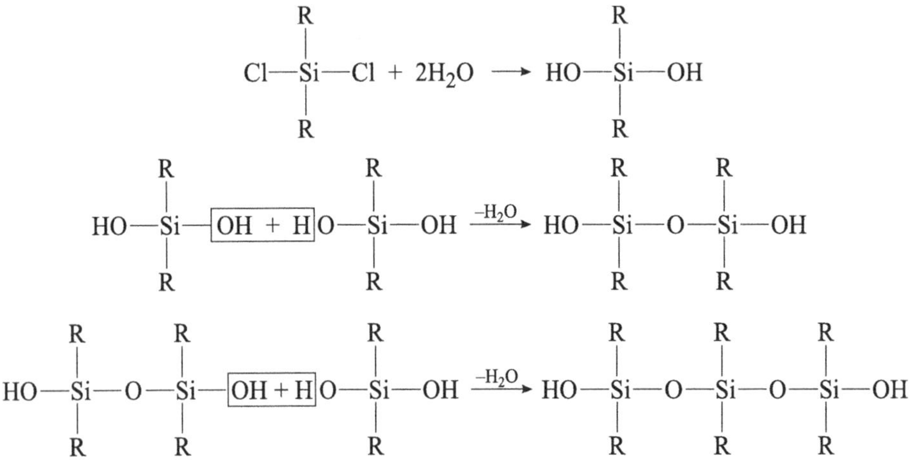

..... and so on

Methyl substituted compounds are mainly used. Hydrolysis of dimethyl chlorosilane gives linear polymers with an –OH group at the both ends. Polymerisation increases the chain length. $(CH_3)_2$ $SiCl_2$ is taken as a chain building unit.

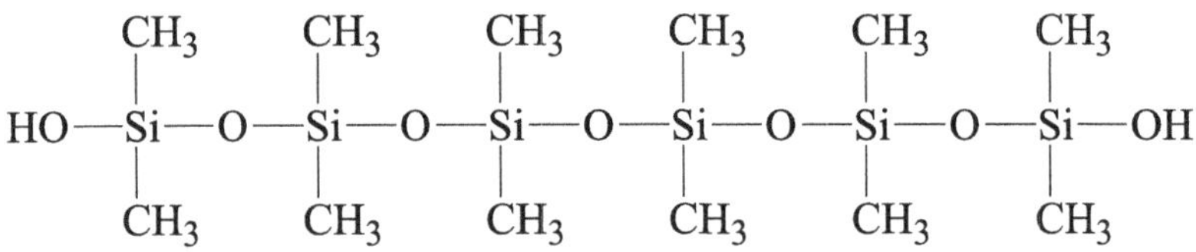

Hydrolysis under controlled conditions can produce cyclic polymers containing rings having 3, 4, 5, and 6 silicon atoms.

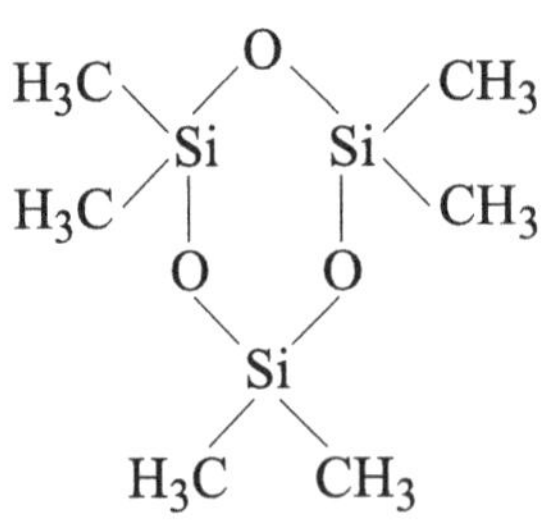

Tris cyclic dimethyl silicoxane

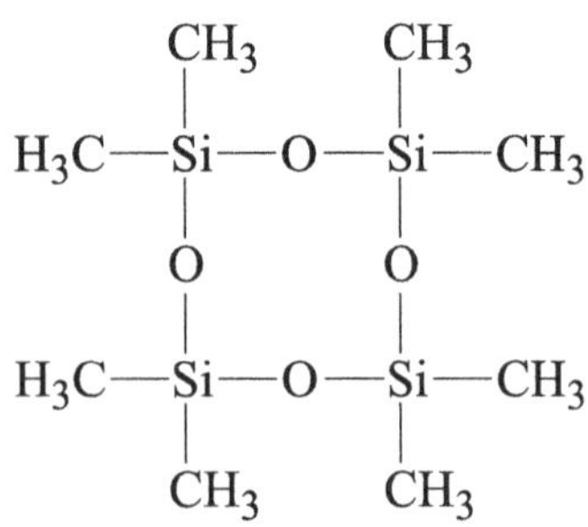

Tetrakis cyclic dimethyl silicoxane

Hydrolysis of Trimethyl Chlorosilane

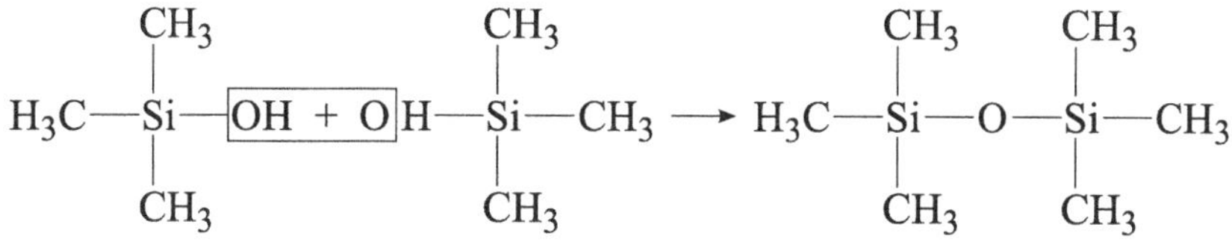

Hexamethyl disiloxane

When $(H_3C)_3$ SiCl and $(H_3C)_2$ $SiCl_2$ are mixed and hydrolysed there occurs no straight chain polymer, there is no –OH group at this end of the chain and the other end is blocked. $(H_3C)_3$ SiCl is, therefore, chain termination unit. Ratio of $(H_3O)_3$ SiCl and $(H_3C)_2$ $SiCl_2$ will determine the chain size.

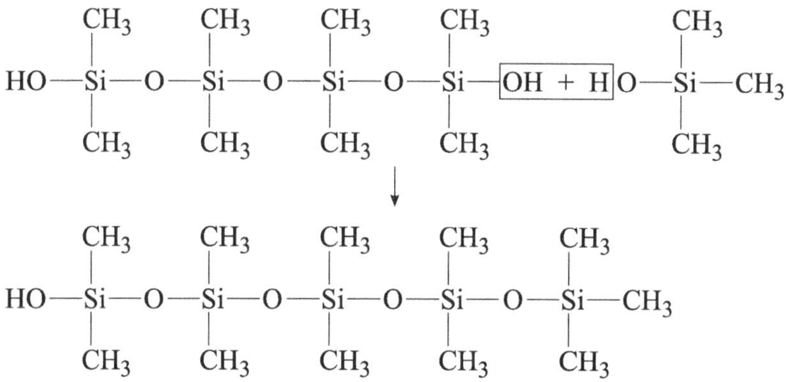

Hydrolysis methyl trichlorosilane H_3C $SiCl_3$ gives a condensed cross-linked polymer, similarly cross-linked polymer chain can be terminated by adding small amount of CH_3SiCl_3.

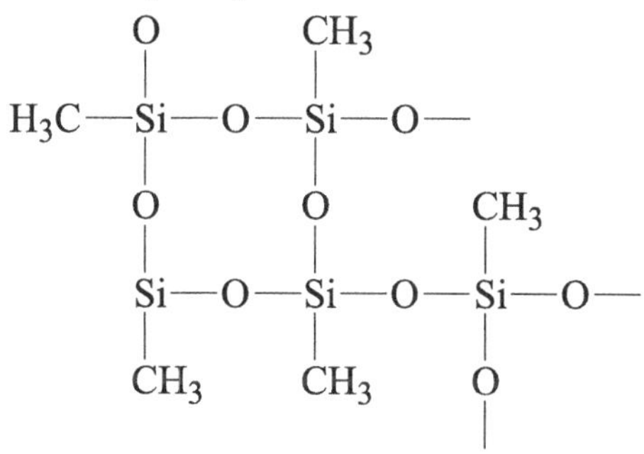

Silicones are expensive but have multivarious useful properties, earlier they were used as electrical insulators. They are resistant to heat, oxidation and chemicals. They are non-stick and water repellent. The inertness is due to very strong Si—O bond and stable Si—O—Si—O—Si skeletal. They are used in non-stick pans. H_2SO_4 Splits Si—O—Si bonds forming Si—O—SO_3H esters. They are used water repellents in masonry, shoe polish and car polish. Silicon rubbers are useful as they can retain elasticity between –90°C to + 250°C. They are good electrical insulators.

Hydrides

Silicon hydrides are covalent. Carbon forms great number of chain and ring compounds due to catenation. Catenation is less strong in the case of silicon, therefore the number of hydrides are limited. Silicon forms saturated hydrides Si_nH_{2n+2} called silanes. The hydrides are straight and branched chains up to 8 Si atoms, Ring compounds are very few.

Compounds like alkenes and alkynes do not exist because of $p\pi{-}d\pi$ bonding. SiH_4 and $SiHCl_3$ is by far the most important silane. Mixture of silanes are produced by the hydrolysis of Hg_2Si with H_2SO_4 or H_3PO_4. Besides SiH_4, other silanes are thermally unstable. Reactive silanes are handled under vacuum.

$$2Mg + Si \xrightarrow[\text{in absence of air}]{\Delta} Mg_2Si$$

$$Mg_2Si + H_2SO_4 \longrightarrow SiH_4\ (40\%) \quad \left.\begin{array}{l} Si_2H_6 \\ Si_3H_8 \\ Si_4H_{10} \\ Si_5H_{12} \\ Si_6H_{14} \end{array}\right] (60\%)$$

Monosilane (SiH_4) are now prepared by the reduction of $SiCl_4$ by $LiAlH_4$, Li H and NaH in ether at nelatively low temperature. They give only one product and not the mixture of silanes.

$$SiCl_4 + Li(AlH_4) \longrightarrow SiH_4 + AlCl_3 + LiCl$$

$$Si_2Cl_6 + 6LiH \longrightarrow Si_2H_6 + 6LiCl$$

$$Si_3Cl_8 + 8NaH \longrightarrow Si_3H_8 + 8NaCl$$

Silanes may be prepared by the direct action of Si or Fe/Si and anhydrous HCl.

$$Si + 2HCl \longrightarrow SiH_2Cl_2$$

$$Si + 3HCl \longrightarrow SiHCl_3 + H_2$$

$$Si + 2CH_3Cl \longrightarrow CH_3SiHCl_2 + C + H_2$$

Silanes have greater reactivity than alkanes. Silanes are strong seducing agents. Silanes hydrolyse in alkaline solution.

$$SiH_4 + (4 + n)\,H_2O \xrightarrow[\text{alkaline}]{\text{slightly}} 2SiO_2nH_2O + 2H_2$$

With alkanes chlorosilicone give hydrosilation reaction in the presence of Pt catalyst.

$$R\,CH = CH_2 + SiHCl_3 \xrightarrow[\text{cat.}]{Pt} R\,CH_2 - CH_2\,SiCl_3$$

The difference between alkanes and silanes is attributed to the following reasons.

(*i*) Pauling's electronegative values of C, Si and H suggest

$$Si^{\delta+} - H^{\delta-}$$

Thus Si invites nucleophilic attack.

(*ii*) The larger size of Si makes the attack easier.

(*iii*) Si contains low energy *d* orbitals which can help to form intermediate compound and lower the activation energy of the reaction.

Germanes are similar to silanes Ge_nH_{2n+2}. It is limited to $n = 5$. Stannanes are much less stable. It can be obtained by reducing $SnCl_4$ by $Li[Al\,H_4]$ or $Na[BH_4]$. Sn_2H_6 is known but is less stable, plumbanes are even less stable. PbH_4 is less stable and difficult to prepare.

Complexes

The elements of group 13 have a high charge, small size and empty orbitals which are suitable for complex formation. Carbon shares a maximum of eight electrons in the second shell. This resembles noble gas configuration and maximum stability, therefore, carbon does not form complexes. The rest of the elements in the group form complexes due to the available *d* orbitals. They form 4 to 6 coordinate complexes.

$$SiF_4 + 2F^- \longrightarrow [SiF_6]^{2-}$$

$$GeF_4 + 2N(CH_3)_3 \longrightarrow [GeF_4[N(CH_3)_3]_2]$$

$$SnCl_4 + 2Cl \longrightarrow [SnCl_6]^{2-}$$

The VB Theory requires 4 covalent and 2-coordinate bonds.

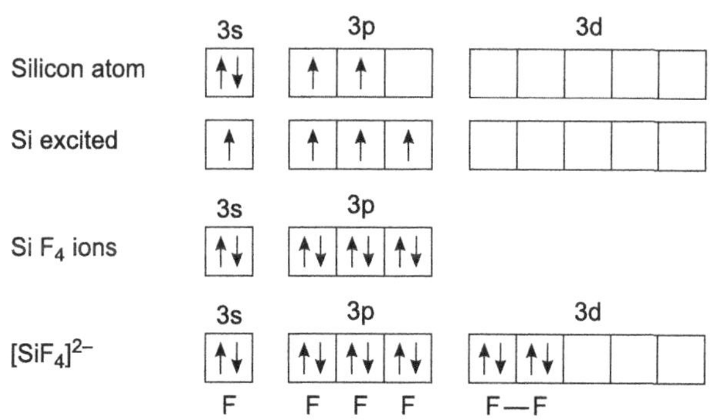

s—p^3—d^2 dybridisation

The sp^3d^2 hybridisation suggests octahedral geometry. The geometry of octahedral is acquired due to VSEPR theory.

$[SiF_6]^{2-}$ complex is stable in water and alkali. $[GeF_6]^{2-}$ and $[SnF_6]^{2-}$ are relatively less stable and are hydrolysed by alkali $[PbF_6]^{2-}$ is hydrolysed by water and alkali both. Ge, Sn and Pb also form chloride complexes and oxalate complexes.

HALIDES

Tetrahelides

All the elements of group 12 form tetrahalides except PbI_4. They are all volatile and covalent. They acquire tetrahedral geometry. Only SnF_4 and PbF_4 have three dimensional structures and high melting. SnF_4 sublimes at 705°C and PbF_4 melts at 600°C. After carbon all other elements contain *d*-orbitals thus Si—F, Si—Cl and Si—O bonds are stronger than the corresponding carbon due to $p\pi$–$d\pi$ bonding.

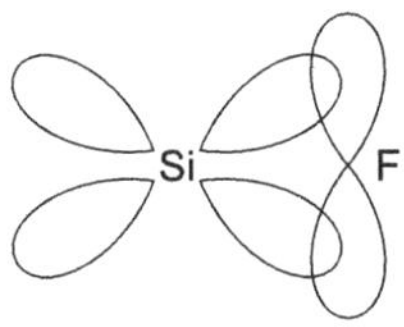

Fig. 12.13 $p\pi$–$d\pi$ bonding in SiF_4

CARBON

C–F bond is one of the strongest bond therefore CF_4 is a least reactive gas.

$$CO_2 + SF_4 \longrightarrow CF_4 + SO_2$$
$$SiC + 2F_2 \longrightarrow SiF_4$$
$$CF_2Cl_2 + F_2 \longrightarrow CF_4 + Cl_2$$

Hexafluoro ethane C_2F_6 and tetrafluoro ethylene C_2F_4 are known. C_2F_4 polymerises under pressure to $(C_2F_4)_n$. PTFE is a white solid plastic. It is chemically inert and a good electrical insulator. It is used as a non-stick coating for cooking pans. Fluoro carbons are used as solvents and lubricants.

$$HCCl_3 + HF \xrightarrow[\text{Catalyst}]{SbCl_4\,F} CF_2ClH$$
$$\downarrow \Delta$$
$$C_2F_4 \xrightarrow[\text{pressure}]{\text{under}} (C_2F_4)n$$

Carbon Tetrachloride CCl_4

Tetrachloromethane (CCl_4) is a very useful organic solvent. It is also used as a fire extinguisher. It is a raw material for the preparation of freon-12.

$$CCl_4 + 2HF \xrightarrow[SbCl_3\text{ catalyst}]{\text{anhydrous}} \underset{\text{Freon}}{CCl_2F_2} + 2HCl$$

Carbon halides are not readily hydrolysed because of the absence of *d*-orbitals in carbon. Silicon halides, on the contrary, are easily hydrolysed because of the presence of 3*d* orbitals. The *d* orbitals are able to coordinate OH^- or water. The empty orbitals in CCl_4 are of high energy therefore CCl_4 can hydrolyse in superheated steam.

$$CCl_4 + H_2O \xrightarrow[\text{Steam}]{\text{Superheated}} \underset{\substack{\text{Carbonyl chloride}\\\text{(phosgene)}}}{COCl_2} + 2HCl$$

Carbonyl chloride (phosgene) is a highly poisonous gas. It can be prepared by the combination of CO and Cl_2 with a carbon catalyst in sunlight.

$$CO + Cl_2 \xrightarrow[h\nu]{\text{Carbon}} COCl_2$$

$COCl_2$ is further used in the manufacture of the polymer polyurethane.

Chloro Fluorocarbons (CFC)

The mixed chloro fluorocarbons carbons $CFCl_3$, CF_2Cl_2 and CF_3Cl are called Freons. CFC's are unreactive and find use as refrigeration fluids. Freons are causing damage to the Ozone layer present in upper atmosphere. Freons prevent the harmful UV radiation from reaching the earth. Less harmful HFC's hydro fluorocarbons are now in use.

Halides of Ge, Sn and Pb

Ge, Sn and Pb form two types of halides MX_4 and MX_2. In the case of Ge(+IV) oxidation state is more stable while in the case of Pb, (+2) state is more stable.

MX_4 are all colourless volatile liquids except GeI_4 and SnI_4 which are orange solids. The colour of GeI_4 and SnI_4 occurs due to the charge transfer spectra, *i.e.*, the transfer of electron from I to Sn. The charge transfer occurs because of similar sizes of Ge and Sn.

$GeCl_4$ and $GeBr_4$ hydrolyse less rapidly. $SnCl_4$ and $PbCl_4$ hydrolyse in dilute solutions but the hydrolysis is incomplete. Complete hydrolysis can be achieved by the halogen acid.

$$Sn(OH)_4 \underset{H_2O}{\overset{HCl}{\rightleftharpoons}} SnCl_4 \underset{H_2O}{\overset{HCl}{\rightleftharpoons}} [SnCl_6]^{2-}$$

The halides of Si, Ge, Sn and Pb can expand the coordination number from 4 to 6 in the presence of excess halo acids, e.g., $[Si\ F_6]^{2-}$, $[GeF_6]^{2-}$ $[SnCl_6]^{2-}$. Pb(+IV) iodide, i.e., PbI_4 is not known because I^- reduces Pb^{4+} to Pb^{+2} (PbI_2).

Dihalides

The increase in the stability of dihalides is in the order

$$CX_2 << SiX_2 < GeX_2 < SnX_2 < PbX_2$$

SiF_2 is a white solid made at high temperature. It can be stored in liquid nitrogen. It polymerizes on warming up to $Si_{16}F_{34}$.

$$SiF_4 + Si \rightleftharpoons 2SiF_2$$

GeF_2 is a white solid prepared by the following reactions:

$$Ge + 2HF \longrightarrow GeF_2 + H_2$$

or

$$GeF_4 + Ge \longrightarrow 2GeF_2$$

Similarly SnF_2 and $SnCl_2$ are white solid obtained by heating Sn or SnO with gaseous HF or HCl. $SnCl_2$ hydrolyses to basic chloride Sn(OH)Cl.

$$SnCl_2 + H_2O \longrightarrow Sn\,(OH)Cl + HCl$$

Sn^{2+} ions are readily oxidised to Sn^{4+} by air, PbX_2 is much more stable than PbX_4. PbX_2 can be prepared by taking the soluble salt of Pb^{2+} and reacting it with halide ion or halogen acid. It is partially hydrolysed by water.

$$Pb^{2+}\ 2H_2O \longrightarrow [PbOH]^+ + H_3{}^+O$$

Cyanides

Cyanide ion (CN^-) form stable complexes with Cr, Mn, Fe, Co, Ni, Cu and Zn. The common complexes of iron are $K_4[Fe(CN)_6]$ and $K_3\ [Fe(CN)_6]$.

The later transition elements form cyanide complexes because they can use filled *d* orbitals for $d\pi$–$p\pi$ back bonding in addition to coordinate bond $M \leftarrow (CN)^-$. This bonding is similar to that in the case of carbonyls (CO). The metal cyanide bond is stronger owing to the negative charge on the CN^- as a π acceptor.

Cyanide complexes with the metals show extreme toxicity (KCN) but these complexes are also a part of enzymes and haemoglobin in the body. Many metal ions Ca^+, Ni^+, Au^+ and Mn^{3+} are unstable in solutions but they are stabilised when complex with cyanide ions. Cyanide complexes are important in the extraction of Ag and Au. These metals dissolve in the presence of NaCN solution in the presence of air and form argenocyanide and aurocyanide complexes.

$$4Ag + 8NaCN + 2H_2O + O_2 \longrightarrow 4Na[Ag(CN)_2] + 4NaOH$$

$$4Au + NaCN + 2H_2O + O_2 \longrightarrow 4Na[Au(CN)_2] + 4NaOH$$

CN^- ion can also act as a reducing agents as well as a complexing agent.

$$2Cu^{2+}\ 4CN^- \longrightarrow (CN)_2 + 2CuCN \xrightarrow{CN^-} [Cu(CN)_4]^{3-}$$

In the above reaction CN^- is oxidised to $(CN)_2$ by Cu^{2+} while Cu^{2+} itself is reduced to Cu^+ ion.

EXERCISES

1. Which of the following compounds is not electron deficient?

   (*a*) B_2H_6 (*b*) BCl_3 (*c*) Al_2Cl_6 (*d*) CCl_4.

2. Draw the layer structure of graphite and indicate the hybridization involved for carbon.

3. Which compound has zero dipole moment?
   (*a*) CO (*b*) CO_2 (*c*) NH_3 (*d*) H_2O
4. Carbon atom in CO_2 is:
   (*a*) *sp* (*b*) sp^2 (*c*) sp^3 (*d*) dsp^2
   Hybridized
5. Which bond has the highest energy
   (*a*) C – C (*b*) Si – Si (*c*) Ge – Ge (*d*) Sn – Sn
6. CO_2 is a gas whereas SiO_2 is a solid. Explain
7. What is the importance of catenation?
8. Draw the structure of:
   (*a*) Chain silicates (*b*) Graphite (*c*) Sheet-silicates
9. What are silicones? How do we prepare cross-linked silicones?
10. What are silanes? Give two methods of its preparation. Why do silanes differ in behavior from alkanes?
11. Carbon forms covalent compounds and lead forms ionic compounds. Explain.
12. How do we prepare CS_2. Discuss its reaction with (*i*) PCl_5 (*ii*) NaOH.
13. (*a*) Carbon has a greater tendency for catenation than silicon.
    (*b*) $PbCl_2$ is more stable than $SnCl_2$. Explain
14. CCl_4 is not hydrolyzed while $SiCl_4$ gets hydrolyzed easily. Explain.
15. How do we prepare silicon carbide? How does it react with alkalies? Discuss its structure.
16. Draw the structure of diamond and graphite and explain their electrical conductivity.
17. Explain the bonding in CO and CO_2.
18. What are the advantages and limitation of CO as a reducing agent?
19. What happens CO_2 in passed through a solution of $Ca(OH)_2$? What happens in case of excess CO_2.
20. Compare the shapes of the following molecules or ions
    (*i*) CCl_4 and $TeCl_4$ (*ii*) CO_2 and NO_2 (*iii*) SiF_4 and ICl_4^-
21. Why SnI_4 is an orange solid and CCl_4 and $SiBr_4$ are colourless liquids?
22. Give the preparation of monosilane and compare its properties with CH_4.
23. Give three examples of freons. Discuss their uses and damage to the environment.
24. Discuss the use of zeolites as water softeners.
25. Give the name and formula of six different types of silicates.
26. How can silicates be detected in qualitative analysis?

27. What are main ores of Sn and Pb and how the metals are extracted?
28. What is lead pollution? What in the effect on humans and how it can be avoided?
29. How do you prepare lead tetra acetate? What is its structure and use?
30. What is red lead and what is its use?

13

Group 15 Elements or Nitrogen Family of Elements

The group 14 contains elements such as N, P, As, Sb and Bi, N_2 is a colourless diatomic gas. The di nitrogen molecule is triply bonded N ≡ N imparting unusual stability. This makes nitrogen useful for inert environment to many chemical studies and as a food preservative. Presently liquid nitrogen is used as a coolant for studying the low temperature reactivity. N_2 gas is currently. Preferred over air to fill car tyres due to inertness.

Element	Electronic Configuration	Ionization Energy (kJ/mol)	Electron Affinity (kJ/mol)	Melting Point (°C)	Boiling Point (°C)
N	$[He]2s^22p^1$	1402	–7	–210	–195.8
P	$[Ne]3s^23p^1$	1012	72	443	280.5
As	$[Ar]3d^{10}4s^24p^1$	947	78	*b*	*b*
Sb	$[Kr]4d^{10}5s^25p^1$	834	103	631	1587
Bi	$[Xe]4f^{14}5d^{10}6s^26p1$	703	91	271	1564

General Characteristics

Phosphorus is known for many allotropes. The most common of these allotropes are white phosphorus which exists in two forms namely α-P_4 (cubic) and β-P_4 (hexagonal). Condensation of phosphorus from the gas or liquid phases contain tetrahedral P_4 (white phosphorus) of the α-form which changes to β-form above –76.9°C, white phosphorus is stored under water. In matches white phosphorus is not used due to highly toxic nature. The material used in the matches is mixture of P_4S_3 and red phosphorus which is much less toxic. Heating white phosphorus in the absence of air gives red phosphorus which is polymeric. Heating white phosphorus at very high pressures converts it to black allotrope which is very stable. At still high pressures, black phosphorus converts to other forms.

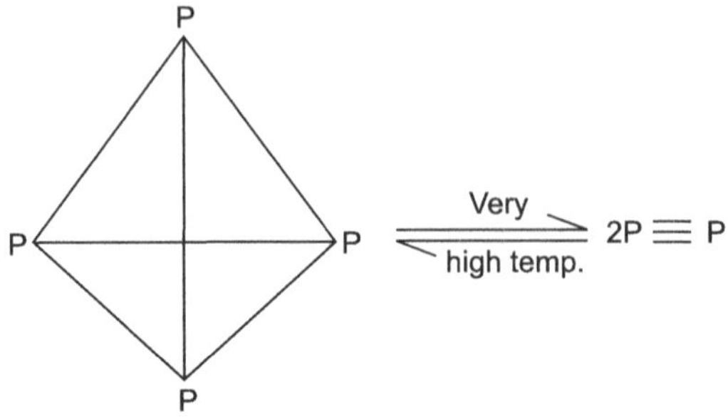

Tetrahedral

At 1800°C the dissociation reaches 50%.

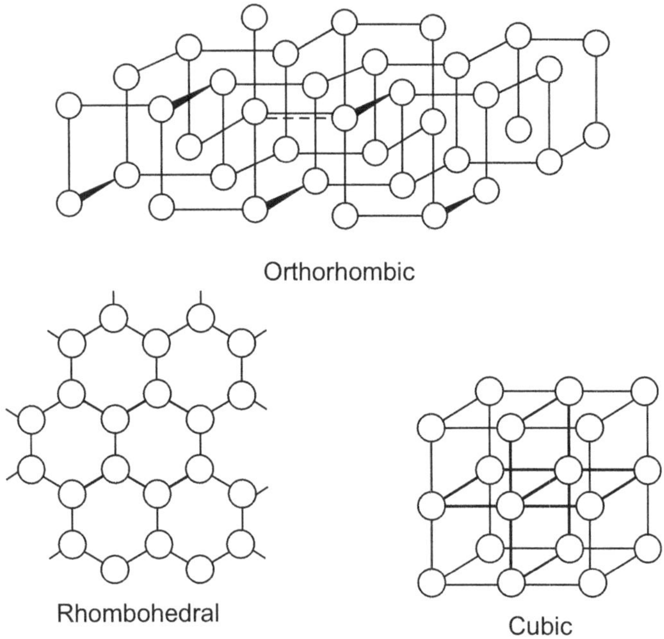

Fig. 13.1 ATP

Arsenic, antimony and bismuth also show many allotropes. Arsenic exists in the gray α-form which is similar to rhombohedral form of phosphorus. Arsenic and phosphorus exist as tetrahedral in the vapour phase. Sb and Bi have similar α forms. These three elements show somewhat metallic appearance and they are brittle and moderately good conductors. Arsenic is the most conducting but its resistivity is 20 times as great as copper.

Bismuth is the heaviest element to show a stable non-radioactive nature. Polonium is the heaviest radioactive element.

Anions

Nitrogen form three anions with lithium and group 2 element namely N_3^- (azide), N^{3-} (nitride) and N^{2-} (nitride). Many other covalent nitrides are also known. $Pb(N_3)_2$ is an explosive.

Phosphates, arsenates, antimonides and bismuthides are also known *e.g.*, Na_3 P, Ca_3As_2 etc. They are ionic, lustrous and have a good thermal and electrical conductivity. Properties are more metallic than ionic.

Occurrence Extraction and Uses

78% of earth's atmosphere contains nitrogen. Nitrates are water soluble therefore they are not present on earth's surface except in desert region. $NaNO_3$ is found in northern Chile and is called Chile salt petres. Large deposits of KNO_3 are found in India called Indian salt petres.

Proteins and amino acids essentially contain nitrogen. Nitrates and other nitrogen compounds are used as fertilizers. NH_4NO_3 is currently used as fertilizer due to large amount of nitrogen. N_2 finds its use in the manufacture of NH_3 and $CaCN_2$. Liquid nitrogen is used as a refrigerant.

Nitrogen is obtained in the laboratory by warming NH_4NO_2 or by oxidizing NH_3 by $Ca(OCl)_2$, bromine water or CuO. It can also be obtained from NaN_3 by thermal decomposition at 300°C. Thermal decomposition of NaN_3 is used in car air bags.

$$NH_4Cl + NaNO_2 \longrightarrow NaCl + \underset{\text{warm}}{NH_4NO_2 + N_2 + 2H_2O}$$

$$4NH_3 + 3Ca(OCl)_2 \longrightarrow 2N_2 + 3CaCl_2 + 6H_2O$$

$$8NH_3 + 3Br_2 \longrightarrow N_2 + 6NH_4Br$$

$$2NaN_3 \xrightarrow{300^\circ C} 3N_2 + 2Na$$

Phosphorus

Phosphorus is quite abundant on earth's surface. It is an essential element in plant and animal life. Bones and teeth contain $Ca_3(PO_4)_2$. Nucleic acids RNA and DNA contain this genetic material. Phosphorus in the form of adrenosine triphosphate ATP and adenosine diphosphate ADP is important for the production of energy in the cells. Water splits one phosphate group from ATP to form ADP with the production of energy in the cells, (33 kJ $mole^{-1}$ of energy is released).

Phosphate can also be used in the form of phosphatic fertilizers.

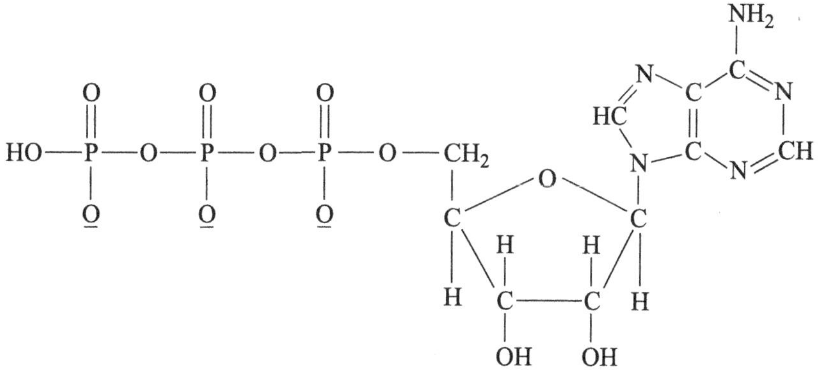

P is extracted from calcium phosphate by reduction with carbon in an electric furnace at 1400 – 1500°C Silica or sand is added to remove calcium silicate slag. P remains in the from of P_4O_{10}, which is reduced by carbon. At this temperature gasesous phosphorus distils off as P_4 with some P_2. The phosphorus so obtained is condensed to white phosphorus by passing through water.

$$2Ca_3(PO_4)_2 + 6SiO_2 \longrightarrow 6CaSiO_3 + P_4O_{10}$$

$$P_4O_{10} + 10C \longrightarrow P_4 + 10CO$$

The elemental phosphorus so produced is used to make very pure H_3PO_4. About 10% is used to make P–S compounds *e.g.*, P_4S_3 used in making matches.

$$P_4 + 5O_2 \longrightarrow P_4O_{10} + 6H_2O \longrightarrow 4H_3PO_4$$

$$P_4 + 10S \longrightarrow P_4S_{10}$$

Arsenic, Antimony and Bismuth

As, Sb and Bi are less abundant. They occur in ores as sulphides in trace amount. They occur as by-products of metallurgical processes of roasting. Most of the compounds of arsenic and antimony are poisonous. The sulphide ores have a specific colour. Arsenic is obtained as As_2O_3 from roasting of CuS, PbS, FeS, CoS and NiS. The oxide is reduced to As by reduction with carbon. Arsenopyrite Fe As S is a white grey ore. As_4S_4 has red-orange colour while As_2S_3 is yellow. These are found in volcanic ores. As can also be obtained from NiAs and $NiAs_2$ or FeAsS by heating to 700°C in the absence of air when As sublimes out.

$$4FeAs\ S \longrightarrow As_4(g) + 4FeS$$

Arsenic compounds are generally obtained from As_2O_3. Rat poison contains arsenic. Sb_2O_3 is obtained from the roasting of ZnS ores.

Sb_2O_3 is reduced to antimony by reduction with carbon.

$$Sb_2O_3 + 3C \xrightarrow{\Delta} 2Sb + 3CO$$

Important ore of antimony is stibnite Sb_2S_3.

$$Sb_2S_3 + 3Fe \longrightarrow 2Sb + 3\ FeS$$

Antimony alloys with Sn and Pb. It is used to prevent steel from rusting Bi_2O_3 is obtained from roasting of PbS, ZnS and CuS and then reduced to bismuth by carbon. It occurs os bismuthinite Bi_2S_3 and bismite Bi_2O_3 Bismuth is a low melting metal it is used in batteries and solders.

General Characteristics

Nitrogen: As in the case of previous groups the first element differs in properties from the rest of the elements in the group. Nitrogen is a colourless, odourless, tasteless gas represeuted by a diatomic molecule N_2. N_2 molecule cotains a triple bond $N \equiv N$ with a short bond length of 1.09Å. Inspite of π bonds N_2 is very stable with a dissociation energy 945 kJ mol^{-1}. It is inert but it reacts with lithium to form Li_3N, CO, CN^- and NO^+ are isoelectronic with N_2 molecule but they are more reactive than nitrogen molecule as they are partly ionic. By passing electric spark through N_2 gas at low pressure it changes to atomic nitrogen which is more active.

Nitrogen Cycle

Nitrogen is contiually let off in the atmosphere through soil, sea and living organisms. This is known as **nitrogen cycle.**

Nitrogen in the soil is present in the form of nitrates, nitrites and ammonium compounds. N_2 is evolved from the soil due to many reasons

1. Plants in the soil absorb these compounds to grow and make protoplasm. These plants are consumed by animals. The animals excrete nitrogeneous compounds *e.g.*, urea and uric acid. Hence N_2 comes back to the soil. Besides, death and decay also gives nitrogen to the soil.
2. The denitifying bacterial in the soil converts nitrates into N_2 and NH_3 gases which are given out to the atmosphere. Ammonia goes to the soil by the rain while N_2 by the denitrifying bacteria:

$$\text{nitrates} \longrightarrow \text{nitrites} \longrightarrow NO_2 \longrightarrow N_2 \longrightarrow NH_3$$

3. Loss of nitorgen compounds occurs in the soil as the water on the soil surface flows into the sea. In water nitrogen supports aquatic plant life.
4. Loss of NO and NO_2 in the atmosphere occurs because of burning of plants, coal and fuel waste of cars. This is a strong factor for 'smog' formation. Finally N_2 returns to the soil along with rains.

Nitrogen is also gained by the soil by **nitrifying bacteria.**

Nitrifying bacteria fixes N_2 from the atmosphere into nitrates and ammonium salts. This process contributes 60% of the nitrogen gain to the soil annually. The nitrogen fixing enzyme is nitorgenase. Presence of transition elements help in the process. Nitrogenase reduces N_2O, RCN and N_3^- to NH_3 and C_2H_2 to C_2H_4.

Fixation of N_2 comes from:

1. NH_4NO_3 as a fertilizer. Haber-Bosch process to make NH_3 and Ostwald's process to convert NH_3 to HNO_3.

$$NH_3 + HNO_3 \longrightarrow NH_4NO_3$$

2. Small amount of nitorgen comes from the deposits of chile salt petre $NaNO_3$.
3. Electric lightning brings about photo chemical changes:

$$N_2 + O_2 \xrightarrow{\text{air}} 2NO$$

$$2NO + O_2 \xrightarrow{h\nu} 2NO_2$$

$$2NO_2 + 2H_2O + O_2 \xrightarrow{\text{rain}} 4HNO_3$$

Phosphorus

Phosphorus is a soft, waxy solid and very reactive at room temperature. It emits light in moist air. It ignites spontaneously in air at 35°C, hence it is kept under water to avoid this. It is highly toxic. Its moelcule exists as a tetrahedral P_4.

Above 800°C P_4 molecule in the gas form breaks up to into P_2 molecule. P_2 molecules have a bond energy of 459.6 Kcal/mol. This value is half of that of N_2 molecules owing to the poor $p\pi$-$p\pi$ overlap. When white phosphorus is heated to about 250°C in the presence of sunlight then white phosphorus changes to red variety. Red phosphorus is polymeric and less reactive. It does not ignite unless heated to 400°C. It is insoluble in organic solvents, when white phosphorus is heated under pressure it changes to black allotrope which is the most stable variety. It is inert and has a layer structure.

Arsenic, Antimony and Bismuth

As, Sb and Bi also show alltropic forms. As vapours contain As_4 molecules yellow allotrope of arsenic is like white phosphorus. Antimony also exhibits yellow allotrope. All these elements are less reactive metallic or α-forms. These elements show layer structures. Another form of antimony is formed under high pressure and exhibits hexagonal close packed structure.

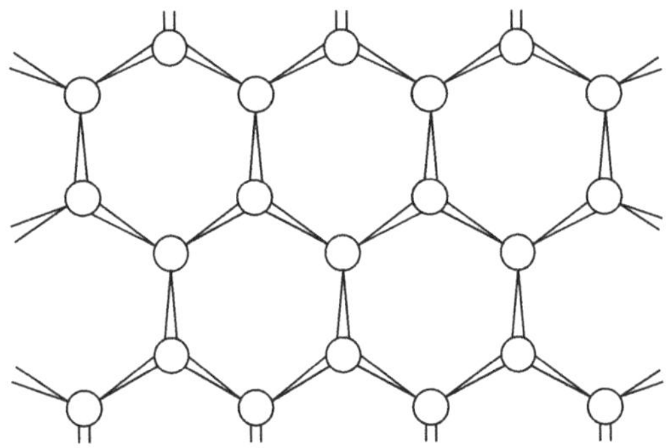

Fig. 13.2 Layer structure of bismuth

This group of elements forms mostly covalent compounds. If the lone pair is donated then a coordination number of four can be obtained *e.g.*, $[NH_3 \longrightarrow H]^+$

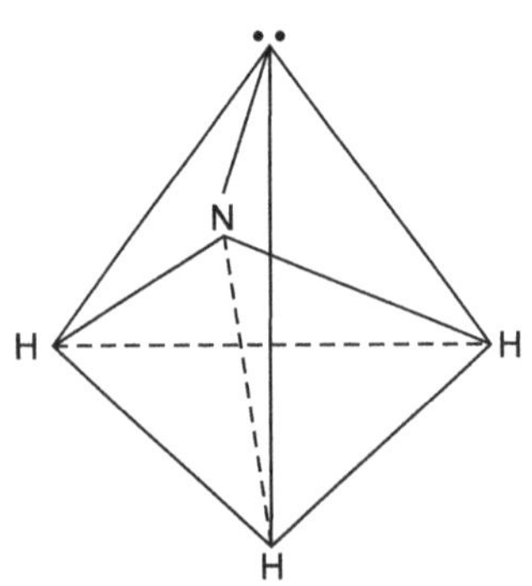

Fig. 13.3 Ammonia molecule

Sb and Bi do not form M^{5+} ions as very high energy is required to remove all the five electrons. Sb and Bi can lose three electrons easily to form M^{3+} ions but I.E for the rest of the elements in the group is too high SbF_3 and BiF_3 can exist as ionic solids but M^{3+} ions are not very stable in solution, $SbCl_3$ and $BiCl_3$ repidly hydrolyse in water to form SbO^+ and BiO^+ ions.

$SbCl_3 + H_2O \rightleftharpoons SbOCl + 2HCl$

$BiCl_3 + H_2O \rightleftharpoons BiOCl + 2HCl$ Bismuthyl chloride

The above reaction can be reversed in 5N HCl.

Nitrogen atom can acquire these electrons to attain noble gas configuration to form N^{3-} ion. This requires 2125 kJ/mol of energy. The nitrides can only be formed with highly electrositive metals.

e.g. Li_3N, Mg_3N_2, Ca_3N_2 Phosphides and bismuthides of sodium are well known but they are not ionic.

General Properties of Group 13 Elements Metallic Character

As in other groups, the metallic character increases in going down the group. N and P are essentially non-metals; As and Sb are metalloids and Bi is a true, metal. The metallic character in this group of elements is further confirmed by:

1. Their appearance and structures.
2. Tendency to form positive ions.
3. The oxides down the group are basic, N and P form acidic oxides; As and Sb form amphoteric oxide and Bi_2O_3 is basic.
4. The increase in metallic conductivity increases on descending the group.

Their resistivity values show increase in metallic properties.

Reactivity

Dinitrogen gas N_2 is unreactive or inert. Phosphorus burns an exposure to air forms P_4O_{10}. Phosphorus is kept under water to aviod burning. Red phosphorus is polymeric but reacts on heating. Arsenic is stable in dry air white it turms black in moist air arsenic sublimes on heating at 615°C forms As_4O_6 and As_4O_{10}. As_4O_{10} and arsenic acid H_3AsO_4 are oxidising agent. Antimony is less reactive and stable towards water and air at room temperature. On heating is air Sb forms Sb_4O_6, Sb_4O_{10} whereas Bi forms only Bi_2O_3.

Some Important Compounds of Group 14 Elements

Hydrides of Nitrogen

The important hydrides of nitrogen are:

1. Ammonia NH_3
2. Hydrazine N_2H_4
3. Hydrazoic acid HN_3

4. Hydroxyl amine NH_2OH

Out of these NH_3 is the most important hydride.

Ammonia NH_3

Preparation

1. It is prepared in the laboratory by heating together ammonium chloride and slaked lime $Ca(OH)_2$.

   $2NH_4Cl + Ca(OH)_2 \longrightarrow CaCl_2 + 2H_2O + 2NH_3$
2. The alternative of preparing NH_3 on a smaller scale is by the action of water on magnesium nitride, Mg_3N_2.

   $Mg_3N_2 + 6H_2O \longrightarrow 3Mg(OH)_2 + 2NH_3$
3. On an industrial scale it is prepared by Haber's process which involves the direct combination of nitrogen (1 volume) and hydrogen (3 volumes) at 425°C and 200 atmospheric pressure in the presence of Fe containing small amounts of K_2O and Al_2O_3 as catalyst and Mo as promoter of catalyst.

   $N_2 + 3H_2 \underset{Mo}{\overset{FeO}{\rightleftharpoons}} 2NH_3 + 93.6 \text{ Kcal mol}^{-1}$

   1 vol + 3 vol, 2 vol.

The process is carried forward by decrease in volume. The best yield is obtained by the following conditions:

(*i*) **Low Temperature:** The reaction being exothermic therefore low temperature *i.e.*, between 400–500°C favours forward reaction.

(*ii*) **High Pressure:** Since the reaction is accompained by the decrease in volume therefore high pressure of 200 atmospheres favours maximum yield.

(*iii*) **High Concentration** of reactants N_2 and H_2 maintains the best yield.

(*iv*) **Catalyst:** iron oxide along with K_2O and Al_2O_3 gives maximum speed to the product formation.

Physical characteristics: Ammonia is a colourless gas having b.pt. –33.35°C and freezing point –77.8°C. It has a pungent small and is highly soluble in water. It is lighter than air and therefore causes tears in the eyes. Liquid nitrogen on evaporation causes intense cooling and this forms the basis of its use in refrigeration plants.

Chemical Characteristics

1. Basic Nature: Ammonia is basic in its aqueous solution which can turn red litmus blue.

$$NH_3 + HOH \rightleftharpoons [NH_4^+] + OH^-$$

The basic character is because of OH^- ions formation.

It behaves as a Lewis base due to the existence of lone pair on nitrogen. Due to the small size of nitrogen lone pair is concentrated on a small region therefore NH_3 acts as a goods base. It can form complexes with the metal ions:

$$Ag^+ + 2NH_3 \rightleftharpoons [Ag(NH_3)_2]^+$$
$$Cu^{2+} + 4NH_3 \rightleftharpoons [Cu(NH_3)_4]^{2+}$$
$$Cd^{2+} + 4NH_3 \rightleftharpoons [Cd(NH_3)_4]^{2+}$$

In the case of Ag+ ions, the white precipitate of AgCl dissolves in excess ammonia because of the soluble complex ion $[Ag(NH_3)_2]^+$. On passing NH_3 through $CuSO_4$ solution a bluish green precipitate of $Cu(OH)_2$ dissolves in excess ammonia solution to give a deep blue solution of $[Cu(NH_3)_4]SO_4$.

(*i*) $CuSO_4 + 2NH_4OH \longrightarrow Cu(OH)_2 + (NH_4)_2SO_4$

(*ii*) $Cu(OH)_2 + (NH_4)_2SO_4 + 2NH_4OH$

$$\downarrow$$

$$\underset{\text{Deep blue}}{[Cu(NH_3)_4]SO_4} + 4H_2O$$

Tetra ammine copper (II) suphate Ammonia molecule is highly associated due to formation of hydrogen bond between the ammonia molecules.

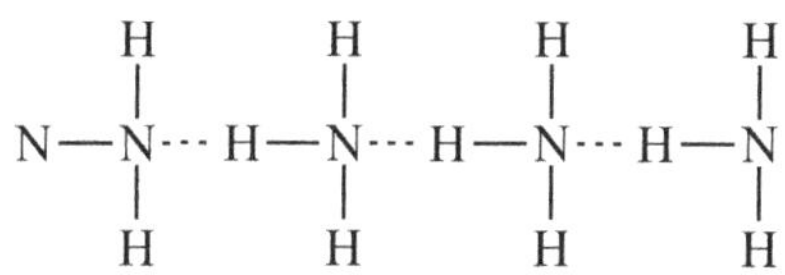

Hydrogen bonding in ammonia leads to the high boiling point and high heat of vapourisation.

2. Reactivity with air: Due to high reactivity NH_3 burns in excess air as given by the reaction:

$$4NH_3 + 3O_2 \longrightarrow 2N_2 + 6H_2O$$

However, in the presence of Pt/Rh at 800°C it forms nitric oxide.

$$4NH_3 + 5O_2 \xrightarrow[\text{Pt}]{800^\circ C} 4NO + 6H_2O$$

Halogens

Ammonia reacts with halogens to give greenish yellow flame of NF_3. With chlorine it forms a number of products such as NH_4Cl, NHCl, $NHCl_2$, NCl and N_2.

(*i*) $4NH_3 + 3F_2 \longrightarrow NF_3 + 3NH_4F$

(*ii*) $NH_3 + 3Cl_2 \longrightarrow NCl_3 + 3HCl$

(*iii*) $8NH_3 + 3Cl_2 \longrightarrow 6NH_4Cl + N_2$

(*iv*) $8NH_3 + 3Br_2 \longrightarrow 6NH_4Br + N_2$

(*v*) $2NH_3 + 3I_2 \longrightarrow NH_3.NI_3 + 3HI$

nitorgen triiodide ammoniate

$NH_3.NI_3$ is a stock-sensitive compound and explodes readily.

$$8NH_3.NI_3 \longrightarrow 5N_2 + 9I_2 + 6NH_4I$$

Reaction with Methane

The reaction of ammonia and methane is used for the preparation of HCN.

$$CH_4 + NH_3 + - O_2 \xrightarrow[\substack{1000 \quad 1200°C \\ \text{at 2 atm.}}]{Pt/Rh} \underset{\substack{\text{hydro} \\ \text{cyanic} \\ \text{acid}}}{HCN} + 3H_2O$$

This is called andrussow process.

Another reaction of CH_4 and NH_3 is known as Degussa process.

$$CH_4 + NH_3 \xrightarrow[1200 - 1300°C]{Pt} HCN + 3H_2$$

With Non-metals or their Oxides or Halides

(*i*) $C + 2NH_3 \xrightarrow{\text{red heat}} NH_4CN + H_2$

(*ii*) $P_4 + 4NH_3 \longrightarrow 4PH_3 + 2N_2$

(*iii*) $10S + 4NH_3 \longrightarrow S_4N_4 + 6H_2S$

On continuation the product $(NH_4)_2S$ is formed

(*iv*) Reaction with metals give nitrides

$3Mg + 2NH_3 \xrightarrow{900°C} Mg_3N_2 + 3H_2$

(*v*) Copper and brass are corroded by moist NH_3 and air.

(*vi*) $PCl_3 + 6NH_3 \longrightarrow P(NH_2)_3 + 3NH_4Cl$

(*vii*) P_4O_{10} reacts with wet and dry ammonia to form a number of polymeric products which are used for water softening (remove Ca^{2+} ions)

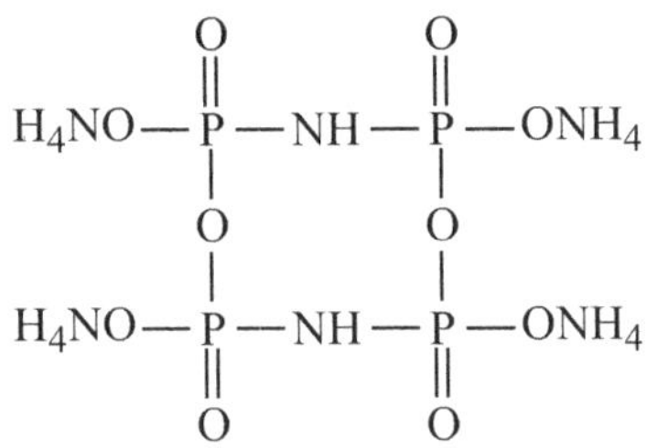

Fig. 13.4

(*viii*) $2PCl_5 + 16NH_3(\text{liq.}) \longrightarrow [(H_2N)_3\,P = N - P\,(NH_2)_3]^+\,Cl^- + 9NH_4Cl$

(*ix*) $6S_2Cl_2 + 16NH_3 \longrightarrow S_4N_4 + 8S + 12NH_4Cl$

$6SCl_2 + 16NH_3 \longrightarrow S_4N_4 + 2S + 14NH_4Cl$

Ammonia as a Solvent

Ammonia is extensively used as a non-aqueous solvent as discussed in chapter 12 of the book. Non-aqueous solvents are invoked to study those inorganic reaction which cannot be studied in aqueous solutions. The most important aspect of non-aqueous ammonia is that it dessolves all the alkali metals to form a blue coloured, paramagnetic and electrically conducting solutions containing solvated electrons.

$$Na + \text{liquid ammonia} \downarrow [Na(NH_3)_n]^+ + e$$

Structure

The structure of ammonia molecule is defined as a pyramid or tetrahedral with one position occupied by the lone pair. It involves SP^3 hybridisation with $\angle H - N - H$ equal to 107.3° less than the tetrahedral 109.5°. It is attributed to the unequal lone pair and bond pair repulsions.

Uses

1. It is used in the manufacture of (*a*) nitric acid and (*b*) Hydrocyanic acid.
2. It is used in the synthesis of S_4N_4.
3. As a coolant in refrigeration.
4. In the manufacture of Na_2CO_3.
5. used in the manufacture of fertilizers.

Hydrazine N_2H_4

It is considered to be a derivative of ammonia wher a hydrogen atom is substituted by —NH_2 group. It behaves a base but weaker than ammonia.

Preparation

1. On a large scale it is obtained from ammonia by oxidation using NaOCl and gelatine.

$$2NH_3 + NaOCl \longrightarrow N_2H_4 + NaCl + H_2O$$

The reaction is Split in Two Steps

Step I: $H_3N + NaOCl \longrightarrow NaOH + NH_2Cl$ chloramine

Step II: $NH_3 + NH_2Cl + NaOH \longrightarrow N_2H_4 + NaCl + H_2O$

Setp I is a fast reaction while step II is a slow reaction. Presence of gelatin helps in the formation of hydrazine and destroys the unwanted reaction

$$N_2H_4 + 2NH_2Cl \xrightarrow{\text{Fast}} 2NH_4Cl + N_2$$

Gelatin removes any heavy metal ion which may catalyse the above reaction.

Finally 70% of hydrazine is obtained whcih on distillation may give 85% solution of hydrazine.

2. Reduction of NO: Solution of K_2SO_3 and nitric oxide is reduced to hyrazine when sodium amalgams is used.

$$K_2SO_3 + 2NO \longrightarrow K_2SO_3.N_2O_2$$

$$K_2SO_3.N_2O_2 + 6H \xrightarrow{Na/Hg} K_2SO_4 + N_2H_4 + H_2O$$

Physical and Chemical Characteristics

Physcial Caracteristics

1. Anhydrous hydrazine is a fuming liquid with a very high dielectric constant.
2. Pure hydrazine or its aqueous solutions are kinetically very stable with respect to decomposition despite the endothermic nature of the compound and positive free energy of formation.

$$N_2(g) + 2H_2(g) \longrightarrow N_2H_4(l)$$

$$\Delta H_f^\circ = 50.6 \text{ kJ mol}^{-1}$$

$$\Delta G_f^\circ = 149.2 \text{ kJ mol}^{-1}$$

3. **Ignition:** When ignited in air hydrazine burns completely with evolution of heat.

$$N_2H_4(l) + O_2(g) \longrightarrow N_2(g) + 2H_2O$$

$$\Delta H = -621.5 \text{ kJ mol}^{-1}$$

Due to the highly exothermic reaction hydrazine is found most suitable as rocket fuel in missiles and space shuttles.

Ozone and H_2O_2 oxides N_2H_4 to hydrazoic acid:

$$3N_2H_4 + 5O_3 \longrightarrow 2HN_3 + 5H_2O + 5O_2$$

$$3N_2H_4 + 5H_2O_2 \longrightarrow 2HN_3 + 10H_2O$$

Physical Properties

Table 13.1

Physical Property	Value
mp. (°C)	2.0
b.p. (°C)	113.5
density at 25° (g cm^{-3})	1.00
dipole moment (D)	1.85
viscosity (centipoise)	0.9
dielectric constant (25°C)	51.7
specific conductivity. K (ohm^{-1} cm^{-1}) at 25°C	2.5×10^{-6}
ΔG_f° (25°C). kJ mol^{-1}	149.2
ΔH_f° (25°C). kJ mol^{-1}	50.6

Chemical Characteristics

1. On heating hydrazine decomposes to NH_3 and N_2

$$3NH_2.NH_2 \longrightarrow N_2 + 4NH_3$$

2. Reduction with sodium in the absence of oxygen

$$2H_2N.NH_2 + 2Na \longrightarrow \underset{\text{sodium hydrazide}}{2H_2N\text{-}NHNa} + H_2$$

3. with HNO_2

It reacts with nitrous acid to form hydrazic acid

$$N_2H_4 + HNO_2 \longrightarrow \underset{\text{hydrazoic acid}}{HN_3} + 2H_2O$$

4. **Reducing Property**

Hydrogine is a powerful seducing agent. It is used as a reducing agent in a number of seactions.

(*i*) Reduces Au_4Cl_3, $AgNO_3$ and $PtCl_4$ to their respective metals.

$$4AuCl_3 + 3N_2H_4 \longrightarrow 4Au + 3N_2 + 12HCl$$

$$4AgNO_3 + N_2H_4 \longrightarrow 4Ag + N_2 + 4HNO_3$$

$$PtCl_4 + N_2H_4 \longrightarrow Pt + N_2 + 4HCl$$

(*ii*) Reduces alkaline fehling solution (CuO) to Cu_2O

$$4CuO + N_2H_4 \longrightarrow 2Cu_2O + 2H_2O + N_2$$

(*iii*) Halogens to halogen acids

$$2Cl_2 + N_2H_4 \longrightarrow N_2 + 4HCl$$

$$2Br_2 + N_2H_4 \longrightarrow N_2 + 4HBr$$

I_2 to HI in the presence of $NaHCO_3$

$$N_2H_4 + 2I_2 \longrightarrow N_2 + 4HI$$

$$HI + NaHCO_3 \longrightarrow NaI + H_2O + CO_2$$

HI formed above can make the reaction go in the reverse direction.

(*iv*) Fe^{3+} salts to ferrous salts

$$4FeCl_3 + N_2H_4 \longrightarrow 4FeCl_2 + N_2 + 4HCl$$

$$2NaIO_3 + 3N_2H_4 \longrightarrow 2NaI + 6H_2O + 3N_2$$

(*v*) MnO_4^- is reduced to Mn^{2+} in dil. H_2SO_4

$$4KMnO_4 + 6H_2SO_4 + 5N_2H_4 \longrightarrow K_2SO_4 + 4MnSO_4 + 16H_2O + 5N_2$$

(*vi*) Hydrogen peroxide oxidise hydrazine to free N_2 and water

$$N_2H_4(l) + 2H_2O_2(l) \longrightarrow N_2(g) + 4H_2O(g) + \text{heat}$$

It should be noted that the product formed by two liquids are gases. This suggests that the reaction is highly exothermic. This reaction is made use of in the propulsion of rockets.

Formation of Complexes

Hydrazine acts as a ligand and forms donor-acceptor complexes with metal ions such as Ni^{2+} and Co^{2+} etc.

$$[Ni(NH_2.NH_2)_4]^{2+} \text{ and } [Co(NH_2.NH_2)_4]^{2+}$$

Structure: Structure of hydrazine resembles that of H_2O_2. It has a dipole moment of 1.85D which rules out trans configuration, $H_2N—NH_2$ group in hydrazine rotates along N—N single bond. Molecular rotation of $–NH_2$ groups is possible over energy barriers as shown in the figure.

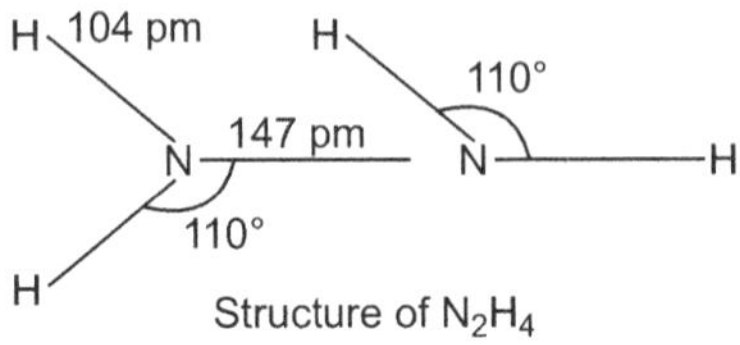

Structure of N_2H_4

In the above structure rotation of NH_2 groups along N—N bond occurs in which H atoms either get eclipsed or staggered. The eclipsed cis-conformation (*c*) is much higher –20 kJ mol^{-1} and the staggered (*b*) trans conformation is some what higher (–3kJ mol^{-1}) in energy than the gauche (*a*) or (*a′*). This is attributed to electron repulions and hydrogen nuclear repulnions are higher in cis as well as trans-hydrazine than in gauche-hydrazine

Hydrazine energy profile shows two minimas of the same energy at angles 100° and 260°, therefore hydrazine exists in a thermodynamic equilibrium of equimolar mixture of gauche conformers with mirror image rotamers (*a*) and (*a′*) to one another.

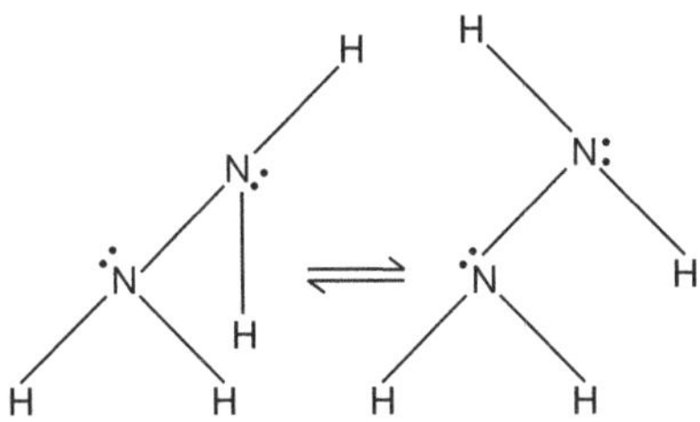

(α) (α')

Reciprocal transformation of the two isomers requires a small activation energy of 3 kJ mol^{-1} lower than thermal energy at room temperature. Consequently the lift period of α and α′ at room temperature is very small. N_2H_4 molecules vibrate with frequency of 10^{12} sec^{-1} at room temperature between *a* and *a′* via trans configuration (*b*) (staggered)

Hydroxylamine NH_2OH

It has been derived from NH_3 in which one hydrogen atom is replaced by an OH group. Its aqueous solutoin is less basic than NH_3 or $H_2N—NH_2$.

$$H_2NOH + HOH \rightleftharpoons NH_3OH^+ + OH^-$$

$$K_{25°} = \frac{[H_3NOH^+][OH^-]}{[NH_2OH]} = 6.6 \times 10^{-9} \text{ mol l}^{-1}$$

$$\text{or } K_{25°} = \frac{[NH_4^+][OH^-]}{[NH_3]} = 1.8 \times 10^{-5} \text{ mol l}^{-1}$$

Preparation: Anhydrous NH_2OH is thermally unstable therefore it is kept in aqueous solution or its salt. The preparation of pure hydroxylamine is difficult and hazardous the salts of hydroxylamine are obtained by of the following methods:

(*i*) **Raschig Synthesis:** It involves the reduction of NH_4NO_2 with SO_2 at 0°C and its hydrolysis

$$2NH_4NO_2 + 4SO_2 + 2NH_3 + 6H_2O \longrightarrow [NH_3OH]_2SO_4 + 2NH_4HSO_4$$

Pure hydroxylamine is further obtained by ammonlysis with liquid ammonia

$$[NH_3OH]_2SO_4 + 2NH_3 \longrightarrow 2NH_2OH + (NH_4)_2SO_4$$

$(NH_4)_2SO_4$ is insoluble in liquid NH_3 is separated by filtration.

(*ii*) It is also prepared by the reduction of aqueous nitrous acid or KNO_2 with $KHSO_3$ or $NaHSO_3$

$$HNO_2 + 2HSO_3^- + H_3O^+ \longrightarrow [NH_3OH]^+ HSO_4^- + HSO_4^-$$

Pure NH_2OH can be obtained ley treating hydroxyl anumonium salt with sodium butoxide NaOBu in butanol.

$$[NH_3OH]^+ HSO_4^- + NaOBu \longrightarrow NH_2OH + NaHSO_4 + BuOH.$$

(*iii*) It can also be prepared by hydrogenation of nitric oxide in acid medium using Pt/C catalyst.

$$2NO + 3H_2 + H_2SO_4 \xrightarrow{Pt/C} [NH_3OH]_2SO_4$$

Physical characteristics: Pure hydroxylamine is thermally unstable and hygroscopic in nature. Its melting point is 33°C

Chemical Characteristics

1. **Stability:** Hydroxylamine is unstable and the solid slowly decomposes above 15°C

$$3NH_2OH \longrightarrow N_2 + NH_3 + 3H_2O$$

$$4NH_2OH \longrightarrow N_2O + 2NH_3 + 3H_2O$$

It absorbs atmospheric oxygen in the presence of alkali to form HNO_2.

$$NH_2OH + O_2 \longrightarrow HNO_2 + H_2O$$

2. **Action of halogens:** It explodes an contact with halogens

$$2NH_2OH + 2Cl_2 \longrightarrow N_2O + H_2O + 4HCl$$

3. It also explodes with $KMnO_4$ and $K_2Cr_2O_7$.

4. **With HNO_2 and HNO_3:** When NH_2OH is heated with HNO_3 it forms nitric oxide

(*a*) $NH_2OH + HNO_3 \longrightarrow 2NO + 2H_2O$

(*b*) nitrous acid and NH_2OH give N_2O

$$NH_2OH + HNO_2 \longrightarrow N_2O + 2H_2O$$

(*c*) with hyponitrous acid $H_2N_2O_2$ it gives N_2.

$$2NH_2OH + H_2N_2O_2 \longrightarrow 2N_2 + 4H_2O$$

5. **Basic character:** NH_2OH dissolves in water to give monoacid base

$$NH_2OH + H_2O \rightleftharpoons [NH_3OH]^+ + OH^-$$

Therefore NH_2OH forms salts with acids *e.g.*, hydroxylamine hydrochloride $NH_2OH.HCl$ hydroxylamine sulphate $NH_2OH.H_2SO_4$

Reducing Property

1. Powerful reducing agent in acid medium

$$2NH_2OH \longrightarrow N_2O + H_2O + 4H \text{ (Basic equation)}$$

2. Reduces Fehling solution to red Cu_2O

$$2CuO + 2H \longrightarrow \underset{\text{red ppt.}}{Cu_2O\downarrow} + H_2O$$

3. **Reduces $AuCl_3$ to Au**

$$AuCl_3 + 3H \longrightarrow Au + 3HCl$$

$$\text{or } 4AuCl_3 + 6NH_2OH \longrightarrow 4Au + 3N_2O + 12HCl + 3H_2O$$

4. **Reduces Fe(III) to Fe(II)**

$$4FeCl_3 + 2NH_2OH \longrightarrow 4FeCl_2 + N_2O + 4HCl + H_2O$$

5. **Halogens to halogen acids**

$$2Cl_2 + NH_2OH \longrightarrow N_2O + H_2O + 4HCl$$

$$2Br_2 + NH_2OH \longrightarrow N_2O + H_2O + 4HBr$$

6. **Na Br O_3 to NaBr**

$$4NaBrO_3 + 12NH_2OH \longrightarrow 4NaBr + 18H_2O + N_2 + 3O_2$$

By a similar reaction

IO_3^- is reduced to I^-.

7. **$HgCl_2$ to Hg**

$$2HgCl_2 + 2NH_2OH \longrightarrow 2Hg + N_2O + 4HCl + H_2O$$

Oxidising property

Hydroxylamine acts as an oxidising agent in alkaline medium.

$$NH_2OH \longrightarrow NH_3 + O$$

(*a*) Sodium arsenite is oxidised to to sodium arsenate

$$Na_3AsO_3 + NH_2OH \longrightarrow \underset{\text{Sodium arsenate}}{Na_3AsO_4} + NH_3$$

(*b*) $Fe(OH)_2$ to $Fe(OH)_3$

$$2Fe(OH)_2 + NH_2OH + H_2O \longrightarrow 2Fe(OH)_3 + NH_3$$

(*c*) Hypronitrous acid to N_2

$$2NH_2OH + H_2N_2O_2 \longrightarrow 4H_2O + 2N_2$$

Reaction with aldehydes and ketones

$$RCH\boxed{O + H_2}NOH \longrightarrow \underset{\text{Aldoxime}}{R - CH{=}NOH}$$

$$R_2C\boxed{O + H_2}NOH \longrightarrow \underset{\text{Ketoxime}}{R_2C = NOH}$$

Uses:

1. It is used as an antioxidant in photographic developers.
2. It reduces Cu^{11} to Cu^{1} in the dyeing of acrylic fibres.
3. It forms oximes with aldehydes and ketones which is useful in the formation of polyamide fibres *e.g.*, nylon,

Structure: Hydroxylamine exists as is and trans isomers. N–O bond length is 147 pm with a single bond.

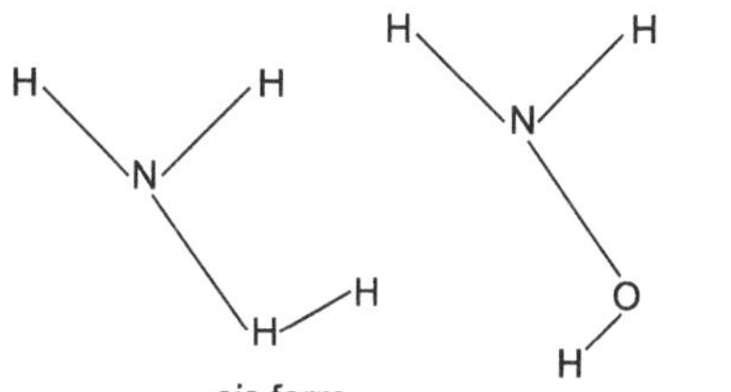

cis form *trans* form

Hydrazoic acid HN_3: Although it is hydride of nitrogen but it does not resemble with other hydrides such as NH_3 and N_2H_4. It is a weak acid comparable in strength to acetic acid.

Preparation: The anhydrous acid is a colourless explosive liquid.

It has enthalphy $\Delta Hf = +269.5$ kJ mol^{-1} at 25°C

The free energy of formation $\Delta Hf° = 327$ kJ mol^{-1} at 25°C

The above value suggests that HN_3 is unstable.

It is prepared by the oxidation of aqueous hydrazine with the help of HNO_2

$$N_2H_4 + HNO_2 \longrightarrow HN_3 + 2H_2O$$

other oxidising agents such as HNO_3, H_2O_2 and $K_2S_2O_8$ can also be used.

Physical Characteristics

It is a colourless volatile liquid with m.pt. = –80°C and b.pt. = 37°C. Due to explosive nature even its dilute solution should be handled with care.

It is a highly poisonous gas.

Chemical Characteristics

1. Decomposition: On standing it decomposes to N_2 and H_2 with the evolution of large amount of heat.

$$2HN_3 \longrightarrow H_2 + 3N_2 \ \Delta H = -51.5 \text{ kJ mol}^{-1}$$

2. Acidic nature: It is a weak acid comparable to acetic acid

$$HN_3(aq) \longrightarrow H^+(aq) + N_3^-(aq)\ Ka = 1.8 \times 10^{-5}$$

3. Azides: Alkali metal and active metals such as Zn, Mg etc. form azide salts.

$$HN_3(aq) + NaOH \longrightarrow NaN_3 + H_2O$$
$$Mg + 2HN_3(aq) \longrightarrow Mg(N_3)_2 + H_2$$
$$AgNO_3 + HN_3 \longrightarrow AgN_3 + HNO_3$$
$$NH_3 + HN_3 \longrightarrow NH_4N_3 \text{ ammonia azide}$$
$$HN_3 + N_2H_4 \longrightarrow N_2H_5^+N_3^- \text{ Hydrazinium azide}$$
$$AlH_3 + 3HN_3 \longrightarrow Al(N_3)_3 + 3H_2$$

Azides of heavy metals like Ba and Pb are highly explosives, alkali metal azides are not explosives on heating they supply N_2 gas.

$$2NaN_3 \longrightarrow 2Na + 3N_2\uparrow$$

4. Reduction: HN_3 is reduced to NH_3 and hydrazine by Na/Hg

$$HN_3 + 4H_2 \longrightarrow 2NH_3$$
$$2HN_3 + 5H_2 \longrightarrow 3HN_2.NH_2$$

In the case of acid solution

$$Sn^{2+} + 3H^+ + HN_3 \longrightarrow Sn^{4+} + NH_4^+ + N_2$$

5. Oxidation: The acid HN_3 can be oxidised to N_2 by HNO_2, Ce^{4+} salts and acid $KMnO_4$

$$HN_3 + HNO_2 \longrightarrow N_2 + N_2O + H_2O$$

with acid $KMnO_4$

$$2HN_3 + O[KMnO_4] \longrightarrow 3N_2 + H_2O$$

6. Decomposition: HN_3 is decomposed by HCl and HI to give N_2

Structure: In HN_3 gas all the three N atoms are in straight line

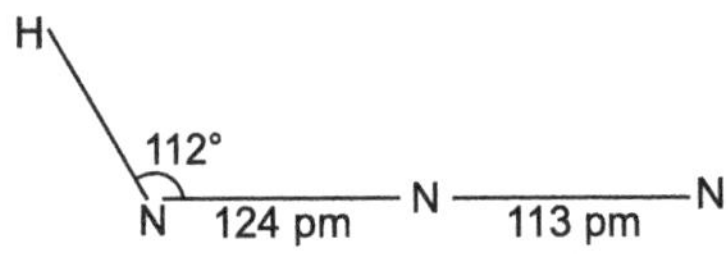

The azide N_3^- is symmetrical and linear (both N–N bonds being 116 pm). This implies that the electronic structure according to VBT is a resonance hydride of these structures

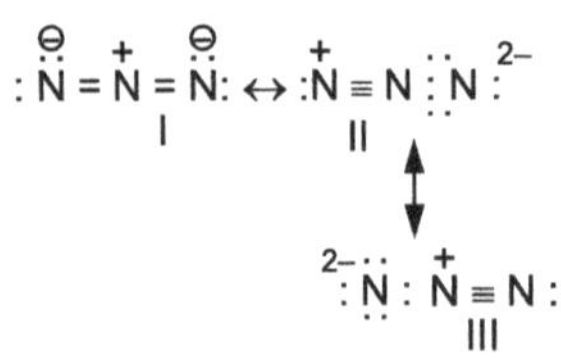

It is isoelectronic with CO_2 and NO_2^+

In structure I the three N atoms evoke *sp* hybridisation to form σ-bonds

N – N – N

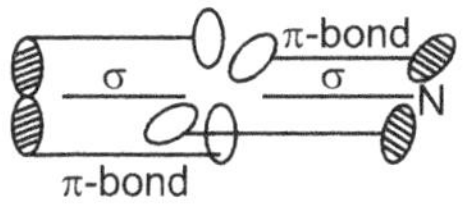

N = N = N

2π-bonds in N_3^- with each terminal N atom containing a lone pair.

Uses: Heavy metal azides are extensively used as explosives *e.g.* $Pb(N_3)_2$.

Phosporus Hydrides

Phosphorus forms a number of hydrides but two are important namely phosphine PH_3 and diphosphine P_2H_4.

Phosphine PH_3: Phosphine is a gas at room temperature.

Preparation

1. It may be prepared by heating white phosphorus with 40% NaOH solution.

$$P_4 + 3NaOH + 3H_2O \longrightarrow \underset{\text{Sodium hypo phosphite}}{3NaH_2PO_2} + PH_3$$

The phosphine gas catches fire in air therefore air is driven out by the current of hydrogen or coal gas. Pure PH_3 is not inflammable but PH_3 is almost invariably mixed with diphosphine P_2H_4 as impurity which is spontaneously inflammable.

2. It is more convenient to prepare PH_3 by the action of water on metal phosphides.

$$Ca_3P_2 + 6H_2O \longrightarrow 3Ca(OH)_2 + 2PH_3$$

$$AlP + 3H_2O \longrightarrow Al(OH)_3 + PH_3$$

3. Pure PH_3 gas can be prepared by the reaction of PH_4I and NaOH solution

$$PH_4I + NaOH \longrightarrow PH_3 + H_2O + NaI$$

Physical Characteristics

Phosphine gas is colourless, poisonous with a smell of rotten fish. It is heavier than air and in slightly soluble in water.

1. It burns in air or oxygen at 100°C to form meta phosphoric acid

$$PH_3 + 2O_2 \longrightarrow \underset{\text{Metaphosporic acid}}{HPO_3} + H_2O$$

Chemical Characteristics

1. It decomposes when heated to 420°C or by passing electric spark.

$$4PH_3 \longrightarrow P_4 + 6H_2$$

2. It burns in chlorine forming PCl_3 and PCl_5

$$PH_3 + 3Cl_2 \longrightarrow PCl_3 + 3HCl$$
$$PH_3 + 4Cl_2 \longrightarrow PCl_5 + 3HCl$$

3. It can reduce oxides of nitrogen

$$PH_3 + 4N_2O \longrightarrow H_3PO_4 + 4N_2$$
$$PH_3 + 4NO \longrightarrow H_3PO_4 + 2N_2$$

4. Passing PH_3 gas through metal salts give the metal precipitate

$$3CuSO_4 + 2PH_3 \longrightarrow \underset{\text{Black ppt.}}{Cu_3P_2\downarrow} + 3H_2SO_4$$

$$3AgNO_3 + PH_3 \longrightarrow \underset{\text{Black ppt.}}{Ag_3P\downarrow} + 3HNO_3$$

5. With halogen acids PH_3 forms phosphonium salts

$$PH_3 + HI \longrightarrow PH_4I \text{ Phosponium Iodide}$$

Uses:

1. It is used in preparing Holme's signals which guide the ships against the presence of rocks at night.
2. It forms smoke screen in warfare
3. It is used to make organophosphorus compounds

Structure: It has sp^3 hybridised pyramidal structure like ammonia but the bond angle H–P–H is only 93° and the P–H bond length of 124 pm. Unlike NH_3 phosphine is not present in liquid state. The properties of NH_3 and PH_3 can be compared

Ammonia	Phosphine
1. Non-Poisonous	Poisonous
2. Ammoniacal smell	Smells like rotten fish
3. Lighter than air	Heavier than air
4. Water soluble	Sparingly soluble in H_2O
5. Turns red litmus blue	Neutral to litmus
6. Hydrogen-bonded molecules	Non hydrogen bond
7. Deep blue colour with $CuSO_4$ solution	Produces black ppt. with $CuSO_4$ soluton.

Diphosphine P_2H_4

Preparation

1. Small amount of diphosphine is formed during the preparation of phosphine.
2. It can be prepared by the action of water on Ca_2P_2 which is formed by heating red phosphorus and lime.

$$Ca_2P_2 + 4H_2O \longrightarrow 2Ca(OH)_2 + P_2H_4$$

General Properties

It is a colourless unstable liquid with boiling point 27°C and melting point –99°C. It catches fire on exposure to air.

It decomposes in the preseuce of light to produce soild $P_{12}H_6$

$$15P_2H_4 \longrightarrow P_{12}H_6 + 18PH_3$$

It does not possess basic properties. It is a strong reducing agent like PH_3.

Structure: Its structure is akin to the structure of hydrazine

Oxides of Nitrogen

There exists *pπ-pπ* bonding between nitrogen and oxygen atoms in all oxides of nitrogen. This does not happen in heavier elements like P, As Sb and Bi, nitrogen forms a wide range of oxides *i.e.*, form oxidation states (+1) to (+VI). The lower oxides are basic and higher oxides are acidic.

Table 13.2 Oxides of nitrogen

Formula	Oxidation number	Name
N_2O	+I	Nitrous oxide
NO	+II	Nitric oxide
N_2O_3	+III	Nitrogen sesquioxide
$NO_2.N_2O_4$	+IV	Nitrogen dioxide, dinitrogen tetroxide
N_2O_5	+V	Dinitrogen pentoxide
(NO_3,N_2O_6)	+VI	Nitrogen trioxide, dinitrogen hexoxide
very unstable		

N_2O nitrous oxides: N_2O is a stable unreactive colourless gas.

Preparation: It is prepared by the decomposition of ammonium nitrate at 280°C

$$NH_4NO_3 \longrightarrow N_2O + 2H_2O$$

It can also be prepared by heating solution of NH_4NO_3 and HCl

$$NH_4NO_3 \xrightarrow{HCl} N_2O + 2H_2O$$

It explodes on strong heating. N_2O is neutral and does not form $H_2N_2O_2$ with water. It helps to form sodium azide with soda mide.

$$N_2O + \underset{\text{Sodamide}}{2NaNH_2} \longrightarrow NaN_3 + NH_3 + NaOH$$

It has no taste and is non-toxic. N_2O is used as an anaesthetic by dentists. It causes hysteric laughter and therefore called laughing gas, N_2O in the presence of dioxygen does not make a person completely unconscious. The patient can die if O_2 is not given Therefore N_2O is not suitable for long operations.

The molecule is triatomic and has a linear structure. The molecule has 16 outershell electrons. In comparison O=C=O has a symmetrical structure while in N_2O the orbitals favour the assymmetrical molecule N–N–O rather than the symmetrical N–O–N.

$$N \xrightarrow{1.126\text{Å}} N \xrightarrow{1.186\text{Å}} O$$

Bond order calculated as N–N = 2.73 and N–O is 1.61.

Nitric Oxide NO

NO is a colourless gas. It is formed during the oxidation of ammonia by Ostwald's process.

1. $4NH_3 + 5O_2 \xrightarrow{Pt} 4NO + 6H_2O$

2. The obsolete birkland Eyde process involving the combination of dinitrogen and dioxygen at 3500°C to form 5% nitric oxide.

$$N_2 + O_2 \longrightarrow 2NO$$

3. NO can also be prepared in the laboratory by the action of dilute HNO_3 on Cu.

$$3Cu + 8HNO_3 \longrightarrow 3Cu(NO_3)_2 + 2NO + 4H_2O$$

Alternatively it can be prepared by the reduction of nitrous with iodide ion:

$$2HNO_2 + 2I^- + 2H^+ \longrightarrow 2NO + I_2 + 2H_2O$$

NO is a neutral oxide. It is an odd electron molecule (11 electrons), it is paramagnetic in the gaseous state. In the solid or liquid state it dimerises forming N_2O_2 which is diamagnetic. NO is a red solid due to assymmetry in the presence of HCl. N–O bond length is 1.15Å which is in-between a double and triple bond. The bonding is siimilar to N_2 and CO which have 10 electrons. According to the molecular orbital theory NO has 11 outer electrons. The extra unpaired electron occupies an antibonding π^* $2p$ orbital. This will make the bond order as 2.5.

If the electron is removed from NO by oxidation then NO^+ nitrosonium ion is formed which has a bond order of 3.0. In N—O bond length is 1.15Å and N—O is 1.06Å in NO^+ ion.

The odd electron NO molecule tends to dimerise to become even electron molecule. It reacts with dioxygen to become NO_2 and with halogens to give nitrosyl halide.

$$2NO + O_2 \longrightarrow 2NO_2$$

$$2NO + Cl_2 \longrightarrow 2NOCl \text{ nitrosyl chloride}$$

NO has a tendency to form nitrosyl complexes *e.g.*, $[Fe(H_2O)_5NO]^{2+}$ in which NO^+ is responsible for the brown ring test for nitrates. Another example is cited in the formation of $Na_2\,[Fe(CN)_5NO].2H_2O$ *i.e.*, sodium nitroprusside. NO is a three electron donor while CO is a two electron donor. Hence 3CO are replaced by 2NO *e.g.*,

$$[Fe(CO)_5] + 2NO \longrightarrow [Fe(CO)_2(NO)_2] + 3CO$$

$$Cr(CO)_6 + 4NO \longrightarrow [Cr(NO)_4] + 6CO$$

In such complexes M–N–O atoms are linear or nearly linear. Lot of complexes where M–N–O angle in the range of 120 – 130° are known. These are weaker than the linear molecules. NO can also act as a bridging ligand similar to CO.

Nitrogen Sesquioxide N_2O_3

It can be obtained by condensing NO and NO_2 molecule in equimolar proportions.

$$NO + NO_2 \xrightarrow[\text{temp.}]{\text{low}} N_2O_3$$

It can also be obtained by reacting NO with calculated amount of oxygen

$$4NO + O_2 \longrightarrow 2N_2O_3$$

It is also called an anhydride of nitrous acid. With alkalis it forms nitrites.

$$N_2O_3 + H_2O \longrightarrow 2HNO_2$$
$$N_2O_3 + NaOH \longrightarrow 2NaNO_2 + H_2O$$

It reacts with concentrated acids forming nitrosyl salts.

$$N_2O_3 + 2HClO_4 \longrightarrow \underset{\text{nitrosyl perchlorate}}{2NO[ClO_4]} + H_2O$$

$$N_2O_3 + 2H_2SO_4 \longrightarrow \underset{\text{nitrosyl hydrogen sulphate}}{2NO[HSO_4]} + H_2O$$

It exists in two different forms. These can be interconverted by irradiation. N–N bond length is 1.864Å is assymmetrical form compoared to N–N bond length 1.45Å in hydrazine. There N–N bond length is weak.

assymmetrical form symmetrical form

Nitrogen Dioxide NO_2 and Dinitrogen Tetraoxide N_2O_4

NO_2 is a reddish brown gas produced by the Ostwald's reaction in the manufacture of nitric acid.

On the smaller scale it is produced by heating lead nitrate.

$$2Pb(NO_3)_2 \longrightarrow 2PbO + 4NO_2 + O_2$$

NO_2 and O_2 are passed through U-tube cooled under ice. NO_2 condenses to a pale liquid b.pt. 21°C finally NO_2 condenses to a colourless solid. NO_2 is an odd electron molecule and is paramagnetic NO_2 dimerises into a calourless N_2O_4. It has no unpaired electrons and behaves as diamagnetic

$$\underset{\substack{\text{brown}\\\text{paramegnetic}}}{2NO_2} \rightleftharpoons \underset{\substack{\text{colourless}\\\text{diamagnetic}}}{N_2O_4}$$

N_2O_4 is mixed anhydride as it forms HNO_2 and HNO_3 on reaction with water.

$$N_2O_4 + H_2O \longrightarrow \underset{\text{nitrous acid}}{HNO_2} + \underset{\text{nitric acid}}{HNO_3}$$

HNO_2 decomposes to give

$$2HNO_2 \longrightarrow NO_2 + NO + H_2O$$

Therefore moist NO_2 or N_2O_4 are acidic.

NO_2 molecule is angular with a bond angle ∠O–N–O of 132°. Bond length N–O is 1.20Å in-between a single and double bond. NO_2 solid shows planar structure.

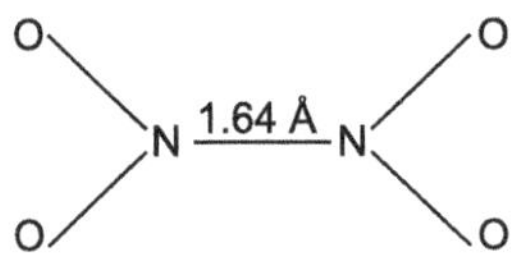

N–N bond is very long and weak.

N–N bond distance is 1.47Å in hydrazine.

Liquid N_2O_4 is a useful non-aqueous solvent.

Self Ionisation of N_2O_4

$$N_2O_4 \rightleftharpoons \underset{\text{acid}}{NO^+} + \underset{\text{base}}{NO_3^-}$$

The acid-base reaction yields N_2O_4

$$NOCl + NH_4NO_3 \longrightarrow \underset{\text{salt}}{NH_4Cl} + \underset{\text{solvent}}{N_2O_4}$$

liquid N_2O_4 is particularly used to prepare anhydrous metal nitrates and nitrato complexes.

$$ZnCl_2 + N_2O_4 \longrightarrow Zn(NO_3)_2 + 2NOCl$$

$$TiBr_4 + N_2O_4 \longrightarrow Ti(NO_3)_4 + 4NO + 2I_2$$

NO_2 is a strong oxidising agent. It reacts with F_2 and Cl_2 to give nitrosyl fluoride and chloride respectively

$$2NO_2 + F_2 \longrightarrow 2NO_2F$$

$$2NO_2 + Cl_2 \longrightarrow 2NO_2Cl$$

$$2NO + 4HCl \longrightarrow 2NOCl + Cl_2 + 2H_2O$$

$$NO_2 + CO \longrightarrow CO_2 + NO$$

N_2O_5

It can be easily prepared by dehydrating HNO_3 with P_2O_5 at low temperatures. It is a colourless diliquescent solid. It is highly reactive and strong oxidising agent. It is an anhydride of nitric acid and sensitive to light.

$$N_2O_5 + H_2O \longrightarrow 2HNO_3$$

$$N_2O_5 + Na \longrightarrow NaNO_3 + NO_2$$

$$N_2O_5 + NaCl \longrightarrow NaNO_3 + NO_2Cl$$

$$N_2O_5 + 3H_2SO_4 \longrightarrow H_3O^+ + \underset{\text{nitronium iron}}{NO_2^+} + 3HSO_4^-$$

In the gas form N_2O_5 decomposes to NO_2, NO and O_2.

$$N_2O_5 \longrightarrow N_2O_3 + O_2$$

Solid N_2O_5 is ionic NO_3^-, NO_2^+. It is called nitronium nitrate. It has the structure.

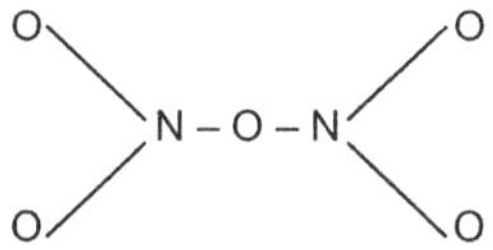

Oxides of Phosphorus Aresenic, Antimony and Bismuth

P_4O_6	III		As_4O_6 III		Sb_4O_6 III		Bi_2O_3 II
P_4O_7	III	V			$(SbO_2)_4$ III	V	
P_4O_8	III	V					
P_4O_9	III	V					
P_4O_{10}		V	As_4O_{10}	V	Sb_4O_{10}	V	

These oxides are limited in comparison to nitrogen owing to their inability to form $p\pi$-$p\pi$ double bonds as in the case of nitrogen.

Phosporus Trioixde P_4O_6

Phosphorus trioxide is dimeric and should be formulated as P_4O_6 instead of P_2O_3. Phosphorus exists as a tetrahedron. It molecular state is P_4. There are 6 oxygen atoms along the edges and each oxygen atom is bonded to two phosphorus atoms.

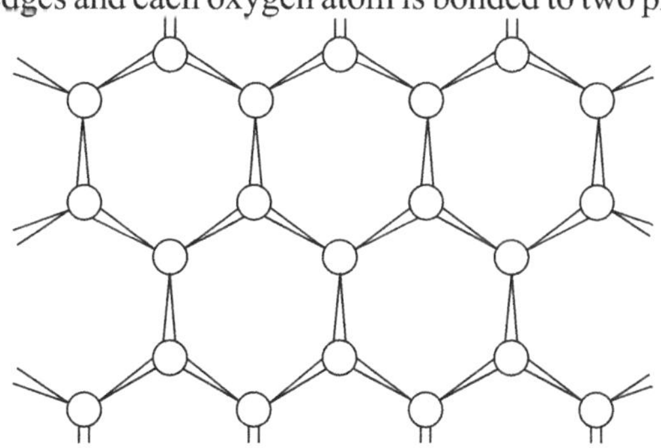

Fig. 13.5 Structure of phosphours trioxide P_4O_6

As_4O_6 and Sb_4O_6 are similary bonded while Bi_2O_3 ionic. The structure of P_4O_6 is shown in the figure above. The $\angle$P–O–P is 127°.

It is prepared by burning yellow phosphorus in the air.

$$P_4 + 3O_2 \xrightarrow[\text{air}]{\text{limited}} P_4O_6$$

P_4O_6 is a soft white solid m.pt. 24°C and b.point 175°C. It burns in air to form higher oxide P_4O_{10}.

$$P_4O_6 + 2O_2 \longrightarrow P_4O_{10}$$

Similary As_4O_6 and Sb_4O_6 are obtained by burning the respective metals in air or dioxygen.

Alternately, As_4O_6 and Sb_4O_6 can be abtained by their sulphide ores in air.

$$2As_2S_3 + 9O_2 \xrightarrow{\text{air}} As_4O_6 + 6SO_2$$

$$2Sb_2S_3 + 9O_2 \xrightarrow{\text{air}} Sb_4O_6 + 6SO_2$$

Both the oxides As_4O_6 and Sb_4O_6 are poisonous. Bi_2O_3 is not dimeric like these oxides. The basic nature of oxides increases on descending the group. P_4O_6 hydrolyses in water forming phosphorous acid.

$$P_4O_6 + 6H_2O \longrightarrow 4H_3PO_3$$

$$As_4O_6 + 12NaOH \longrightarrow 4Na_3AsO_3 + 6H_2O$$

$$As_4O6 + 12HCl \longrightarrow 4AsCl_3 + 6H_2O$$

As_4O_6 and Sb_4O_6 are both amphoteric as they react with acids as well as bases. Copper arsenites exist as schecle's green and paris green but they are not used now due to their toxic nature.

Pentoxides of P, As and Sb

These are very important oxides P_4O_{10} is dimeric $(P_2O_5)_2$. The basic structure is same as that of P_4O_6. In addtion each P atom in P_4O_6 forms three bonds to O atoms. There are 5 electrons in the valency shell of P atom; three electrons are used in bonding leaving a lone pair on P atoms. The lone pair on each of the four P atoms form a coordinate bond to an oxygen atom.

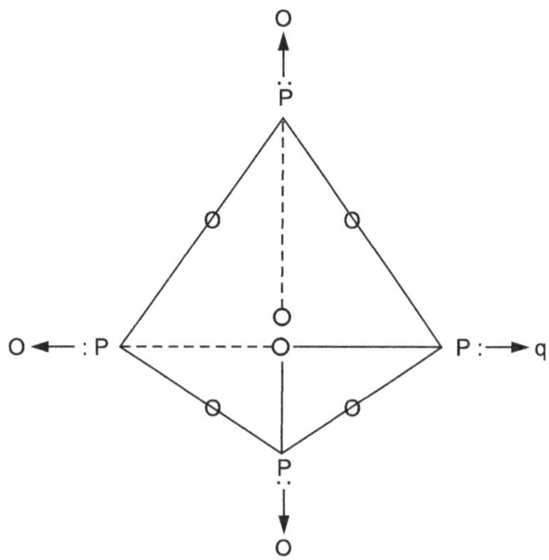

Fig. 13.6

The bridging bonds P–O are 1.60Å and the coordinate bonds at the corners are 1.43Å. The bonds at the corner are much shorter than a single bond and they represent a double bond. This is different from the > C = C < double bond in ethene in two respects.

1. It is *p* orbital overlop with *d* orbital and not *p-p* overlop.

2. Since both the electrons came from one atom therefore it forms a dative bond.

In the gas phase As_4O_{10} has a structure similar to P_4O_{10} but in the crystal form it contains equal number of $[AsO_4]$ tetrahedra and $[AsO_6]$ octahedra joined by sharing corners.

+5 oxidation state is the highest for this group of elements. They can be easily reduced to the lower oxidation state therefore they are used as strong oxidising agents.

P_4O_{10} is a wonderful dehydrating agent as it absorbs water from the air and becomes sticky. It is used mainly for drying the solvents. It hydrolyses with water.

$$P_4O_{10} + 6H_2O \longrightarrow 4H_3PO_4 \text{ Phosphoric acid}$$

It reacts with aclohols, ethers forming the corresponding phosphate esters.

$$P_4O_{10} + 6EtOH \longrightarrow 2O = P(OEt)(OH)_2 + 2O = 9(OEt)_2(OH)$$

$$P_4O_{10} + 6Et_2O \longrightarrow 4O = P(OEt)_3$$

As_4O_{10} dissolves slowly in water forming H_3AsO_4 arsenic acid. It is a tribasic acid. Lead arsenate and calcium arsenate are used as inseticides. Sb_4O_{10} is insoluble in water. Some antimonates are well known $[Sb(OH)_6]^-$. Bismuth does not form a pentoxide. Highest oxidation states decrease on descenting the group *i.e.*, higher oxidation are more acidic.

Uncommnon Oxides

The oxides P_4O_7, P_4O_8 and P_4O_9 are reported. These oxides have +III and +V phosphorus. Heating P_4O_6 under vacuum gives red phosphorus plus a mixture of P_4O_7, P_4O_8 and P_4O_9. These oxides represent strctures between P_4O_6 and P_4O_{10}. On hydrolysis they yeild a mixture of oxyacids in both oxidation states.

$$\left.\begin{matrix} P_4O_8 \\ P_4O_9 \end{matrix}\right] + H_2O \longrightarrow \underset{\text{Phosphoric acid}}{H_2PO_4} + \underset{\text{Phosphorus acid}}{H_2PO_3}$$

Oxyacids

Nitrogen has two important oxyacids namely nitrous acid and nitric acid.

Nitrous acid: This acid is unstable except in very dilute solutions.

Preparation: It can be prepared by the action of an acid on a nitrite of a metal.

$$Ba(NO_2)_2 + H_2SO_4 \longrightarrow BaSO_4 + 2HNO_2$$

Alkali metal nitrites can be obtained by heating the corresponding nitrates either by itself or by lead.

$$2NaNO_3 \xrightarrow{\Delta} 2NaNO_2 + O_2$$

$$NaNO_3 + Pb \xrightarrow{\text{heat}} NaNO_2 + PbO$$

It acts as a weak oxidising agent. It oxidises Fe^{2+} to Fe^{3+}; I^- to I_2 and itself gets reduced to N_2O or NO. It can be oxidised to HNO_3 or NO_3^- by $KMnO_4$ and Cl_2.

$$2KI + 2HNO_2 \longrightarrow I_2 + 2NO + 2KOH$$

$$KNO_2 + Cl_2 + H_2O \longrightarrow KNO_3 + 2HCl$$

Nitrites are a very useful reagent to make diazornum compounds which are further converted to azodyes and other pharmaceutical products.

$$PhNO_2 + HNO_2 \longrightarrow \underset{\text{Phenyl diagonium chloride}}{PhN_2Cl} + 2H_2O$$

Nitrites can also be converted to NH_2OH which is a very useful reagent. $NaNO_2$ is used as a food additive for sausages but it considered to be slightly poisonous (5 – 10 g per day). During cooking of meat nitrites react with amines to form nitrosamine R_2–N–N=O which is cancer causing.

The NO_2^- ion is a good ligand and forms coordination complexes in which both N and O can contain lone pairs to form M ←NO_2 nitro and M←ONO nitrito complexes *e.g.*, $[Co(NH_3)_5(NO_2)]^{2+}$, $[Co(NH_3)_5(ONO)]^{2+}$, Co^{2+} ions when treated with NO_2^- are oxidised to Co^{3+} *e.g.*, precipitate of potassium cobalitnitrite $K_3\,[Co(NO_2)_6]$.

Nitric acid HNO_3

Pure nitric acid is a colourless liquid on exposure to sunlight it turns slightly brown due to decomposition to NO_2 and O_2.

$$4HNO_3 \xrightarrow{hv} 4NO_2 + O_2 + 2H_2O$$

It disssociates 100% in aqueous solution into H_3^+O and NO_3^- ions. Its salts known as nitrates are soluble in water. NO_3^- ion is a planer triangle. Other elements in this group form tetrahedral oxoacids *e.g.*, PO_4^{3-}, AsO_4^{3-}. This is attributed to the small size of nitrogen and the limit of eight electron in the outershell.

It is a powerful oxidising agent when hot and concentrated. It become a stronger oxidising agent in acid solution. Cu and Ag dissolve in HNO_3 but Au dissolves in aqua regia (3 vol. HCl and, vol. HNO_3) due to oxidising power of HNO_3 and complex formation of Cl^- ions.

Preparation

1. HNO_3 was in the beginning made from KNO_3 by the action of concentrated H_2SO_4.

$$2KNO_3 + H_2SO_4 \longrightarrow K_2SO_4 + 2HNO_3$$

2. HNO_3 was synthesized by Birkland and Eyde process in an electric are furnace

$$N_2 + O_2 \xrightarrow[\text{arc}]{\text{electric}} 2NO$$

$$2NO + O_2 \longrightarrow 2NO_2$$
$$4NO_2 + 2H_2O + O_2 \longrightarrow 4HNO_3$$

This method entails high cost of electricity.

3. **Ostwald's Process:** This process involves catalytic oxidation of ammonia followed by the oxidation of NO to NO_2 and finally conversion of NO_2 to HNO_3.

$$4NH_3 + 5O_2 \xrightarrow[\text{5 at 850°C}]{\text{Pt/Rh}} 4NO + 6H_2O$$

NO and air are cooled and the mixture of gases is absorbed by water.

$$2NO + O_2 \longrightarrow 2NO_2$$
$$2NO + H_2O \longrightarrow HNO_2 + HNO_3$$
$$2HNO_2 \longrightarrow H_2O + NO_2 + NO$$
$$3NO_2 + H_2O \longrightarrow 2HNO_3 + NO$$

NO is recycled to give NO_2 as in the first step.

Overall Reaction

$$NH_3 + 2O_2 \longrightarrow HNO_3(60\%) + H_2O$$

Distillation of 60% HNO_3 raises the concentration to 68% since a constant boiling mixture is formed. Further increase in concentration to 98% can be achieved by dehydrating over concentrated H_2SO_4.

Nitrating Agent

On the other hand when HNO_3 is mixed with concentrated H_2SO_4, nitronium ion is formed NO_2^+ which acts as a nitrating agent for benzene and toluene.

Nitration of Benzene

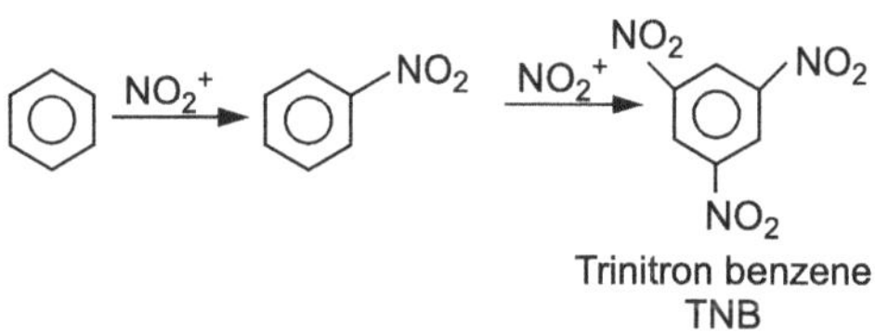

NItration of Toluene

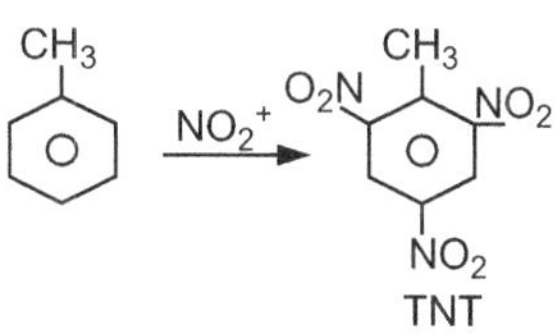

Organo nitrates are less stable and behave as explosive *e.g.*, nitroglycerine, nitro cellulose, TNT etc.

Oxidising Agent

It is manily used to oxidise cyclohexanol, cyclohexanone to adipic acid used in the manufacture of nylon-66

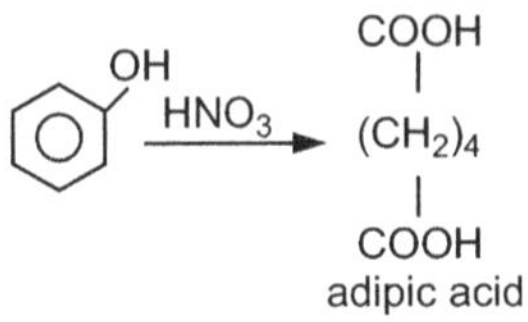

$$\begin{matrix} COOH \\ | \\ (CH_2)_4 \\ | \\ COOH \end{matrix} + (CH_2)_6(NH_2)_2 \rightarrow \text{Nylon} - 66$$

It can oxidise *p*-xylene to terephthalic acid

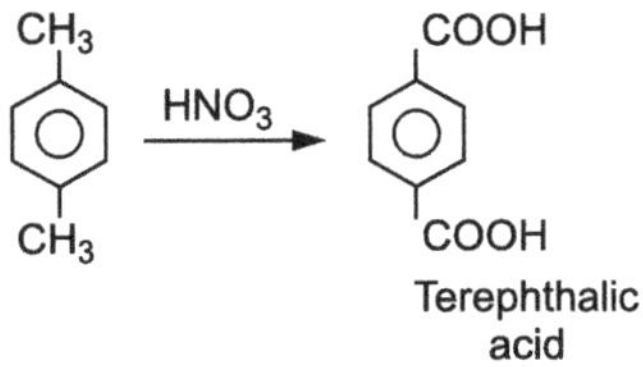

This in turn is used in the manufacture of terylene. The structure of NO_3^- ion is planar triangular.

Reaction with Metals

With copper metal HNO_3 reacts both in concentrated as well as dilute solutions to get reduced to NO_2 and NO.

$$Cu + 3HNO_3 \xrightarrow{\text{conc.}} NO_2 + Cu(NO_3)_2 + H_2O$$

$$3Cu + 8HNO_3 \xrightarrow{\text{dilute}} 2NO + Cu(NO_3)_2 + 4H_2O$$

Ammonia is produced when Cu/Al/Zn alloy reacts with NaOH.

$$Cu/Al/Zn + NaOH \longrightarrow H$$

$$NO_2^- + 9H \longrightarrow NH_3 + 3H_2O$$

$$NO_2^- + 7H \longrightarrow NH_3 + 2H_2O$$

NO_2^- is +V oxidation state which is finally reduced to –111 oxidation state (NH_3)

Oxyacids of Phosphorus

The phosphorus forms two series of oxyacids, with phosphorus having oxidation state +III and +V. The +III oxyacids are reducing agents as they can be easily oxidised to +V state while the +V oxidation state oxyacids are oxidising agents as they can be reduced to +III state. In all these oxyacids P is 4-coordinate and tetrahedrally disposed. *p*π-*p*π back bonding in P–O bonds gives rise to P = O bonds. The –OH groups are ionisable and therefore acidic while the

P—H bonds in acids give reducing property. The PO_4^{3-} ions can polymerize to give isopolyacids and their salts. The more important phosphoric acid series are discussed first.

Orthophosphoric acid H_3PO_4

H_3PO_4 contains three replaceable H-atoms therefore it is tribasic. It dissociates as mentioned.

$$H_3PO_4 \rightleftharpoons H^+ + H_2PO_4^- \quad K_1 = 7.5 \times 10^{-3}$$
$$H_2PO_4^- \rightleftharpoons H^+ + HPO_4^{2-} \quad K_2 = 6.2 \times 10^{-8}$$
$$HPO_4^{2-} \rightleftharpoons H^+ + PO_4^{3-} \quad K_3 = 1.0 \times 10^{-12}$$

This can give rise to these salts.

1. NaH_2PO_4 sodium dihydrogen phosphate give slight acidic behaviour
2. Na_2HPO_4 Disodium monohydrogen phophate which is basic in water.
3. Na_3PO_4 Trisodium phosphate which make the aqueous solution quite basic.

NaH_2PO_4 and Na_2HPO_4 are obtained from H_3PO_4 and Na_2CO_3 but Na_3PO_4 is produced from H_3PO_4 and NaOH. These salts are anhydrous and hydrated.

H_3PO_4 reacts with aclohols to form trialkoxy phosphate ester.

$$(HO)_3 - P = O + \underset{\text{alcohol}}{3Et\,OH} \longrightarrow \underset{\substack{\text{Triethyl phosphates} \\ \text{(ester)}}}{(EtO)_3 - P = O} + 3H_2O$$

Test for PO_4^{3-} Ions

Qualitatively the phospate salt solution is mixed with dilute HNO_3 and then the ammonium molybdate solution is added. A yellow precipitate of ammonium 12–molybdo phosphate is obtained slowly. Arsenic gives a similar yellow precipitate but on slight heating. Orthophosphates of titanium, zirconium and hafnium are insoluble in acids, therefore zirconyl nitrate is used to remove phosphate present in the solution.

Phophates are quantitatively estimated by adding Mg^{2+} salt solution and NH_4OH to a solution of the phosphates. White precipitate of magnesium ammonia phosphate is filtered and estimated. On ignition $Mg_2P_2O_7$ magnesium pyrophosphate obtained can be weighed and estimated.

$$Mg^{2+} + NH_4^+ + PO_4^{3-} \longrightarrow Mg(NH_4)PO_4\downarrow$$
$$2Mg(NH_4)PO_4 \longrightarrow Mg_2P_2O_7 + 2NH_3 + H_2O$$

Structure of Orthophosphoric Acid H_3PO_4

As discussed earlier, it is tribaric 3(OH) groups and P = O bond due to $p\pi$–$p\pi$ back donated bond.

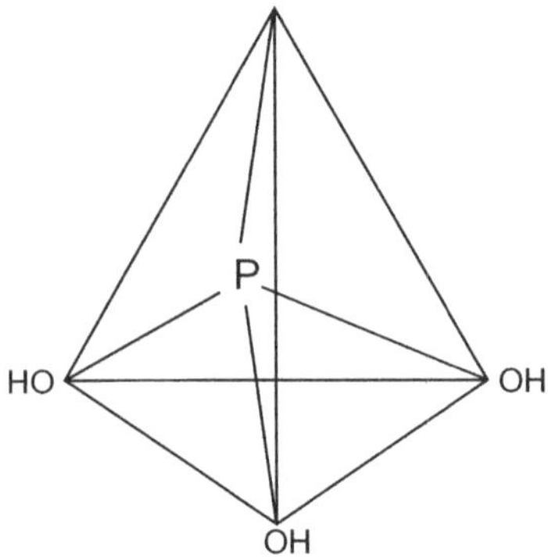

H_3PO_4

Preparation

1. Orthophosphoric acid can be obtained from phospate rock and H_2SO_4.

$$\underset{\text{Tricalciumphosphate}}{Ca_3(PO_4)_2} + 3H_2SO_4 \longrightarrow 2H_3PO_4 + 3CaSO_4$$

$$\underset{\text{rock phasphate}}{[3Ca_3(PO_4)_2 + CaF_2]} + 10H_2SO_4 \longrightarrow 6H_3PO_4 + 10CaSO_4 + 2HF$$

Phosphoric acid produced by the above mentioned process is impure. Pure H_3PO_4 is prepared by taking molten phosphorus and heating with air and steam in a furnace.

$$P_4 + 5O_2 \longrightarrow \underset{\substack{\text{Phosphorus}\\\text{pentoxide}}}{P_4O_{10}}$$

P_4O_{10} obtained is immediately hydrolysed to form H_3PO_4.

$$P_4O_{10} + 6H_2O \longrightarrow 4H_3PO_4$$

Concentrated H_3PO_4 is viscous as it hydrogen bonded in aqueous soluton. Concentrated acid is 85% by weight. The acid produced by this method is used in the laboratory kraft cheese is Na_2HPO_4.

It can be very easily prepared by reacting P with concentrated HNO_3.

$$P_4 + 20HNO_3 \longrightarrow 4H_3PO_4 + 20NO_2 + 4H_2O$$

Action of Heat on Orthophosphoric Acid

It loses water stepwise

$$\underset{\substack{\text{orthophosphoric}\\\text{acid}}}{2H_3PO_4} \xrightarrow[-H_2O]{220°C} H_4P_2O_7 \xrightarrow{320°C} \underset{\substack{\text{metaphosphoric}\\\text{acid}}}{(HPO_3)_n}$$

Metaphosporic acid usually polymerises partecularly $(NaPO_3)_n$.

Polyphospates: Isopolyphosphate and isopoly acids arise out of polymerising PO_4 units in acid medium. They consist of chain of tetrahedra giving simple unbranched chains.

The hydrolysis of P_4O_{10} in steps gives an insight to the range of phosphoric acids.

$$P_4O_{10} + 6H_2O \longrightarrow 4H_3PO_4 \text{ final stage reaction.}$$

Polyphosphates are straight chain structures. From the structures of these isopoly acids we can estimate the replaceable H atoms from the number of OH groups.

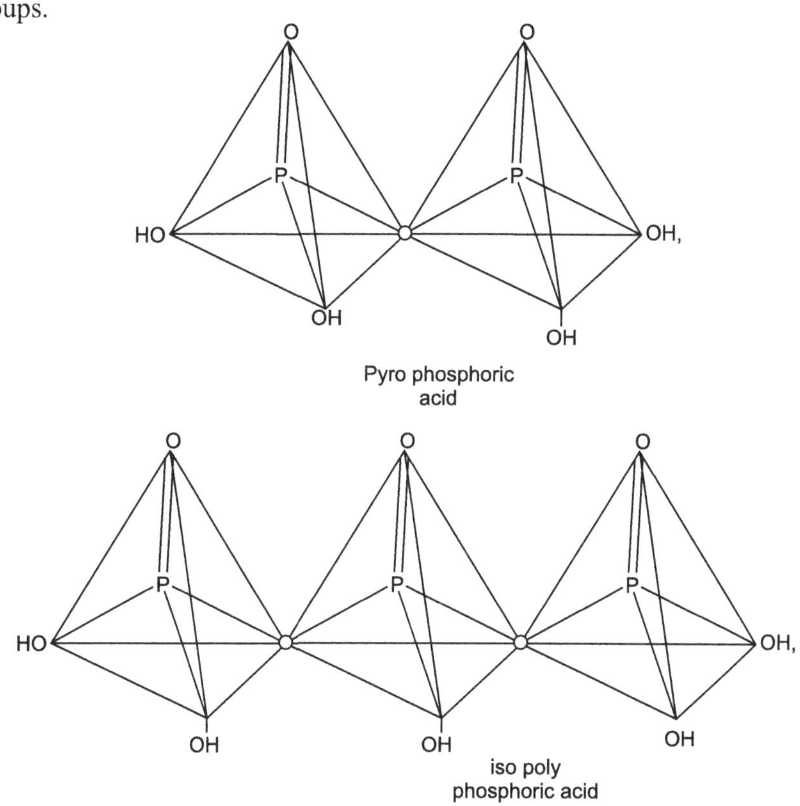

Fig. 13.7 Sturcture of pyrophosphoric acid and tripolyphosporic acid

Pyrophosphoric acid is tetrabasic tripolyphosphoric acid is pentabasic acid tetrapolyphosphoric acid is hexa basic while tetra metaphosphoric acid is tetrabasic.

$$P_4O_{10} + 6H_2O \longrightarrow 4H_3PO_4 \text{ overall reaction}$$

Polyphosphates up to the chain of ten PO_4 have been isolated. Out of these the first four are well know.

1. H_3PO_4 orthophosphoric acid
2. $H_4P_2O_7$ pyrophosphoric acid (dipolyphoric acid)
3. $H_5P_3O_{10}$ tripolyphosphoric acid
4. $H_6P_4O_{13}$ tetrapolyphosphoric acid

One of long chain poly phosphates is called Graham's salt.

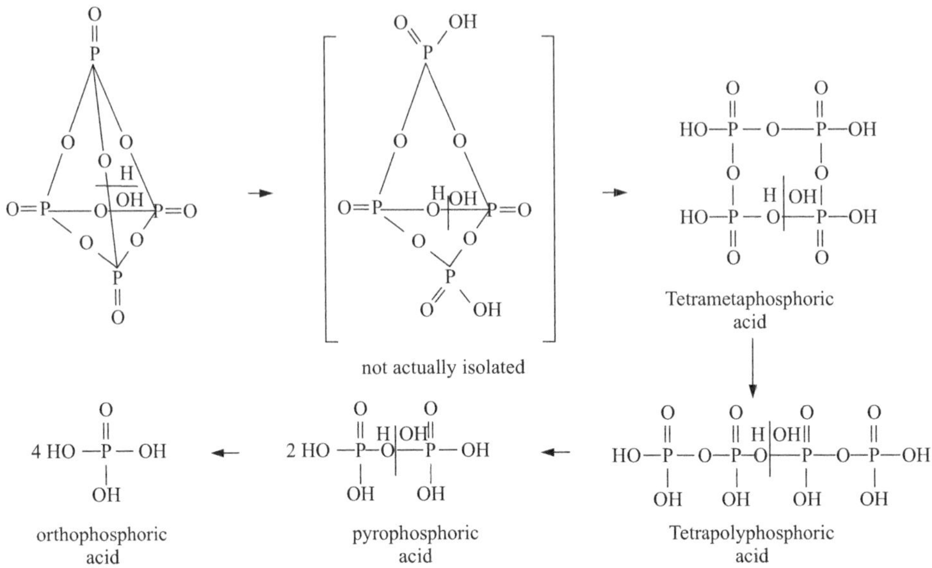

Fig. 13.8 Stepwise hydrolysis of P_4O_{10} lee P. 514

$Na_2H_2P_2O_7$ when mixed $NaHCO_3$ is used in fermenting bread in place of using yeast. $Na_2P_2O_7$ when added to soap solutions or soap powder acts as a water softener.

Sodium tripoly phosphate solution dissolves grease and improves the quality of detergents. It is prepared by:

1. Fuse proper amounts of Na_2HPO_4 and NaH_2PO_4. Recrystallization from water yields $Na_5P_3O_{10}.6H_2O$.

2. Fuse Na_2O and P_4O_{10}. On cooling sodium pyrophosphate crystallises out but on slow cooling this changes to $Na_5P_3O_{10}$ structures of some polyphosphate ions are given below:

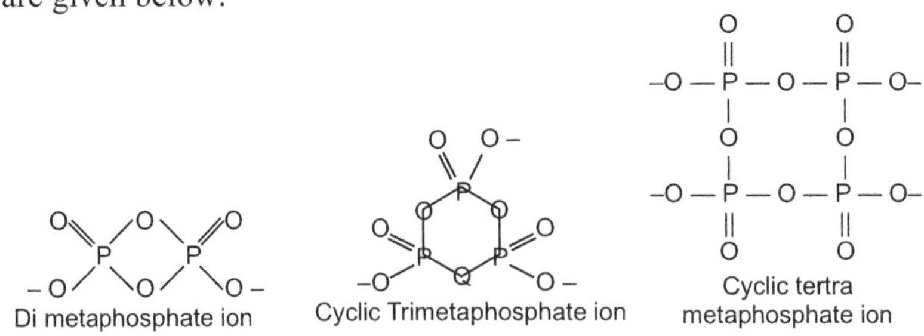

Fig. 13.9

A few large rings PO_4 units up to 8 tetrahedra have been isolated *e.g.* $Na_8[P_8O_{24}]$. On heating NaH_2PO_4 to 640°C trimetaphosphate is obtained ($Na_3P_3O_9$). Hydrolysis of trimetaphosphate with NaOH gives chain of tripolyphosphate compound.

$$3NaH_2PO_4 \xrightarrow{\text{heat}} Na_3P_3O_9 + 3H_2O$$
$$Na_3P_3O_9 + 2NaOH \longrightarrow Na_5P_3O_{10} + H_2O$$

On treating P_4O_{10} with a solution of cold NaOH sodium tetrametaphosphate is formed ($Na_4P_4O_{12}$).

Hypophosphoric Acid $H_4P_2O_6$

It has one less oxygen in comparison to pyrophosporic acid $H_4P_2O_7$. Here P is in +IV oxidation state. It is prepared by the oxidation of red phosphorus by NaOCl or yellow phosphorus by water and air. It contains no P—H bonds therefore it does not behaves as a reducing agent. There are four P—OH bonds hence it is tetrabasic. It contains P—P bond which is stronger than P—O—P bond, so the hydrolysis is slow.

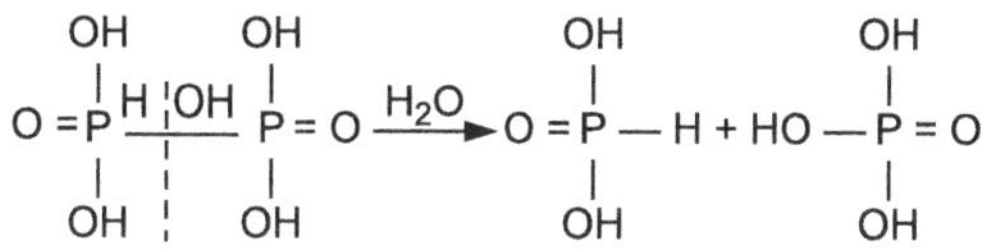

H_3PO_3 Phosphorus acid H_3PO_4 Phosphoric acid

Phosphorus Acid and Related Acids

These acids represent the oxidation state +III. They do contain P–H bonds therefore they act as reducing agents. They are prepared by the hydrolyies of P_4O_6. They can yield pyro and orthophosphorus acids which are dibasic as well as reducing agents.

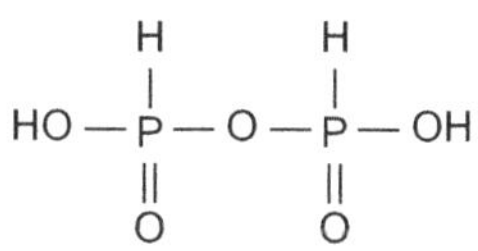

Pyrophosphorous acid

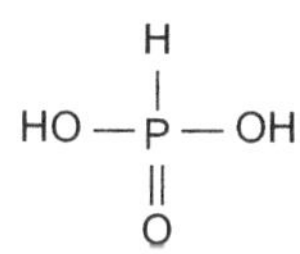

ortho phosphorous acid

Orthophosphorus Acid H_3PO_3

It contains two acidic H atoms *i.e.* 2O—H groups and P – H hydrogen atom. It is dibasic because only two of the three hydrogen atoms can ionise.

$$H_3PO_4 \rightleftharpoons H^+ + H_2PO_3^- \quad Ka_1 = 1.6 \times 10^{-2}$$
$$H_2PO_3^- \rightleftharpoons H^+ + HPO_3^{2-} \quad Ka_2 = 7 \times 10^{-7}$$

Due to this ionisation, two salts can be formed *e.g.* NaH_2PO_3 and Na_2HPO_3. These phosphates act as strong reducing agents in basic solutions. In acid solution we get H_3PO_3 which is a moderate reducing agent.

Metaphosphorus Acid $(HPO_2)_n$

It can be prepared from phosphine at low pressure.

$$PH_3 + O_2 \xrightarrow{25\text{ mm Hg}} H_2 + HPO_2$$

In this P atom will form three bonds or a double bond. It prefers to polymerise instead of forming a double bond. It may form a ring structure but the exact structure is uncertains.

Hypophosphorus Acid H_3PO_2

Here the oxidation state is + 1. It has one oxygen atom less than H_3PO_3.

It is prepared by alkaline hydrolysis of phosphorus.

$$P_4 + 3NaOH + 3H_2O \longrightarrow 3NaH_2PO_2 + PH_3$$

It is monobasic and a strong reducing agent. Its salts are called hypophosphites NaH_2PO_2 and are used to bleach wood and to make paper.

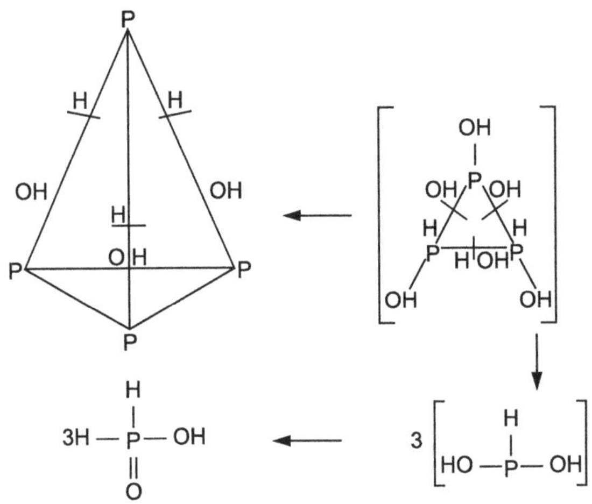

Fig. 13.10 Hypophosphorus acid

Phosphates: Phosphates are useful in many fields.

1. **Fertilizers:** Superphosphate of lime, triple superphosphate and the common ammonium phosphate are used as fertilizers. Here, cent per cent purity does not matter.
2. **Detergents:** Sodium tripoly phosphate and sodium pyrophosphate in powder and liquid state respectively are used as fillers in detergents.
3. **Foods:** They are used in cola drink to give acid taste and root beer.
4. **Emulisfier:** Used as emulsifiers in processed cheese, dried milk and other sausages.
5. (*a*) **Metals:** Some metals are treated with phosphoric acid at 95°C. Zn^{2+}, Mn^{2+} and Cu^{2+} are used for rust proofing.

(*b*) Pickling or removing scale and oxide film from the metal surface *e.g.* iron and steel.

(*c*) Aluminium parts are dipped in a both of H_3PO_4 with a small amount of nitric acid and $Cu(NO_3)_2$ to give extra shining metal surface protected by Al_2O_3.

6. **Water softening:** Na_3PO_4 and calgon (sodium hexametaphosphate) are effectively used as water softeners.
7. **Buffers:** Phosphate buffers NaH_2PO_4 and Na_2HPO_4 are used for removing H_2S in the petroleum in dustry.
8. **Matches:** Phosphorus sulphide are used in making matches.
9. **Platicizers:** Organophosphorus compounds are used as plasticizers *e.g.* Triaryl phosphates.
10. **Insecticides *e.g.* Triethyl phosphate**
11. **Petrol:** Tritolyl phosphate is used as petrol additives.
12. **Tooth Paste:** $CaHPO_4.2H_2O$ is used in tooth paste and $Ca_2P_2O_7$ in fluoride tooth paste.
13. **Baking powder:** $Ca(H_2PO_4)_2$ mixed with $NaHCO_3$ is used as a baking powder.
14. **Fabrics:** To mae the fabric flame proof ammonium phosphate and urea phosphate are used.

Some of the uses of phosphates are criticized *e.g.* as water softeners they also contribute to water pollution. The phosphates go through sewage disposal into rivers and lakes where they consume dissolved oxygen and kill aquatic animals like fish and others.

Halides of Phosphorus, Trihalides

Trihalides of N, P, As, Sb and Bi are well known. Trihalides of nitrogen *i.e.* NCl_3 is explosive while NF_3 is quite stable, NBr_3 and NI_3 are formed as ammeniates $NBr_3.6NH_3$ and $NI_3.6NH_3$ but they are unstable. The trihalides are covalent and have a tetrahedral structure similar to NH_3. The rest of the trihalides are all very stable.

The trihalides have a tetrahedral structure with one position occupied by the lone pair except BiF_3 Which is ionic other trihalides are all covalent.

The trihalides are prepared by the direct combination of the elements.

$$P_4 + 6Cl_2 \longrightarrow 4PCl_3$$
$$N_2 + 3Cl_2 \longrightarrow 2NCl_3$$
$$As_4 + 6Cl_2 \longrightarrow 4AsCl_3$$
$$2Sb + 3Cl_2 \longrightarrow 2SbCl_3$$
$$2Bi + 3Cl_2 \longrightarrow 2BiCl_3$$

The trihalides hydrolyse quickly with water but the products vary depending on the elements.

$$NCl_3 + 6H_2O \longrightarrow NH_4OH + 3HOCl$$

$$PCl_3 + 3H_2O \longrightarrow H_3PO_3 + 3HCl$$

$$AsCl_3 + 3H_2O \longrightarrow H_2AsO_3 + 3HCl$$

$$SbCl_3 + H_2O \longrightarrow \underset{\text{anti monyl chloride}}{SbO^+Cl^-} + 2HCl$$

$$BiCl_3 + 3H_2O \longrightarrow \underset{\text{bis muthyl chloride}}{BiO^+Cl^-} + 2HCl$$

NF_3 does not hydrolyse with water, dilute acids and alkalis because it is unreactive PF_3 is less prone to hydrolysis. PF_3 can donate its lone pair to form coordinate bond *e.g.* $Ni(PF_3)_4$. It forms σ bond as well as a π bond due to back bonding from a filled metal orbital to the empty *d* orbital on P similar to CO ligand.

$$Ni(CO)_4 + 4PF_3 \longrightarrow Ni(PF_3)_4 + 4CO$$

Many other trifloro phosphine complexes are known.

Reaction with NH_3

The trihalides react with NH_3 *e.g.* $PCl_3 + 6NH_3 \longrightarrow P(NH_2)_3 + 3NH_4Cl$

PF_3 is a colourless, odourless gas which is very toxic. The trihalides show acceptor properties. It can accept a lone pair from another ion to form a complex ion $[SbF_5]^{2-}$ and $[Sb_2F_7]^-$. They do react with organometallic reagents.

PCl_3 is the most important trihalide it is instrumental in preparing PCl_5

$$PCl_3 + Cl_2 \text{ or } S_2Cl_2 \longrightarrow PCl_5$$

PCl_3 is widely used as a reagent to convert alcohols to alkyl halides and carboxylic acids to acid chloride.

$$PCl_3 + 3C_2H_5OH \longrightarrow 3C_2H_5Cl + H_3PO_3$$

$$PCl_3 + 3RCOOH \longrightarrow 3RCOCl + H_3PO_3$$

PCl_3 is oxidised by O_2 or P_4O_{10} to give phosphoryl chlodie ($POCl_3$).

$$2PCl_3 + O_2 \longrightarrow 2POCl_3$$

$$6PCl_3 + P_4O_{10} + 6Cl_2 \longrightarrow 10POCl_3$$

$POCl_3$ is used to prepare trialkyl and triaryl phosphates.

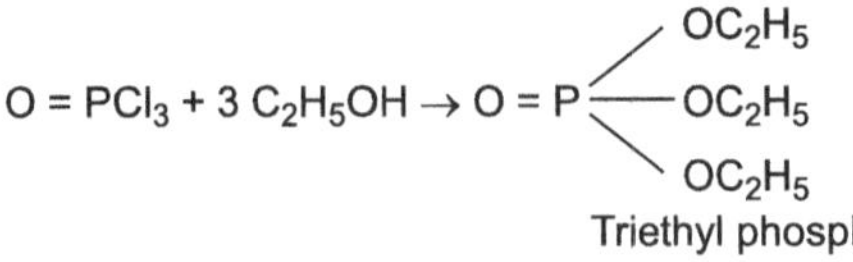

Triethyl phosphate

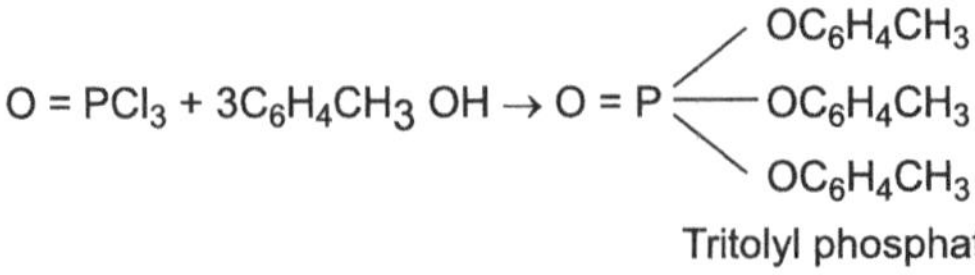

Tritolyl phosphate

These phosphate derivatives are used in making insecticides, plasticizers, solvent extraction and petrol additives.

Penta Halides

Nitrogen is not able to form pentahalides owing to eight electrons in the second shell. The remaining elements have available *d* orbitals for the valency expansion *e.g.*

$$PF_5, AsF_5, SbF_5, BiF_5$$

$$PCl_3, \underset{\text{Doubtful}}{(AsCl_5)}, 5bCl_5$$

$$PBr_5$$

$$PI_5$$

$AsCl_5$ has a temporary existance BiF_5 is very reactive and explodes with H_2O into O_2 and F_2O. It oxidises UF_4 to UF_6 and BrF_3 to BrF_5.

Pentahalides are prepared as mentioned

$$3PCl_5 + 5AsF_5 \longrightarrow PF_5 + 5AsCl_3$$

$$PCl_3 + Cl_2\ (CCl_4) \longrightarrow PCl_5$$

$$2As_2O_3 + 10F_2 \longrightarrow 4AsF_5 + 3O_2$$

$$2Sb_2O_3 + 10F_2 \longrightarrow 4SbF_5 + 3O_2$$

$$2Bi + 5F_2 \longrightarrow 2BiF_5$$

The pentahalides of P, As Sb and Bi ahve a trigonal bipyramidal structure expected from VSEPR theory.

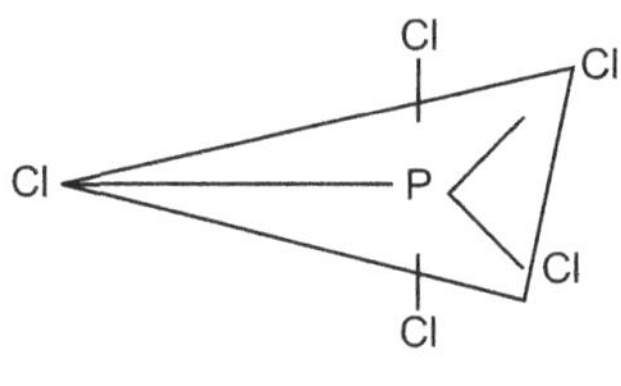

Fig. 13.11 Gageous PCl_5

Valence Bond Theory

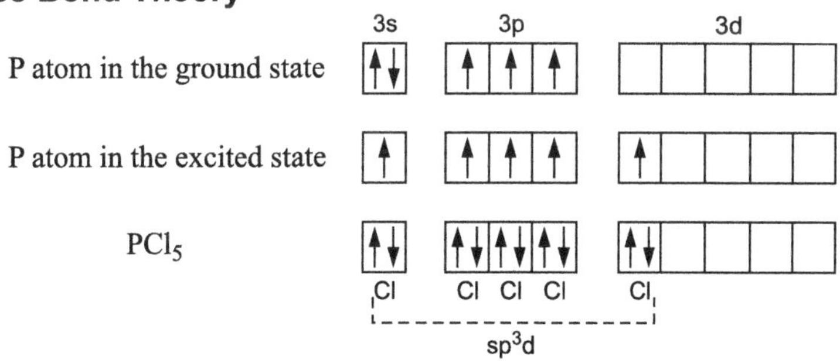

sp^3d hybridisation suggests trigonal bipyramidal structure. It is not a regular structure the bond angles Cl—P—Cl inside the triangle are 120° while the apex bond angles are 90°. The P—Cl bond lengths are 1.58Å while the P—Cl bond length at equatorial position are 1.53Å. NMR studies suggest that all the five P—Cl bonds are equivalent while the electron diffraction studies gives on the spot position wheras *nmr* gives the average picutre. This suggests interchange of axial and equatorial positions or resonance.

PCl_5 becomes ionic in the crystal form.

$$2PCl_5 \rightleftharpoons [PCl_4]^+ [PCl_6]^-$$

In the gas and liquid from PCl_5 stays covalent. PCl_5 happens to be the most important pentahalide PCl_5 hydrolyses violently with water.

$$PCl_5 + 4H_2O \longrightarrow \underset{\text{orthophosporic acid}}{H_3PO_4} + 5HCl$$

In quimolar amounts the reaction is gentle with the formation of phosphoryl chloride

$$\underset{\text{1 vol.}}{PCl_5} + \underset{\text{1 vol.}}{H_2O} \longrightarrow \underset{\text{phophoryl chloride}}{POCl_3} + 2HCl$$

PCl_5 acts as a chlorinating agent to convert carboxylic acids to acid chlorides and alcohols to alkyl chloride

$$PCl_5 + 4R-\overset{\overset{\displaystyle O}{\|}}{C}-OH \longrightarrow 4RCOCl + H_3PO_4 + HCl$$

$$PCl_5 + 4R\text{–}OH \longrightarrow 4R\text{–}Cl + H_3PO_4 + HCl$$

Reaction with P_4O_{10} and SO_2

$$6PCl_5 + P_4O_{10} \longrightarrow \underset{\text{phophoryl chloride}}{10POCl_3}$$

$$PCl_5 + SO_2 \longrightarrow POCl_3 + \underset{\text{Thionyl chloride}}{SOCl_2}$$

Reaction with NH_4Cl: With NH_4Cl, PCl_5 forms a variety of phosphonitrilic polymers

$$nPCl_5 + nNH_4Cl \longrightarrow \underset{\text{Ring compound}}{(NPCl_2)_n} + 4nHCl \text{ and } \underset{\text{chain compound}}{Cl_4\,P-(NPCl_2)_n\,NPCl_3}$$

Parallel PH_5 are not known. Hydrogen is not strongly electronegative to invoke orbitals in P atom but PHF_4 and PH_2F_3 are prepared.

Phosphorus oxy-trichloride or phosphoryl chloride $POCl_3$

Preparation

(*i*) It is prepared by the action of O_2, O_3 or SO_3 on PCl_3.

$$2PCl_3 + O_2 \longrightarrow 2POCl_3$$

$$PCl + SO_3 \longrightarrow POCl_3 + SO_2$$

(*ii*) action of H_2O (calculated amount), P_4O_{10}, $H_2C_2O_4$, H_3BO_3 and B_2O_3 on PCl_5

$$PCl_5 + H_2O \longrightarrow POCl_3 + 2HCl$$
$$6PCl_3 + P_4O_{10} \longrightarrow 10POCl_3$$
$$PCl_5 + H_2C_2O_4 \longrightarrow POCl_3 + CO + CO_2 + 2HCl$$
$$3PCl_5 + H_3BO_3 \longrightarrow 3POCl_3 + B_2O_3 + 6HCl$$
$$3PCl_5 + B_2O_3 \xrightarrow{\text{heating}} 3POCl_3 + 2BCl_3$$

Properties: It is a colourless fuming liquid of specific gravity 1.721 at 0°C and boiling point = 107.2°C and melting point 1.35°C.

1. It hydrolyses slowly by H_2O

$$POCl_3 + 3H_2O \longrightarrow H_3PO_4 + 3HCl$$

2. It replace –OH group from the organic compounds by the –Cl groups

$$3R\text{–}OH + POCl_3 \longrightarrow 3RCl + H_3PO_4$$

3. In cases such as H_3PO_4 and $H_4P_2O_7$; $POCl_3$ forms metaphosphoric acid (HPO_3)

$$POCl_3 + H_3PO_4 \longrightarrow 2HPO_3 + 3HCl$$
$$POCl_3 + 2H_4P_2O_7 \longrightarrow 5HPO_3 + 3HCl$$

Structure of $POCl_3$

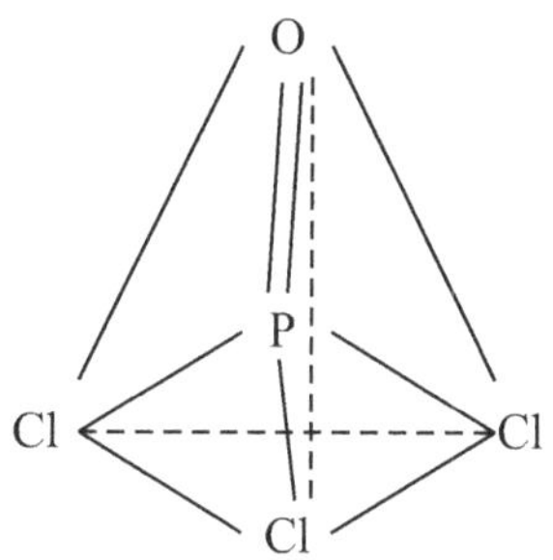

Fig. 13.12 Tetrahedral shape of $POCl_3$

Phosphazenes and Polyphosphazenes

Limited catenation is observed in nitrogen and phosphours. The chain in nitrogen is three in azide ion (N_3^-) and two in phosphorus *e.g.* P_2H_4 and $Me_2(S)$—P—P(S) Me_2.

In ring compounds 4, 5 or 6 phosphorus atoms are joined. N and P may also bond together giving compounds known as phosphazenes. In these compounds P exists in +V state and N is +III state. These are unsaturated compounds.

Monophosphazenes are prepared by reacting azide with PCl_3, POR_3 or $P(C_6H_5)_3$.

1. $PCl_3 + C_6H_5N_3 \longrightarrow Cl_3\,P = N\,C_6H_5 + N_2$
2. $P(C_6H_5)_3 + C_6H_5N_3 \longrightarrow (C_6H_5)_3\,P = N\,C_6H_5 + N_2$

Diphosphazene

$$3PCl_5 + 2NH_4Cl \longrightarrow [Cl_3P = N\text{–}PCl_2 = N\text{–}PCl_3^{+}]Cl^{-} + 8HCl$$

N and P may catenate together to form series of polymers.

$$nPCl_5 + nNH_4Cl \xrightarrow{150°C} (NPCl_2)_n + 4nHCl + Cl_3P.(NPCl_2)_n.NPCl_3$$

The above reaction creates 3, 4, 5, 6... linear chains. Rings where $n = 3$ and 4 may contain 6 or 8 atoms. The former are flat and the later may exist in 'chair' and 'boat' form.

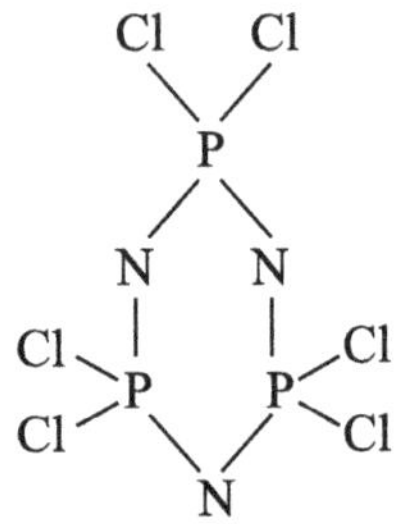

Chair

Fig. 13.13

Cyclophosphazenes

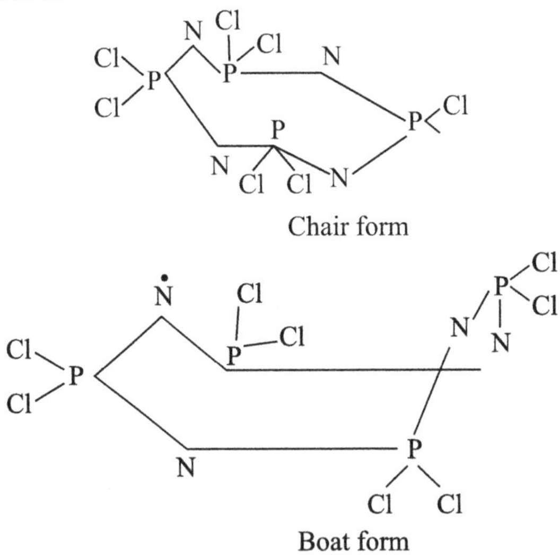

Boat form

Fig. 13.14

These compounds range from P_2NCl_7, $P_3N_2Cl_9$ and $P_4N_3Cl_{11}$ originally they were called phosphorilic halides or poly chloro phosphazenes

Phosphorus Sulphides

A series of phosphorus sulphides are formed when P and S are heated together at 100°C *e.g.* P_4S_3, P_4S_5, P_4S_7 and P_4S_{10} depending upon the amounts of P and S. The other phosphorus sulphides are prepared by other reactions *e.g.* P_4S_4 and P_4S_9.

P_4S_{10} is quite akin to P_4O_{10} but the absence of parallel P_4S_6 related to P_4O_6 is surprising. The other sulphides have no correlation to oxides. In the suphides P atoms are tetrahedrally disposed with S atoms forming a bridge between P atoms. The structure of phosphorus sulphides are drawn as under:

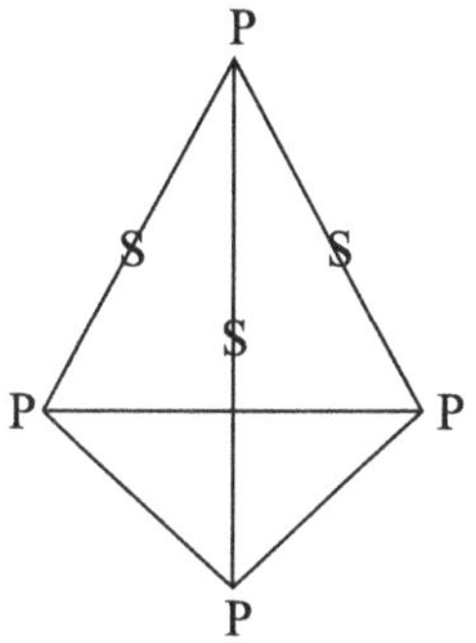

Tetra phosphorus
tri sulphide
(P_4S_3)

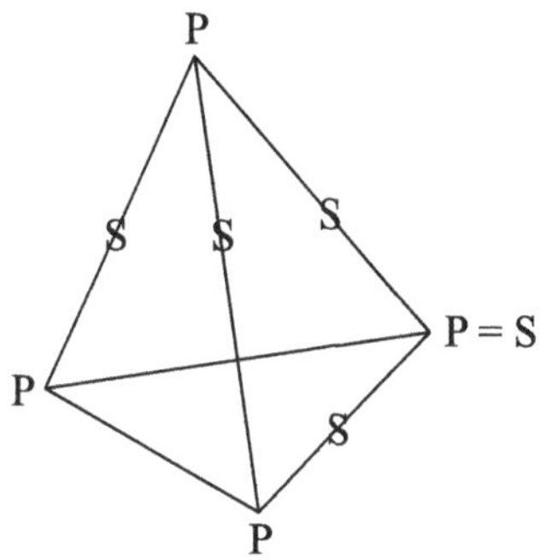

Tetra phosphorus
panta sulphide
P_4S_5

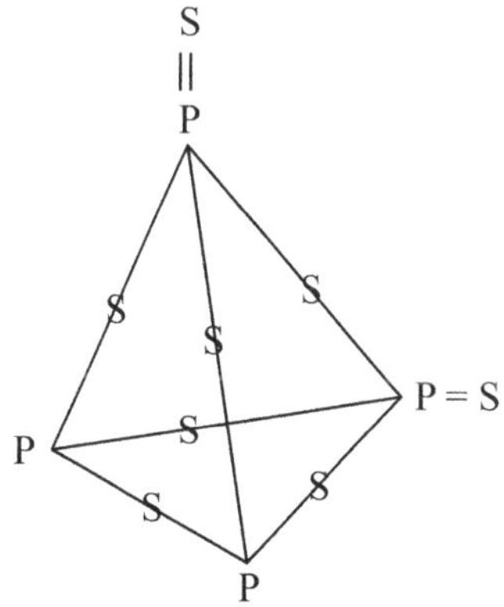

Tetra phosphorus
heptasulphide
P_4S_7

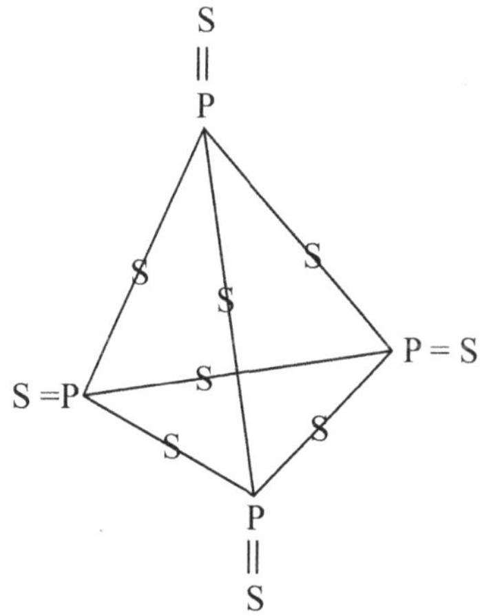

Tetra phosphorus
decasulphide
P_4S_{10}

Phosphorus Trisulphide P_4S_3

It is the most stable sulphide. It is prepared by heating red phosphorus in a limited amount of sulphur at 180°C in an inert atmoshphere. It is soluble in CS_2

and toluene. P_4S_3 is used in making matches. The constituents of matches are P_4S_3, $KClO_3$, fillers plus glue. The friction produced between match and the sandpaper starts the reaction between P_4S_3 and $KClO_3$ which generates enough heat to burn match stick into flame.

Tetraphosphorus Decasulphide P_4S_{10}

It is produced by reacting the liquified white phosphorus at 300°C with slight excess of sulphur.

$$\underset{\text{(liquid)}}{\underset{\text{White}}{P_4}} + 10S \xrightarrow{300°C} P_4S$$

It is hydrolysed by water forming H_3PO_4 similar to P_4O_{10}.

$$P_4S_{10} + 16H_2O \longrightarrow 4H_3PO_4 + 10H_2S$$

P_4S_{10} undergoes hydrolysis in alcohols and phenols as well to give dialkyl or diaryl dithiophosphoric acid.

$$P_4S_{10} + 8EtOH \longrightarrow 4(EtO)_2.P.S\text{–}SH + 2H_2S$$

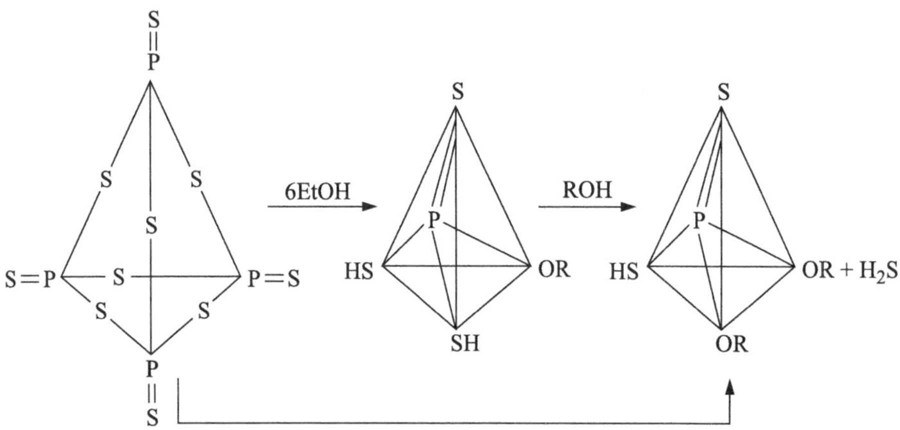

Structure of thio phosphoric acid

The zine salts of dialkyl and diaryl thio phosphates $[(RO)_2 - P.S]_2$ Zn are used as lubricant in gear box. The methyl and ethyl derivatives are used to manufacture pesticides *e.g.* malathion and parathion.

$$(EtO)_2.P\text{–}S.SH + Cl_2 \longrightarrow (EtO)_2P\text{–}(S)\text{–}Cl + HCl + S$$

$$(EtO)_2P.S.Cl + NaOC_6H_4.NO_2 \longrightarrow (EtO)_2.P.S.O.C_6H_4NO_2 \text{ Parathion.}$$

Parathion is an effective insecticide. It attacks the nervous system of insects. It is not toxic to mamals as the digetive system breaks the insecticide before it enters the body.

EXERCISES

1. Discuss briefly the allotropic forms of phosphorus and their structures.
2. Give names and formulae of oxides and oxy acids of nitrogen.
3. Describe the structure of
   (*i*) N_2O_3 (*ii*) N_2O (*iii*) $H_2N_2O_2$ (*iv*) HNO_3
4. Detail the method of preparation and structure of the following:
   (*i*)P_4O_{10} (*ii*) H_3PO_4 (*iii*) HPO_3 (*iv*) $H_4P_2O_7$
5. Describe the preparation of hydrazine justify by giving examples of its:
   (*a*) Reducing action (*b*) Donor molecule.
   Elaborate the structure of hydrazine.
6. Discuss preparation, reducing action and structure of hydroxylamine.
7. Define the terms 'acid' and 'base' by giving examples assuming NH_3 as a solvent.
8. Give reasons
   (*i*) HNO_2 is a weak acid w.r.t HNO_3.
   (*ii*) NO molecule is paramagnetic.
   (*iii*) H_3PO_3 is a dibasic acid.
   (*iv*) HPO_2 in unknown
   (*v*) Structure of HNO_3 and meta phosphoric acid quite different.
9. Discuss the following reactions in liquid ammonia. Give at least two examples
   (*i*) Acid base neutralization (*ii*) Solvolytic reaction
   (*iii*) Complex formation
10. Give reaction of hydrazine with $AlCl_3$ and $AgNO_3$
11. Balance the following reactions:
    (*i*) $CS_2 + PCl_5 \longrightarrow$ (*ii*) $PCl_3 + 3H_2O \longrightarrow$
    (*iii*) $H_2PO_3 + AgNO_3 + H_2O \longrightarrow$ (*iv*) $H_3PO_3 + 3PCl_5 \longrightarrow$
12. Nitrogen molecules have the formula N_2 whilst phosphorus has the formulae P_4.
13. Explain the commercial preparation of Nitrogen and phosphorus.
14. Evoke the molecular orbital theory to describe the bonding in N_2 and NO. What is their bond order.
15. What is effect of heat on:
    (*i*) $NaNO_3$ (*ii*) NH_4NO_3 (*iii*) $NH_4Cl + NaNO_2$
    (*iv*) $Cu(NO_3)_2.2H_2O$ (*v*) $Pb(NO_3)_2$ and (*vi*) NaN_3
16. How the following react with water:
    (*i*) Li_3N (*ii*) CuNCN (*iii*) AlN (*iv*) N_2O_5 (*v*) NCl_3

17. How NH_3 and HNO_3 are manufactured? How is HNO_3 concentrated? What are its uses?
18. Discuss the preparation, properties and structures of the oxides of nitrogen.
19. Describe the conditions under which the following react and give the products formed:
    (*a*) Copper and nitric acid (*b*) N_2O and sodamide
    (*c*) CaC_2 and nitrogen (*d*) $CuSO_4$ and CN^- ions
    (*e*) Ammonia and acidified NaOCl
    (*f*) HNO_2 and iodide ions.
20. What is preparation of hydrazine and hydrazine sulphate?
21. Give reasons:
    (*i*) NF_3 is stable whilst NCl_3 and NI_3 are explosive.
    (*ii*) NF_3 has donor properties whereas PF_3 forms many complexes
    (*iii*) PF_5 is known but NF_5 is not
22. What are the main ingredients in fertilizers? How do we prepare these? What are their uses in plants?
23. Why P—O bond length in $POCl_3$ is 1.45Å whereas the single bond phosphorus and oxygen is 1.83Å?
24. What are the uses of phosphates in industry?
25. Give examples of phosphazenes. How are they made and what are their structures?
26. Give reactions of the following compounds with water:
    (*a*) As_4O_6 (*b*) As_4O_{10} (*c*) $SbCl_3$
    (*d*) Mg_3Bi (*e*) Na_3As
27. Identify the substance A and give the reactions involved:
    Substance (A) is a gas of vapour density 8.5. On oxidation at high temperature with a Pt catalyst it gave a colourless gas (B) which turns brown in air forming a gas (C). (B) and (C) were condensed together to give substances (D), which reacted with water forming an acid (E). Acid (E) with acidified solution of KI evolved gas (B), but (E) was treated with a solution of NH_4Cl to give colourless gas (F). (F) did not support combustion but magnesium continued to burn in it. However, (F) reacted with CaC_2 in an electric furnace forming a solid (G) which was hydrolyzed with water forming a solution of substance (A), which turned Nester's segment yellow. Identify substances (A) to (G).

14

Group 16 Elements or Oxygen Family of Elements

This group includes O, S, Se, Te and Po elements. Their electronic configurations and available oxidation states are listed in the table below:

Table 14.1 Electronic Structures and Oxidation States

Elements		Electronic structure		Oxidation states*				
Oxygen	O	[He]	$2s^2 2p^4$	–II	(–I)			
Sulphur	S	[Ne]	$3s^2 3p^4$	–II		(II)	IV	VI
Selenium	Se	[Ar]	$3d^{10} 4s^2 4p^4$	(–II)		II	IV	VI
Tellurium	Te	[Kr]	$4d^{10} 5s^2 5p^4$			II	IV	VI
Polonium	Po	[Xe]	$4f^{14} 5d^{10} 6s^2 6p^4$			II	IV	

*The most important oxidation states are shown in hyphen. Other well-characterized but less important states are shown in normal type. Oxidation states that are unstable, or in doubt, are given in parentheses.

General Characteristics

The first two are essentially non-metals, Se and Te are metalloids and Po is a radioactive metal. All these elements together are known as chalcogens *i.e.*, elements available in their ores. The common ores of the elements are oxides or sulphides.

Some of these elements are a source of important industrial chemicals *e.g.*, H_2SO_4, Na_2SO_3 and H_2O_2. These elements show increasing metallic behaviour with the increase in atomic number. Oxygen is an important elements as it reacts with almost all other elements. Most of the oxides are covalent.

S, Se and Te are moderately reactive. They combine directly with most other elements *i.e.*, metals and non-metals. They usually form dioxides with oxygen.

They react with acids except those which are oxidising agents. Po being a true metal dissolves in H_2SO_4, HF, HCl and HNO_3.

Oxygen has a smaller size with high electronegativity. It has no '*d*' orbitals therefore it can form double bonds. Other elements can also form double bonds but it becomes weaker with increase in atomic number. Sulphur shows a tendency to form chains and rings. Sulphur shows a very wide range of allotrophic forms, O and S contain electrons in *s* and *p* orbitals, Se, Te and Po contain electrons in *d* orbitals also. S shows the highest oxidation state of +VI *e.g.* H_2SO_4. Se, Te and Po compounds are quite toxic.

Electronic Configuration Vs Oxidation States

The general configuration of group 14 elements is ns^2np^4. In order to complete their octet oxygen being strongly electronegative can gain two electrons, form O^{2-} ion or by sharing two electrons to form covalent bonds. Sulphur is moderately electronegative. With more electropositive elements sulphur forms sulphides, and selenium and tellurium form S^{2-}, Se^{2-}, Te^{2-} ions. These compounds are 50% ionic and 50% covalent. The covalent compounds containing a covalent bond pair are H_2O, F_2O, Cl_2O, H_2O and SCl_2. S, Se, Te show oxidation state of IV and VI also which are more stable than the +11 state.

Occurrence

Oxygen is the most abundant elements in the group. It occurs in the free state in the atmosphere as dioxygen molecule O_2. Photosynthesis is the main source of supply of dioxygen in the atomosphere. Chlorophyll in plants with sun's energy gets converted to glucose

$6CO_2 + 6H_2O + \text{Sun's energy} \rightarrow C_6H_{12}0_6 + 60_2$. Besides, oxygen occurs in oxide ores of the metals as well as oxosalts such as carbonates, sulphates, nitrates and borates. Oxygen comprises of 89% weight of ocean water. Oxygen is also present in upper atmosphere as ozone O_3.

Sulphur is 16th most abundant elements on earth's crust. It is available in free state as sulphur deposits and in the combined state as sulphide ores and gypsum $CaSO_4$, $2H_2O$. In the early times it was used to make explosive gun powder. Presently it is used to make H_2SO_4.

The other three elements Se, Te and Po are available in very small amounts.

Extraction

Dioxygen is produced by the factional distillation of liquid air. O_2 is mainly used in the steel industry. Oxygen obtained from the liquid air contains traces of N_2 and noble gas argon. O_2 is mixed with anaesthetic for surgical operations. Oxy-acetylene flame is used for welding. In the laboratory it is obtained from the thermal decomposition of $KClO_3$ in the presente of MnO_2 as catalyst.

$$2KClO_3 \xrightarrow[150°C]{MnO_2} 2KCl + 3O_2$$

Oxygen can also be prepared frome the catalytic decomposition of hypochlorites

$$2HOCl \xrightarrow{Co^{2+}} 2HCl + O_2$$

Oxygen can be obtained by the electrolysis of water plus trace amount of H_2SO_4 or $Ba(OH)_2$.

Uses: All the elements react with oxygen at room temperature or by heating with the exception of noble metals *e.g* Pt, Au and W plus the noble gases. The reactions of oxygen are spontaneous and exothermic. O_2 is essential for respiration by animals and plants.

$$C_6H_{12}O_6 + 6O_2 \xrightarrow{\text{respiration}} 6CO_2 + 6H_2O + \text{energy}$$

In the human beings complex formed by haemoglobin and dioxygen transports O_2 to the different parts of the body. The three most importants uses of oxygen are:

1. The conversion to steel is quicker.
2. Large size of pig iron.
3. Metal nitrides are not formed when air is used.

Other Uses of Dioxygen are

1. The preparation of TiO_2, a white pigment used in paints, paper and plastics.
2. Oxidation of NH_3 to HNO_3.
3. Manufacture of ethylene oxide.
4. Oxidant in rockets.

Sulphur

Sulphur can be obtained from the natural source where it is present in the form of H_2S (20%). H_2S has an objectionable smell. Burning of sulphur produces SO_2 with acid smell. H_2O and SO_2 react together to produce S.

$$2H_2S + 3O_2 \longrightarrow 2SO_2 + 2H_2O$$
$$SO_2 + 2H_2O \longrightarrow 2H_2O + 3S\downarrow$$

In the form of under ground deposits of sulphur, sulphur is mined by Frasch process. The important suphide ores are:

Table 14.2 Some Important Sulphide Ores

MoS_2	Molybdenite
FeS_2	Pyrites (fool's gold)
FeS_2	Marcasite
FeAsS	Arsenopyrites
$(Fe, Ni)_9S_8$	Pentlandite
Cu_2S	Copper glance or chalcocite
$CuFeS_2$	Copper pyrites or chalcopyrite
Cu_5FeS_4	Bornite or peacock mineral
Ag_2S	Silver glance or argentite
ZnS	Zinc blende or sphalerite
ZnS	Wurtzite
HgS	Cinnabar
PbS	Galena or lead glance
As_2S_3	Orpiment
As_4S_4	Realgar
Sb_2S_3	Stibnite
Bi_2S_3	Bismuthinite

Sulphur element can be obtained from sulphates by the action of carbon.

$$2CaSO_4 + C \xrightarrow{1200^\circ C} 2SO_2 + 2CaO + CO_2$$

Uses: Sulphur is mostly used to produce SO_2, SO_3 and finally to H_2SO_4. The H_2SO_4 obtained is used to make fertilizers *e.g.* $(NH_4)_2SO_4$. A part of the sulphur is utilized in making sulphites, HSO_3^- and SO_2 which are used for bleaching action. The sulphur element is used to make carbon disulphide which is a good solvent. CS_2 is further used to make CCl_4 and viscose rayon. Besides, sulphur is used in the vulcanisation of rubber. With alkenes it reacts to form mustard gas. Sulphur is also used in the making of gun powder and fungicides.

Extraction of Selenium and Tellurium

Se and Te are produced as an anode mud in the electrolysis of copper. The sludge has trace amount of Pt, Ag and Au. Se and Te are also obtained during roasting of sulphide ores with the help of electrostatic precipitators. Se is used in Xerox photocopier where it is used as photoreceptor to capture the image. In the human body the presence of Se in trace amount is necessary.

Selenium and its compounds are toxic in large amounts. Tellurium is used is making steel and non-ferrous alloys *e.g.*, hardened lead. The foul smelling breath and sweat contain Se an Te.

Polonium Po

Polonium was discovered in thorium and uranium ores by Marie curie. Po is one of the decay products in the radioactive series.

$$^{210}_{82}\text{Pb} \xrightarrow{\beta} {}^{210}_{83}\text{Bi} \xrightarrow{\beta} {}^{210}_{84}\text{Po} \xrightarrow{\beta} {}^{206}_{82}\text{Pb}$$

Presently Po is obtained by the irradiation of Bismuth by a neutron.

$$^{210}_{83}\text{Bi} \xrightarrow{^{1}_{0}n} {}^{211}_{83}\text{Bi} \longrightarrow {}^{211}_{83}\text{Po} + {}^{0}_{-1}e$$

All the isotopes of polonium are strongly radioactive. Out of the isotopes of Po, $^{211}_{84}$Po is the most stable one. The studies of Po are complicated in the aqueous solutions by the α-emission.

Allotropy

Oxygen is a gas at room temperature while other elements in the group are solids. Oxygen exists in two gas forms *e.g.* dioxygen O_2 and ozone O_3. S, Se, Te and Po have complicated structures.

The bonding in the dioxygen O_2 appears simple. It is expected to have all paired electrons *i.e.* must be a diamagnetic molecule but actually it has unpaired electrons and so it is paramagnetic.

$$:\ddot{O}: + :\ddot{O}: \longrightarrow :\ddot{O}::\ddot{O}: \text{ or } O{=}O$$

Diamagnetic

The paramagnetic nature of dioxygen O_2 is explained best by the molecular orbital theory (MOT).

Liquid dioxygen is pale blue and the solid dioxygen is blue. The colour is attributed to the unpaired electrons in the outermost shell in the oxcited state. The transition to the oxcited state is ferbidden in the gaseous state. This is best explained by the molecular orbital thoery which shows unpaired electron in the antibonding $\pi^*2p_y^{\,1}$ and $\pi^*2p_x^{\,1}$ molecular orbitals.

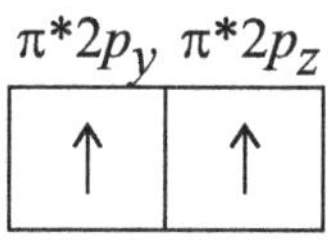

Due to singlet excited state, O_2 is more reactive than normal ground state. Paramagnetic dioxygen is involved in Diels-Alder reaction and other biological oxidations.

Ozone O_3

The moleclue is triatomic allotrope of oxygen. It decomposes to O_2. O_3 has an angular structure with an O—O—O bond angle of 116°48′ and O—O bond length of 1.25Å. Which is in between a single (1.48Å) and a double bond

(1.21Å). The central oxygen atom has one lone pair and the terminal O atom has two lone pairs.

I II

The actual structure is a resonance hybrid of structure I and structure II.

Ozone present in the gas mixture can be extimated by passing the gas through KI solution buffered at pH 9.2.

$$O_3 + 2KI + H_2O \longrightarrow I_2 + 2KOH + O_2$$

The iodine liberated can be tetrated against $Na_2S_2O_3$ solution, O_3 can be used as a disinfectant. It can destroy bacteria and viruses. It is preferered over chlorine because chlorine has unpleasant smell. It is used in swimming pools for the same reason.

Ozone is thermally unstable and decomposes to O_2. The decomposition of O_3 is exothermic and often explosive.

Ozone is a powerfuly oxidising agent. The gas decomposes catalytically to O_2 and the change in volume measured.

$$\underset{\text{2 volumes}}{2O_3} \longrightarrow \underset{\text{3 volumes}}{3O_3}$$

$\Delta G = -163 \text{ kJ mol}^{-1}$.

Ozone reacts readily in comparison to oxygen.

$$3PbS + 4O_3 \longrightarrow 3PbSO_4$$

$$2NO_2 + O_3 \longrightarrow N_2O_5 + O_2$$

$$S + H_2O + O_3 \longrightarrow H_2SO_4$$

Potassium ozonide KO_3 is an orange solid and it contains paramagetic O_3^- ion. O_3 give addition reactions as it adds to the unsaturated organic compounds.

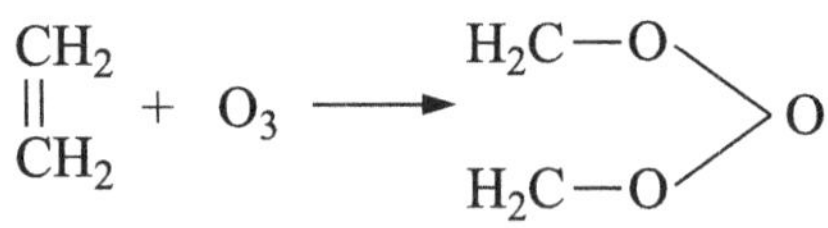

Ethylene ozonide

Ozone can be cleaved to aldehydes and ketones in solution or oxidised by air to carboxylic acids.

Allotropy in Sulphur

Sulphur has many allotropic forms compared to other elements. These allotropic forms arise out of polymerised sulphur and from the crystal strcutures adopted. The two most important forms are α rhombic state at room temperature and β or monoclinic sulphur stable above 95.5°C. The two forms are interchangeable with slow heating or cooling. Sulphur in volcanic areas is naturally occuring rhombic form (Yellow crystals). Third varriety of sulphur looks like mother of pearls and is called *r*-monoclinic. It can be obtained from chilling of concentrated solution in solvents like CS_2, toluene or ethyl alcohol. All these three forms show puckered rings S_8 with a crown.

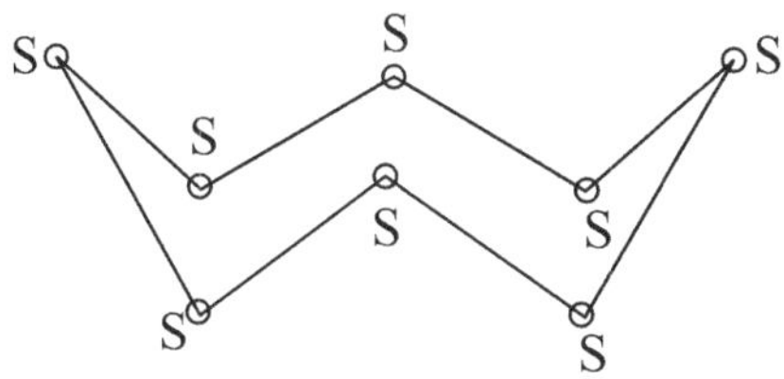

Fig. 14.1 S_8

The three forms differ in overall packing of the rings which affects its densities.

α-rhombic 2.069g cm^{-1}

β-monoclinic 1.94 – 2.01g cm^{-1}

r-monoclinic 2.19g cm^{-1}.

Engle's Sluphur

It is prepared by pouring $Na_2S_2O_3$ solution on concentrated HCl and finally taking out S from toluene. Alternatively if can be made from the following reaction:

$$H_2S_4 + S_2Cl_2 \longrightarrow \underset{\text{engel's sulphur}}{S_6} + 2HCl$$

Other rings exhibit S_7, S_9, S_{10}, S_{11}, S_{12}, S_{18} and S_{20}. They can be obtained by the reactions.

$$H_2S_8 + S_2Cl_1 \longrightarrow S_{12} + 2HCl$$
$$H_2S_8 + S_4Cl_2 \longrightarrow S_{12} + 2HCl$$

In these the ∠S—S—S is between 102° to 108° and S—S bond length 2.04 to 2.06Å.

On pouring liquid sulphur into water plastic sulphur is obtained. Several other forms are produced by molten sluphur are fibrous, laminar and rubber-like sulphurs called crystex.

At higher temperatures longer sulphur chain break into shorter chains.

Sulhpur boils at 444°C. This decreases its viscosity. At 200°C it is S_8 rings with 1 to 2% S_2 molecule. At 600°C sulphur contains S_2 molecule which is stable up to 2200°C.

Allotropes of Se, Te and Po

Selenium exhibits six allotropes. These are mainly used in electronic gadgets. Their uses include LEDs and Xerox photocpiess. Red non-metallic forms are three, contain Se_8 rings. They differ in the packing of rings. Amorphous red form has polymeric chains. There are two grey forms, the grey metallic form is more stable. Black vitreous form of Se is made of irregular rings.

Tellurium has only one allotropic form. It is semi-metallic silvery white. It is quite akin to grey Se. Po is a true metal it has two allotropes; α-form is cubic while β-form is rhombohedral. These forms are metallic.

From S, Se to Te there exists a decrease in the allotric forms and increase in the metallic behaviour. O and S are insulators, Se and Te are semi-conductors and Po is metallic. The structures are simple diahamic molecule in oxygen, rings and chains in Se Te and Po are simply metallic.

Uniqueness of Oxygen

Oxygen has a high value of electronegativity (3.5), therefore if forms mostly ionic compounds.

Hydrogen bonding is a common experience in oxygen compounds. Oxygen has a limitation of eight electrons in the second shell. Its higher coordination number is four. In the rest of the elements in the group the maximum covalency is 6 by using *d* orbitals. Oxygen uses $p\pi$ orbtials to form strong double bonds. The double bonds become weaker as the atomic number increases.

Type of Oxides

Almost all elements react with dioxygen to form oxides. These oxides are classified on the basis of structure of oxides or the chemical properties. On the baiss of structures.

Oxides are classified into:

1. Normal oxides
2. Peroixdes
3. Superoxides
4. Sub oxides.

Normal Oixdes

Oxides in which oxidation number of oxygen is taken as (–11) are normal oxides *e.g.*, H_2O, MgO and Al_2O_3. They all contain M–O bonds except H_2O.

Peroxides $(O-O)^{2-}$

These oxides contain more oxygen than desired by the oxidation number of M.

They are ionic and contains O_2^{2-} ion *e.g.*, Na_2O_2, BaO_2. Other peroxides contain —O—O— covalent bond *e.g.*, H_2O_2, H_2SO_5 and $H_2S_2O_8$.

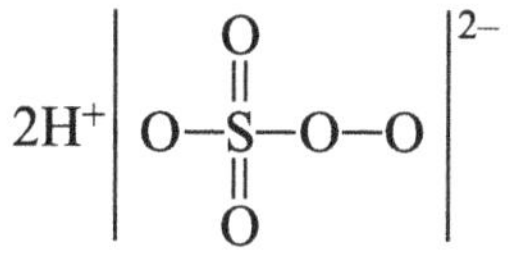

and

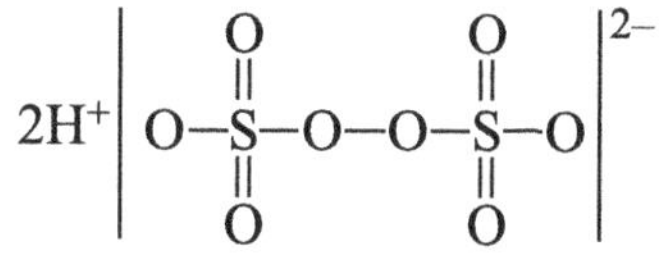

They are all strong oxidising agents and are easily hydrolysed.

$$H_2SO_5 + H_2O \longrightarrow H_2SO_4 + H_2O_2$$

Superoxides KO_2

These are even stronger oxidising agents with acids and water they give H_2O_2 and oxygen.

$$KO_2 + 2H_2O \longrightarrow KOH + H_2O_2 + -O$$

KO_2 is used in breathing masks used in space and sub marines. KO_2 produces O_2 and removes CO_2.

$$4KO_2 + 2CO_2 \longrightarrow 2K_2CO_3 + 3O_2$$

$$4KO_2 + 4CO_2 + 2H_2O \xrightarrow[\text{excess } CO_2]{} 4KHCO_3 + 3O_2$$

Sodium superoxide can be obtained by reacting Na_2O_2 with O_2 at high temperature and pressure.

$$Na_2O_2 + O_2 \longrightarrow 2NaO_2$$

Bonding Peroxides and Superoxides

The bonding in these is described with the help of **molecule orbital theory.**

Peroxides O_2^{2-}: |—O—O—|$^{2-}$ ion contains 18 electrons. These electrons are distributed in molecular orbitals like this:

$$\sigma 1s^2, \sigma^* 1s^2, \sigma 2s^2, \sigma^* 2s^2, \sigma 2p_x^{\,2}, \left| \begin{matrix} \pi 2p_y^2 \; \pi^* 2p_y^2 \\ \pi 2p_3^2, \pi^* 2p_3^2 \end{matrix} \right. \longrightarrow \text{increasing energy.}$$

The bond order this way is one *i.e.,* a single bond.

Superoxides O_2^-: The superoxide O_2^- ion contains 17 electrons. They are distributed among molecular orbtials as shown.

$$\sigma 1s^2, \sigma^* 1s^2, \sigma 2s^2, \sigma^* 2s^2, \sigma 2p_x^{\,1}, \left| \begin{matrix} \pi^2 p_y^2 \\ \pi 2p_z^2 \end{matrix} \right| \begin{matrix} \pi^* 2p_y^2 \\ \pi^* 2p_3^1 \end{matrix}$$

The bond order obtained is 1.5. The stability of the peroxides and superoxides increase with the increase in the size of metal ions. Therefore large cations can be stabilised by large anions. This will give high coordination number and high lattice energy.

Suboxides: In this the element contains less oxygen than desired. These involve M—M bonds in addition to element oxygen bond (M—O) *e.g.*, $O=C=C=C=O$

Classification of Oxides on the Basis of Acid-base Properties

Basic Oxides: Most of the metal oxides are basic. The metal oxides are ionic and contain oxide O^{2-} ion. The oxides of more electropositive elements of group I and group II and the lanthanides are basic. A good amount of energy is needed to form ionic oxides. O_2 molecule is first broken into atoms for which energy is required and then two electron are added to form O^{2-} is also large. These ionic oxides have high lattice energy too. The ionic oxides have high melting points *e.g.* $Na_2O = 1275°C$, $MgO = 2800°C$. When water reacts with them they are converted to OH^- ions.

$$Na_2O + H_2O \longrightarrow 2NaOH$$

Metal oxides M_2O_3 and MO_2 do not react with water *e.g.* Ti_2O_3 and Bi_2O_3. These oxides react with acids to forms salts and hence they are basic. In case of metal ions with more oxidation states *e.g.* CrO, Cr_2O_3, CrO_3, PbO, PbO_2, $(Sb_2O_3)_2$ and $(Sb_2O_5)_2$, the lowest oxidation state is ionic and most basic *e.g.*, CrO is basic Cr_2O_3 is amphoteric and CrO_3 is acidic.

Amphoteric Oxides: BeO_2, Al_2O_3, Ge_2O_3, SnO, PbO and ZnO react both with strong acids and strond bases. These are called amphoteric oxides.

$$Al_2O_3 + 6HCl \longrightarrow 2AlCl_3 + 3H_2O$$

$$Al_2O_3 + 2OH^- + 3H_2O \longrightarrow 2[Al(OH)_4]^-$$

$$PbO + 2HNO_3 \longrightarrow Pb(NO_3)_2 + H_2O$$

$$PbO + OH^- \longrightarrow [PbO.OH]^-$$

Acidic Oxides

The non-metal oxides are mostly covalent. They show low melting and boiling points. They occur as discrete molecules *e.g.*, CO_2, NO, SO_2, Cl_2O etc. B_2O_3 and SiO_2 do occur as giant molecules. They are all acidic. They are considered as anhydrides of the corresponding acids.

$$B_2O_3 + 3H_2O \longrightarrow \underset{\text{ortho Boric acid}}{2H_3BO_3}$$

$$N_2O_2 + H_2O \longrightarrow 2HNO_3$$

$$P_4O_{10} + 6H_2O \longrightarrow \underset{\text{ortho Phosphoric acid}}{4H_3PO_4}$$

$$SO_3 + H_2O \longrightarrow H_2SO_4$$

SiO_2 does not react with water but it does react with NaOH. This exhibits its acidic nature. Oxides N_2O_3 and N_2O_5; SO_2 and SO_3 the oxide with higher oxidation state is the most acidic.

$$N_2O_3 + H_2O \longrightarrow 2HNO_2$$
$$N_2O_5 + H_2O \longrightarrow 2HNO_3$$

In N_2O_3; N is (+III) oxidation state and N_2O_5 has N in (+V) oxidation state. Higher the oxidation state of the central atom, weaker the O—H bond facilitating easy release of H^+ ions.

Neutral Oxides: Some of the covalent oxides hae neither acidic or basic behaviour but they stay neutral *e.g.* N_2O, NO and CO.

Dioxides MO_2: It includes SO_2, SeO_2, TeO_2 and PoO_2, SO_2 in available on a vast scale.

Sulphur Dioxide: SO_2 can be produced by burning S in air or by burning H_2S in air.

$$S + O_2 \longrightarrow SO_2$$
$$2H_2S + 2O_2 \longrightarrow 2H_2O + SO_2$$

SO_2 is also formed during the roasting of sulphide ores *e.g.*

$$2ZnS + 3O_2 \xrightarrow{\Delta} 2ZnO + 2SO_2$$
$$2CuS + 5/2O_2 \xrightarrow{\Delta} Cu_2O + 2SO_2$$

While burning large amount of waste *i.e.* coal or fossil fuel SO_2, CO and CO_2 pollutants are set free in the environment.

Properties: SO_2 is a colourless gas which cause continuous coughing due to its acidic nature. It has a b.pt. of −10°C and melting point −75.5°C. It is highly soluble in water. SO_2 in solution is present as a hydrate of SO_2 *e.g.* $SO_2.6H_2O$ with a trace amount of H_2SO_3. Above 5 ppm it is poisonous to inhale.

Test for SO_2: SO_2 has a choking acidic smell. It turns $K_2Cr_2O_7$ green due to the formation of Cr^{3+} salts.

$$K_2Cr_2O_7 + 3SO_2 + H_2SO_4 \longrightarrow \underset{\text{green}}{Cr_2(SO_4)_3} + K_2SO_4 + H_2O$$

It turns iodate paper blue (starch and I_2 reaction)

$$2KIO_3 + 5SO_2 + 4H_2O \longrightarrow I_2 + 2KHSO_4 + 3H_2SO_4$$

SO_2 vapours cause environmental pollution by producing 'acid-rain'.

Methods for Quantitative Estimation of SO_2

1. SO_2 is oxidised to H_2SO_4 by H_2O_2

$$SO_2 + H_2O_2 \longrightarrow SO_3 + H_2O$$
$$SO_3 + H_2O \longrightarrow H_2SO_4$$

2. SO_2 reacts with $K_2(HgCl_4)$ to form a complex with the dye pararosaniline which is later estimated colourimetrically

$$K_2(HgCl_4) + 2SO_2 + 2H_2O \longrightarrow K_2\,[Hg(SO_3)_2] + 4HCl$$

3. SO_2 produced is oxidised to SO_3 by the contact process:

$$2SO_2 + O_2 \underset{\text{or } V_2O_5}{\overset{\text{Pt (500°C)}}{\rightleftharpoons}} 2SO_3 \;\Delta H = -98 \text{ kJ mol}^{-1}$$

SO_3 is used for other important products

$$2SO_2 + Na_2CO_3 + H_2O \longrightarrow \underset{\text{sodium hydrogen sulphite}}{2NaHSO_3} + CO_2$$

$$2NaHSO_4 + Na_2CO_3 \longrightarrow \underset{\text{Sodium sulphite}}{2Na_2SO_3} + H_2O + CO_2$$

$$Na_2SO_3 + S \xrightarrow{\Delta} \underset{\text{Sodium thiosulphate}}{Na_2S_2O_3}$$

SO_2 can also be used as a non-aqueceus solvent. It undergoes self-ionisation:

$$2SO_2 \rightleftharpoons \underset{\text{Thioxyl ion}}{SO^{2+}} + \underset{\text{Sulphite ion}}{SO_3^{2-}}$$

It has a V-shaped structure

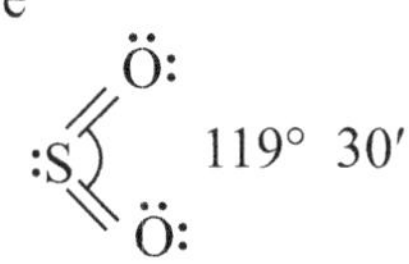

SeO_2, TeO_2 and PoO_2 are prepared by burning in air or oxygen at room temperature.

$$Se + O_2 \longrightarrow SeO_2$$
$$Te + O_2 \longrightarrow TeO_2$$
$$Po + O_2 \longrightarrow PoO_2$$

The structures of these dioxides are similar to SO_2 but the solid forms infinite chains.

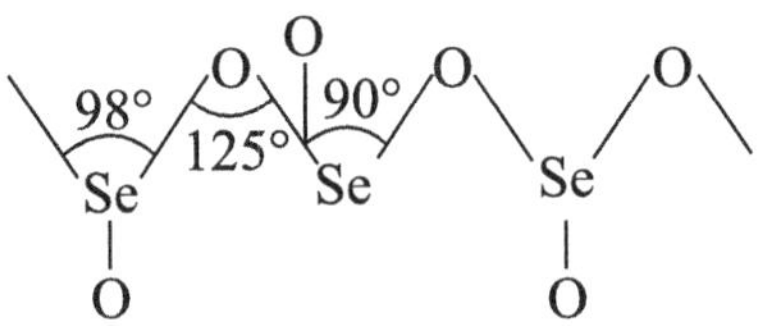

Structure of SeO_2

Trioxides: SO_3 is the most important trioxide. It is manufactured by the contact process using V_2O_5 catalyst.

$$2SO_2 + O \overset{V_2O_5}{\rightleftharpoons} 2SO_3$$

It reacts with water producing heat,

$$SO_3 + H_2O \longrightarrow H_2SO_4 \quad \Delta H = -ve$$

SO_3 dissolves in concentrated H_2SO_4

$$SO_3 + H_2SO_2 \longrightarrow H_2S_2O_7 \text{ or } H_2SO_2\,(SO_3)$$

It has a plane triangular structure

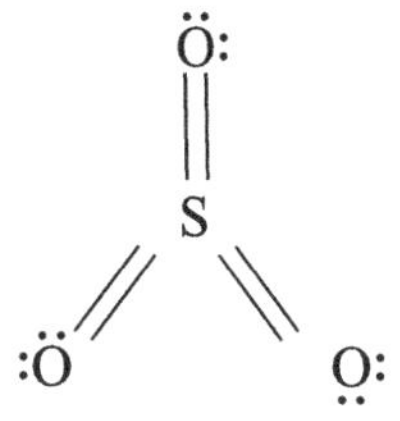

S—O bond

Three σ-bonds

Three π-bonds

SO_3 in the solid form exists in three form (*i*) SO_3 ice-like and a cyclic trimer (m. pt. 16.8°C). SO_3 when allowed to stand for a long time with a trace of water gives β-SO_3 or α-SO_3 bundle of white silky needles with helical chains similar to chain phosphate.

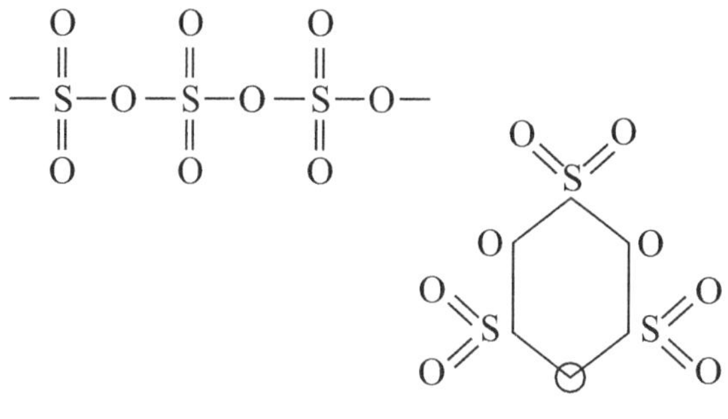

Fig. 14.2 Chain and Cyclic trimer

SO_3 is a powerful oxidising agent

$$2HBr + SO_3 \longrightarrow Br_2 + H_2O + SO_2$$

$$P_4 + 10SO_3 \longrightarrow P_4O_{10} + 10SO_2$$

It is used in the manufacture of H_2SO_4. It is strong sulphonating agent. It sulphonates alkyl benzene. Chain alkyl benzene sulphonates are active ingredients of detergents.

SO_3 is used to form sulphamic acid with urea.

$$\underset{\text{urea}}{NH_2.CO.NH_2} + SO_3 + H_2SO_4 \longrightarrow \underset{\text{sulphamic acid}}{2\,NH_2SO_3H} + CO_2$$

It is solid at room temperature. It is used to clean sugar and mechineries.

SeO_3 is prepared by passing silent electric discharge over Se and O_2. TeO_3 is obtained from H_6TeO_6 telluric acid by strong heating SeO_3 and TeO_3 are anydrides of selenic acid (H_2SeO_4) and telluric acid (H_6TeO_6).

Oxyacids of Group 14 Elements

Oxyacids of sulphur

Oxyacids of sulphur are many and they are important than those of Se and Te. Sulphurous acid is a lower acid of sulphur having oxidation state of sulphur as (+IV) and sulphuric has sulphur in higher oxidation state of (+VI). Their salts are called sulphites and sulphates respectively. Their oxoanions have strong π bonds *i.e.* *p*π–*d*π bonding. They have little tendency to polymerize unlike phosphates and silicates.

The oxo acids of sulphur are of four types:

1. Sulphurous acid
2. Sulphuric acid
3. Thionic acid
4. Peroxoacids

1. Sulphurous acid series

(*a*) Sulphurous acid

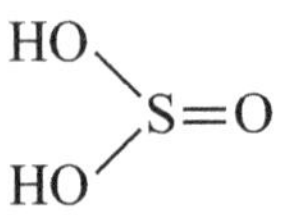

S = +1V oxidatoin sate

(*b*) Pyrosulphurous acid $H_2S_2O_5$

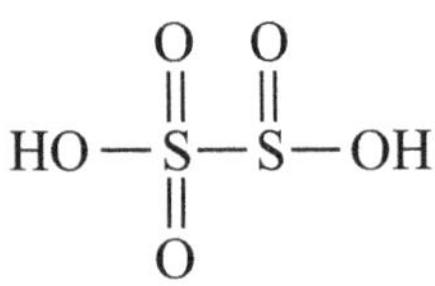

S = (+V)

S = (+III)

(*c*) Dithionous acid $H_2S_2O_4$

S = (+III)

2. Sulphuric acid series

(*a*) Sulphuric acid H_2SO_4

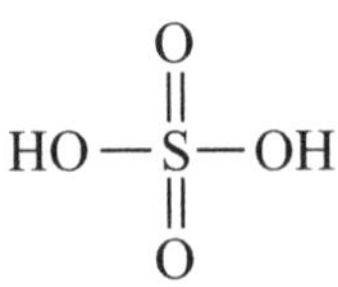

S = (+VI)

(*b*) Dithiosulphuric acid $H_2S_2O_3$

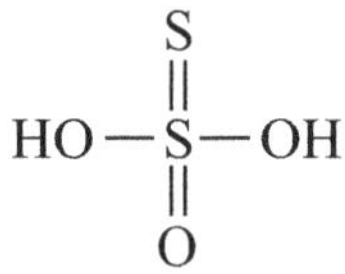

S = (+VI)

S = (–II)

(*c*) Pyrosulphuric acid $H_2S_2O_7$

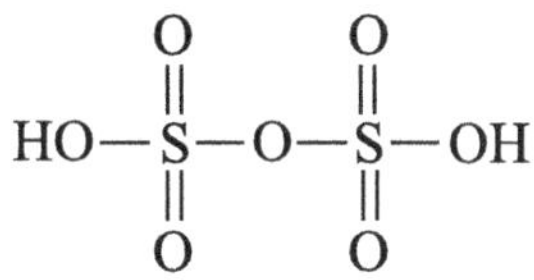

S = (+VI)

3. Thionic acid series

(*a*) Dithionic acid $H_2S_2O_6$

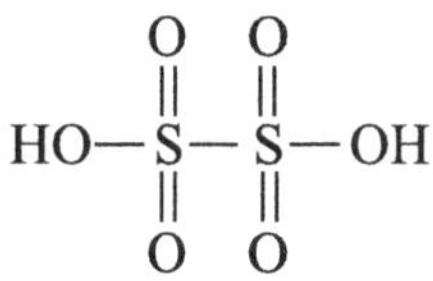

S = (+V)

(*b*) Polythionic acid H_2SnO_6

$n = 1$ to 12

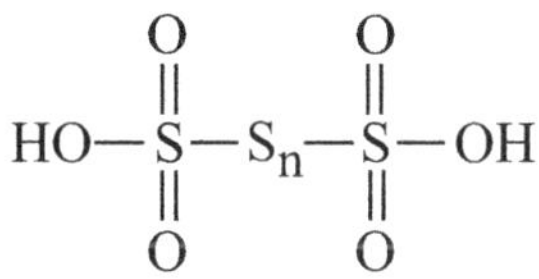

S = (+V)

S = (O)

4. Peroxo acids

(*a*) Peroxo monosulphuric acid H_2SO_5

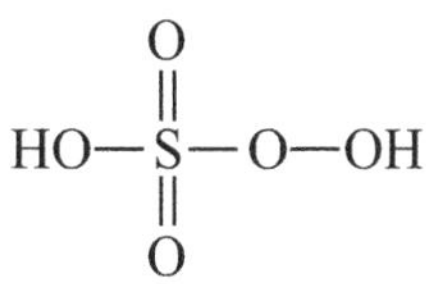

S = (+VI)

(*b*) Peroxo disculphuric acid $H_2S_2O_8$

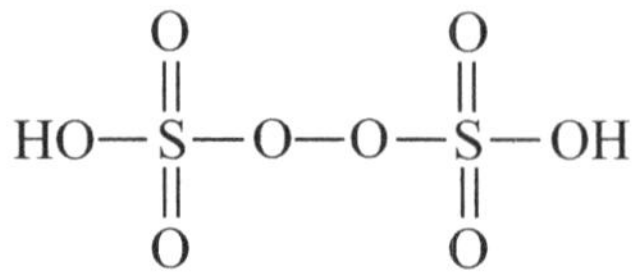

$S = (+VI)$

Sulphurous Acid Series

SO_2 is highly soluble in water. It is mostly present as a hydrate of sulphur dioxide ($SO_2.xH_2O$). The aqueous solution is acidic therefore small amount of H_2SO_3 may exist in solution. The salts are known as sulphites. Many sulphites are sparingly soluble in water *e.g.* $BaSO_3$, $CaSO_3$ and Ag_2SO_3 whereas alkali metals and ammonium sulphites are soluble. In dilute solution hydrogen sulphites HSO_3^- are formed *e.g.* $RbHSO_3$ and $C_5HSO_3^-$. $NaHSO_3$ leads to internal dehydration to disulphites $S_2O_5^{2-}$.

$$2NaHSO_3 \rightleftharpoons Na_2S_2O_5 + H_2O$$

Na_2SO_3 is used to bleach wood pulp in the paper industry. Na_2SO_3 is a good reducing agent, reduces corrosion in pipes and used in photography as a developer. With dilute acids it forms SO_2 which provides a reducing atmosphere.

$$Na_2SO_3 + 2HCl \longrightarrow 2NaCl + SO_2 + H_2O$$

Presence of sulphites and HSO_3^- can be determined by reaction with I_2 where the excess I_2 is determined by $Na_2S_2O_3$.

$$NaHSO_3 + I_2 + H_2O \longrightarrow NaHSO_4 + 2HI$$

$$2Na_2S_2O_3 + I_2 \longrightarrow 2\,NaI + \underset{\text{sodium tetrathionate}}{Na_2S_4O_6}$$

Structure: SO_3^{2-} ion has a pyramidal structure *i.e.* tetrahedral with one lone pair. The $\angle$ O—S—O is 106° and bond length 1.51Å. The π electrons are delocalised giving bond order 1.33.

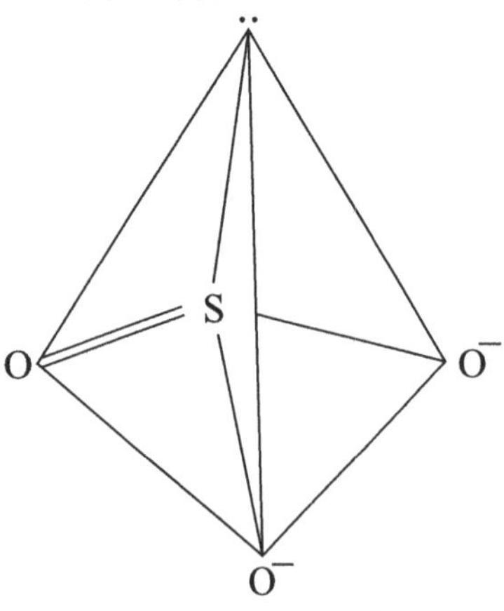

Fig. 14.3 Structure of SO_3^{2-} ion.

On heating HSO_3^- with SO_2 solution disuphites are formed (—S—S—) bond.

$$2RbHSO_3 \xrightarrow{\Delta} Rb_2S_2O_5 + H_2O$$

or $Na_2SO_3 + SO_2(aqueous) \longrightarrow Na_2S_2O_5$

$Na_2S_2O_5$ is called sodium disulphite or sodium pyrosulphite or sodium metabisculphite. The free disulphite acid is not isolated. With acid disulphite gives SO_2.

$$Na_2S_2O_5 + HCl \longrightarrow NaHSO_3 + NaCl + SO_2$$

The oxidation of sulphites give sulphates and with sulphur, thiosulphates are obtained.

$$SO_3^{2-} + H_2O_2 \longrightarrow SO_4^{2-} + H_2O$$

$$SO_3^{2-} + S \longrightarrow S_2O_3^{2-}$$

Reduction of SO_3^{2-} solution with SO_2 and zinc dust yields dithionites with S in (+III) state.

$$2HSO_3^- + SO_2 \xrightarrow{Zn} \underset{\substack{\text{dithionite} \\ \text{or} \\ Na_2S_2O_4}}{S_2O_4^{2-}} + SO_3^{2-} + H_2O$$

Sodium dithionite is extracted by adding NaCl to the mixture, $Na_2S_2O_4$ is a strong reducing agent. It is used as a food preservative in squashes.

Sulphuric Acid Series

The most commercially important chemical is sulphuric acid H_2SO_4. The most important process to procure it is contact process in which SO_2 is oxidised by air in the presence of a catalyst vanadium pentoxide V_2O_5.

$$2SO_2 + \underset{\text{(air)}}{O_2} \xrightleftharpoons{V_2O_5} 2SO_3$$

Earlier platinished asbestos was used which was very easily poisoned.

$$SO_3 + H_2O \longrightarrow H_2SO_4$$

$\Delta H = -ve$.

The reaction is violent. It produces dense H_2SO_4 which is difficult to condense. SO_3 is absorbed by H_2SO_4 to produce pyro sulphuric acid $H_2S_2O_7$ called oleum.

$$H_2SO_4 + SO_3 \longrightarrow \underset{\text{oleum}}{H_2S_2O_7}$$

Oleum is diluted with water to get concentrated H_2SO_4.

$$H_2S_2O_7 + H_2O \longrightarrow \underset{\text{concentrated}}{2H_2SO_4}$$

Alternatively H_2SO_4 can also be produced by lead chamber process but this method is now obsolete.

In this method a homogeneous catalyst NO_2 is used which oxidises SO_2 in the presence of water.

$$NO_2 + SO_2 + H_2O \longrightarrow H_2SO_4 + NO$$
$$2NO + O_2 \longrightarrow 2NO_2$$

It produces impure acid to the extent of 78% pure and is unable to produce concentrated acid.

Properties: Pure H_2SO_4 melts at 10.5°C giving viscous acid. H_2SO_4 is strongly hydrogen bonded in the obsence of water. It does not react with metals to produce H_2. Some metals reduce H_2SO_4 to SO_2. Pure H_2SO_4 can form an azeotropic mixture with water (98.3% H_2SO_4 + 1.7% H_2O). It boils at 338°C.

Pure H_2SO_4 is a sulphonating agent *i.e.* it can insert $-SO_3H$ group to benzene or toluene (Aromatic alkanes).

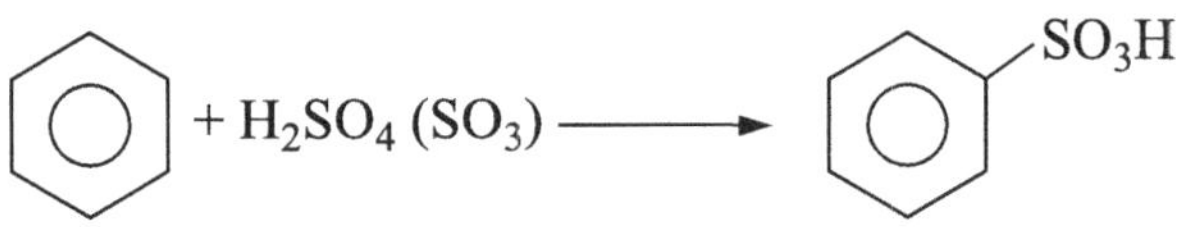

Benzene
sulphonic acid

With H_2O: Concentrated H_2SO_4 and water makes an explosive mixture in all proportions. To make dilute H_2SO_4 it is safe to carefully add H_2SO_4 into water with constant stirring.

Oxidising Agent: Concentrated H_2SO_4 is a strong oxidising agent.

$$2NaBr + H_2SO_4 \longrightarrow 2HBr + Na_2SO_4$$
$$2HBr + H_2SO_4 \longrightarrow 2H_2O + Br_2 + SO_2$$

Copper is lower in electro chemical series therefore it does not react with acids. Noble metals such as Cu dissolve in conc. H_2SO_4 due to the oxidising action of H_2SO_4.

$$Cu + 2H_2SO_4 \longrightarrow CuSO_4 + 2SO_2 + 2H_2O$$

Dehydrating Agent

Concentrated H_2SO_4 absorbs H_2O therefore it is an effective drying agent in dessiccators. It can dehydrate HNO_3 forming nitronium ion NO_2^+ which is a very good nitrating agent.

$$HNO_3 + 2H_2SO_4 \longrightarrow NO_2^+ + H_3O^+ + H_3O^+ + 2HSO_4^-$$

H_2SO_4 can remove water from ethyl alcohol to produce ethers.

$$2C_2H_5OH + H_2SO_4 \longrightarrow C_2H_5.O.C_2H_5 + H_2SO_4.H_2O$$

It removes water from sugar, paper and cloth with a residue of carbon.

$$C_{12}H_{22}.11.H_2O \xrightarrow[H_2SO_4]{\text{conc.}} 12C + 12H_2O$$

In dilute aqueous solution H_2SO_4 acts as a strong acid.

$$H_2SO_4 + H_2O \longrightarrow H_3SO_4^{+} + HSO_4^{-}$$

Structure of SO_4^{2-} Ion

SO_4^{2-} ion has a tetrahedral structure with 4S—O bonds of length 1.49Å. The bond order of S—O bonds is 1.5. It has 4σ bonds between S—O atoms with 2π bonds delocalised over one S and 4 oxygen atoms.

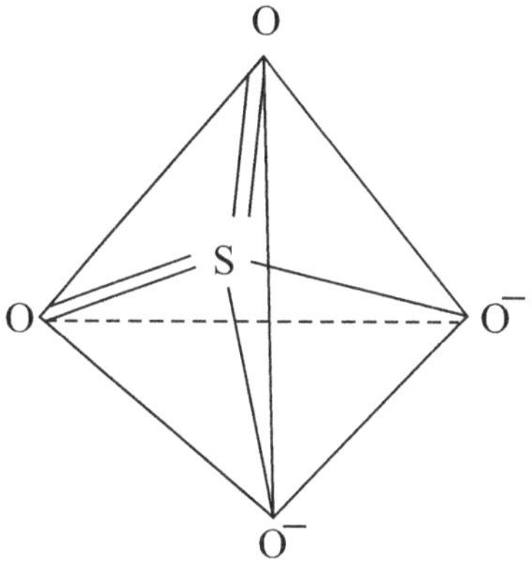

Fig. 14.4 Structure of SO_4^{2-} ion

Thio sulphuric acid and thio sulphates $H_2S_2O_3$: The free acid is not isolated as it decomposes in water to a mixture of S, H_2S, H_2Sx, SO_2 and H_2SO_4.

It can be prepared in the absence of water (ether) at low temperature (–78°C)

$$H_2S + SO_3 \xrightarrow{\text{ether}} H_2S_2O_3 \text{ (ether)}$$

The salts of are called thiosulphates. They are isolated by boiling alkaline or neutral shulphite solutions with sulphur and also by oxidation of polysulphides with air.

$$Na_2SO_3 + S \xrightarrow[\text{water}]{\text{boiling}} Na_2S_2O_3$$

$$2Na_2S_3 + 3O_2 \xrightarrow[\text{in air}]{\text{heating}} 2Na_2S_2O_3 + 2S$$

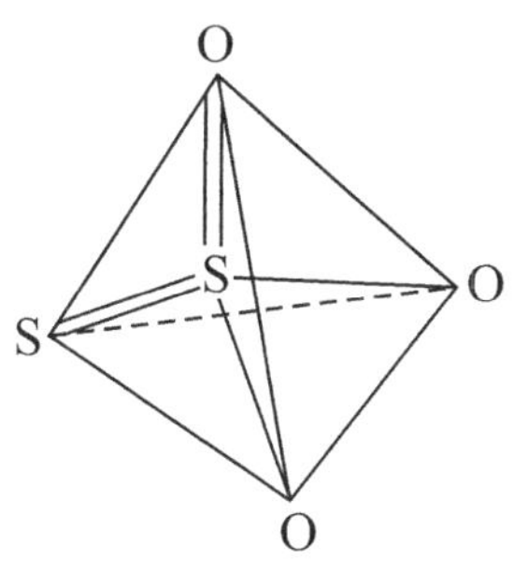

Crystals of sodium thiosulphate $Na_2S_2O_3.5H_2O$ are colourless hexagonal and are called 'hypo', m.pt. 48°C. It is readily soluble in water. It is titrated against iodine solution using starch indicator.

$$2Na_2S_2O_3 + I_2 \longrightarrow 2NaI + \underset{\text{(Sodium tetrathionate)}}{Na_2S_4O_6}$$

$Na_2S_2O_3$ is used to remove excess Cl_2 in bleaching indeustry. It removes th taste of chlorine from chlorinated drinking water.

$$Na_2S_2O_3 + 4Cl_2 + 5H_2O \longrightarrow 2NaHSO_4 + 8HCl$$

Cl_2 is a stronger oxidising agent than I_2 therefore there is no tetrathionate ion formed. Hypo is used in film photography and prints. Parts of the film exposed to light decomposes to Ag forming a negative film. After developing the film is dipped in hypo solution where a soluble complex with silver salts is formed. When no emulsion is left in the solution it can be taken out as it is not affected by the exposed light.

$$Na_2S_2O_3 + \underset{\text{flim}}{2AgBr} \longrightarrow Ag_2S_2O_3$$

$$Ag_2S_2O_3 + 2Na_2S_2O_3 \longrightarrow Na_3\,[Ag(S_2O_3)_2]$$

Pyrosulphates $S_2O_7^{2-}$: The can be prepared by heating hydrogen sulphates or by dissolving SO_3 in concentrated H_2SO_4.

$$2NaHSO_4^- \xrightarrow[-H_2O]{\Delta} Na_2S_2O_7$$

$$H_2SO_4 + SO_3 \longrightarrow \underset{\text{Pyro sulphuric acid}}{H_2S_2O_7}$$

$$H_2SO_4 + 2SO_3 \longrightarrow \underset{\text{Trisulphuric acid}}{H_2S_3O_{10}}$$

Thionic Acids $H_2S_xO_6$

The simplest of thionic acids is dithionic acid *i.e.* $H_2S_2O_6$. It is dibasic acid the salts are dithionates *e.g.* $Na_2S_2O_6$. Here, the sulphur has (+ve) oxidation state. It can be prepared by oxidising a sulphite. On a large scale it is obtained by oxidising a cooled aqueous SO_2 solution with MnO_2 or Fe_2O_3.

$$2MnO_2 + 3SO_2 \longrightarrow MnS_2O_6 + MnSO_4$$

Dithionates are easily soluble in $H_2O.BaS_2O_6$ salt can be readily isolated. Dithionates are stable to mild oxidation but the strong oxidising agents like $KMnO_4$ or Cl_2/H_2O oxidise them to sulphates. In the same way, mild reducing agents have no effect but strong reducing agents reduce them to dithionites and sulphites.

$$2Na_2S_2O_6 + 2Na/Hg \longrightarrow Na_2S_2O_4 + Na_2SO_3$$

Dithionic acid is similar in structure to ethane

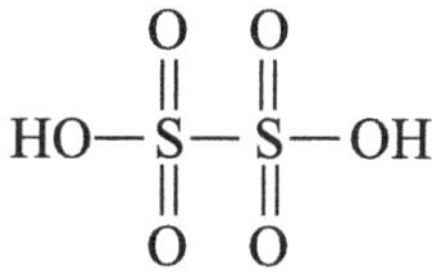

S—S bond length is 2.15 Å; S—O bonds are 1.43 Å; the angles S—S—O are 103° (close to tetrahedral).

Polythionate salts are known. They are prepared by H_2S reaction on SO_2 the series of ions are formed $S_3O_6^{2-}$, $S_4O_6^{2-}$, $S_5O_6^{2-}$ and $S_6O_6^{2-}$.

They are named according to the number of S atoms. They are formulated as:

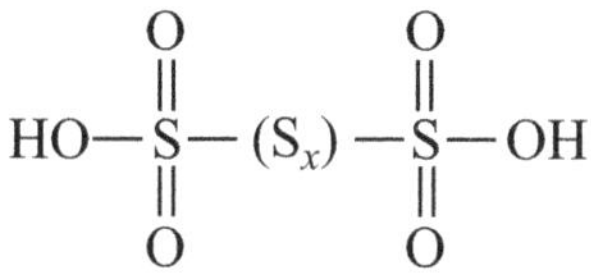

The tri, tetra, penta and hexa prefixes are used with the name of thionic acid.

Peroxo acids: These compounds essentially contain —O—O— linkage. The peroxo acids are:

1. Peroxo monosulphuric acid H_2SO_5
2. Peroxo disulphuric acid $H_2S_2O_8$.

Selenium and Tellurium do show oxoacids but no peroxoacids.

Peroxo mono sulphuric acid is obtained by oxidising H_2SO_4 with H_2O_2

$$H_2SO_4 + H_2O_2 \longrightarrow H_2SO_5 + H_2O$$

$$ClSO_3H + H_2O_2 \longrightarrow (OH)\,(HOO)\,SO_2 + HCl$$

It has colourless crystals with m.pt. 45°C. It can detonote if not handled carefully.

Peroxo Disulphuric Acid $H_2S_2O_8$

It can be prepared from the electrolysis of H_2SO_4 or sulphates at high current denisty.

Anode: (*i*) $2HSO_4^- - 2e \longrightarrow H_2S_2O_8$

Anode: (*ii*) $2SO_4^{2-} - 2e \longrightarrow \underset{\text{Disuphate}}{S_2O_8^{2-}}$

It has a strong tendency to convert to SO_4^{2-} ion therefore it can be used as a powerful oxidising agent. It can convert Mn^{2+} to MnO_4^- and Cr^{3+} to CrO_4^{2-}. $K_2S_2O_8$ and $(NH_4)_2\ S_2O_8$ are its important salts. $(NH_4)_2S_2O_8$ can start the polymerisation of vinyl acetate to produce rayon and PTFE (polytetra fluoro ethylene). In the case of PVC and styrene and rubbers $R_2S_2O_8$ is used as an initiator.

Hydrolysis of peroxo disulphuric acid yields peroxo mono sulphuric acid known as Caro's acid,

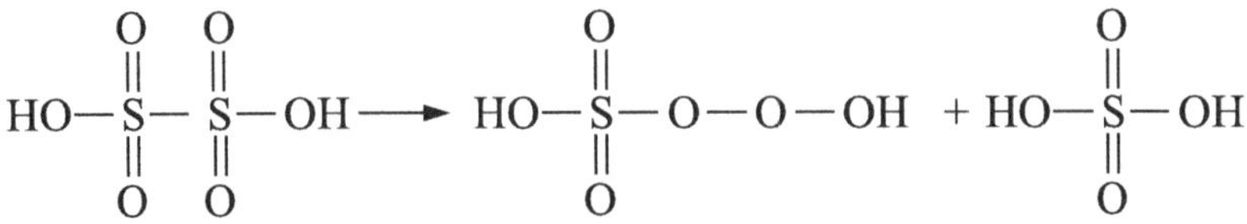

Caro's acid   Sulphuric acid

With PCl_5 Caro's acid can be coverted to chhoro sulphuric acid

$$HO.SO_3.OH \xrightarrow{PCl_5} ClSO_3OH$$

Chloro sulphuric acid

Oxo acids of Se and Te: Se forms selenious acid (H_2SeO_3) and selenic acid (H_2SeO_4). They are prepared when SeO_2 disrolves in water. In can form two salts *e.g.* selenites and selenates. H_2SeO_3 can be changed to H_2SeO_4 by refluxing with H_2O_2. On heating selenates we can get $Se_2O_7^{2-}$ pyroselenates, H_2SeO_4 are isomorphous with sulphates, H_2SeO_3 and H_2SeO_4 are moderate oxidising agents.

TeO_2 and water can give H_2TeO_3 tellurous acid. With bases it can form tellurites and polytellurites H_6TeO_6 telluric acid is different from sulphuric acid. It exists as octahedral molecules in the solid form. It is a strong oxidising agent and forms two series of salts *e.g.* $NaTeO(OH)_5$ and $Li_2TeO(OH)_4$. The acid is obtained by the action of $KMnO_4$ on Te or TeO_2.

Halides of Group 14 Elements

Halogens like fluorine is more electronegative than oxygen. The more electronegative part of the compound is named later. Therefore fluorine and oxygen compounds are called oxygen fluorides. Whereas chlorine and oxygen compounds are called chlorine oxides because oxygen is more electronegative. The compounds of group 14 elements with halogens are listed in the Table 14.3.

Table 14.3 Compounds with the Halogens

	MX_6	MX_4	MX_2	M_2X_2	M_2X	Others
O			OF_2	O_2F_2		O_3F_2, O_2F_2
			Cl_2O		ClO_2	Cl_2O_6, Cl_2O_7
					BrO_2	BrO_3
						I_2O_4, $I_2O_4O_4$, I_2O_5
S	SF_6	SF_4	SF_2	S_2F_2		SSF_2, S_2F_4, S_2F_{10}
		SCl_4	SCl_2	S_2Cl_2		
				S_2Br_2		
Se	SeF_6	SeF_4				

		$SeCl_4$		Se_2Cl_2		
		$SeBr_4$		Se_2Br_2		
Te	TeF_6	TeF_4				
		$TeCl_4$	$TeCl_2$			
		$TeBr_4$	$TeBr_2$			
		TeI_4	TeI_4			
Po		$PoCl_4$	$PoCl_2$			
		$PoBr_4$	$PoBr_2$			
		PoI_4	(PoI_2)			

Fluorine is the most electronegative element hence it exposes the other elements to the maximum valency. S, Se and Te show a maximum valency of six with fluorine *e.g.* SF_6, SeF_6 and TeF_6. All these have ortahedral structure and are available as gases.

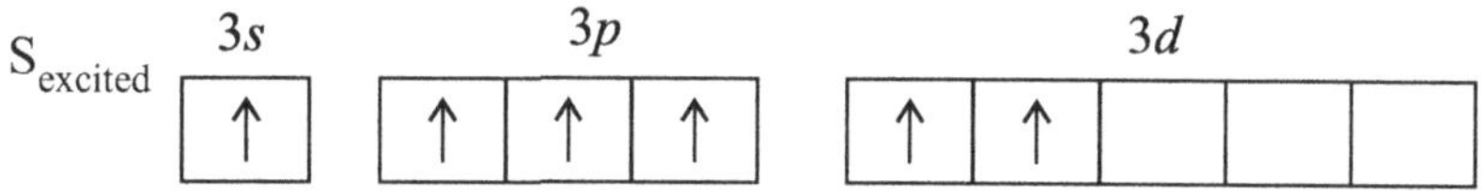

SF_6 is a colourless odourless, non-inflammable gas insoluble in water. It is used as an insulator because it is inert.

SeF_6 is a little more reactive while TeF_6 is easily hydrolysed by water.

$$TeF_6 + 6H_2O \longrightarrow 6HF + H_6TeO_6$$

In the case of TeF_6; $|TeF_7|^-$ and $|TeF_8|^{2-}$, TeF_6 does form these anions, tetrafluoride are difficult to form because they easily convert to SF_6.

SF_4 is also gasesous. SeF_6 is liquid and TeF_6 is a solid.

These are prepared as given:

$$S + F_2 \text{ (mixed with inert } N_2) \longrightarrow SF_4 \text{ and } SF_6$$

$$3SCl_2 + 4NaF \longrightarrow SF_4 + S_2Cl_2 + 4NaCl$$

$$S + 4CaF_2 \longrightarrow SF_6 + 4CoF_2$$

$$S + 4CoF_2 \longrightarrow SF_6 + 4CoF_2$$

$$SeCl_4 + 4AgF \longrightarrow SeF_4 + 4AgCl$$

$$TeO_2 + 2SeF_4 \longrightarrow TeF_4 + 2SeOF_2$$

SF_4 is more reactive towards water

$$SF_4 + 2H_2O \longrightarrow SO_2 + 4HF$$

SF_4 also acts as a fluorinating agent

$$3SF_4 + 4BCl_3 \longrightarrow 4BF_3 + 3Cl_2 + 3SCl_2$$

$$5SF_4 + I_2O_5 \longrightarrow 2IF_5 + 5OSF_2$$

It is a selective fluorinating agent for organic chemicals.

$$RCOOH \xrightarrow{SF_4} RCF_2$$

$$R_2C = O \longrightarrow R_2CF_2$$
$$RCHO \longrightarrow R - CHF_2$$
$$R - OH \longrightarrow R - F$$

The tetrahalides possess a trigonal bipyramidal structure with one equatorial position occupied by the lone pair.

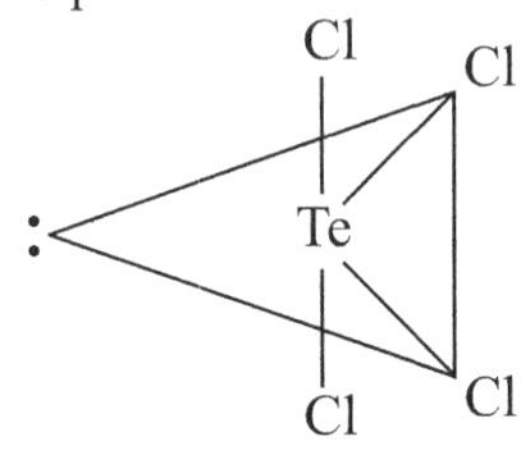

Fig. 14.5 Structure of $TeCl_4$

$TeCl_4$ reacts with HCl to produce a complex ion $(TeCl_6)^{2-}$

$$TeCl_4 + 2HCl \longrightarrow H_2[TeCl_6]$$

Polonium forms series of complex compounds *e.g.* $(NH_4)_2\,(PoX_6)$, $Cs(PoX_6)$ and $Cs_2\,[PoX_6]$.

Tetrabromides: Terabronides of Se, Te and Po are known. $SeBr_4$ is known to hydrolyse easily.

$$SeBr_4 + 4H_2O \longrightarrow [Se(OH)_4] + 4HBr$$
$$[Se(OH)_4] \longrightarrow H_2SeO_4 + H_2O$$

$SeBr_4$ is also unstable.

$$2SeBr_4 \longrightarrow Se_2Br_2 + Br_2$$

Only Te and Po form tetraiodides SCl_2 is an important dihalide. It is a red liquid with a foul smell with m.p. –122°C and b.pt. 59°C.

On heating S and Cl_2; S_2Cl_2 is obtained. It can be saturated with more Cl_2 to get SCl_2. With hydrogen polysulphides a wide range of dichloro sulphones are formed.

$$H_2S_2 + SCl_2 \longrightarrow S_2Cl_2 + 2HCl$$
$$H_2S_x + 2SCl_2 \longrightarrow S_{(x+2)}\,Cl_2 + 2HCl$$

SCl_2 adds to double bond *e.g.* in alkenes

$$SCl_2 + 2CH_2 = CH_2 \longrightarrow \underset{\text{Mustard gas}}{S(CH_2 - CH_2 - Cl)_2}$$

Mustard gas is a volatile liquid. When sprayed it remains close to the ground. It causes blistering of the skin and finally death. It reacts with the proteins in the body. It is used in the chemical war.

The dimeric monohalides when S and halogens combine directly *e.g.* S_2F_2, S_2Cl_2. Similarly Se_2Cl_2 and Se_2Br_2 are formed. These monohalides hydrolyse slowly and tend to disproportionate.

$$2S_2F_2 + 2H_2O \longrightarrow 4HF + SO_2 + 3S$$
$$2SeCl_2 \longrightarrow SeCl_4 + Se$$

S_2Cl_2 is a yellow liquid b.pt. 138°C. It is used in vulcanizing rubber and in making chlorohydrins. It is also used in making sulphur rings with S equal to 7 to 20 atoms.

$$H_2S_8 + S_2Cl_2 \longrightarrow S_{10} + 2HCl$$
$$H_2S_x + 2S_2Cl_2 \longrightarrow \underset{\text{dichloro sulphone}}{S_{(x+4)}Cl_2} + 2HCl$$

Structure of S_2Cl_2 is similor to that of H_2O_2 with a bond angle of 104° the distortion occurs due to two one pairs on sulphur.

Oxohalides: The oxohalides are of two types:

(*i*) Thionyl compounds SOX_2

(*ii*) Sulphuryl compounds SO_2X_2

Thionyl Compounds

S and Se are known to form oxohalides.

Thionyl chloride $SOCl_2$ is a colourless fuming liquid m.pt. 78°C. It is prepared by reacting PCl_5 and SO_2.

$$PCl_5 + SO_2 \longrightarrow \underset{\text{Phosphoryl chloride}}{POCl_3} + \underset{\text{Thionyl chloride}}{SOCl_3}$$

Thionyl chloride hydrolyses readily

$$SOCl_2 + H_2O \longrightarrow SO_2 + 2HCl$$

Thioniyl chloride is an important reagent to convert —COOH to acid chlorides.

$$RCOOH + SOCl_2 \longrightarrow \underset{\text{acid chloride}}{RCOCl} + SO_2$$

$SOBr_2$ is prepared from $SOCl_2$ and HBr and SOF_2 is obtained from $SOCl_2$ and SbF_3.

The structure of $SOCl_2$ is tetrahedral with one position occupied by the lone pair.

Sulphuryl Chloride SO_2Cl_2

Sulphuryl halides known are SO_2F_2, SO_2Cl_2, SO_2FCl and SO_2FBr. It is the result of reaction of SO_2 and Cl_2 in the presence of a catalyst.

$$SO_2 + Cl_2 \xrightarrow[\text{catalyst}]{\text{Pt}} SO_2Cl_2$$

It is a colourless furning liquid b.pt. 69°C. It is a very good chlorinating agent, SO_2F_2 is not hydrolysed while SO_2Cl_2 does hydrolyse with water. SO_2Cl_2 has a distorted tetrahedral structure. This seem to be derivatives of H_2SO_4 where the two 2 — O — H group are replaced by chlorine. In case one —O—H is replaced it forms $ClSO_3H$ chloro sulphonic acid, FSO_3H and $BrSO_3H$ are also isolated.

Hydrides: All the elements in the gourp 14 form hydrides *e.g.* H_2O, H_2S, H_2Se, H_2Te and H_2Po. Water is a colourless liquid at room temperature while all other are colourless foul smelling gases. Except H_2Te all others can be prepared from the combination of elements. It is easier to prepare H_2S, H_2Se, H_2Te from the action of mineral acids on sulphides, selenides and tellurides of metals.

$$FeS + H_2SO_4 \longrightarrow H_2S + FeSO_4$$

$$FeSe + 2HCl \longrightarrow FeCl_2 + H_2Se$$

$$Al_2Te_3 + 6H_2O \longrightarrow 3H_2Te + 2Al(OH)_3$$

H_2Te is also made from electrolysis of dilute H_2SO_4 with Te as cathode.

$$\underset{\text{Alloy}}{(Mg + Po)} + 2HCl \longrightarrow H_2Po + MgCl_2$$

H_2S, H_2Se and H_2Te are water soluble and burn in air with a blue flame.

$$2H_2S + 3O_2 \longrightarrow 2H_2O + 2SO_2$$

H_2S is used as a qualitative reagent for tests of metals. It is a weak dibasic acid. It forms hydrogen sulphide and sulphide salts and therefore a dibasic acid.

$$H_2S + NaOH \longrightarrow \underset{\text{Hydro sulphide}}{NaHS} + H_2O$$

$$NaHS + H_2S \longrightarrow \underset{\text{Sulphide}}{Na_2S} + H_2O$$

Metal sulphides hydrolyse in water to give a basic solution.

$$Na_2S + H_2O \longrightarrow NaHS \quad \underset{\text{base}}{NaOH}$$

Heavy metals sulphides are insoluble in water. NH_3 and H_2S form NH_4HS and $(NH_4)_2S$ in the solution. $(NH_4)_2Sx$ or yellow ammonium sulphide is used to precipitate metal sulphides in the mixture analysis (qalitative).

Water: Watre is most abundantly present in the oceans which is 70% of earth's surface. Ocean water contains so many dissolved salts. The fresh water covers less than 3% of earth's water. The preparation of pure drinking water is quite a job. Water is also needed in the industry. The drinking water should be free from certain inpurities or the impurities should be under toleration limit. The limit of contaminants in the drinking water is tabulated below:

Table 14.4 EEC Limits for Contaminants in Drinking Water

	(mg litre^{-1})	**Source of contaminant and suspected problems**
Al	200	$Al_2(SO_4)_3$ is added to water in some places to clarify it. Al may be related to Alzheimer's disease (senile dementia)
Pb	50	Water pipes made of lead are the problem. Pb damages the brains of children
NO_3^-	50	Nitrate are added as fertilizers to make crops grow. This gets into the water supply. Nitrates affect the level of O_2 in the blood of babies and cause blue baby syndrome. Nitrates be related to stomach cancer.
Trihalomethanes	100	Water supplies are treated with chlorine to kill bacteria. Over-chlorination may allow chlorotorm to be formed by reaction with peat. This is a possible cause of cancer of the gut and bladder.
Pesticides	0.1	For individual pesticides
	0.5	For all pesticides together DDT is now banned in many places. Its harmful effects occur because it accumulates and undergoes biological amplification in the food chain, and UV light converts DDT into polychlorinated biphenyls which are toxic

Utrapure water is needed for laboratory and industry. This can be obtained by distillation. Distillation is an expensive method. During distillation water tends to absorb CO_2 which makes it acidic. Deionised water is cheaply produced by ion-exchange resins *e.g.* organic (R or Ar) acid with $—SO_3H$ group. This removes metal ions from the water solution and replace by H^+ ions.

$$\text{resin} - SO_3H + Na^+ \longrightarrow \text{resin} - SO_3Na + H^+$$

$$2\ \text{resin} - SO_3H + Ca^{2+} \longrightarrow (\text{resin} - SO_3)_2\ Ca + 2H^+$$

In order to remove negative ions from water resin having a positive group or basic groups $-NR_4^+OH^-$ may be used.

$$\text{resin} - NR_4{}^+OH^- + Cl^- \longrightarrow \text{resin} - NR_4{}^+\ Cl^- + OH^-$$

The resin gets exhausted after sometime and can be regenerated by treating with H_2SO_4 and Na_2CO_3 alternatively.

Purification of drinking water

Drinking water must contain not more than 0.5 grams of dissolved salts for the purpose of taste. The maximum permissible limit is 3 times of 0.5 grams. Suspended and colloidal partides in water are removed by treatment with $Al(OH)_3$ or $Fe(OH)_3$ *e.g.* alum. Further purification of water is done by ion-exchangers. To kill the bacteria present in water it is treated with chlorine or ozone. The untreated water can cause water borne diseases.

To obtain drinking water from the sea water is very expensive as the sea water contains a large number of salts. Water for irrigation can be produced by the desalination of sea water. These methods are expensive, therefore resorted to when the water shortage is servere.

Distillation, ion-exchange, electrodialysis, reverse osmosis and freezing out of ice are methods which can give pure water.

Table 14.5 Some Properties of H_2O, H_2Se and H_2Te

	Enthalpies of formation (kJ mol^{-1})	Bond angle	Boiling point (°C)
H_2O	–242	H—O—H = 104°28°	100
H_2S	– 20	H—S—H = 92°	– 60
H_2Se	+ 81	H—Se—H = 91°	– 42
H_2Te	+ 154		– 2.3

Properties of H_2O, H_2S, H_2Se and H_2Te.

Besides water, other hydrides have unpleasant odour. The order of decrease in the stability of hydrides is

$H_2O < H_2S < H_2Se < H_2$ Te.

This is attributed to less orbital overlap with the hydrogen orbital.

Structure of H_2O

H_2O is sp^3 hydribised with a $\angle$ HOH = 104°28′. The closing in of the bond angle is due to the presence of lone pairs. The strcutre is tetrahederal.

Water has an exceptionally high boiling point of 100°C as compared to the other hydrides owing to its tendency to form hydrogen bonds.

The strucuture of liquid water is not certain but it probably consists of two or three hydrogen bonded molecules put together. The structure of ice is hexagonal. With the increase in pressure it forms more dense structures. Nine different forms of ice are known. A normal covalent bond has strength of ~100 kJ/mole whereas a hydrogen bond is ~ 20 kJ/mole. Soild ice is less dense than the liquid. Hence lakes and the sea freeze from the top. Therefore water underneath ice is difficult to cool. The maximum density of water occurs at 4°C. Above 4°C the density decreases.

Hydrogen Peroxide H_2O_2

Oxygen and sulphur differ from the other elements in the property of **Catenation.** These elements catenate to form polyoxides and polysulphides *e.g.*

H—O—O—H

H_2S_2 H — S — S — H

H_2S_3 H — S — S — S — H

H_2S_4 H — S — S — S — S — H

Pure H_2O_2 is a colourless liquid resembling water. It has more hydrogen bonding than water therefore it has a high boiling point b.pt. 152°C m.pt. –0.4°C. It has a higher density of 1.4g cm^{-1}.

It is used as a mild bleaching agent *e.g.* oxidises or bleaches textile and paper/ wood pulp. It is used in the treatment of sewage water and also as a rocket feul. It is used in the preparation of peroxoborates, initiates polymerization reaction in PVC, poly urethanes and epoxy resins. It bleaches hair, feathers, fats and waxes. It is useful oxidising agent in the laboratory as well as an antiseptic.

With Cl_2 it acts as a reducing agent.

$$H_2O_2 + Cl_2 \longrightarrow 2HCl + O_2$$

It is an unstable compound which disproportionates with the rise in temperature.

$$2H_2O_2 \longrightarrow 2H_2O + O_2$$

Earlier H_2O_2 was obtained on electrolysis of H_2SO_4 or $(NH_4)_2SO_4$:

$$2SO_4^{2-} \xrightarrow{\text{electrolysis}} S_2O_8^{\,2-} + 2e$$

$$H_2S_2O_8 + H_2O \longrightarrow H_2SO_5 + H_2SO_4$$

$$H_2SO_5 + H_2O \longrightarrow H_2SO_4 + H_2O_2$$

In the laboratory it can be prepared by the addition of acid to a peroxide.

$$BaO_2 + H_2SO_4 \longrightarrow BaSO_4 + H_2O_2$$

Ionic perorides like Na_2O_2 give H_2O_2 with water.

$$Na_2O_2 + H_2O \longrightarrow Na_2O + H_2O_2$$

H_2O_2 acts as a strong oxidising agent. In acid medium the reaction is slow and in basic medium it is fast. It oxidises fessocyanide to ferricyanide.

$$[FeCN)_6]^{4-} + 2H^+ + H_2O_2 \longrightarrow [Fe(CN)_6]^{3-} + 2H_2O$$

Similarly Fe^{2+} salts are converted to Fe^{3+} salts.

With stronger oxidising agents H_2O_2 is oxidised therefore it acts as a reducing agent.

$$2KMnO_4 + 3H_2SO_4 + 5H_2O_2 \longrightarrow 2MnSO_4 + K_2SO_4 + 5O_2 + 8H_2O$$

$$KIO_4 + H_2O_2 \longrightarrow KIO_3 + O_2 + H_2O$$

In these reactions O_2 is always evolved.

Structure: H_2O_2 has skew structure. The —O—O— bond is perhaps due to repulsion between the OH groups. Both liquid and soild forms have a similar structure.

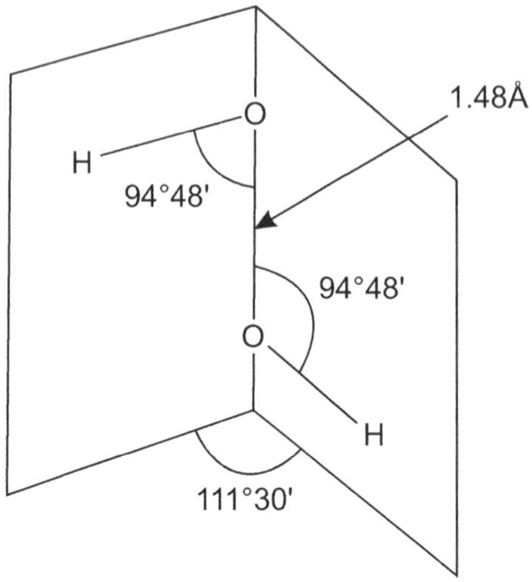

Fig. 14.6 Structure of H_2O_2

Other Hydrides: Hydrides of group 14 yield H^+ ions therefore they all behave as weak acids the acid strength increases from H_2O to H_2Te. According to Fajan's rule *i.e.* the large the negative ion greater is the tendency to covalency. This suggests that H_2Te is the most covalent. In H_2O, H_2S, H_2Se and H_2Te the acidic behaviour depends upon the enthalpy of formation, ionisation energy, electron affinity and enthalpy of hydration. In these hydrides the enthalpy of formation out wighs all other factors. The values are –120, –10, +43 and +77 kJ mol^{-1} respectively. Thermodynamically the last two hydrides are unstable.

$$H_2Te \text{ (hydracid)} + H_2O \rightleftharpoons H_3O^+ + HTe \text{ (hydracid)}$$

If the hydrogen atom is more acidic then more stable will be the salts *e.g.* oxides, sulphides, sdenides and tellurides

Hydrides of Sulphur

The most common hydride of sulphur is hydrogen sulphide or H_2S in which sulphur has –2 oxidation state. Sulphur also forms polysulphides known as sluphanes some important sulphanes are H_2S_2, H_2S_3, H_2S_4, H_2S_5 and H_2S_6.

Hydrogen Sulphide H_2S

Preparation

1. H_2S is prepared by the action of dilute sulphuric acid on FeS.

$$FeS + H_2SO_4 \longrightarrow FeSO_4 + H_2S\uparrow$$

2. H_2S can also be prepared by heating the solution of thio acetamide.

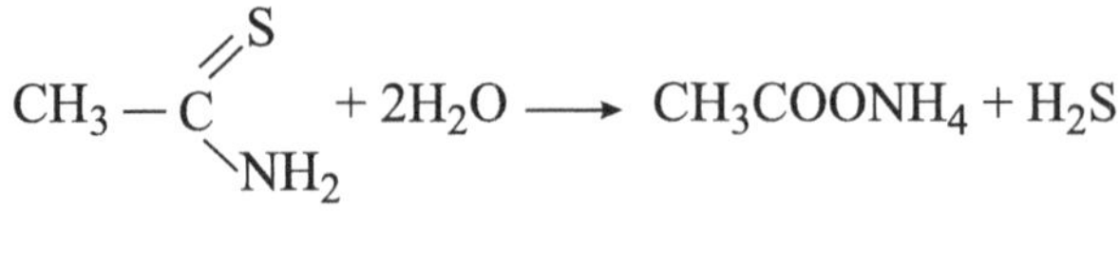

Properties

(*a*) Physical characteristics: It is a colourless gas with a smell of rotten eggs. H_2S is quite poisonous in nature. It causes headache when inhaled even in small quantity.

(*b*) Chemical Characteristics

1. Acidic behaviour: Aqueous solution of H_2S behaves as a weak diprotic acid:

$$H_2S \rightleftharpoons H^+ + HS^-$$
$$HS^- \rightleftharpoons H^+ + S^{2-}$$

2. Reaction with SO_2: It reacts with SO_2 in moisture to produce sulphur

$$SO_2 + 2H_2S \longrightarrow 2H_2O + 3S$$

3. Reaction with O_2

$$2H_2S + 3O_2 \longrightarrow 2H_2O + 2SO_2$$

4. Reducing action: Hydrogen sulphide is a very active reducing agent because it produces nascent hydrogen.

$$H_2S \longrightarrow 2H + S\downarrow$$

It takes up oxygen and thus behaves as a reducing agent

$$H_2S + \frac{1}{2}O_2 \longrightarrow H_2O + S\downarrow$$

It can reduce Cl_2 to halogen acids

$$H_2S + Cl_2 \longrightarrow 2HCl + S\downarrow$$
$$H_2S + Br_2 \longrightarrow 2HBr + S\downarrow$$
$$H_2S + I_2 \longrightarrow 2HI + S\downarrow$$

It can reduce ferric salts to ferrous satls

$$2FeCl_3 + H_2S \longrightarrow 2FeCl_2 + 2HCl + S\downarrow$$
$$Fe_2(SO_4)_3 + H_2S \longrightarrow 2FeSO_4 + H_2SO_4 + S\downarrow$$

Sulphuric acid to SO_2

$$H_2S + 2H_2SO_4 \longrightarrow 2H_2O + 2SO_2 + S\downarrow$$

Nitric acid to NO_2

$$H_2S + 2HNO_3 \longrightarrow 2H_2O + 2NO_2 + S$$

Hydrogen peroxide to water

$$H_2S + H_2O_2 \longrightarrow 2H_2O + S$$

Acidified $KMnO_4$ to Mn^{2+} ion

$$2KMnO_4 + 3H_2SO_4 + 5H_2S \longrightarrow K_2SO_4 + 2MnSO_4 + 8H_2O + 5S\downarrow$$

Acidified potassium dichromate to Cr^{3+} ion

$$K_2Cr_2O_7 + 4H_2SO_4 + 3H_2S \longrightarrow K_2SO_4 + Cr_2(SO_4)_3 + 7H_2O + 3S\downarrow$$

H_2S in used as a qualitative reagent to give coloured precipitates of metal sulphides. In acidic medium it gives precipitates of sulphides of Hg^{2+}, Pb^{2+}, Cu^{2+}, Cd^{2+}, As^{3+}, Sb^{3+} and Sn^{2+} which belong to group II of mixture analysis. In alkaline medium it precipitates sulphides of Ni^{2+}, Co^{2+}, Mn^{2+} and Zn^{2+} which belong to group IV of the mixture analysis.

Structure: The structure of H_2S to similar to the structure of H_2O. It shows sp^3 hybridised S with a bond angle ∠H—S"H of 92.5° and a bond length between H—S of 132 pm.

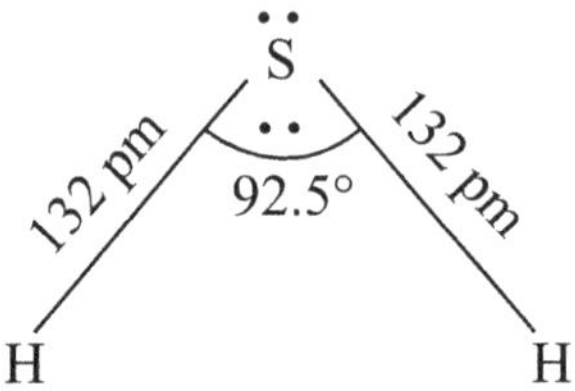

Fig. 14.7 Structure of H_2S molecule

Hydrogen Polysulphides or Sulphanes

Sulphur forms only a few hydrogen poly sulphides because sulphur has a limited tendency of catenation due to weak S—S bond.

Formula	Name	Structure
H_2S_2	Hydrogen disuphide	H—S—S—H
H_2S_3	Hydrogen trisulphide	H—S—S—S—H
H_2S_5	Hydrogen pentasulphide	H—S—S—S—S—H

They are prepared by slowly pouring alkali polysulphides solutions into equal volume of HCl cooled in a freezing mixture. The heavy yellow oily liquid is formed which upon fractional distillation under reduced pressure yields individual sulphanes.

They can also be prepared by the reactions such as

$$2H_2S + SnCl_2 \longrightarrow 2HCl(g) + H_2S_x\downarrow + Sn\ (liquid)$$

The yellow oily liquid possesses more powerful odour than H_2S.

These poly sulphides are unstable and decompose to H_2S and S on keeping or warming *e.g.*

$$H_2S_2 \longrightarrow H_2S + S$$

Sulphur and Nitrogen Compounds

S and N compounds exist as ring and chain compounds. Nitrogen and sulphur are diagonally opposite elements in the periodic table. Some of their properties are closely related. They have similar charge densities, electronegativities (N = 3.0;

S = 2.5) so covalent bonding is expected from them. The sulphur and nitrogen compounds are not routine compounds therefore are explained by unusual theories. Their structures are uncommon. The most important tetrasulphur tetranitride S_4N_4 is a beginning for other sulphur and nitrogen compounds S_4N_4 may be prepared by:

1. $6SCl_2 + 16NH_3 \longrightarrow S_4N_4 + 2S + 14\ NH_4Cl$

2. $6S_2Cl_2 + 16NH_3 \xrightarrow{CCl_4} S_4N_4 + 8S + 12NH_4Cl$

3. $6S_2Cl_2 + 4NH_4Cl \longrightarrow S_4N_4 + 8S + 16HCl$

Properties: It is a solid m.pt. 178°C. It changes colour with the change in temperature known as thermochrome. At room temperature it is orange yellow while at 100°C it turns red. It is stable in air but may detonate on heating or grinding.

Its structure is a heterocyclic ring. It differs from S_8 ring which is crown shaped.

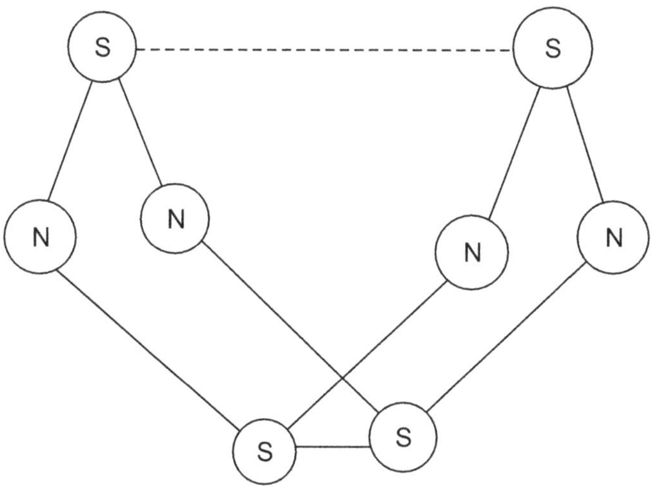

Fig. 14.8 Structure of S_4N_4

S_4N_4 is slowly, hydrolysed by water. It reacts quickly with warm NaOH which breaks the ring structure.

$$S_4N_4 + 6NaOH + 3H_2O_2 \longrightarrow Na_2S_2O_3 + 2Na_2SO_3 + 4NH_3$$

S_4N_4 can be reduced by $SnCl_2$ in CH_3OH to give tetrasulhur teraimide $S_4(NH)_4$.

Vapours of S_4N_4 under reduced pressure when passed through silver wood produce S_2N_2.

$$S_4N_4 + 4Ag \longrightarrow S_2N_2 + 2Ag_2S + N_2$$

EXERCISES

1. What are the various oxides of sulphur? Discuss the structures of SO_2 and SO_3.
2. Discuss the structures of the various oxy acids of sulphur:
   (*i*) H_2SO_4 (*ii*) $H_2S_2O_3$ (*iii*) $H_2S_2O_6$ (*iv*) H_2SO_5
3. Explain:
   (*i*) SF_6 is known but SCl_6 is not known.
   (*ii*) SF_5Cl is more reactive than SF_6.
   (*iii*) SF_6 does not behave as a Lewis acid.
4. Describe the structure of the following compounds:
   (*i*) S_2Cl_2 (*ii*) SCl_2 (*iii*) SF_4 (*iv*) SF_6
5. Why is OF_2 smaller than OH_2? Explain.
6. How does SO_2 react with H_2S? Elaborate.
7. Draw the structure of S_4N_4. How does S_4N_4 react with:
   (*i*) Ag wool at 300°C (*ii*) $SnCl_2$ in ethanol
8. H_2S is both an oxidant and a reductant. Explain.
9. What is the laboratory preparation of SO_2? How does moist SO_2 act as a bleaching agent?
10. What is action of SO_2 on lime water?
11. Explain why sulphur exhibits +4 and +6 oxidation states but oxygen does not.
12. Why H_2S is a better reducing agent than H_2O. Explain.
13. Describe the electron dot structures of:
    (*i*) $H_2S_2O_3$ (*ii*) $H_2S_2O_6$
14. Give equations for the preparation of oxygen from:
    (*i*) H_2O (*ii*) H_2O_2 (*ii*) Na_2O (*iv*) $NaNO_3$
    (*v*) KCl_3 (*vi*) HgO
15. How do you obtain oxygen commercially? What are its main uses?
16. On what basis may oxides be classified?
17. Describe the application of molecular orbital theory on (*i*) O_2 (*ii*) superoxide O_2^- (*iii*) peroxide O_2^{-2} ion.
    Explain bond order and magnetic properties.
18. Compare the oxide of Na, Ca with S and N. Discuss their melting points, nature of bonding and their reactions with water, acids and bases.
19. Why oxygen molecule is O_2 whilst the sulphur molecule is S_8.
20. How do we prepare H_2O_2? What is its structure in the gas phase? Write its equations with:

(*i*) Acidified solution of $KMnO_4$
(*ii*) Aqueous HI
(*iii*) Acidified solution $K_4[Fe(CN)_6]$

21. What are the main sources of sulphur? Describe the allotropic forms of sulphur.
22. Describe the Frasch process for mining sulphur. What changes occur when sulphur is heated?
23. Explain the differences in bond angles and the boiling points of H_2O and H_2S.
24. How does the π-bonding occur in O_2, O_3, SO_3 and SO_4^{2-}.
25. Describe the contact process in the manufacture of sulphuric acid. Give its uses.
26. Describe the preparation, properties and structure of SO_2, SO_3, H_2SO_5 and $H_2S_2O_8$.
27. Give the differences in the structure of gaseous and solid SO_3.
28. Give the following:
    (*i*) Reaction between SO_3 and H_2SO_4
    (*ii*) Action of heat on $NaHSO_3$
    (*iii*) Structures of SO_4^{2-} and $S_2O_3^{2-}$
    (*iv*) Reaction between $Na_2S_2O_3$ and I_2
    (*v*) H_2SO_3 and sulphites reduction.
29. Give the preparation of $Na_2S_2O_3$. What is the use of $Na_2S_2O_3$ or hypo in photography and volumetric analysis?
30. Enlist the fluorides of sulphur. How are they prepared? Describe structures and uses.
31. Explain why:
    (*i*) SF_6 in unreactive towards H_2O but TeF_6 reacts.
    (*ii*) SF_6 is known but OF_6 is not known.
32. Compare and contrast H_2SO_4, H_2SeO_4 and H_2TeO_4

15

Group 17 Elements or Halogen Family of Elements

The elements existing in this group are fluorine, chlorine, bromine, iodine and astatine. They are symbolised, their electronic configuration and the oxidation state are tabulated in the Table 15.1.

Table 15.1 Electronic Structures and Oxidation States

Element		Electronic configuration		Oxidation states*
Fluorine	F	[He]	$2s^2\,2p^5$	–I
Chlorine	Cl	[Ne]	$3s^2\,3p^5$	–I + I + III + IV + V + VI + VII
Bromine	Br	[Ar]	$3d^{10}\,s4^2\,4p^5$	–I + I + III + IV + V + VI
Iodine	I	[Kr]	$4d^{10}\,5s^2\,5p^5$	–I + I + III + V + VII
Astatine	At	[Xe]	$4f^{14}\,5d^{10}\,6s^2\,6p^5$	

General Discussion

The name halogens is derived from the Greek word halogens which means salt. They are all very reactive with non-metals. Out of these fluorine is the most reactive element. All these elements acquire the general configuration of $ns^2\ np^5$ in the outermost shell. They are one electron short of noble gas configuration. These elements try to acquire the noble gas configuration by accepting an electron by forming an ionic bond. All halogens are therefore, oxidising agents and all halides X^- are reducing agents. With electronegative elements they form covalent compounds. In brief, with metals they form ionic and with non-metals convalent compounds. Halogen group has close knit members in the group. Flourine which contains no *d*-orbitals has unique properties because

1. Fluorine has a very small size and therefore hold its electrons tightly.
2. Fluorine has no *d*-orbital participating in bonding.

chlorine and bromine exhibit similar properties as they have similar sizes.

The ionic radius of chloride Cl^- is bigger than F^- ion. The ionic radius of bromine is slightly greater than that of Cl^- ion. This results in their close electronegativity values which makes the polarity of their bonds being not very different. Most of these elements show oxidation state (±1). if halogen is electronegative in a compound then it is –1 otherwise (+1). Except F all other element can expand their valencies to the presence of *d*-orbitals. Fluorine has electrons in the second shell where *d*-orbitals are missing therefore its covalent bonds are very few. Flourine is a very strong oxidising agent because of small size and high electro-negativity. It can expose the combining elements to the maximum oxidation state *e.g.*, IF_7. PtF_6, SF_6, BiF_5 etc.

All the halognes exist as diatomic molecules. They are coloured.

F is light yellow

u is yellow-green

Br is dark red brown

Cl is violet.

On descending the group energy levels get closer. I_2 is solid at room temperature and exists as balck flakes having a metallic lustre, liquid I_2 conducts little bit and undergoes self ionisation.

$$3I_2 \rightleftharpoons I_3^+ + I_3^-$$

Halogens produce important chemicals having commercial use.

Occurrence

All the halogens are very reactive therefore they exist in the combined state on the earth. The element astatine At is radioactive and has a short half life. Fluorine and chlorine are quite abundant while bromine and iodine are rare.

The mineral fluorspar CaF_2 is the most important source of fluorine. Another important mineral of fluorine is fluoroapatite $3Ca_3(PO_4)_2.CaF_2$. It is also an important source of phosphorus. Fluoroapatite contains lot of SiO_2 which reacts with HF to form hydrofluosilicic acid $H_2[SiF_6]$. Another mineral cryolite $Na_3[AlF_6]$ is quite rare and is used in the extraction of aluminium.

Chlorine is abundantly available as NaCl. It can produce Cl_2 electrolytically. NaCl is abundant is sea water. Sea water also contains other halogen salt like NaBr, NaI, KI, $MgCl_2$ etc. Bromides occur in sea water. They are also present in sea-weeds. They have higher concentration of iodides *e.g.*, $NaIO_3$ and $NaIO_4$. $NaIO_3$ and $NaIO_4$ are present as impurities in $NaNO_3$ crystals.

Extraction of Halogens

Fluorine: Fluorine is most reactive. It has important ores like CaF_2, fluoropatite and cryolite. Fluorine is obtained from CaF_2 by the action of concentrated H_2SO_4 to give aqueous HF.

$$CaF_2 + H_2SO_4 \longrightarrow CaSO_4 + 2HF$$

HF forms a cooling mixture of KHF_2. Anhydrous HF may be electrolysed giving F_2 and H_2.

$$KF + HF = KHF_2$$

$KF + KHF_2$ on electrolysis gives F_2 and H_2. Many difficulties are confronted to isolate F_2:

1. HF etches glass and causes skin wounds. Wounds are slow to heal because F^- ions remove Ca^{2+} ions from the tissues.
2. HF is very toxic like HCN.
3. HF is less ionised therefore to make a good electrolyte it is mixed with KF to give $K^+|HF_2|^-$.
4. Water should not be present with fluorine otherwise it will be oxidized to dioxygen.
5. H_2 liberated at cathode should be separated from F_2 lest it may recombine to give HF.
6. Fluorine can catch fire with traces of grease or silicon.
7. Only steel cathodes and carbon anodes should be used with teflon in solution because graphite reacts with fluorine to form fluorocarbons. It stops graphite from conducting and explosion may occur. Cylinders of F_2 are used commercially but now they are replaced by ClF_3(b.p 12°C) which is easier to transport.

$$3F_2 + Cl_2 \xrightarrow{300°C} 2ClF_3$$

Cryolite is chiefly used for the extraction of Al but it also give F_2.

Fluorocarbons are interesting compounds CF_4 is corresponding to methane. CF_4 is called perfluoromethane. It can be heated with no fear of burning. Fluorocarbons are inert to concentrated HNO_3 and H_2SO_4 and to strong oxidising agents like $KMnO_4$ or O_3 and likewise to strong reducing agents *e.g.*, $LiAlH_4$. Fluorocarbons pyrolyse at very high temperature to break C—C bonds to form $F_2C = CF_2$

$$2CHClF_2 \xrightarrow{1000°C} F_2C = CF_2 + 2HCl$$

Polytetrafluoro ethene PTFE is called Teflon.

Freons are chlorofluorocarbons *e.g.*, $CClF_3$ CCl_2F_2 and CCl_3F. They are used cooling agents in refrigerators. Amount of F^- ions in drinking water (~1 ppm) take care of the dental problems. In nuclear reactors 75% of fluorine produced UF_6 which is used to enrich unranium.

$$U + 4HF \longrightarrow UF_4$$

$$UF_4 + F_2 \longrightarrow UF_6$$

or $$UF_4 + ClF_3 \longrightarrow UF_6$$

Chlorine

1. Chlorine was prepared first by the oxidation of HCl by MnO_2 in the laboratory.

$$4HCl + MnO_2 \longrightarrow MnCl_2 + 2H_2O + Cl_2$$

Cl_2 so prepared must be purified. To remove HCl it is passed through water. Finally to remove water it is passed through concentrated H_2SO_4. Traces of H_2O can be further removed by passing over CaO and P_4O_{10}.

2. On commercial scale it can be obtained by electrolysis of fused NaCl and aqueous NaCl.
3. Fused NaCl

   $$\text{Fused } 2NaCl \xrightarrow{\text{electrolysis}} 2Na + Cl_2$$

4. $2NaCl + 2H_2O \xrightarrow{\text{electrolysis}} 2NaOH + Cl_2 + 2H_2$
5. According to Deacon process HCl is oxidised by air.

   $$4HCl + O_2 \longrightarrow 2H_2O + 2Cl_2$$

This process is now modified in which 1, 2 dichloromethane is pyrolysed to vinyl chloride.

$$H_2C - Cl - CH_2Cl \xrightarrow{450^\circ} CH_2 = CH - Cl + HCl$$

Presently $CuCl_2$ catalyst is used which lowers the temperature condition The by product HCl is used for the oxidation of HCl.

Chlorine gas is toxic causes cougting and lung damage. Chlorine finds use in producing organo-chloro compounds and in bleaching. Some inorganic chloro compounds are also important.

Chloro organic compounds are used in plastic industry *e.g.,* 1, 2 dichloro ethane, vinyl chloride (PVC), benzene haxachloride, chloro-phenols, DDT, dichloromethane etc. For bleaching of textiles, wood and paper pulp we use bleaching powder ($CaOCl_2$), sodium hypocholorite (NaOCl), chlorine dioxide (ClO_2), sodium chlorate ($NaClO_3$), metal and non-metal chlorides.

Bromine (Br_2)

Bromine is obtained from sea water and brine lakes. Water from brine lakes is concentrated and then treated by H_2SO_4 to maintain pH of 3.5. Then Cl_2 gas is passed through it.

$$2Br^- + Cl_2 \longrightarrow 2Cl^- + Br_2$$

The more electronegative replaces the less electro negative element, Br_2 is more volatile and is passed through Na_2CO_3 to give a mixture of NaBr and $NaBrO_3$. The solution is then acidified to give bromine.

$$3Br + 3Na_2CO_3 \longrightarrow 5NaBr + NaBrO_3 + 3CO_2$$

$$5NaBr + NaBrO_3 + 3H_2SO_4 \longrightarrow 5HBr + HBrO_3 + 3Na_2SO_4$$

$$5HBr + HBrO_3 \longrightarrow 3Br_2 + 3H_2O$$

Bromine is used to make organic bromo compounds, 1, 2 dibromo ethane is a lead scavenger. CH_3Br is used to kill earth-worms.

Bromocompounds are also used as pesticides, AgBr is used in photosensitive film. KBr is a sedative.

Iodine (I_2)

I_2 is obtained from chile salt petre which $NaNO_3$ with small amounts of $NaIO_3$ and $NaIO_4$. $NaNO_3$ is dissolved in water to crystallise out $NaNO_3$. The remaining liquid contains the iodates.

One part of this liquid is treated with $NaHSO_3$ to give I^- which is mixed with unreacted liquid to give I_2 which is purified by sublimation.

$$2IO_3^- + 6HSO_3^- \longrightarrow 2I^- + 6SO_4^{2-} + 6H^+$$

$$5I^- + IO_3^- + 6H^+ \longrightarrow 3I_2 + 3H_2O$$

I_2 is absorbed by KI to form KI_3 which is more soluble. I_2 is an effective antiseptic iodoform (CHI_3). AgI is used in photo films and seeding of clouds. Iodine deficiency in the human body causes goitre. Tincture iodine is I_2, KI and alcohol. Iodide and iodates are used in volumetric titration. $K_2[HgI_4]$ or nesler's reagent is used to detect the presence of ammonium ion (NH_4^+).

Astatine

Astatine exists as twenty odd isotopes which are all radioactive. The most stable of all the isotopes is ^{210}At. It has a half life of 8.3 hours. It was prepared in 1940 by bombarding Bi with α-particles.

$$^{209}_{83}Br + ^{4}_{2}He \longrightarrow ^{211}_{85}At + 2^{1}_{0}n$$

The chemistry of ^{211}At is studied by taking trace amount of 10^{-14} mole. Astatine resembles closely in properties to iodine.

Physical characteristics of Halogens

Size of halogen atoms and ions is given in the Table 15.1

Table 15.2 Ionic and Covalent Radii

	Covalent radius (Å)	Ionic radius X^- (Å)
F	0.72	1.33
Cl	0.99	1.84
Br	1.14	1.96
I	1.33	2.20

Ionization Energy

The ionization energies decrease with the increase in size of the halogen atoms. Generally speaking, the ionization energies of the halogens are quite high because of their relatively small size. It is the tendency of the atoms to lose electron to give positive ions.

Table 15.3 Ionization and Hydration Energy, Electron Affinity.

	First ionizatoin energy (kJ mol^{-1})	Electron affinity (kJ mol^{-1})	Hydration energy X^- (kJ mol^{-1})
F	1681	– 33	–513
Cl	1256	–349	–370
Br	1143	–325	–339
I	1009	–296	–274
At	—	0270	—

Fluorine has exceptionally high ionization energy due to its very small size. It always reflects an oxidation state of (–1). In F_2 molecule the oxidation state is zero. In most of the compounds fluorine gains an electron to become F^- except in covalent compounds.

Hydrogen reflects a high ionization energy of 1311 kJ mol^{-1} in the form of H^+. Cl, Br and I have ionization energies lower than that of hydrogen although they do not easily form X^+ ions.

Iomization energy is defined as the energy required to pull off an electron from the outermost shell of a neutral gaseous atom in its ground state of configuration. Generally the ions exist in solid crystalline form or in solutions. This brings into play lattice energy or the hydration energy. H^+ is very small in size therefore in solids it must have high lattice energy and in solution a high hydration energy. (~ 1091 kJ mol^{-1}). The negative ions also show hydration energy. In a way H^+ are formed only when the lattice or hydration energy is greater than ionization energy. The condition for positive ion is opposite *i.e.*, halide ion X^+ must be large and consequently should have low hydration and lattice energies. Therefore these X^+ ions are not formed. In few compounds I^+ is stabilized by forming a complex *e.g.*, $[I(\text{pyridine})_2]^+NO_3^-$.

Energy is evolved when halogens form negative ions. Here the energy is evolved when a halogen gains an electron.

$$Cl + e \longrightarrow Cl^-$$

Hence the halogens form halide ion.

Table 15.4 Electonegativity and Electrode Potential

	Pauling's electronegativity	Standard electrode potential E° (volts)
F	4.0	+2.87
Cl	3.0	+1.40
Br	2.8	+1.09
I	2.5	+0.62
At	2.2	+0.3

Oxidation States and the Type of Bonds

Because of high electronegativities values the two halogen atoms form a covalent bond. The compounds between halogens and non-metals are mostly covalent. With metals they usually form ionic bonds due to large electro negativity difference between the two. As shown by the large electron affinity values halide ions are quite easily formed. The E° values decrease down the group and so energy evolved an forming the halide ions should also decrease. Some metal halides where metal ions have small size and high charge form unexpectedly covalent halides *e.g.*, $BeCl_2$ and $AlCl_3$. This is attributed to their chain structure.

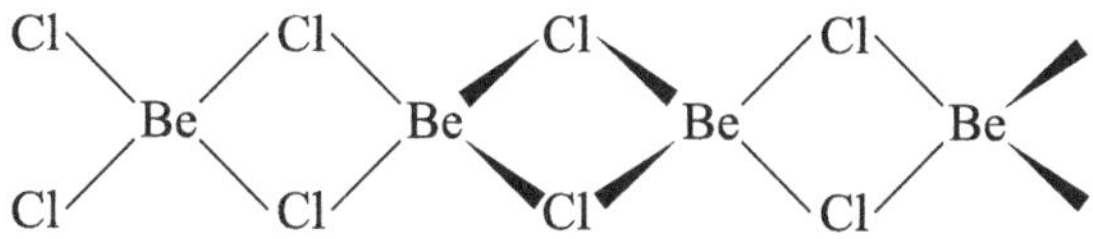

$BeCl_2$ structure

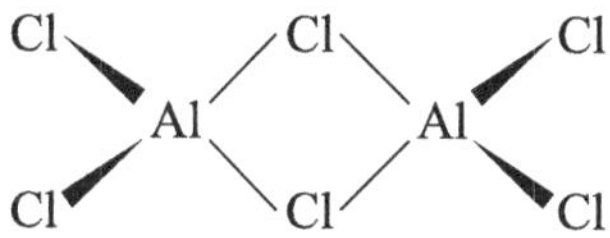

$AlCl_3$ structure

Fluorine is the most electronegative and has the oxidation number (–1) always. Cl, Br and I the covalency of one is quite common. Here the oxidation state may be (–1) or (+1) Cl. Br and I also exhibit higher valencies with oxidation numbers (+III), (+V) and (+VII). These valencies are formed as a results of promoting electrons from filled *p* and *s* levels to the empty *d* levels. The unpaired electrons so formed can form three, five or seven covalent bonds. The higher oxidation states are quite common in interhalgens and halogen oxides.

Electronic can configuration of halogens

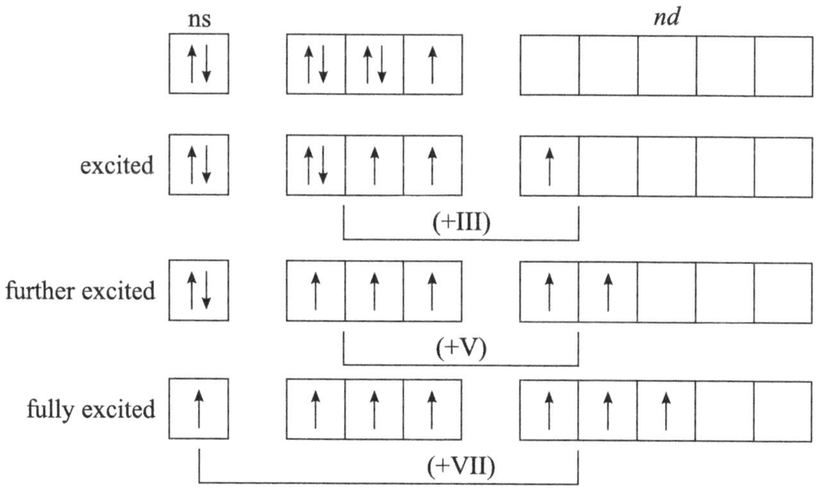

These oxidation states of (+IV) and (+VI) occurs in their oxides ClO_2, BrO_2, Cl_2O_6 and BrO_3.

Melting and Boiling Points

The melting and boiling points increase with the increase in atomic numbers. Fluorine and chlorine are gases, bromine is a liquid and iodine is a solid at room temperature. Only two elements are liquids at room temperature namely mercury and bromine. In very hot climates Cs and Tl are also liquid. I_2 solid sublimes without melting.

Table 15.5 Melting and Boiling Points

	Melting point (°C)	Boiling point (°C)
F_2	–219	–188
Cl_2	–101	–34
Br_2	–7	60
I_2	114	185

Bond Energy

All the halognes except At are diatomic molecules. Bond energy of X_2 decreases with increase in the size of atoms. This is attributed to the less effective overlap of orbitals. Bond energy of F_2 deviates from the trend.

Table 15.6 Bond Energy and Bond Length

	Bond energy (free energy of dissociation) (kJ mol^{-1})	Bond length X_2 (Å)
F	126	1.43
Cl	210	1.99
Br	158	2.28
I	118	2.66

Bond energy of F_2 is 126 kJ mol^{-1} which compared to other hologens Table 15.5 is very low and hence high chemical reactivity. Other element in the first row of the periodic table like N—N, O—O, P—P and S—S show weak bonds. Two reasons have been cited.

1. In Cl_2, Br_2 and I_2 *pd* hybridizations occurred which gives rise to multiple bonding. This actually makes the bond stronger than F_2 where no *d*-orbitals are available.
2. Fluorine atom is very small in size *i.e.,* 1.48Å. Therefore, some inter nuclear repulsion is experienced. The large electron-electron repulsion between two F atoms actually weakens the bonds (coulson model is largely accepted).

Oxidizing Power

The tendency of a neutral gaseous atom in its ground state of configuration is called electron affinity. Electron affinity of chlorine obtains maximum value. Oxidation is removal of electron or deelectronation. Oxdizing agent on the contrary gain electrons. The strength of the oxidizing agent is called oxidation potential. This is best represented by the Born-Haber cycle.

Oxidation potential is the energy altered between the element in its standard state to its hydrated ions. Thus iodine change from $\frac{1}{2}I_2$ (solid) to I^- (hydrated). The total oxidation potential is equal to the sum of energy as the enthalpies of fusion, vapourization and dissociation minus the energy evolved as the electron affinity EA and the energy of hydration. In a similar way oxidation potential of bromine and chlorine can be calculated *e.g.*,

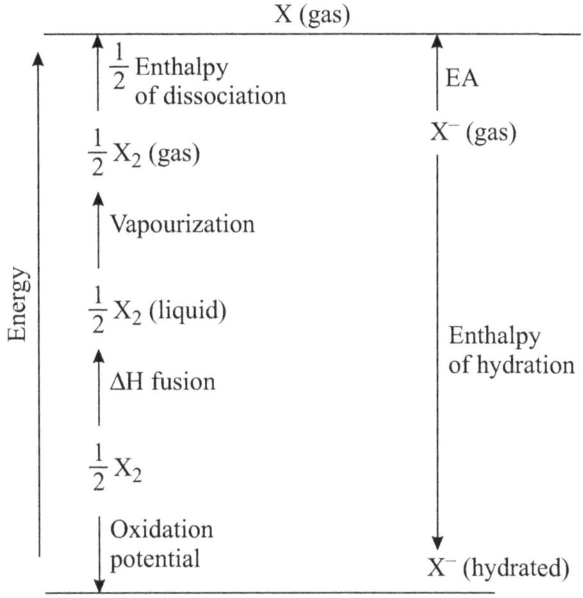

Table 15.7 Enthalpy (ΔH°) Values for $\frac{1}{2}X_2 \rightarrow X_{(hydrated)}$ (all values in kJ mol^{-1})

1/2 Enthalpy of fusion	1/2 Enethalpy of vaporization	1/2 Enthalpy of dissociation	Electron affinity	Enthalpy of hydration	Sum of ΔH°
—	—	+159/2	–333	–513	–836
—	—	+243/2	–349	–370	–597.5
1	+30/2	+193/2	–325	–339	–552.5
+15/2	+42/2	+151/2	–296	–247	–466

Since the halogens are gases therefore ΔH fusion and ΔH vapourization are ignored.

Electron affinity value for chlorine is the maximum *i.e.,* gaseous Cl atom accepts electron most readily, therefore Cl is the strongest oxidizing agent. The values in the table show that F is the most electronegative element. Accordingly F should be strongest oxidizing agent. This is explained by:

1. F_2 dissociation is low due to the weak F—F bond.
2. F_2 has a high hydration energy due to exceptionally small F^- ion.

Fluorine is a stong oxidizing agents as it can displace Cl^- ion both in solution form and dry. Chlorine gas can displace Br^- ion from solution Commonly halogen with less atomic number can oxidize halides with higher atomic number.

Chemical Characteristics

Reaction with water

Reaction with water: all the halogens are water soluble. The halogens react with water to different extents.

Fluorine can oxidize water to dioxygen. This reaction is spontaneous and exothermic. In so doing fluorine is reduced to F^- ions.

$$F_2 + 3H_2O \longrightarrow 2H_3O^+ + 2F^- + \frac{1}{2}O_2 \qquad \Delta G° = -795 \text{ kJ mol}^{-1}$$

Chlorine and water reaction is possible but the reaction is slow due to high activation energy.

$$Cl_2 + 3H_2O \longrightarrow 2H_3O^+ + 2Cl^- + \frac{1}{2}O_2$$

Cl_2 may give an alternative disproportionate which occurs swiftly.

$$\underset{(0)}{Cl_2} + H_2O \longrightarrow \underset{(-1)}{HCl} + \underset{(+1)}{HOCl}$$

Table 15.8 Concentrations in Saturated Solutions at 25°C.

	Solubility (ml l^{-1})	Concentration X_2 (hydrated) (mol l^{-1})	Concentration HOX (mol l^{-1})
Cl_2	0.091	0.061	0.030
Br_2	0.21	0.21	1.1 × 10–3
I_2	0.0013	0.0013	6.4 × 10–6

Br_2 and I_2 may undergo similar disproportionation to a very limited extent *i.e.*, Cl_2 may contain 2/3 hydrated X_2 and 1/3 OCl^-. Br and I_2 may exhibit still small extent of OBr^- and OI^- (negligible).

I_2 is a very weak oxidizing agent $\Delta G°$ here is positive *i.e.*, energy may be supplied to make it to oxidize water.

$$I_2 + H_2O \longrightarrow 2H^+ + 2I^- + \frac{1}{2}O_2 \qquad \Delta G = +105 \text{ kJ mol}^{-1}$$

The reverse reaction may occur spontaneously at –150 kJ mol^{-1}. The atmospheric oxygen oxidises iodide to iodine.

Titration

Iodine and sodium thiosulphate titration, the brown colour of iodine because pale yellow (minimum iodine) and starch indicator is added to give deep blue colour which disappears on addition of more of $Na_2S_2O_3$ solution to become colourless.

$$I_2 + 2S_2O_3^{2-} \longrightarrow 2I^- + S_4O_6^{2-}$$

If the solution in the titration flask is allowed to stand the blue colour appears again. This is attributed to the reverse reaction by atmospheric oxygen.

$$2I^- + \frac{1}{2}O_2 + 2H^2 \longrightarrow I_2 + H_2I$$

The end point is noted when the blue colour just disappears and the solution remains colourless for 30 seconds.

Reactivity of Halogens

Fluorine is the most reactive of all the elements. All elements react with fluorine except the few lighter noble gases such as He, Ne and Argon. Fluorine reacts with Xe because of the low ionization potential of Xe element. Fluorine gives vigorous reaction with many elements often explosive. Metals such as Cu, Ni, Fe and Al form a protective fluoride coating on their surface. If the metals are in powdered form (large surface area) or the reaction mixture is heated then the reaction tends to be vigorous. In the other halogens the reactivity decreases in the order Cl > Br > I. Chlorine and bromine react less vigorously with elements in comparison to fluorine. Iodine in still less reactive and does not combine with S and Se. In addition, fluorine and chlorine oxidize the elements than do bromine and iodine exposing the elements to higher oxidation states *e.g.*, PCl_5, SF_6, PBr_5 etc.

Reasons for Great Reactivity of Fluorine

1. The low dissociation energy of F—F bond.
2. The increasingly strong bond with other elements.

Table 15.9 Some Bond Energies of Halogen Compounds (all values in kJ mol^{-1})

	HX	BX_3	AlX_3	CX_4	NX_3
F	566	645	582	439	272
Cl	431	444	427	347	201
Br	366	368	360	276	243
I	299	272	285	238	*

The above mentioned properties of fluorine arise out of exceptionally small size of fluorine. The weak F—F bond is because of lone pair repulsion between the two fluorine atoms. Strong bonds are attributed to high coordination number and high lattice energy. The bond energies given in the Table 15.9 show that halogens form very strong bonds even stronger than C—C bond.

* Unstable and explosive

Reaction of Halogens

1. **With Water**

$$2F_2 + 2H_2O \longrightarrow 4H^+ + 4F^- + O_2 \text{ vigorous}$$
$$2I_2 + 2H_2O \longrightarrow 4H^+ + 4I^- + O_2$$

The above reaction tends to go in reverse direction because HI is one of the strongest reducing agents.

$$X_2 + 2H_2O \longrightarrow H^+ + X^- + HOX$$

This reaction is maintained in Cl, Br and I, the order being Cl > Br > I.

2. **With Hydrogen:** The general reaction is maintained by all halogens.

$$X_2 + X_2 \longrightarrow 2HX$$

3. **Metals:** Most of the metals form halides. In the case of fluorine it is vigorous.

$$nX_2 + 2M \longrightarrow 2MXn$$

4. **Carbon Monoxide**

$$X_2 + CO \longrightarrow \underset{\text{carbonyl halides}}{COX_2}$$

5. **Non-Metals**

(*a*) Phosphorus

$3X_2 + 2P \longrightarrow 2PX_3$ Trihalides of As, Sb and Bi

$5X_2 + 2P \longrightarrow 2PX_5$ F, Cl and Br give pentahalides AsF_5, SbF_5, $SbCl_5$ and BiF_5

(*b*) Sulphur

$$X_2 + 2S \longrightarrow S_2X_2 \text{ (Cl and Br)}$$
$$2Cl_2 + S \longrightarrow SCl_4 \text{ (only Cl)}$$
$$3F_2 + S \longrightarrow SF_6 \text{ (only F)}$$

6. **H_2S**

$$X_2 + H_2S \longrightarrow \underset{\substack{\text{oxidise } S^{2-} \text{ to } S^\circ \\ \text{(all halogens)}}}{2HCl + S}$$

7. **SO_2**

$$X_2 + SO_2 \longrightarrow SO_2X_2 \text{ (only F and Cl)}$$

8. **NH_3**

$$3X_2 + 8NH_3 \longrightarrow N_2 + 6NH_4Cl \qquad \text{(given by F, Cl and Br)}$$

9. **Inter halogens**

$$X_2 + X_2 \longrightarrow 2XX' \text{ Inter halogens}$$
$$X_2 + XX' \longrightarrow X'X_3 \text{ higher inter halogens}$$

Important Compounds of Halogens

Hydrogen Halide HX

The anhydrous HX is referred to as hydrogen halides and the aqueous solutions as halogen acids.

$$H_2 + X_2 \longrightarrow 2HX$$

The reactivity of halogens towards hydrogen decrease down the group. H_2 and F_2 react violently. Reaction H_2 and Cl_2 is slow in the dark and faster in daylight and explosive in the sunlight. With iodine reaction is slow at room temperature.

Hydrogen Fluoride HF

On industrial scale it is prepared by heating CaF_2 and concentrated H_2SO_4.

$$CaF_2 + H_2SO_4 \longrightarrow CaSO_4 + 2HF$$

Calcium fluoride used should be free from SiO_2 as SiO_2 reacts with HF produce.

$$SiO_2 + 4HF \longrightarrow SiF_4 + 2H_2O$$
$$SiF_4 + 2HF \longrightarrow H_2[SiF_6]$$

HF so produced is purified by washing, cooling and fractional distillation, Vapours of HF are toxic while solutions of HF are corrosive. HF is handled in metal apparatus as the glass attacks HF to form fluorosilicate.

$$\underset{\text{glass}}{SiO_2} + 6HF \longrightarrow \underset{\substack{\text{fluorosilicate}\\ \text{ion}}}{[SiF_6]^{2-}} + 2H^+ + 2H_2O$$

Uses of HF

1. HF is used to form CFC (freons) which are used as refrigerating fluids and propellant in aerosols. CFC damages ozone layer

$$CCl_4 + 2HF \xrightarrow[SbCl_3]{\text{anhydrous}} CCl_2F_2 + 2HCl$$

2. Used to make cryolite (Na_3AlF_6) for extraction of aluminium metal

$$3NaF + AlF_3 \longrightarrow \underset{\text{Cryolite}}{Na_3AlF_6}$$

3. Used in anhrydrous form to make alkyl benzene which is converted to alkyl benzene sulphonate detergents.
4. Used in uranium processing where UF_6 intermediate is used.
5. Aqueous HF is used for etching of glass and pickling steel and making metal fluorides.

$$B_2O_3 + 6HF \xrightarrow[H_2SO_4]{\text{Conc.}} 2BF_3 + 3H_2SO_4 + H_2O$$

$$Al_2O_3 + 6HF \longrightarrow 2AlF_3 + 3H_2O$$

Hydrogen Chloride HCl Preparation

1. It is prepared on a very large scale by salt cake method.

   Stage I: $NaCl + H_2SO_4 \xrightarrow{150°C} HCl(g) + NaHSO_4$

   Stage II: $NaHSO_4 + NaCl \xrightarrow{550°C} HCl(g) + Na_2SO_4$

   The reaction is endothermic. Na_2SO_4 is used for paper making.

2. Good amount of HCl has become available as a by-product in heavy organic chemical industry.

$$\begin{array}{c} CH_2Cl \\ | \\ CH_2Cl \end{array} \xrightarrow[\text{halogenation}]{\text{Dehydro}} CH_2{=}CHCl + HCl$$

It can also be obtained from chlorinated fluorocarbons.

3. Pure HCl can be obtained from the direct combination of the gaseous mixture of H_2 and Cl_2. The process is strongly exothermic and may turn explosive. Therefore the reaction is allowed to proceed in a hydrogen chlorine flame.
4. On a smaller scale it is prepared by treating NH_4Cl with concentrated H_2SO_4. It is preferred as NH_4HSO_4 is soluble and the reaction goes on in uninhibited.

$$2NH_4Cl + H_2SO_4 \longrightarrow 2HCl + (NH_4)_2SO_4$$

HCl gas is highly soluble in water and that forms hydrochloric acid. HCl acid is used to make metal chlorides, manufacture of dye stuffs and in the sugar industry.

Hydrogen Bromide and Hydrogen Iodide HBr and HI

HBr and HI are prepared by the action of concentrated H_3PO_4 on metal bromides and iodides. Care is taken to use H_3PO_4 which is a non-oxidizing acid. Concentrated H_2SO_4 is a strong oxidizing agent and has the capability to oxidize HBr and HI to Br_2 and I_2

$$H_3PO_4 + NaBr \longrightarrow HBr + \underset{\text{Sodium hypophosphite}}{NaH_2PO_2}$$

$$H_3PO_4 + NaI \longrightarrow HI + NaH_2PO_2$$

In the laboratory Br_2 or I_2 with red phosphorus in water

For HBr

$$\underset{\text{red}}{2P} + 3Br_2 \longrightarrow 2PBr_3 \xrightarrow{6H_2O} 6HBr + 2H_3PO_3$$

For HI: What is added to a mixture of red phosphorus and iodine.

$$\underset{\text{red}}{P} + 3I_2 \longrightarrow 2PI_3 \xrightarrow{6H_2O} 6HI + 2H_3PO_3$$

Only HF exists as a liquid at room temperature (b.pt. 19.9°C) while HCl, HBr and HI exist as gases. The boiling points increase steadily for HCl to HI.

The high boiling point of HF is attributed to the existence of hydrogen bond. The hydrogen bond is formed between F atom of one molecule and the hydrogen atom of the other molecule. The molecules of HF are represented as $(HF)_n$ forming zig-zag chains in liquid and solid. Some hydrogen bond also exists in the gas form which are shown as $(HF)_2$ and cyclic $(HF)_n$.

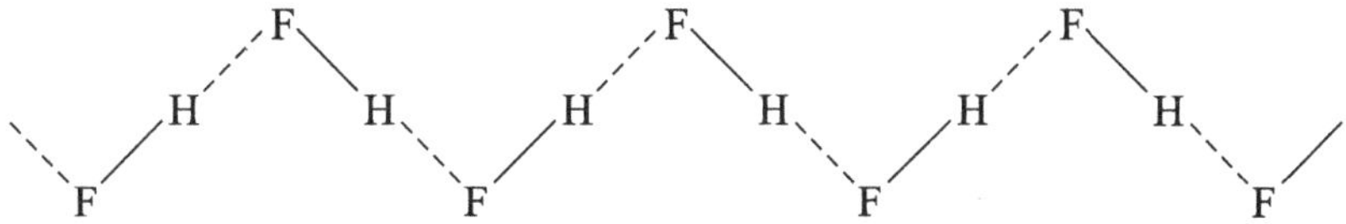

HCl, HBr and HI are monomeric in the gas and liquid form but HCl and HBr form weak hydrogen bonds in the solid form.

Hydrogen bonds are weak with bond energy between 5 to 35 kJ mol^{-1} while normal covalent bonds —C—C— have bond energy 347 kJ mol^{-1} but still they are significant.

Table 15.10 Some Physical Properties of Hydrogen Halides

Property	Hf	HCl	HBr	HI
Boiling point, K	293.4	188	206	237
Melting point, K	190	159	187	222
Limization concentration in water	7×10^{-7}	10^{7}	10^{9}	10^{11}
Heat of formation (DHP) kJ mol^{-1}	–271.1	–92.3	–36.4	+26.5
Bond Energy, kJ mol^{-1}	574	428	362	295

Heat of Formation

The heat of formation of hydrogen halides given in the table 15 increase with the increasing atomic mass of the halogen which points to corresponding decrease in the quantities of energy released during the formation of hydrogen halides. Thus as we go down the group the formation of hydrogen halides become more difficult which tallies with the experimental data.

Reducing Properties

From the heat of formation data it is evident that the stability of hydrogen halides decrease down the group. HF and HCl are stable up to 1200°C while HBr dissociate slightly (only 1%) and HI dissociates 20% at 440°C. This implies that the reducing nature of hydrogen halides increase an we move down the group.

$$\xrightarrow[\text{Decreasing reducing action}]{\text{HI HBr HCl HF}}$$

Therefore HF has no appreciable reducing nature whereas HI is a very strong reducing agent. HI is very easily oxidized even by the atmospheric oxygen as it acquires brown colour of iodine an exposure.

$$4HI + O_2 \longrightarrow 2H_2O + 2I_2$$

The general reaction during reduction is

$$2X^- \longrightarrow X_2 + 2e$$

The ease of oxidation of halide ion is expected to increase in the order of increase in the size of halide ions. Therefore HF is a poor reducing agent

(F^- electron is difficult to remove). From the discussion it is concluded that HF is a poor reducing agent and HI is a strong reducing agent (I^- ion is large and removal of electron is easy).

Acidic Strength

All the hydrogen halides are soluble in water and their aqueous solutions are acidic. This is further supported by the dissociation constants. The order of acidic strength is HI > HBr > HCl > HF. This order may appear strange in view of the order of decreasing polarity of H–X bonds. The positive charge an H atom in HF molecule is greater than in HCl, HBr and HI molecules. The tendency to release H^+ to water is less than HCl, HBr and HI. The anomaly becomes clear on studying the following cycle.

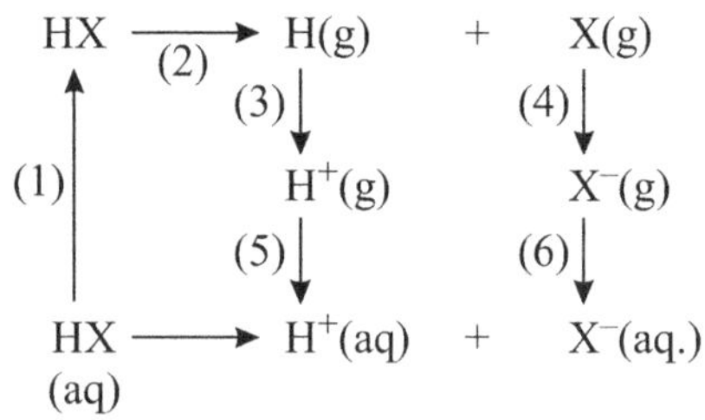

This cycle is such that ionization is related to a number of atomic and molecular species.

Energy of dehydration of HX molecules (step I), energy of dissociation of gaseous HX into atoms (step 2), ionization energy of H atom (step 3), electron affinity of the X atom (step 4) and energies of hydration of H^+ and X^- (steps 5 and 6) the main factor contributing to the lower acidify of HF is its bond dissociation energy.

Oxides of Halogens

Halogens do not combine directly with oxygen. The halogen-oxygen compounds are prepared indirectly. Many of the halogen-oxygen compounds are unstable as they have positive energies of formation except OF_2. However, the halogen-oxygen compounds that are stable are shown in the Table 15.11. The bonds in these compounds are largely covalent because of similar electronegativities (F = 4.0; Cl = 3.0; Br = 2.8; I = 2.5 and O = 3.5). Since fluorine is more electro-negative than oxygen therefore fluorine oxygen compounds are called oxygen fluorides rather than oxides of fluorine. In other halogens oxygen is more electronegative therefore they are called oxides (*e.g.*, oxides of chlorine, bromine and iodine). There oxides are powerful oxidizing agents. There compounds are reactive and unstable to heat. Iodine oxides are most stable while bromine oxides are least. The higher oxides are more stable than the lower ones.

Table 15.11 Halogen-Oxygen Compounds

Fluorine	Chlorine	Bromine	Iodine
OF_2	Cl_2O	Br_2O	I_2O_4
O_2F_2	ClO_2	BrO_2	I_2O_5
	Cl_2O_6	BrO_3	I_4O_9
	Cl_2O_7		

(*a*) Oxygen-fluorides

(*i*) Oxygen difluoride (OF_2): The oxidation state of fluorine is always –1. It is prepared by the action of F_2 on NaOH.

$$2F_2 + 2NaOH \longrightarrow 2NaF + H_2O + OF_2$$

OF_2 is a yellow, poisonous gas which slowly dissolves is water.

$$OF_2 + H_2O \longrightarrow O_2 + 2HF$$

It is thermally stable at 25°C. It is a non-explosive because of negative heat of formation. It is used as a rocket fuel.

Structure

It contains V-shaped structure with a bond angle equal to 103°18′. It has sp^3 hybridized structure with two orbitals overlapping with one electron of fluorine while the remaining two orbitals containing lone pairs as shown.

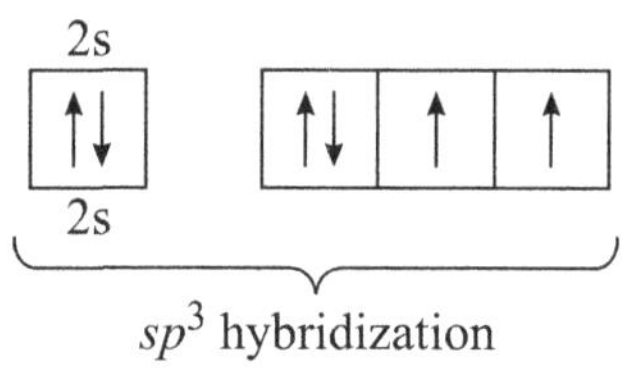

i.e., tetrahedral with two lone pairs.

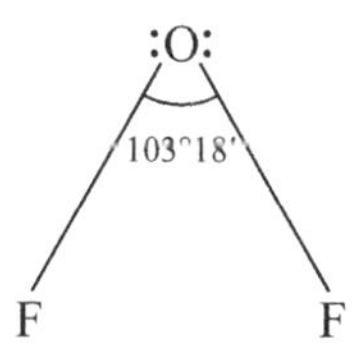

Structure of OF_2

The bond angle is distorted due to the lone pairs.

(*ii*) Diooxygen difluoride (O_2F_2): It is prepared by passing electric discharge through fluorine and oxygen cooled at –185°C. It is a yellow orange solid. It is highly unstable. It disproportionates in its elements at –50°C. It is very strong fluorinating and oxidizing agent.

Structure

Its structure is similar to that of H_2O_2. The O—O distance is shorter that than in H_2O_2 and is of the same size that in oxygen.

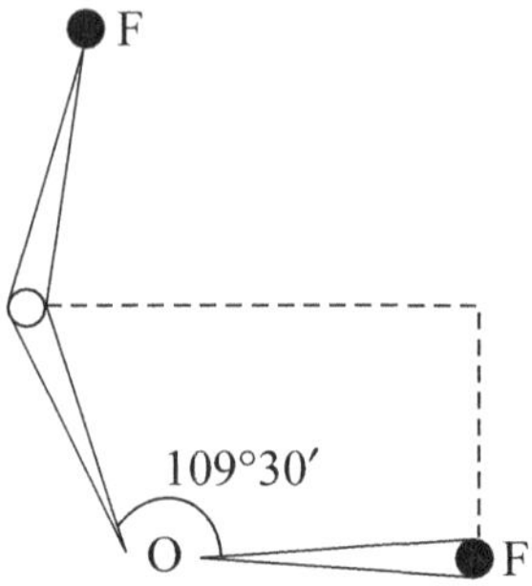

(*b*) Oxides of Chlorine

(*i*) Chlorine monoxide (Cl_2O): Here, chlorine is in +1 oxidation state. It is prepared by the action of chlorine on yellow mercury oxide at 0°C.

$$HgO + 2Cl_2 \longrightarrow HgCl_2 + Cl_2O$$

It is a yellwish-red gas which decomposes explosively. It is an active oxidising agent which dissolves in water forming HClO

$$Cl_2O + H_2O \longrightarrow 2HClO$$

Gaseous mixture of Cl_2O and NH_3 explodes violently.

$$3Cl_2 + 10NH_3 \longrightarrow 2N_2 + 6NH_4Cl + 3H_2O$$

Commercially Cl_2O is used to make $Ca(OCl)_2$ which is an effective bleaching agent for textiles and wood pulp.

It is regarded as an anhydride of HClO. It possesses an angular structure.

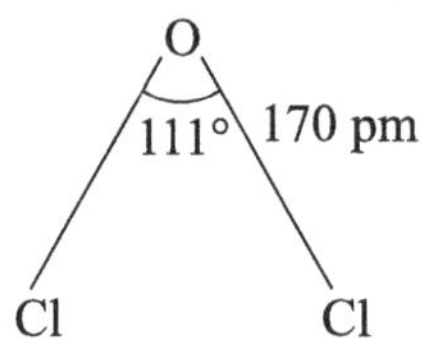

(*ii*) Chlorine dioxide (ClO_2): In this compound chlorine is in +4 oxidation state. It is prepared by the action of sodium chlorite and chlorine plus air.

$$2NaClO_2 + Cl_2 \xrightarrow{\text{air}} 2NaCl + 2ClO_2\uparrow$$

It is a yellow gas which explodes violently. Because of explosive nature it is always prepared in low concentration. It is prepared by reducing $NaClO_3$ with NaCl, HCl and SO_2.

$$\underset{+5}{ClO_3^-} + Cl^- + 2H^+ \longrightarrow \underset{+4}{ClO_2} + \frac{1}{2}Cl_2 + H_2O$$

It is a powerful chlorinating and oxidising agent. Its bleaching power is 30 times stoanger than chlorine. With alkali it forms chlrites and chlorates.

$$2ClO_2 + 2NaOH \longrightarrow \underset{\text{Chlorite}}{NaClO_2} + \underset{\text{Chlorate}}{NaClO_3} + H_2O$$

It is a mixed anhydride. ClO_2 contains three electron bonds coupled with electron pair bond. Its structure is a resonance hybrid of the following structures:

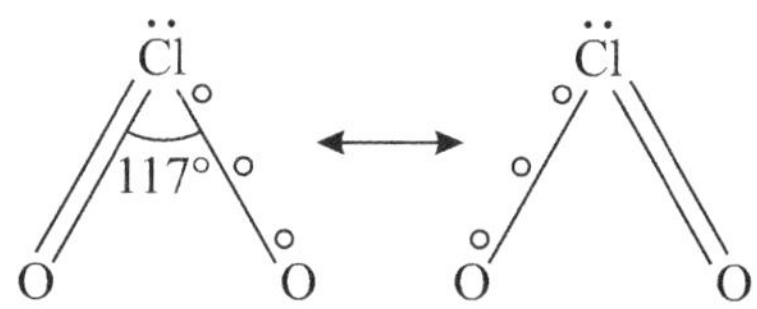

The three electron bond assigns it a paramagnetic nature and a shorter bond length. Because of the presence of odd electron it has a tendency to form a dimer because the odd electron is delocalized as in the case of NO_2.

It is used for bleaching wood pulp and for water treatment.

***(iii)* Dichlorine hexaoxide (Cl_2O_6):** If contains chlorine in the +6 oxidation state. It is prepared by the action of O_3 on chlorine dioxide.

$$2ClO_2 + 2O_3 \longrightarrow Cl_2O_6 + 2O_2$$

It is red oily liquid which explodes on contact with grease. It is a strong oxidising agent. It reacts with akalies to form chlorate and perchlorate.

$$Cl_2O_6 + 2NaOH \longrightarrow \underset{\text{Chlorate}}{NaClO_3} + \underset{\text{Perchlorate}}{NaClO_4} + H_2O$$

***(iv)* Dichloro heptoxide (Cl_2O_7):** In this chlorine has a +7 oxidation state. It is prepared by the dehydration of $HClO_4$ with P_4O_{10} at –10°C and then distilling the product under vacuum. It can explode an heating.

$$4HClO_4 + P_4O_{10} \longrightarrow 2Cl_2O_7 + 4HPO_3$$

With H_2O it changes back to perchloric acid.

$$Cl_2O_7 + H_2O \longrightarrow 2HClO_4$$

Hence it is considered an anhydride of perchloric acid. It is the most stable chlorine oxide and a strong oxidising agent. It has the structure $O_3Cl–O–ClO_3$ with tetrahedral chlorine as shown

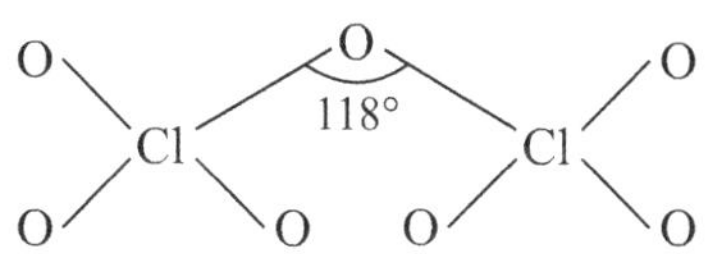

Oxides of Bromine

Bromine monoxide (Br_2O): In this compound Br has +1 oxidation state. It is prepared by the action of Br_2 on HgO in a carbon tetrachloride solution

$$Br_2 + HgO \xrightarrow{CCl_4} Br_2O + Hg$$

Br_2O is a dark brown solid moderately stable at −60°C whereas it decompses above −40°C. It dissolves in water to form weak, unstable acid called hypobromous acid.

$$Br_2O + H_2O \longrightarrow 2HOBr$$

The alkaline hydrolysis of Br_2O leads to desproportionation:

$$Br_2 + 6OH^- \longrightarrow 5BrO_3^- + Br^- + 3H_2O$$

It can oxidise I_2 to I_2O_5 and benzene to 1, 4 quinone. Reaction with F_2 gives $FBrO_2$.

Bromine Dioxide (BrO_2)

BrO_2 contains Br in the +4 oxidation state. It is prepared by passing electric arc through bromine and oxygen gases at low temperature and pressure.

$$Br_2 + 2O_2 \xrightarrow{LT/P} 2BrO_2$$

It is stable below – 40°C where it exists as a yellow solid. On treatment with alkalies it forms a bromide bromate mixture.

$$6BrO_2 + 6NaOH \longrightarrow NaBr + \underset{\text{sodium bromate}}{5NaBrO_3} + 3H_2O$$

***(iii)* Bromine trioxide BrO_3:** In this bromine has +6 oxidation state. It is prepared by the action glow discharge on Br_2 and O_2 mixture at 0°C.

$$Br_2 + 3O_2 \longrightarrow 2BrO_3$$

It is stable below −80°C.

Oxides of Iodine

Iodine pentoxide (I_2O_5): It contains iodine in +5 oxidation state and is obtained as a white hygroscopic solid by heating HIO_3 at 200°C in dry air.

$$2HIO_3 \xrightarrow[\text{Dry.air}]{200°C} I_2O_5 + H_2O$$

It dissolves in water to give iodic acid HIO_3 therefore it is termed as an anhydride of iodic acid. It is the most stable oxide of iodine. It is a very strong oxidising agent and is used for the estimation of CO in the atmosphere.

$$I_2O_5 + 5CO \longrightarrow 5CO_2 + I_2$$

Structure

Each iodine is sp^2 hybridised therefore it acquires a planar structure. The infrared spectra data proposes the structure.

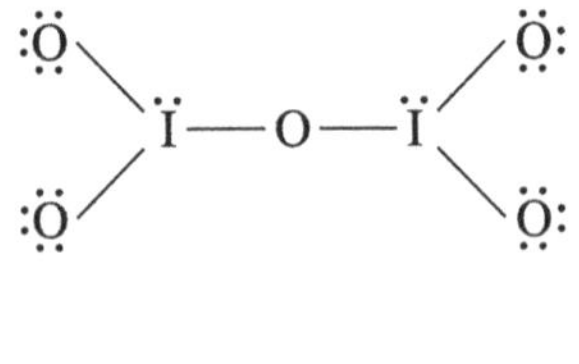

I_2O_5 is an important reagent to oxidise CO to CO_2 in the absence of water. It is used in the detection and estimation of toxic carbon monoxide when it is mixed with other gases. It can also oxidise NO, C_2H_4 and H_2S. It reacts with SO_3 to produce IO_2^+ iodyl salt. With concentrated H_2SO_4 it yields iodosyl compounds IO^+ (HSO_4).

Some other Oxide of Iodine

The less stable oxides of iodine I_2O_4, I_4O_9 and I_2O_7 do exist but they decompose at about 100°C to I_2 and I_2O_6. I_2O_4 contains 10 units of IO_3 cross linked units. The structure of I_4O_9 is formulated as $I(IO_3)_3$. I_2O_7 is an orange polymer formed by dehydrating HIO_4 with column.

OXO-Acids of Halogens

Fluorine does not form oxo-acids because of its strong oxidising property, chlorine, bromine and iodine form four series of oxo-acids with the formulae HXO, HXO_2, HXO_3 and HXO_4. Some of these acids are quite unstable therefore they are available in solutions or as their salts. Only $HClO_4$, HIO_3, H_5IO_6 and HIO_4 are available in the pure form.

Table 15.12 OXO-acids of Halogens

Oxidation state of halogen	Chlorine	Bromine	Iodine	Name of acid	Name of salt	Thermal stability and acid strenght
+1	HClO	HBrO	HIO	Hypohaloous	Hypo-halite	Increases
+3	$HClO_2$	—	—	Halous	Halite	Increases
+5	$HClO_3$	$HBrO_3$	HIO_3	Halic	Halate	Increases
+7	$HClO_4$	$HBrO_4$	HIO_4	Perhalic	Perhalate	Increases
			H_5IO_6			

In general:

(*i*) Acidic character of oxo acids increase with the increase in their oxidation number *e.g.*

$$HClO_4 > HClO_3 > HClO_2 > HClO$$

Reason

(*i*) The strength of the acid is related to the strength of the conjugate base. If the acid is strong its conjugate base has to be weaker and vice versa. The conjugate bases of the oxoacids of chlorine are $[OCl_4]^-$, $[ClO_3]^-$, $[ClO_2]^-$ and $[ClO]^-$ respectively. The oxoanions of chlorine in which central Cl atom has a highest oxidation state have more number of oxygen atoms form an extension of π bonds. Due to the delocalization between electronegative oxygen and Cl atoms stablises the ion and decrease its tendency to accept a proton. This causes a weak conjugate base resulting into an increase in acid strength.

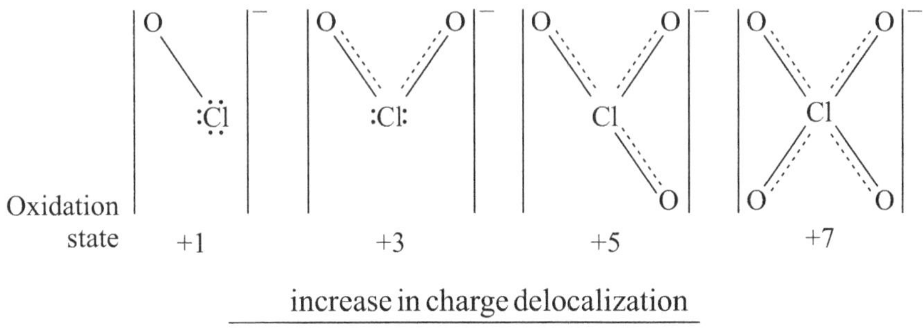

(*ii*) The acid strength of oxo-acids of different halogens having the same oxidation state decrease with increases in the atomic number of the halogens *e.g.,* oxy-acid of Cl are more acidic than the corresponding bromine and in turn more acidic than the corresponding oxo-acid of iodine.

As the atomic number of halogen increases its electronegativity decreases but the size increases which makes the acid weak.

(*iii*) All the halogen oxoacids as well as their salts act as powerful oxidizing agents. The oxidizing action does not increase with the increasing number of oxygen atoms. The rates of oxidation depend on the reaction conditions.

(*a*) **Hypohalous acids and hypohalites:** HOCl, HOBr and HOI are thermally unstable therefore they are not available is pure form. They are generally available in aqueous solutions. They are prepared by shaking HgO in water alongwith the desired halogens.

$$2HgO + 2X_2 + H_2 \longrightarrow HgO.HgX_2\downarrow + 2HOX$$

All these acids have halogen in the +1 oxidation state. They are weak acids, the acid strength decreases with the increase in the atomic number of halogen atom *i.e.,* the acid strength is in the order:

$$HClO > HBrO > HIO$$

HIO is more commonly stated as iodine hydroxide, HClO is the strongest acid out of these. HClO is also most stable. Correspondingly salts of HClO are most stable and are used in bleaching cotton, linen and paper pulp. They kill micro organisms and therefore used as disinfectants and to purify water.

All hypohalites disproportionate at elevated temparature.

$$3OX^- \longrightarrow \underset{\text{halate}}{XO_3^-} + \underset{\text{halide}}{2X^-}$$

All oxo-halides are sp^3 hybridised and their structure is shown below:

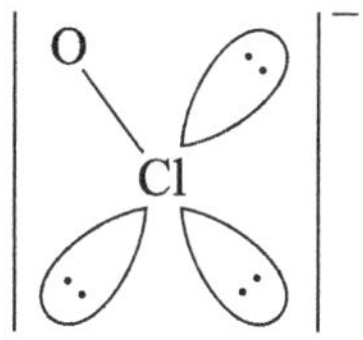

(*b*) **Halous acids and halites:** They contain halogen in the +3 state. The only halous acid to be well established is $HClO_2$. The aqueous solution of $HClO_2$ is prepared by the action of barium chlorite and sulphuric acid

$$Ba(ClO_2)_2 + H_2SO_4 \longrightarrow BaSO_4 + 2HClO_2$$

Its acid strength is greater than HClO but the oxidizing power is similar to that of hyphoalous acid.

Its salts are prepared by the action of ClO_2 on a peroxide.

$$Na_2O_2 + 2ClO_2 \longrightarrow \underset{\text{sodium chlorite}}{2\,NaClO_2} + O_2$$

Chlorites are very good oxidizing agents. Chlorites decompose on heating to chlorates and chlorides.

$$3NaClO_2 \longrightarrow 2NaClO_3 + NaCl$$

The structure of chlorite ion is:

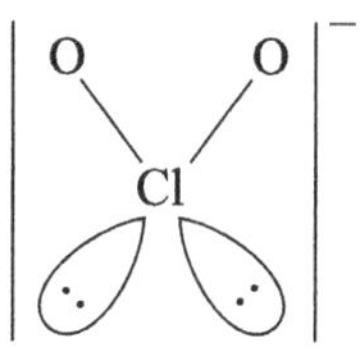

It is an angular molecule represented by two resonance structure.

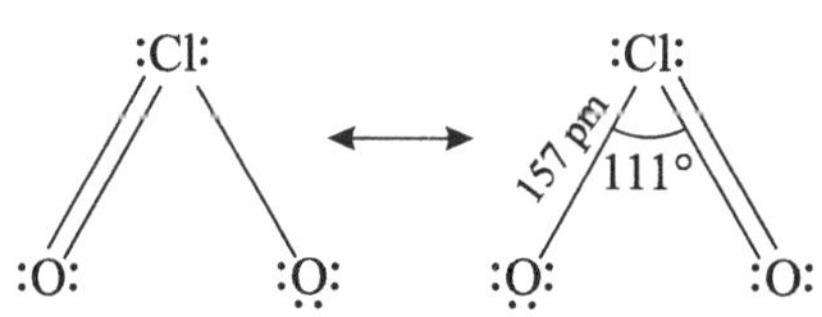

(*c*) **Halide acids and halates:** $HClO_3$, $HBrO_3$ and HIO_3 have halogen atom in the +5 oxidation state. $HClO_3$ and $HBrO_3$ are known in aqueous solution only while HIO_3 is very stable and obtained in the form of white crystals. $HClO_3$ and $HBrO_3$ are easily produced by the action of barium chlorate with H_2SO_4.

$$Ba(ClO_3)_2 + H_2SO_4 \longrightarrow BaSO_4\downarrow + 2HClO_3$$

However, HIO_3 is prepared by the action of concentrated nitric acid on iodine.

$$10HNO_3 + I_2 \longrightarrow 2HIO_3 + 4H_2O + 10NO_2$$

They are strong acids they are stable and their stability increases with increase in atomic number. They are strong oxidizing agents but less than halous and hypohalous acids.

Their salts are conveniently prepared by the action of Cl_2 on hot conc. hydroxides or by the electroylsis of hot chloride solution.

$$6NaOH + 3Cl \longrightarrow NaClO_3 + 5NaCl + 3H_2O$$

Chlorate ion ClO_3^- has a pyramidal structure.

Cl atom

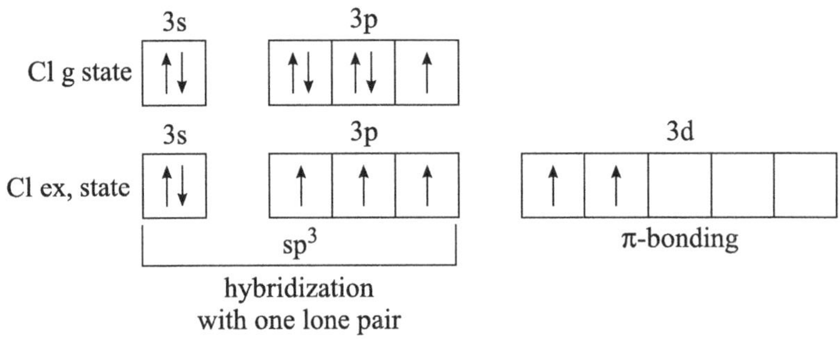

All Cl–O bonds are of equal strength due to resonance. Actual Cl—O bonds are shorter due to *pπ-dπ* bonding.

ClO_3^- is described by the three resonating structures as given:

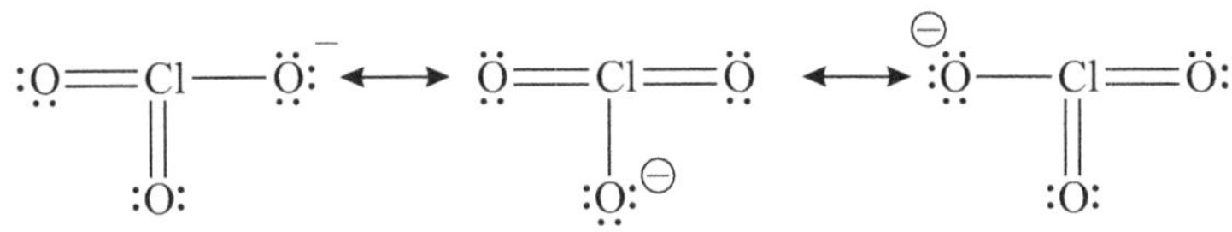

$$\angle OClO = 106°$$
$$Cl - O = 148 \text{ pm}$$

ClO_3^- and $HClO_3$ are powerful oxidizing agents. Chlorates form explosive mixtures with reducing agents like carbon, sulphur and organic compounds. $KClO_3$ is used in fire works matches and explosives due its oxidizing action.

(*d*) **Perhalic acids and perhalates:** $HClO_4$, $HBrO_4$ and HIO_4 contain halogen in the +7 oxidation state. The first two are fairly unstable.

$HClO_4$

1. It is prepared by distilling chloric acid under reduced pressure.

$$3HClO_3 \longrightarrow HClO_4 + 2O_2 + Cl_2 + H_2O$$

2. It can also be obtained by heating $KClO_3$ with concentrated H_2SO_4 at 10–20 mm Hg

$$KClO_4 + H_2SO_4 \longrightarrow KHSO_4 + HClO_4$$

Properties

Physical Characteristics

$HClO_4$ is a colourless liquid with boiling point of 90°C. It decomposes at its boiling point. It becomes black on standing and explodes when kept in a sealed tube.

$$4HClO_4 \longrightarrow 4ClO_2 + 3O_2 + 2H_2O$$

If fumes in moist air and forms hydrates $HClO_4.xH_2O$ where $x = 1, 2, 2.5, 3, 3.5$.

Chemical Characteristics

It behaves as an oxidizing agent. It catches fire in the presence of paper and wood.

1. It is one of the strongest oxidizing agent and is known to ionise as:

$$HClO_4 \longrightarrow H^+ + ClO_4^-$$

2. Metals dissolve in it with the evolution of hydrogen

$$Zn + 2HClO_4 \longrightarrow Zn(ClO_4)_2 + H_2$$

Perchlorates can also be prepared by heating chlorates under special conditions.

$$4KClO_3 \xrightarrow[\text{in silica flask}]{480°\text{C}} 3KClO_4 + KCl$$

Most of the perchlorates are highly soluble except these of potassium, rubidium, caesium and ammonium. Perchlorates are deliquescent. Magnesium perchlrate acts as a drying agent. Perchlorates are used in matches and explosive due to their oxidizing behaviour. The perchlorate ion has a tetrahedral structure.

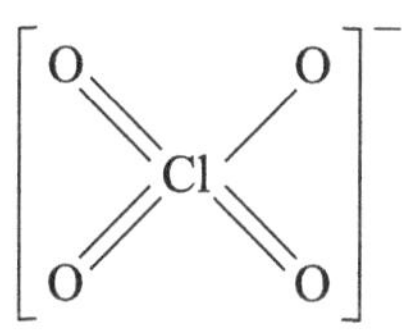

The ClO_4^- ion has four equivalent resonance structures:

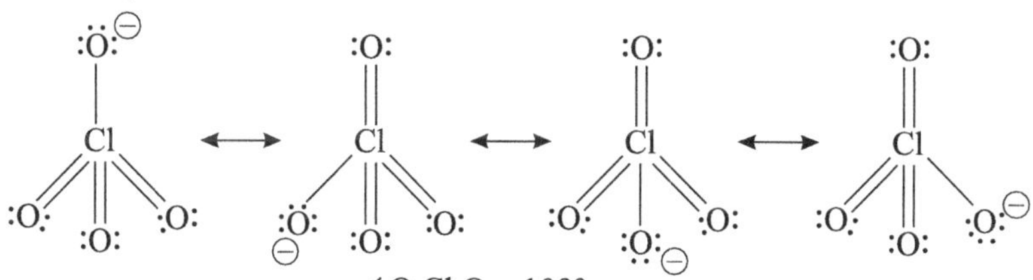

$\angle$O Cl O = 109°

Cl—O = 144 pm

They are more dangerous than chlorates. $KClO_4 + NH_4ClO_4$ + carbon or organic materials are used as explosives and as rocket fuels. They are used in fireworks and flares.

Perbromic Acid and Perbromates

Perbromic acid and its salt were first prepared in 1968. They were prepared by the methods give below:

1. It was prepared by the oxidation of BrO_3^- with aqueous XeF_2 $BrO_3^- + XeF_2 + H_2O \longrightarrow BrO_4^- + Xe + 2HF$. The yield by this method is ~10%.
2. By the oxidation of BrO_3^- by using F_2 gas.

   $$BrO_3^- + F_2 + 2OH^- \longrightarrow BrO_4^- + 2F^- + H_2O$$

   On acidification of perbromates we get perbromic acid which is a monobasic acid stable in aqueous solution. The perbromate are stronger oxidizing agents then perchlorates but periodate are slow in their oxidizing action.

Per-iodic Acids and Periodates

Wide range of periodic acids and periodates are known but the most important of these are meta-periodic acid HIO_4 and para-periodic acid H_2IO_6 and their salts. The large size of iodine permits 6 oxygen atoms to surround it whereas Cl can accommodate only 4 oxygen atoms. Periodic acids and periodates are powerful oxidizing agents and thus are used as analytic reagents.

Preparation

HIO_4

1. It can be obtained by heating para-periodic acid at 100°C in vacuum.

$$2H_5IO_6 \xrightarrow[-3H_2O]{80°C} \underset{\text{Dimero periodic acid}}{H_4I_2O_9} \xrightarrow{100°C} 2HIO_4 + H_2O$$

Salts of meta periodic acid *e.g.* KIO_4 and $NaIO_4$ are prepared by thermal decomposition of iodates.

$$4KIO_3 \longrightarrow 3KIO_4 + KI$$

KIO_4 acts as an oxidizing agent in acid medium.

$$2IO_4^- + 16H^+ + 14e \longrightarrow I_2 + 8H_2O$$

Structure

X-ray studies show four oxygen atoms round the iodine at the centre of a regular tetrahedron. Para-periodic acid is prepared by the action of H_2SO_4 on barium para-periodate.

$$B_5(IO_6)_2 + 5H_2SO_4 \longrightarrow 5BaSO_4 + 2H_5IO_6$$

Another method of preparation involves of passing Cl_2 through silver para-periodate.

$$4Ag_3IO_6 + 10Cl_2 + 10H_2O \longrightarrow 4H_5IO_6 + 20AgCl\downarrow + 5O_2$$

Physical Characteristics

It has white monoclinic crystalline structure and is hygroscopic. It is soluble in water and alcohol. It melts at 146°C with decomposition.

$$2H_5IO_6 \xrightarrow{419\,K} 2HIO_3 + 4H_2O + O_2$$

If the acid is heated under vacuum it gives the following decompsition reaction.

$$\underset{\text{Para-periodic acid}}{H_5IO_6} \xrightarrow{80°C} \underset{\text{Dimeso periodic acid}}{H_4I_2O_9} \xrightarrow{100°C} \underset{\text{meta per iodic acid}}{HIO_4} \xrightarrow{140°C} \underset{\text{iodic acid}}{HIO_3} + \frac{1}{2}O_2$$

It oxidises most of the organic substances. It does not form H_2O_2 or any reaction of H_2O_2 and therefore it does not contain —O—O— linkage. It is a weaker acid than HIO_3.

Structure

Its structure resembles more with H_3TeO_6 rather $HClO_4$. X-ray studies reveal that six oxygen atoms are symmetrically disposed octahedrally round the iodine atom.

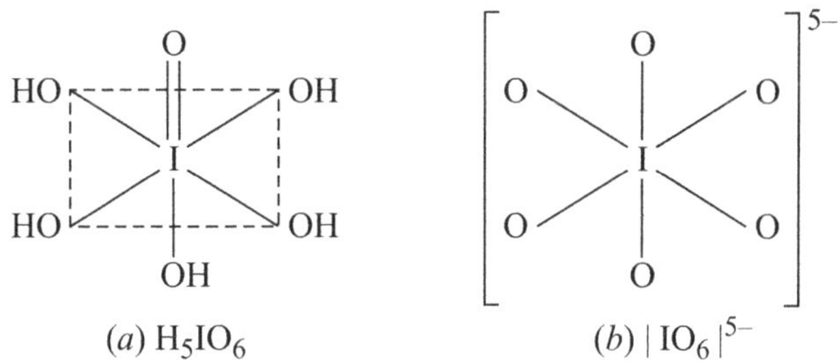

Fig. 15.1 Structure of H_5IO_6 and IO_6^{5-} ion

Halogen-halogen Compounds or Interhalogens

They are a special and unique class of compounds. They are compounds formed by the combination of two different halogens under suitable temperature and pressure conditions. Some important inter-halogens are listed in the Table 15.13.

Table 15.13 Some Inter-halogen Compounds

(Type) AX (Shape) Linear	AX_3 T-shaped	AX_5 Square pyramid	AX_7 Pentagonal
Chlorine fluoride, ClF	Chlorine trifluoride, ClF_3	Chlorine pentafluoride, ClF_5*	—
Bromine fluoride, BrF	Bromine trifluoride, BrF_3	Bromine pentafluoride, BrF_5	—
Iodine fluoride, IF*	Iodine trifluoride, IF_3*	Iodine pentafluoride, IF_5	Iodine heptafluoride, IF_7
Bromine chloride, BrCl	—	—	—
Iodine chloride, ICl	Iodine trichlride, ICl_3	—	—
Iodine brominde, IBr	—	—	—

They are of two types:

(*i*) Neutral molecules *e.g.* ICl, ICl_3 and IF_7

(*ii*) Plyhalide anions I_3^- and polyhalogn cations *e.g.* ICl_2^+.

These compounds are formed because of large electronegativity and the large difference in size of the halogens. ΔG difference of these compounds are negative indicating high thermodynamic stability. The only heptahalide IF_7 formation is explained in the terms of very large size of iodine and very small size of fluorine atoms. This explains that the large size of central iodine atom can accommodate large number of small fluorine atoms.

General Properties

(*a*) All inter-halogens are generally covalent compounds due to small electronegativity difference between them. However, the more electronegative halogen atom acquires a negative charge.

(*b*) Their properties are similar to the component halogens but they are more reactive than the halogens because there exists a weak covalent bond between the dissimilar electronegative elements.

(*c*) ICl and ICl_3 have conductivities values of the order of 10^{-3} ohm^{-1} cm^{-1} *i.e.,* they give cnducting solutions in many solvents. The ionizations that occur are:

$$2ICl \rightleftharpoons I^+ + ICl_2^-$$

$$2ICl_3 \rightleftharpoons ICl_2^+ + ICl_4^-$$

(*d*) All inter-halogens are strong oxidizing agents.

(*e*) Neutral nter-halogen molecules and uninegative inter-halogen ions are diamagnetic which shows that the valence-electrons are either shared or unshared pairs.

(*f*) All inter halogens undergo hydrolysis in which the heavier and less electronegative halogen atom behaves as if it represents the positive

* These inter-halogen compounds are unstable.

oxidation state and the lighter and more electro negative helogens acts in the negative oxidation state *e.g*

$$IF_7 + 6HOH \longrightarrow H_2IO_6 + 7HF$$
$$ICl + HOH \longrightarrow HOI + HCl$$

(*g*) The interhalogens differ in polarity due to difference in their electro negativity from one anther and due to their atomic radius. Thermal stability is decreased by size differences but increase by their polarity, *e.g.*, ClF is thermally more stable to IBr.

(*h*) Halogen fluorides act as strong fluorinating agents and are also very reacttive *e.g.* ClF_3 is as strong a fluorinating agent as F_2.BrF_3 and IF_5 are fluorinating agents particularly in case of the elements in the intermediate oxidation states.

Preparation of Iodine Containing Interhalogens

IF is quite unstable therefore its method of preparation is not known with certainty.

(*i*) ICl

(*a*) It is prepared by mixing liquid chlorine with solid iodine in the freezing mixture.

$$Cl_2 + I_2 \xrightarrow[\text{mixutre}]{\text{freezing}} \underset{\text{crystal}}{2ICl}$$

(*b*) Direct combination of Br_2 and I_2

$$Br_2 + I_2 \longrightarrow 2IBr$$

(*ii*) ICl_3

(*a*) Prepared by the action of Cl_2 on ICl at 100°C

$$ICl + Cl_2 \xrightarrow{373K} ICl_3$$

(*b*) By treating liquid Cl_2 and excess iodine at –80°C

$$3Cl_2 + I_2 \xrightarrow{-80^\circ C} 2ICl_3$$

(*iii*) IF_5

(*a*) Prepared by the action of iodine on AgF.

$$10AgF + 5I_2 \longrightarrow 2IF_5 + 10AgI\downarrow$$

(*b*) I_2O_5 is treated with fluorine

$$2I_2O_5 + 15F_2 \longrightarrow 4IF_5 + 5O_2F_2$$

(*c*) By passing fluorine diluted three times with nitrogen over iodine in quartz vessel.

$$\underset{(N_2)}{I_2} + 5F_2 \longrightarrow 2IF_5$$

(*iv*) IF_7

(*a*) By the action of F_2 on IF_5 at 250°C

$$IF_5 + F_2 \longrightarrow IF_7$$

(*b*) By heating IF_5 at 500°C

$$7IF_5 \longrightarrow 5IF_4 + I_2$$

Structure of Interhalogen

(*a*) Their structures are of interest as some of them do not obey octet rule *e.g.*, IF_5, IF_7.

(*b*) All interhalogen neutral molecules are binary in nature. However, the tertiary compounds are very unstable and the compounds redistribute into binary compounds.

(*c*) The large iodine atom forms heptafluoride while bromine forms pentafluoride and chlorine the trifluorides. Thus, R_{large}/R_{small} ratio increases the number of small atoms per molecule.

(*d*) The general formula of interhalogens is represented by AX_n where n is always an odd number 1, 3, 5 or 7 which is further explained on the basis of hybridization,

(*i*) AX_n where $n = 1$ such as IF, ICl, IBr, ClF, BrF and BrCl

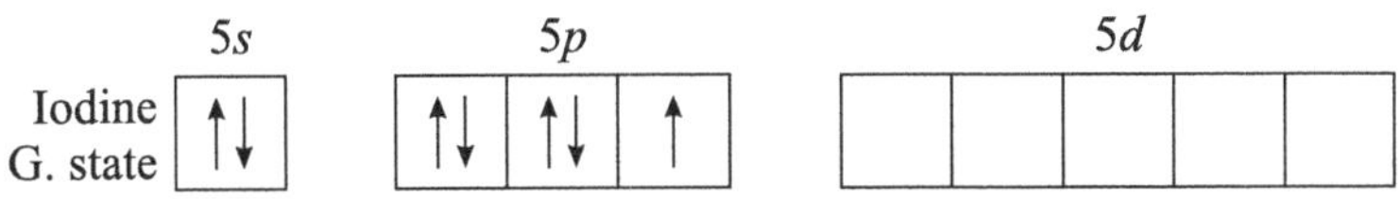

It is the presence of unparied electron in the iodine atom which encourages the formation of a covalent bond by pairing with another unpaired electron of the second halogen atom resulting into a linear structure. It is quite similar to iodine molecule formed from iodine atoms.

***(ii)* AX_3 type compounds:** This is exemplified by compounds such as IF_3, ICl_3, ClF_3 and BrF_3. The additional covalent bond are formed due to unpairing of one paired electron to a higher *d*-orbital. The result is three unpaired electrons and hence three covalent bonds.

Electronic Configurations Chlorine Atom

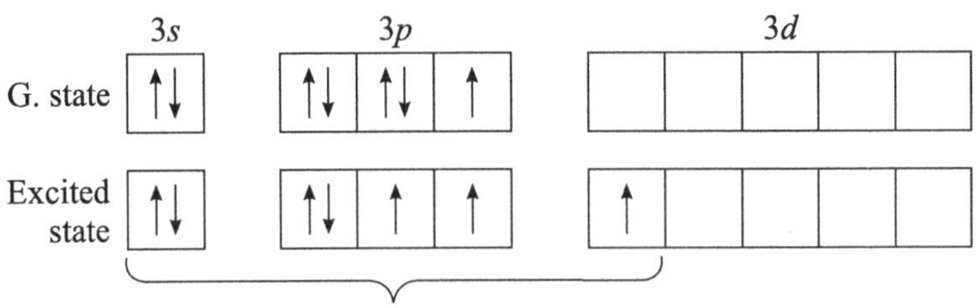

sp^3d hybridization suggesting trigonal bipyramidal structure with two apex positions occupied by lone pairs

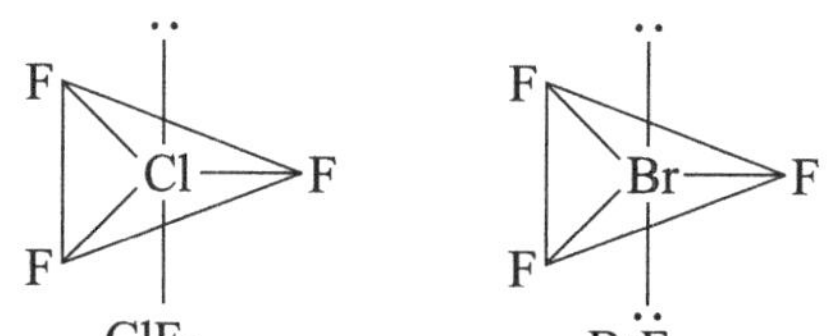

The three unpaired electrons in chlorine atom (excited form) form bonds with three F atoms (pairing electrons) assuming sp^3d hybridization with two positions occupied by lone pairs. However, ICl_3 is different as it exists as dimer unit I_2Cl_6 in the crystalline state.

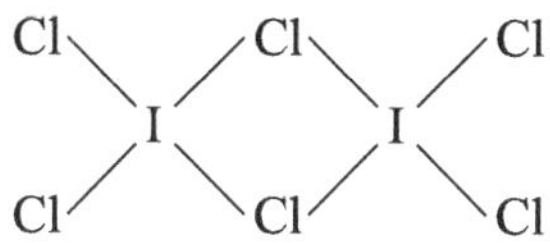

Planar I_2Cl_6

***(iii)* AX_5 type Compounds:** Here $n = 5$. It includes ClF_5, BrF_5 and IF_5. In there are more electron pair of Cl, Br and I in $3p$, $4p$ and $5p$ are promoted to the higher energy d-orbitals resulting in 5 d-orbitals having impaired electron in each.

IF_5

Iodine excited

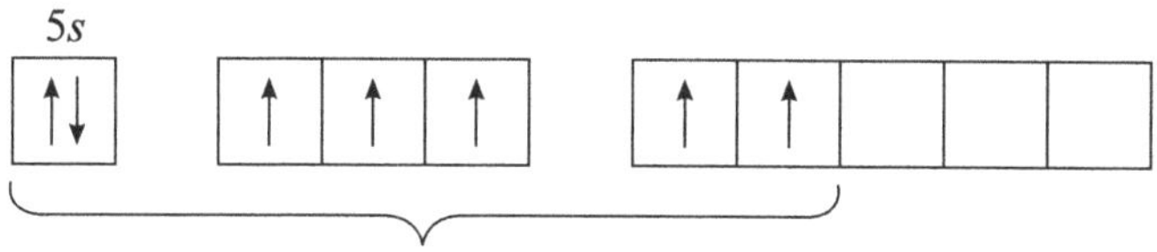

sp^3d^2 hybridization

sp^3d^2 hybridization points to octahedral structure with one position occupied by the lone pair.

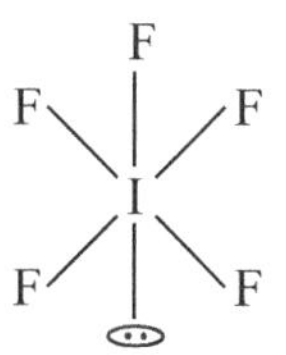

Five unparied electrons in the excited iodine atom form 5 covalent bonds with five F-atoms with one lone pair that remains. The geometry suggested is square bipyramidal or octahedral with four base positions occupied by the F-atoms.

***(iv)* AX_7 type Compounds:** AX_7 where $n = 7$ includes only IF_7. Here, one more electron pair in 5 s-orbital of iodine is promoted to the higher energy d-orbital leaving seven orbitals each containing an unpaired electron.

IF_7

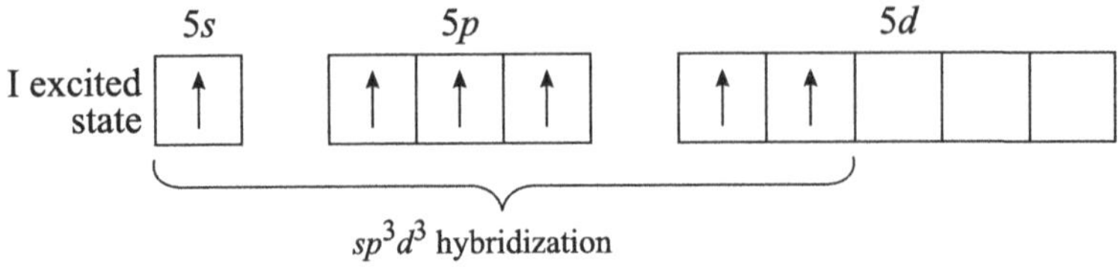

This gives us seven covalent bonds with seven F-atoms. sp^3d^3 hybridization suggests pentagonal bipyramidal structure.

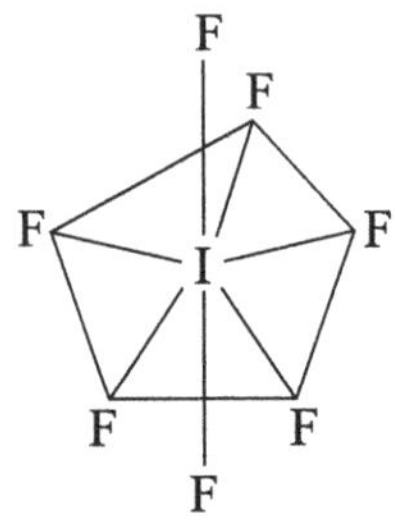

Fig. 15.2 IF_7 pentagonal bipyramidal

Electro Positive Iodine

The ionization energy decreases down the group and the atomic size increase likewise. Therefore though halogen have high eletronegativity yet due to large size there is a tendency to lose electron easily in the case of iodine atom. Thus iodine is able to acquire the electropositive character.

The ionization energy value of iodine is 1000 kJ mol^{-1} which is comparable to mercury metal which has IE equal to 1004 kJ mol^{-1}. Consequently iodine forms cationic compounds. A number of cationic compounds of iodine are given below:

***(i)* ICl:** Iodine monochloride is a good conductor of electricity showing its tendency to form I^+ ions.

(ii) Conductivity measurements of ICl revealed that ICl ionises in nitrobenzene or acetic anlydride as:

$$2ICl \longrightarrow I^+ + ICl_2^-$$

IBr reflects similar behaviour.

***(iii)* ICl_3**: Molten ICl_3 is a god conductor of electricity and ionises as:

$$2ICl_3 \longrightarrow ICl_2^+ + ICl_4^-$$

(iv) Large number iodine compounds have been characterised which contain I^+ that is stabilised by coordination.

$$[py - I - py]^+ X^-$$

Here *py* stands for pyridine and X stands for NO_3^-, ClO_4^-. Electrolysis of these compounds yields I at cathode.

(*v*) Hypoiodous acid (HOI): It is amphoteric and acts as a base and hence may be called iodine hydroxide. Compounds like ICl may be considered salt of this base.

(*vi*) Compounds *e.g.*, ($py - I - ONO_2$) have been prepared. Here iodine has a coordination number of two.

Iodine (III) Compounds

A good number of compounds having I^{3+} have been isolated. These compounds are ionic and contain oxoanion to stabilise. They are prepared by treating iodine with an oxidizing agent and the anlydrous acid.

Some Examples

(*a*) $I_2 \xrightarrow[CH_3COOH]{ClO_3^- \text{(oxid. agent)}} \underset{\text{iodine triacetate}}{I(OCOCH_3)_3}$

(*b*) $I_2 \xrightarrow[HClO_4]{O_3 \text{(oxid. agent)}} \underset{\text{iodine triperchlorate}}{I(ClO_4)_3}$

(*c*) $I_2 \xrightarrow[H_3PO_4]{\text{Fum. } HNO_3 \text{(oxid. agent)}} \underset{\text{iodine phosphate}}{IPO_4}$

There compounds contain I^{3+} cations. On electrolysis of iodine triacetate using Ag cathode, iodine is liberated at cathode which forms AgI.

$$I^{3+} + 3e + Ag \longrightarrow AgI$$

Cations of Iodine

There is considerable evidence of the presence of I_2^+, I_4^{2+} and I_3^+ cations of iodine under special conditions. Here iodine acts as an electrophile and can be stabilized in nucleophilic solvents. These can be isolated in solvents which are highly acidic *e.g.* H_2SO_4, $H_2S_2O_7$ and fluorosulphuric acid or melts KCl/$AlCl_3$ which are weakly basic, stable compounds of these have been isolated and their structures studied by X-ray crystallography.

Polyhalide Ions

Polyhalide ions are prepared by reacting halogens or inter halogens with halide ions. The most common example of polyhalide formation is the disolution of iodine in KI resulting in the formation of I_3^- ion.

$$KI + I_2 \longrightarrow K^+ + I_3^-$$

Important polyhalide ion have been tabulated below:

Table 15.14 Some important polydalide Ions

Cl_3^-, Br_3^-, I_3^-, ICl_2^-, IBr_2^-
ICl_4^-, BrF_4^-, I_5^-
IF_6^-, I_7^-

The reason for triiodite ion formation I_3^- is explained like this. When a negative iodide ion I^- comes closer to big iodine molecule whose outer electrons are far from the positive nucleus then the electronic effect induces a dipole within the iodine molecule. Thus, there exists a negative iodide ion and a positive pole of the large iodine molecule leading to the formation of I_3^- ion.

$$I_2 + I^- \longrightarrow I_3^-$$

Structure of Polyhalide Ion

X-ray studies show that structure of Br_3^-, I_3^- or ICl_2^- is linear and can be explained:

(*i*) In ICl_2^- the number of electrons involved are:

$$\begin{aligned} I &= 7 \\ Cl &= 2 \\ \text{–ve charge} &= \underline{1} \\ &\ \underline{10} \end{aligned}$$

This suggests sp^3d hybridisation which gives trigonal bipyramidal geometry with 3 pairs on non-bonded electrons around the iodine atom in one plane as shown.

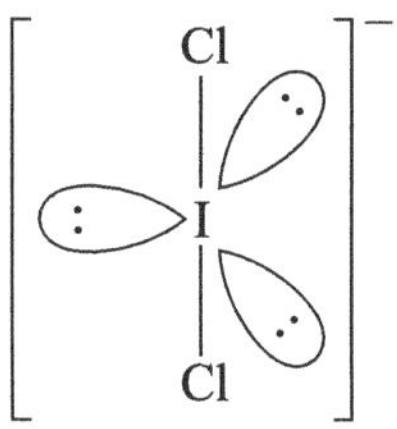

(*ii*) In ICl_4^- the number of electrons are

$$\begin{aligned} I &= 7 \\ \text{Negative charge} &= 1 \\ \text{4Cl atoms} &= \underline{4} \\ \text{Total} &= \underline{12 \text{ electrons}} \end{aligned}$$

This involves sp^3d^2 hybridization with six electron pairs round the iodine atom which are oriented towards the corners of an octahrdon. Iodine is placed at the centre of a square formed by 4Cl-atoms with two lone pairs or non-bonded electrons oriented perpendicular to the plane of the ion ICl_4^- is a planar structure as confirmed by X-ray studies of $KICl_4$.

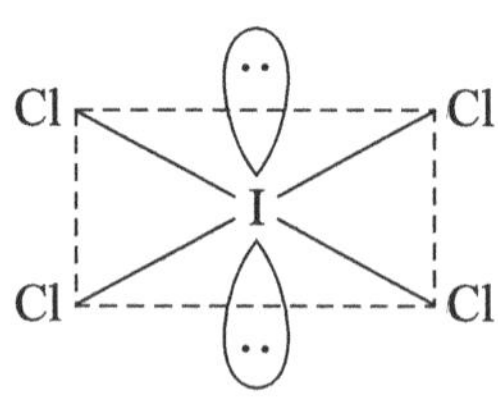

Structure ICl_4^-

Pseudohalogens and Pseudohaldies

Psuedo halogen has two electronegative atoms and its molecular form resembles the halogens. Psuedohalogens give anions which resemble halide ions in their properties. Some common pseudohalides ions and dimeric psuedohalogen species are listed in the table below:

Table 15.15

Pseudopalide	Formula	Pseudohalogen	Formula
Cyanide	CN^-	Cyanogen	$(CN)_2$
Cyanate	OCN^-	Oxoxyanogen	$(OCN)_2$
Thioeyanate	SCN^-	Thiocyanogen	$(SCN)_2$
Selenocyanate	$SeCN^-$	Selenocyanogen	$(SeCN)_2$
Tellurocyanate	$TeCN^-$	—	—
Azide	N_3^-	—	—
Azidothicarbonate	$SCSN_3^-$	Azidocarbondisuphideq	$(SCSN_3)_2$
Isocyanate	ONC^-	—	—

Some typical reactions of the psuedohalogens and pseudohalide ions are summarised below:

(*i*) The original psuedohalogens are volatile compounds. They are formed by the coming together of two radical X–X. They react with alkalies like halogens:

$$(CN)_2 + 2OH^- \longrightarrow CN^- + OCN^- + H_2O$$
$$Cl_2 + 2OH^- \longrightarrow Cl^- + OCl^- + H_2O$$

(*ii*) Inter-psuedohalogen and psueohalogen-halogen compounds are known *e.g* cyanogen bromide reacts with sodium azide to give NaBr and cynogen azide.

$$CNBr + NaN_3 \longrightarrow CNN_3 + NaBR$$

Cyanogen bromide in turn is obtained from Br_2 and HCN.

(*iii*) Psuedohalogens react with metals to give salts containing X^- ions *e.g.*, silver (*i*), mercury (*i*) and load (*ii*) salts correspond to AgCl, Hg_2Cl_2 and $PbCl_2$ in being insoluble in water.

(i*v*) Psuedohalide ions react with H_2 to give HX acids which are weak compared to the corresponding halogen acids.

HN_3 pKa = 4.4

HCN pKa = 8.9

HCl pKa = –7.4

(*v*) Psuedohalide form ions analogous to the polyhalide ions

(NH_4) $(SCN)_3$ and $K(SeCN)_3$ or KCl_3

(*vi*) Psuedohalide ion is oxidised to the parent psuedohalogen by a suitable oxidising agent.

$$2Fe^{3+} + 2SCN^- \longrightarrow 2Fe^{2+} + \underset{\text{Psuedohalogen}}{(SCN)_2}$$

Psuedohalogens

Preparation

1. Cyanogen $(CN)_2$ is obtained by heating AgCN or $Hg(CN)_2$ with $HgCl_2$

$$2AgCN \longrightarrow 2Ag + (CN)_2$$

$$Hg(CN)_2 + HgCl_2 \longrightarrow Hg_2Cl_2 + (CN)_2$$

2. Thio cyanogen $(SCN)_2$ is made by the action of Br_2 on Ag SCN.

$$2AgSCN + Br_2 \longrightarrow 2AgBr\downarrow + (SCN)_2$$

3. **Selenocyanogen (Se CN)$_2$:** It is prepared by displaceing Ag Se CN by I_2.

$$2Ag\ SeCN + I_2 \longrightarrow 2AgI + (Se\ CN)_2$$

4. Azido carbon disulphide $(S\ CSN_3)_2$ is obtained from $KSCSN_3$ by oxidation with H_2O_2.

$$CS_2 + KN_3 \longrightarrow KSCSN_3 \xrightarrow{H_2O_2} (SCSN_3)_2$$

Fluorocarbons

These compounds of fluorine are very amazing. Their importance is gained because of their interesting uses.

Fluorocarbon are obtained from hydrocarbon by substituting H-atom by fluorine *e.g.*,

$$CH_4 \longrightarrow CF_4 \text{ tetrafluoromethane}$$

Preparation

1. Tetrafluoromethane is prepared by the direct combination between carbon and gaseous fluorine

$$C + 2F_2 \longrightarrow CF_4$$

This reaction is dangerous because of the risk of explosion.

2. By the action of fluorine with hydrocarbons in the presence of an inert gas like N_2.

OR

By the action of fluorine, hydrocarbons in the presence of inert solvent CCl_4.

3. By the action of SbF_3 on chloro-derivatives.

$$CCl_4 + SbF_3 \longrightarrow \underset{\substack{\text{dischlorodifluoromethane} \\ \text{or Freon-12}}}{CCl_2F_2} + SbCl_3$$

4. Passing the vapours of the hydrocarbons over CoF_3 at 200 – 300°C

$$RH + CoF_3 \longrightarrow RF + CoF_2$$

where R is alkyl radical.

Properties

1. They are dense colouless liquids with slight unpleasant odour.
2. They have low surface tensions and refractive indices.
3. They have low boiling points quite comparable to inert gases. The boiling points and freezing points do not differ much from the parent hydrocarbons.
4. They have higher fluidities and densities.
5. They are stable to heat and non-inflammable.
6. They are resistant to oxidation.
7. They resist the action of most of reagents including alkalies up to quite a high temperature.

Uses

Fluorocarbons find wide application in industry due to non-inflammability, inertness and stability to heat.

The low reactivity is attributed to:

1. High C—F B.E. which is 485 kJ mol^{-1} while C—H and C—Cl bond energy 404 and 330.5 kJ mol^{-1} respectively.
2. Fluorocarbons are sparingly soluble in water due to low coordination to carbon and fluorine (no expansion of their octet).
3. Resistance to oxidation by oxygen. Here fluorine in C–F bond is a protector of carbon from oxygen.
4. Fluorine has a very samll size, therefore H is easily substituted by F atom. Substitution by other halogens is not difficult.

***(i)* Carbon tetrafluoride (CF_4):** It is used as a monometric liquid at low temperatures.

***(ii)* Dichlorodifluoromethane (CCl_2F_2):** It is used in refirgerators because of its low boiling point and low-corrosion.

***(iii)* Chlorofluorocarbons:** They polymerise to give stable oils, waxes and greases. They are used as lubricants known as fluoroubes. They find application as heat transfer media and dielectric media.

(*iv*) Tetrafluoroethylene (C_2F_2): They quickly polymerise to form polytetra fluoro ethlyene $(C_2F_4)_n$ known as Teflon.

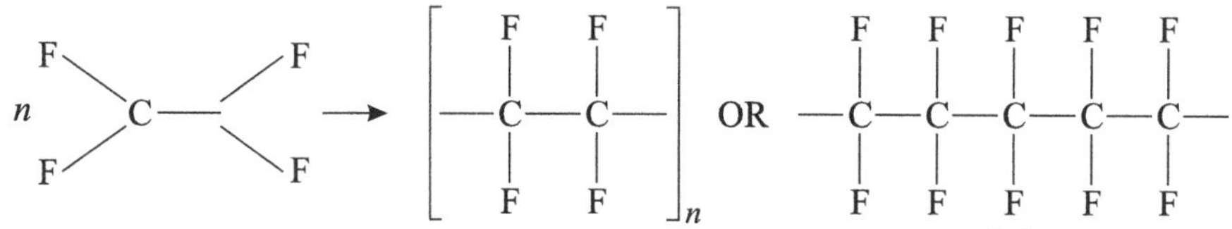

Teflon

Teflon is chemically inert so much so even aqua regia cannot affect it. It is extensively used as a non-stick coating to protect the metals against corrosion. It is used in handling of fluorine gas. It sticks to nothing and nothing sticks to it.

EXERCISES

1. Give the method of preparation of fluorine. Indicate the precautions necessary for it. What is the role of aqueous or dry HF in the electrolysis of aqueous NaF.
2. Why fluorinating agents are preferred over F_2. Given equations to show their use.
3. What are uses of F_2. Given equations on reactions of HF and (*i*) SiO_2 (*ii*) KF (*iii*) CCl_4 (*iv*) Graphite and uranium.
4. Enlist the salts used as sources of chlorine, bromine and iodine. How do these element extracted from these salts?
5. How do we prepare halogen acids HF, HCl, HBr, and HI in aqueous solutions? Explain why HF is weaker than HI.
6. Give the main uses of Cl_2.HCl(g) can be obtained from NaCl and H_2SO_4 but why HBr(g) and HI(g) are not prepared in a similarly. Explain.
7. Give equations for the reactions of Cl_2 with:

   (*a*) Hydrogen (*b*) CO_2 (*c*) P (*d*) S

   (*e*) SO_2 (*f*) Br^-(aq.) (*g*) NaOH.
8. What happens when aqueous silver nitrate is added to the solution of NaF, NaCl, NaBr and NaI? What changes will take place when ammonia is added to each of these mixtures.
9. Discuss the uniqueness of fluorine bond with other halogens.
10. Draw the structures of the following:

    (*i*) OF_2 (*ii*) Cl_2O (*iii*) O_2F_2 (*iv*) I_2O_5

    Why the bond angle in OF_2 is different from Cl_2O.
11. (*a*) Describe the analytical use of I_2O_5.

    (*b*) Why O—F bond is longer in O_2F_2 than in OF_2 and the O—O bond in O_2F_2 is shorter than H_2O_2.
12. Describe the preparation of the following compounds:

    (*i*) NaOCl (*ii*) $NaClO_2$ (*iii*) $NaClO_3$ (*iv*) HIO_4. Give at least one use of each of these.
13. Give four different types of oxoacids of halogens. Write the formula of acid in its stable state.
14. I_2 is insole in water but when KI is added to it, it dissolves quickly. Give reasons.
15. What is the shape of I_3^- ion? Why solid CS_3 is stable but solid NaI_3 is not?
16. What are inter halogens? Draw the shapes of the following molecules and ions:

(*a*) ClF (*b*) BrF_3 (*c*) IF_5^- (*d*) IF_7
(*e*) I_3^- (*f*) ICl_4 and (*g*) I_5^-.

17. (*a*) Draw the structures and explain the hybridizations of:
    (*i*) ICl_2^- (*ii*) IF_7
    (*b*) Why inter halogens are more reactive than halogens.
18. Give the preparation of Cl_2O_7 and draw its structure.

16

Noble Gases or Zero Group Elements

He, Ne, Ar, Kr, Xe and Rn constitute noble gases or zero group. Prior to the name noble gases they were named inert gases and rare gases. Both the names are a misnomer as these gases are found to be selective in their reactions. Inertness was attributed to the stable octet in their outermost shell except in helium. Stablest configuration of the elements is not the only factor for their inert behaviour. The other factors for their reactivity are ionization energies and electron affinity. Radon is a radioactive element. The elements have zero electron affinity and very high ionization energies. As we move down the group the atomic radii increases and the ionization energy becomes attainable so much so Kr and Xe show chemical reactivity and are known to form compounds.

Their electronic configurations are shown in the Table 16.1

Table 16.1 Electronic Structures

Element	Symbol		Electronics structure
Helium He	He		$1s^2$
Neon	Ne	[He]	$2s^2\ 2p^6$
Argon	Ar	[Ne]	$2s^2\ 2p^6$
Krypton	Kr	[Ar]	$3s^2\ 3p^6$
Xenon	Xe	[Kr]	$4d^{10}\ 5s^2\ 5p^6$
Radon	Rn	[Xe]	$4f^{14}\ 5d^{10}\ 6s^2\ 6p^6$

All there elements show a closed shell ns^2np^6 configuration in the outermost shell.

The first chemical compound of the noble gases was reported by N. Bartlett in 1962. The preparation of the xenon compound Xe PtF_6 was based on the chance discovery of the compound between O_2 and PtF_6.

$$O_{2(g)} + PtF_{6(g)} \longrightarrow \underset{\text{Dioxygen hexa fluorollatinate (V)}}{O_2PtF_6}$$

The compound is formulated as $O_2^+(PtF_6)^-$. PtF_6 is a very strong oxidizing agent and could even attract electron from the molecular oxygen. The ionization energy of the O_2 gas is 1165.6 kJ mol^{-1} which is quite comparable to xenon gas which is 1170 kJ mol^{-1}. The molecular dimensions of O_2 and Xe are quite similar Xe = 4Å; O_2 = 4Å. This encouraged Bartlett to proceed ahead with the reaction of Xe(g) and PtF_6(g) at room temperature to obtain orange yellow crystalline compound of $XePtF_6$ as predicted.

$$Xe(g) + PtF_{6(g)} \longrightarrow \underset{\text{(Solid)}}{Xe^+ \mid PtF_6 \mid^-}$$

All the noble gas ordinarily neither lose or gain electron therefore they exist as monoatomic gases at STP.

Occurrence and Discovery of Noble Gases

He, Ne, Ar, Kr and Xe occur in the atmosphere. They were first obtained by Cavendish in 1784. Ar is quite abundant and can be recovered by fractional distillation of liquid air. In the air Ar exists because of the reaction:

$$^{40}_{19}K \longrightarrow {}^{40}_{18}Ar + \underset{\beta\text{-decay}}{{}^{0}_{-1}e}$$

In the nuclear reaction solar energy transform hydrogen to deuterium and then form helium. Xenon and krypton can also be obtained by fractional distillation of liquid air.

In 1901 Dorn discovered Radon by the nuclear reaction.

Flow diagram

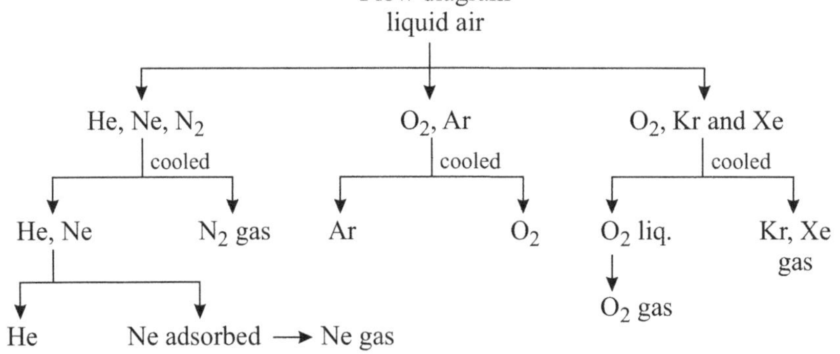

General Properties of Noble Gases

1. **Atomicity:** All the noble gases are monoatomic the cp/cr value is 1.66 which is theoretical value for monoatomic molecules.
2. All the gases are colourless, odourless and tasteless.
3. **Solubility:** The solubility in water increases with increase in their atomic size from He to Xe. He is the least soluble. Increase in size increases the dipole character of water.

4. **Liquefaction:** Because of the completely filled outermost orbital of noble gases they exhibit weak van der woals force between the molecules hence it is difficult to liquify them but the gas with larger size like Xe can be liquified.
5. **Atomic Radius:** Due to the completely filled valence shell inter-electronic repulsion is more which increases the atomic radius from He to Xe. The atomic radius of noble gases is greater than the halogens.
6. **Reactivity:** In general they are unreactive but due to the large size of Xe the ionization potential is attainable.
7. **Melting and boiling points:** They are relatively low because of weak van der woals forces between the molecules.
8. **Ionization potential:** Due to the completely filled valence orbitals the ionization potentials are higher but they decrease down the group due to the increase in size.

$$He > Ne > Ar > Kr > Xe > Rn$$

9. **Electron affinities:** The electron affinities of all the noble gases is almost zero due to saturated valence shells.

Spectral Properties

He which is a core of sun was established by green and yellow lines in the absorption spectra while Neon consists of orange and red lines.

Uncommon Properties of Helium

1. Below 2.1 K new form of He known as He (11) is obtained while above 2.1 K it is He (1).
2. Viscosity of He is $\frac{1}{100}$th of hydrogen gas.
3. Thermal capacity of He is 800 times that of copper at room temperature.
4. He solidifies at 0.9 K and 25 atmosphere pressure.
5. At absolute zero He acquires a superliquid character.

Chemical Properties

For quite a long time it was believed that the noble gases are chemically inert. There was some evidence that some molecular ions are formed in the discharge tubes and the formation of clathrate compounds. The molecular ions observed in the discharge tube are He_2^+, HeH^+, HeH^{2+} and Ar_2^+. They are formed only momentarily.

Neutral molecules such as He_2 are unstable *e.g.,*

$$He_2 \longrightarrow He + He$$

According to the molecular orbital theory the bond order of He_2 is determined by

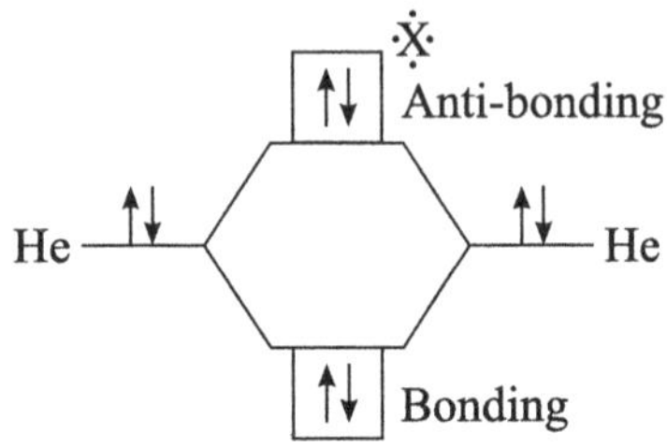

$$\text{B.O.} = \frac{1}{2}\ \text{(Bonding-antibonding)}$$

i.e., $$= \frac{1}{2}(2-2) = \text{O}$$

The value of the bond order shows that no bond like He – He is possible or He is a monoatomic gas or molecule. The same applies to the other noble gases.

Clathrate Compounds

Common compounds have covalent and ionic compounds. Clathrate compounds are those in which atoms or molecules of appropriate size are trapped in cavities of the other compounds. They are also called cage compounds. Normally the gases are trapped but there exists no chemical bond. It is said to be guest and host arrangement represented by *n*HGm where *n* is number of host molecules and *m* is the number of guest molecules. Ar, Kr and Xe are known to form clathrates in aqueous solution of quinol at 10 to 40 atmospheres. The gases are trapped in cavities of 4.0Å diameter in β-quinol structure.

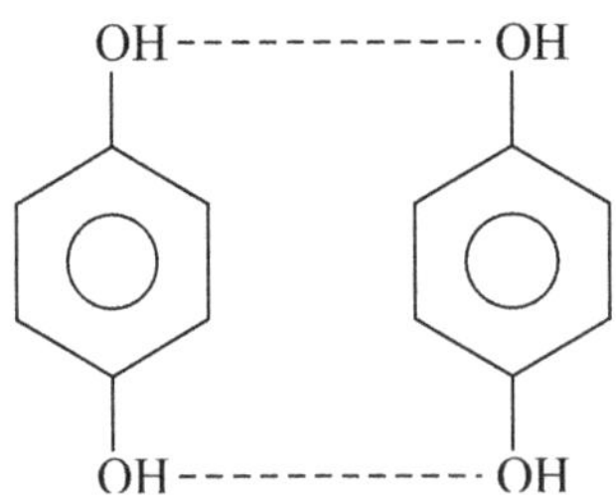

Hydrogen bonded quinol molecules

β-Quinol

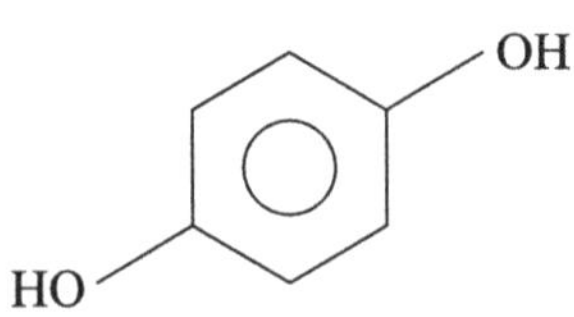

When a clathrate is dissolved the hydrogen bonded β-quinol structure breaks down and the trapped noble gas escapes. Other molecules that are known to form clathrates are O_2, SO_2, H_2S, MeCN and CH_3OH. Smaller molecules such as He and Ne cease to form clathrates as they are small enough to escape out of the cavities. The composition of clathrates is Quinol 3 : 1 trapped molecules. Normally all the cavities are not occupied.

Ar, Kr and Xe can be trapped in the frozen water under high pressure. These clathrates are also called noble gas hydrates. They have the composition of $6H_2O$: 1 gas atom or molecule. Heavier noble gases can be trapped in zeolites in which noble gas atom is 20% by weight, clathrates are useful in storing radioactive isotopes of Kr and Xe produced in nuclear reactors.

Reactivity of Xenon

Soon after the preparation of Xe and PtF_6 compound *e.g.*, $Xe^{\oplus}[PtF_6]^{\ominus}$ orange red crystals it was found that Xe and F_2 reacted at 400°C to give colourless volatile solid XeF_4. This is considered iso structural with $[ICl_4]^{\ominus}$. After this discovery there was a rapid surge in the chemistry of noble gases particularly Xe compounds.

The ioninzation energy values of He, Ne and Ar are much higher than Xe therefore formation of similar compounds is not feasible. The IE value of Kr is slightly higher than that of Xe therefore Kr does form KrF_2. Rn is radioactive but due to low ionization energy as compared to Xe, may form compounds similar to that of Xe. But Rn has all radioactive isotopes. Only RnF_2 and a few complexes are known.

Xenon Compounds

Xenon forms compounds with the most electronegative fluorine and oxygen. These electro negative atoms spread xenon to the +2 and +8 oxidation state compounds. Xenon directly reacts with F_2 while its oxygen compounds are derived from the fluorides. $XeCl_2$ and $XeCl_4$ find some mention in the directory of xenon compounds. The important compounds of xenon are given in Table 16.2.

Table 16.2 Structure of Xe compounds

Formla	Name	Oxidation state	m.p.(°C)	Structure
XeF_2	Xenon difluoride	(+II)	129	linear (RnF_2 and $XeCl_2$ are similar)
XeF_4	Xenon tetrafluoride	(+IV)	117	Square planar ($XeCl_4$ is similar)
XeF_6	Xenon hexafluoride	(+VI)	49.6	distorted octahedron
XeO_3	Xenon trioxide	(+VI)	explodes	pyramidal (tetrahedral with one corner unoccupied)
XeO_2F_2		(+VI)	30.8	Trigonal bipyramid (with one position unoccupied)
$XeOF_4$		(+VI)	–46	Square pyramidal (octahedral with one position unoccupied)
XeO_4	Xenon tetroxide	(+VIII)	–35.9	Tetrahedral
XeO_3F_2		(+VIII)	–54.1	Trigonal bipyramid
$Ba_2[XeO_6]^{4-}$	barium perxenate	(+VIII)	dec. > 3000	Octahedral

Xe and fluorine directly react when the two gases are heated at 400°C in a nickel vessel giving xenon fluorides depending upon F_2 and Xe ratios.

Fluorides of Xenon

(*i*) Xenon difluoride XeF_2

Preparation:

1. It can be prepared in the laboratory by heating xenon and fluorine gases at 400°C in the ratio of 2 : 1 *i.e.,* excess of xenon otherwise XeF_4 may form. XeF_2 is isolated by vacuum sublimation.

$$\underset{2\ :\ 1}{Xe + F_2} \xrightarrow[\text{1 atom}]{400°C} XeF_2$$

2. It may be prepared by exposing xenon and fluorine to sunlight-taken in a dry glass vessel at room temperature.
3. It may also be prepared by exposing to sunlight mixture of xenon and oxygen difluoride (OF_2) or by taking Xe and O_2F_2 at –118°C.

Physical Characteristics: It is a white crystalline solid which can be sublimed. Its melting point is 140°C.

Chemical Characteristics

Xe–F bond has lowest bond energy among the known florides. Due to this XeF_2 is an oxidizing agent as well as a mild fluorinating agent.

1. Oxidising agent: Its aqueous solution is a powerful oxidizing agent.

$$2HCl + XeF_2 \longrightarrow Xe + Cl_2 + 2HF$$

It can oxidise Co^{2+} to Co^{3+}, Cr^{3+} to Cr^{6+}, Ag^{1+} to Ag^{2+} and BrO_3^- to BrO_4^-.

2. **Substitution reactions:** XeF_2 can be substituted by highly electronegative ligands like HSO_3F, $HClO_4$, CF_3COOH etc. Similar reaction with HCl or BCl_3 have failed.

$$F - Xe - F + HSO_3F \longrightarrow F - Xe - SO_3F + HF$$

$$F - Xe - F + HSO_3F \longrightarrow FSO_3 - Xe - SO_3F + HF$$

3. **It oxidizes I_2 to IF:**

$$I_2 + XeF_2 \xrightarrow{BF_3} Xe\uparrow + 2I^+F^-$$

$$IF + XeF_2 \xrightarrow{BF_3} Xe\uparrow + IF_3$$

4. **XeF_2 reacts with sulphur trioxide to give:**

$$XeF_2 + 2SO_3 \longrightarrow Xe\uparrow + \underset{\text{Fluoride of disulpher hexaoxide}}{S_2O_6F_2}$$

5. **H_2 reduces XeF_2 to xenon**

$$XeF_2 + H_2 \longrightarrow Xe\uparrow + 2HF$$

6. **XeF_2 acts as a fluorinating agent**

$$2NO + XeF_2 \longrightarrow 2NOF + Xe\uparrow$$

Above reaction includes fluorination and oxidation both. In the first step ionization of XeF_2 to XeF_2^+ takes place followed by the electron transfer to give $Xe–F^-$ which is very weakly bonded and hence a potential source of fluorination.

7. It reacts with Lewis acids SbF_5 or AsF_5 to form adducts.

$$XeF_2 + SbF_5 \longrightarrow XeF_2.SbF_5$$

$$XeF_2 + AsF_5 \longrightarrow XeF_2.AsF_5$$

It has been established that these adducts comprise of $XeF^+.AsF_6^-$ and $XeF^+.SbF_6^-$

Likewise $XeF_2 + 2SbF_5 \longrightarrow XeF_2.2SbF_5$ that is $XeF^+.SbF_{11}^-$. Here XeF^+ ion is strongly bonded to $Sb_2F_{11}^-$ through a fluorine bridge.

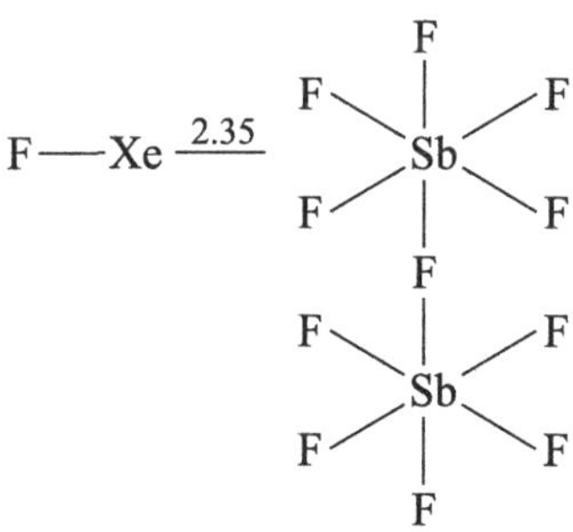

XeF_2 and AsF_5 react in a different stoichiometry

$$2XeF_2 + AsF \longrightarrow 2XeF_2.AsF_5 \longrightarrow [Xe_2F_3]^+[AsF_6]^-$$

The ion $Xe_2F_3^+$ has the linear geometry around xenon and angular geometry at the central fluorine atom

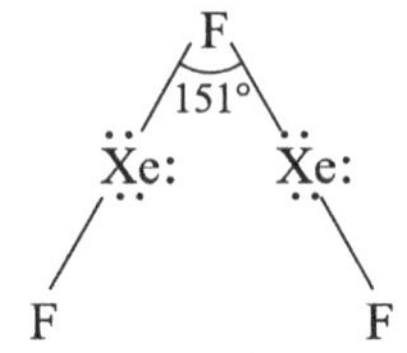

$|Xe_2F_3|^+$ ion

XeF_2 dissolves in hydrofluoric acid without any reaction and without any change in conductivity.

Structure

1. Infrared and Raman spectra show that XeF_2 has a linear structure in the vapour state.
2. Solid crystal structure of XeF_2 shows linear units.

Bonding

Xenon difluoride may be represented by Lewis electron dot formula as:

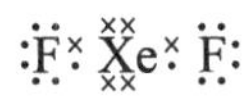

The actual structure is a resonance hybrid of the following structures.

Valence Bond Theory

According to this theory the valence shell electrons of xenon are excited and promoted to the high energy 5*d* vacant orbitals.

e.g.,

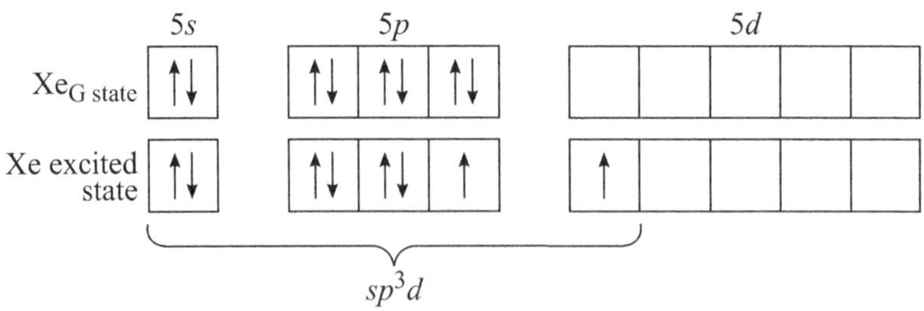

Here, the electron is promoted from the 5*p* level of Xe to the empty 5*d* level. Thus, the two unpaired electrons form bonds with fluorine atoms. The five electron pairs point to the corners of a trigonal bipyramid. Out of these, three lone pairs occupy the equatorial positions and the remaining two bond Pairs occupy the optical positions. The VBT predicts two strong covalent bonds. The geometry predicted is a linear geometry. The three lone pairs are farther removed due to repulsion.

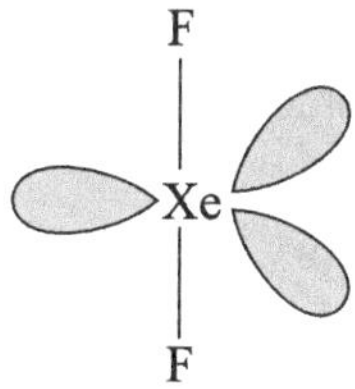

Fig. 16.1

Objections to VBT

1. It is difficult for the 5*p* electrons to jump to high energy 5*d* orbitals. What is the source of supply of such a high energy? How such a high energy is compensated?
2. The formation of two strong covalent bonds for F – Xe – F is improbable because the overlap of orbitals is most suited when the orbitals have similar energies 5*d* orbitals in Xe seem to have too high energy to participate in the proposed sp^3d hybridization.

Rundle offered a simple molecular orbital approach to take care of these objections.

Molecular Orbital Model

XeF_2 compound has three atomic orbitals or it forms a three centre bond. According to the molecular orbital three atomic orbitals give as many molecular orbitals. The three molecular orbitals. Each molecular orbital is represented by two touching circles. Over all there are four electrons and three molecular orbitals.

Molecular Orbitals

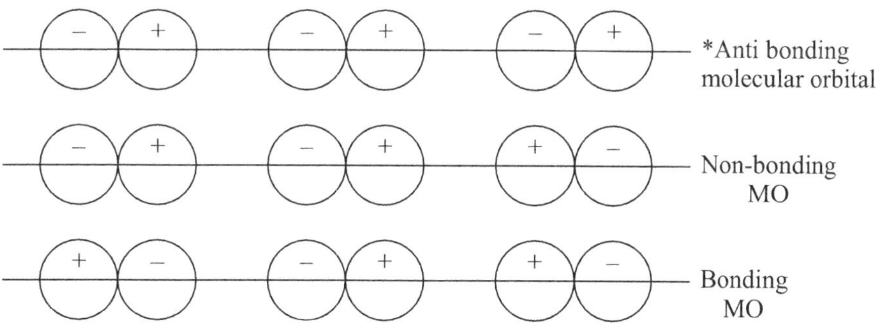

The four electrons are because of 2 $5p_z$ electrons of Xe and 2 $2p_z$ electrons of two fluorine atoms. The order of energy of molecular orbitals is:

Bonding MO < non-bonding MO < antibonding MO.

Due to strongly electro negative nature of fluorine atoms the bonds may acquire some ionic character. The bonding may be described as three centre four electrons σ bonding. Because of the linear structure of XeF_2(F – Xe – F) best overlap of orbitals is expected. The bonds are akin to the bonds described for B_2H_6.

In the molecular orbital theory four electrons are to be distributed among three molecular orbitals energywise as shown:

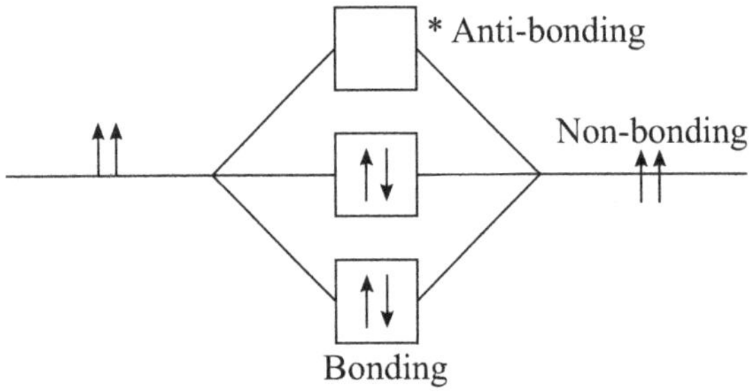

Bond order according to the MO theory is

$$B.O. = \frac{1}{2}(2 - 0) = 1$$

Sigma bond is spread over the F – Xe – F system which gives a bond order of 0.5 for each Xe – F bond. This accounts for the existence of weak Xe – F bonds in XeF_2.

The objections to the valency bond theory get automatically removed as:

1. There is no question of promotion of electrons to the higher energy *5d* orbitals from *5p* orbitals.
2. The two covalent bonds formed F – Xe – F are weak and therefore they can be substituted by the other more electronegative ions.

(*ii*) Xenon (*iv*) Compounds

Xenon Tetrafluoride XeF_4 Prepared

1. It is prepared by heating the mixture of Xe and F_2 in the ratio of 1 : 5 at 400°C at 5-6 atmospheres in a nickel crucible

$$Xe + 2F_2 \xrightarrow[6\text{ atom}]{400°C} XeF_4$$

   It is purified by fractionation.

2. Another method of preparing XeF_4 is by electric discharge in a vessel at –80°C containing Xe and F_2 in molar ratio of 1 : 2 at 2 – 15 atmospheres.

Properties

Physical: It is colourless crystalline solid. It has a melting point of 117.1°C and can be easily sublimed.

Chemical

1. Reduced by H_2 gas at 130°C
$$XeF_4 + 2H_2 \longrightarrow Xe + 4HF$$
2. ***Fluorinating agent:*** It is a stronger fluorinating agent than XeF_2.
$$Pt + XeF_4 \longrightarrow Xe + PtF_4$$
$$2SF_4 + XeF_4 \longrightarrow Xe + 2SF_6$$
3. ***Reduced by mercury***
$$XeF_4 + 2Hg \longrightarrow Xe + 2HgF_2$$
4. ***With*** $\mathbf{BCl_3}$ ***it gives*** $\mathbf{BF_3}$
$$3XeF_4 + 4BCl_3 \longrightarrow 4BF_3 + 3Xe + 6Cl_2$$
5. ***Addition Compounds:*** It forms addition compounds with SbF_5.
$$XeF_4 + SbF_5 \longrightarrow XeF_4.SbF_5 \longrightarrow [XeF_3]^+ [SbF_6]^-$$
$$XeF_4 + 2SbF_5 \longrightarrow XeF_4.2SbF_5 \longrightarrow |XeF_3|^+ [Sb_2F_{11}]^-$$
However, XeF_3^+ ions are represented by T-shaped structure.
6. XeF_4 can be hydrolysed in a controlled manner to yield different products.
$$XeF_4 + H_2O \xrightarrow{-80^\circ C} \underset{\text{Xenon oxydifluoride}}{XeOF_2} + 2HF$$
$$XeF_4 + H_2O \longrightarrow \underset{\text{Xenon trioxide}}{XeO_3}$$
7. XeF_4 forms 1 : 1 stable adducts with strong lewis acid *e.g.*, $Me_4N^+F^-$ (tetra methyl ammonium fluoride) and alkali metal fluorides.
$$XeF_4 + M^+F^- \longrightarrow M^+(XeF_5)^-$$
M^+ may be Na^+, K^+, Pb^+, Cs^+, Me_4^+

XeF_5^- anion shows the unusual pentagonal planar structure where axial positions are occupied by lone pairs.

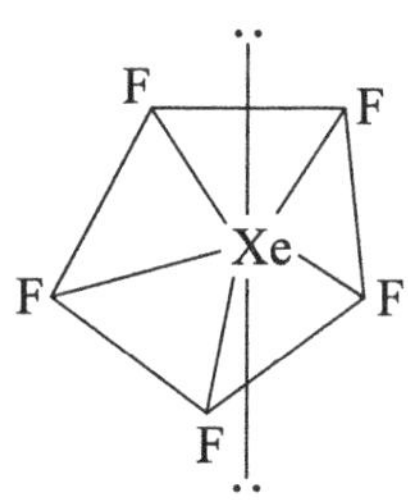

Structure of XeF_5^- anion

Bonding and Structure

1. Spectroscopic and electron diffraction studies have proposed square planar structure of XeF_4.

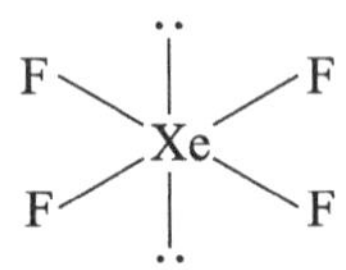

2. **VBT:** Valence bond representation of XeF_4 can be explained in which two paired electrons from the $5p$ sub-level get promoted to the higher energy $5d$ orbitals.

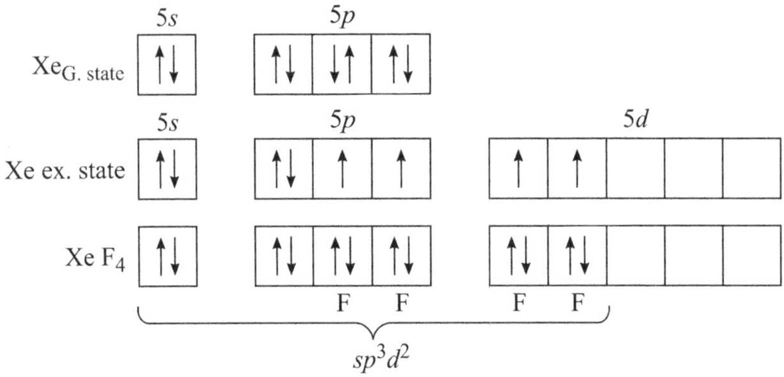

This suggests sp^3d^2 hybridization of orbitals. The four singly occupied hybridised orbitals use one unpaired electron each for bond formation. The remaining two hybridired orbitals contain lone pairs. The overall geometry of the molecule emerges as octahedral geometry with two trans positions occupied by the lone pairs.

Molecular Orbital Theory

Just like in XeF_2 where three p atomic orbitals *i.e.,* one full $5p_z$ atomic orbital and two F $2p_z$ orbitals (one electron each) combine resulting in three centred orbitals *i.e* F – Xe – F. As per molecular orbital theory a total of four electrons are filled *i.e* two in bonding MO and two in the non-bonding MO giving rise to F – Xe – F with two single bonds.

$$\text{Bond order} = \frac{1}{2}(B - A) = \frac{1}{2}(2 - 0) = 1$$

This suggests that each Xe – F bond contains bond order of 0.5 each. Another $5p$ orbital of xenon at right angle to the first can form a similar set of three centre four electron bond F – Xe – F forming XeF_4.

(*iii*) Xenon Hexa Fluoride XeF_6

Preparation

1. It is prepared by the direct combination of Xe and F_2 in the ratio of 1 : 20 at 300°C and 60 atmospheres.

$$\underset{1}{Xe} + \underset{:\ 20}{3F_2} \xrightarrow[60\,atm.]{300^\circ C} XeF_6$$

2. It can also be prepared by the fluorination of XeF_4 by O_2F_2 below – 80°C.

$$XeF_4 + O_2F_2 \xrightarrow{-80^\circ C} XeF_6 + O_2$$

3. Alternatively F_2 three parts and xenon 1 part are passed through electric discharge at low temperature.

Physical Characteristics

It is a colourless solid but the vapours are greenish-yellow. It melts at 49.5°C and hydrolyses rapidly. It acts as a stronger oxidising and fluorinating agent as compared to XeF_2 and XeF_4.

Chemical Characteristics

1. It can be reduced to Xe by H_2, HCl and NH_3

$$XeF_6 + 3H_2 \longrightarrow Xe + 6HF$$
$$XeF_6 + 6HCl \longrightarrow Xe + 3Cl_2 + 6HF$$
$$XeF_6 + 8NH_3 \longrightarrow Xe + 6NH_4F + N_2$$

2. It cannot be stored in a glass or quartz vessel due to following reactions:

$$2XeF_6 + SiO_2 \longrightarrow 2XeOF_4 + SiF_4$$
$$2XeOF_4 + SiO_2 \longrightarrow 2XeO_2F_2 + SiF_4$$
$$2XeO_2F_2 + SiO_2 \longrightarrow 2XeO_3 + SiF_4$$

3. **XeF_6 reacts with water**

$$XeF_6 + 3H_2O \longrightarrow XeO_3 + 6HF$$

XeO_3 is highly explosive.

4. **Reacts with fluoride ion acceptor**

$$XeF_6 + PtF_5 \longrightarrow XeF_6.PtF_5 \longrightarrow |XeF_5|^+|PtF_6|^-$$
$$XeF_6 + AsF_5 \longrightarrow XeF_6.AsF_5 \longrightarrow XeF_5^+AsF_6^-$$
$$XeF_6 + SbF_5 \longrightarrow XeF_6.SbF_5 \longrightarrow XeF_5^+SbF_6^-$$
$$XeF_6 + BF_3 \longrightarrow XeF_6.BF_3. \longrightarrow XeF_5^+BF_4^-$$

XeF_5^+ formed in the reactions shows square pyramid structure.

5. **Reacts with fluoride ion donors**

$$C_sF + XeF_6 \longrightarrow CsF.XeF_6 \xrightarrow{+CsF} Cs_2XeF_8$$

or

$$2Cs^+[XeF_8]^{2-}$$

$$2NOF + XeF_6 \longrightarrow |NO|_2^+|XeF_8|^{2-}$$

Bonding and Structure

1. In the solid state the crystal of XeF_6 contains $[XeF_5^+]F^-$ in which XeF_5^+ cations represent a square pyramid structure bridged by two free F^- ions. Determination of structure of XeF_6 gaseous molecule is a little difficult. The structure established is not a regular perfect octahedron because the two fluorines are distorted in the octahedral arrangement owing to the non-bonding electron pair which occupies a position in the centre of an octahedron and forces apart two fluorines at the corner of this face.

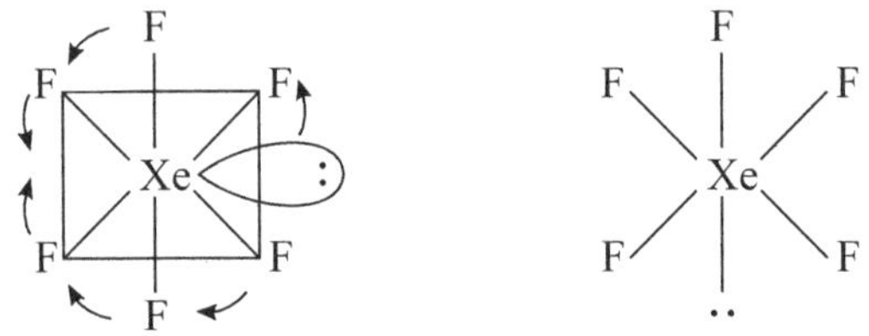

2. **Valence Bond Theory:** According to the theory the three paired electron in 5*p* sub-level of Xe get promoted to the high energy 5*d* sub-level

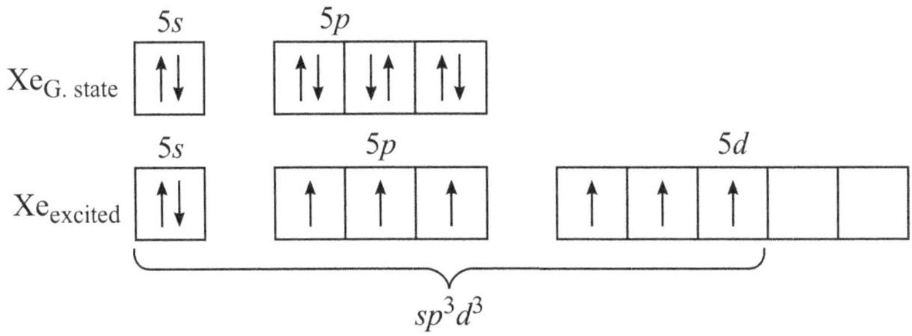

One 5*s*, three 5*p* and three 5*d* orbitals produce sp^3d^3 hybrid orbitals suggesting pentagonal bipyramidal structure. The six unpaired electrons form six sigma bonds with six fluorine atoms. The seventh orbital contains a lone pair that causes distortion and projects an irregular geometry.

3. **Molecular Orbital Theory:** In continuation to XeF_2 and XeF_4 the third 5*p* orbital of xenon (each *p*-orbital at right angles to one another) unpairs and may form a third set of three centre bond in the case of XeF_6. The shape of *p*-orbitals predicts that XeF_2 is linear, XeF_4 square planer and XeF_6 octahedral. The octahedral XeF_6 structure gets distorted due to the presence of a lone pair from one face to another. The F – Xe – F angle gets distorted. In reality neither valence bond nor molecular orbital theory predicts the actual structure of XeF_6. However, valence bond theory gives a clearer picture compared to molecular orbital theory.

(*iv*) Xenon Octafluoride XeF_8

There exist some reports about the existence of XeF_8 but yet the existence of XeF_8 has not been confirmed. Coulsion suggested the following shape of XeF_8 in which two fluorine atoms in XeF_6 are replaced by fluorine molecules (F_2).

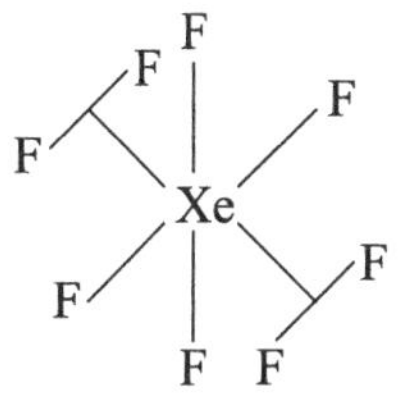

There exists no evidence of the above structure. It is only a conjecture.

Fluoride of Krypton

(*i*) Krypton difluoride KrF_2

Preparation

It is prepared by irradiating the mixture of Kr and F_2 at low temperature (–193°C). Irradiation may be done by γ-rays, UV light or electric discharge.

$$Kr + F_2 \xrightarrow[\text{discharge} - 193°C]{\text{electric}} KrF_2$$

Here, fluorine acts as an oxidising agent *i.e* Kr° changes to Kr^{2+}.

Physical Characteristics

KrF_2 is a colourless solid. It is a powerful oxidising and fluorinating agent.

Chemical Characteristics

1. It is decomposed by water

$$KrF_2 + H_2O \longrightarrow Kr + \frac{1}{2}O_2 + 2HF$$

2. Forms adducts with SbF_5

$$KrF_2 + SbF_5 \longrightarrow KrF_2.SbF_5 \longrightarrow KrF^+SbF_6^-$$
$$KrF_2 + 2SbF_5 \longrightarrow KrF_2.2SbF_5 \longrightarrow KrF^+Sb_2F_{11}^-$$
$$2KrF_2 + SbF_5 \longrightarrow 2KrF_2.SbF_5 \longrightarrow Kr_2F_3^+.SbF_6^-$$

The structures of there compounds are analogous to that of xenon compounds.

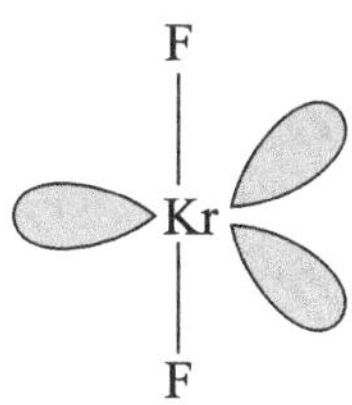

Just like XeF_2 the structure of KrF_2 molecule is linear.

Radon Compounds

Radon has the lowest ionization energy (10.70 eV) out of noble gases. It is expected to be most reactive but the radioactive nature of the element prevents further studies.

Xenon Oxyfluorides

(*i*) Xenon Oxytetrafluoride $XeOF_4$

Preparation

1. It can be prepared by the hydrolysis of XeF_6.

$$XeF_6 + H_2O \longrightarrow XeOF_4 + 2HF$$

2. It may be prepared by the reaction of SiO_2 on XeF_6

$$2XeF_6 + SiO_2 \longrightarrow 2XeOF_4 + SiF_4$$

3. By heating a mixture of Xe(1 part) + fluorine (4 parts) and a large excess of oxygen at 230°C

$$Xe + 2F_2 + \frac{1}{2}O_2 \xrightarrow{230°C} XeOF_4$$

$XeOF_4$ is purified by vacuum distillation.

Physical Characteristics

It is a colourless solid which melts at –46°C.

Chemical Characteristics

It reacts with water or SiO_2 to give XeO_2F_2 and XeO_3

$$XeOF_4 + H_2O \longrightarrow XeO_2F_2 + 2HF$$

$$XeO_2F_2 + H_2O \longrightarrow XeO_3 + 2HF$$

$$2XeOF_4 + SiO_2 \longrightarrow \underset{\text{Xenon dioxygen difluoride}}{XeO_2F_2} + SiF_4$$

$$2XeO_2F_2 + SiO_2 \longrightarrow 2XeO_3 + SiF_4$$

3. Reacts with lewis acid like SbF_5 to form $XeOF_3^+$ cation

$$XeOF_4 + SbF_5 \longrightarrow XeOF_3^+SbF_6^-$$

excess

$$XeOF_4 + 2SbF_5 \longrightarrow XeOF_3^+Sb_2F_{11}^-$$

$XeOF_3^+$ cation shows AX_4E type structure with one lone pair, where the oxygen and fluorine atoms occupy axial positions in the trigonal bipyramid geometry as predicted by the VSEPR theory.

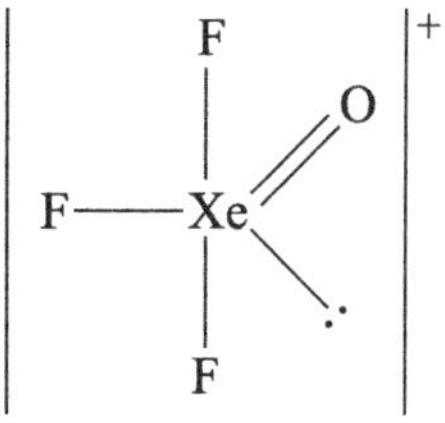

With NOF (nitrosyl fluoride) it yields a crystalline solid $NO^+(XeOF_5)^-$ $XeOF_5^-$ exhibits a pentagonal pyramidal structure with oxygen at the apical position. Here the Xe atom is above the plane of the fluorine atoms.

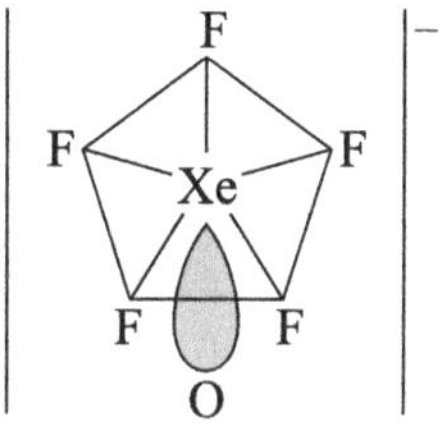

Bonding and Strurture

1. The structure of $XeOF_4$ molecule is square pyramid.
2. **VBT:** According to the theory three *p*-electrons are supposed to be promoted to the 5*d* sub level as illustrated:

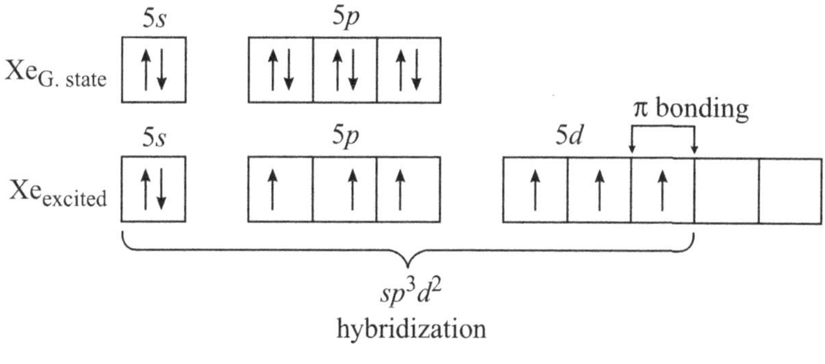

Effective hybridization is reduced to sp^3d^2 an one of *d* orbitals is π-bonding. The five unpaired electrons (three in the *p* orbitals and two in the *d* orbitals) form four σ bonds with four fluorine atoms and one σ bond with oxygen atom. The sixth hybrid orbital is occupied by the lone pair. The remaining electron in the *d* orbital is π bond forming. Therefore, the ultimate geometry is square pyramid which is an uncommon geometry.

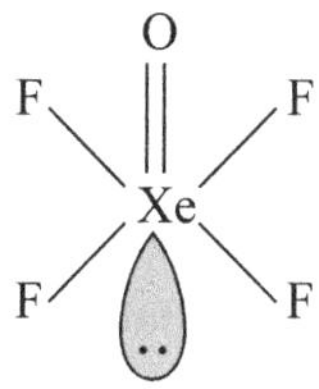

(*ii*) Xenon Dioxygen difluoride XeO_2F_2 Preparation

It is prepared by mixing XeO_3 and $XeOF_4$ at low temperature (–78°C). The product is purified by fractional distillation.

$$XeO_3 + XeOF_4 \xrightarrow{-78°C} XeO_2F_2 + O_2 + F_2$$

Physical Characteristics

It is a colourless solid which melts at 30°C.

Chemical Characteristics

1. It hydrolyses to give
$$XeO_2F_2 + H_2O \longrightarrow XeO_3 + 2HF$$
2. It undergoes slow decomposition to give XeF_2 and O_2
$$XeO_2F_2 \longrightarrow XeF_2 + O_2$$
3. With SbF_5 and SbF_3 in HF it gives a white solid in vacuum
$$XeO_2F_2 + 2SbF_5 \longrightarrow \underset{\text{white solid}}{XeO_2F^+ \cdot Sb_2F_{11}^-}$$

The white solid formed slowly decomposes to

$$XeO_2F^+.Sb_2F_{11}^- \longrightarrow XeF^+.Sb_2F_{11}^- + O_2$$

XeO_2F^+ shows a tetrahedral structure with one position occupied by a lone pair.

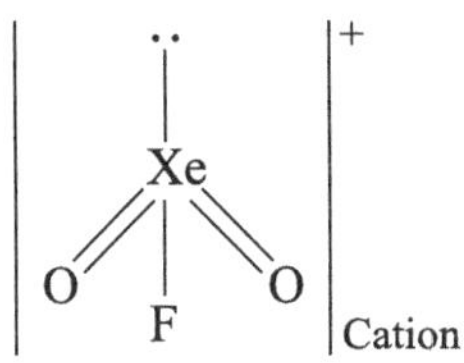

Bonding and Structure

1. It has a see-saw structure.
2. The valence bond theory suggests trigonal bipyramidal geometry with one equatorial position occupied by the lone pair.

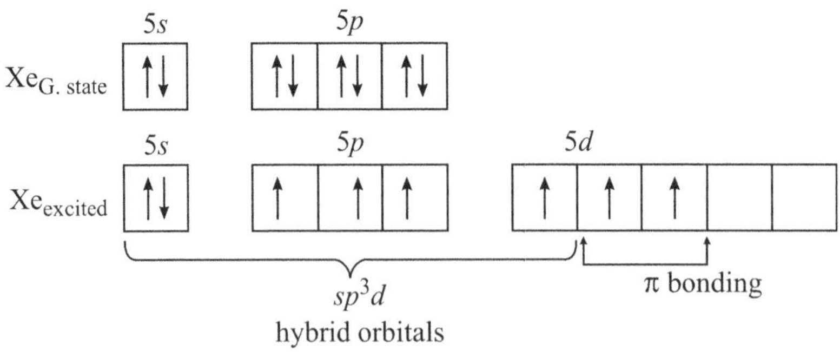

The three 5*p* electrons of Xe are promoted to the high energy 5*d* orbitals. Two *d* orbitals form π bonds are excluded from the hybridization. The remaining one 5*s*, three 5*p* and one 5*d* orbitals hybridise to give sp^3d hybridization which suggests trigonal bipyramidal geometry. The four singly occupied orbitals form four σ bonds; two with two fluorine and two with oxygen atoms, the fifth hybrid orbital contains lone pair. Therefore, the structure is as shown.

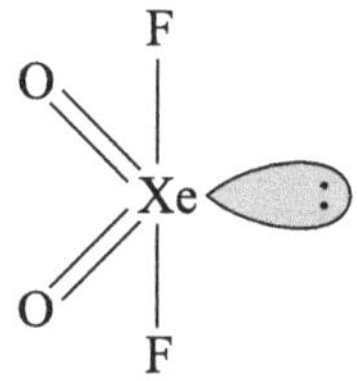

(*iii*) Xenon Oxygen Difluoride $XeOF_2$ Preparation

It is prepared by the hydrolysis of XeF_4 at –80°C.

$$XeF_4 + H_2O \xrightarrow{-80°C} XeOF_2 + 2HF$$

Structure

1. It has a T-shaped structure.
2. **Valence bond approach:** In this compound $XeOF_2$ we assume that two electrons from the 5*p* sub level are promoted to 5*d* sub level.

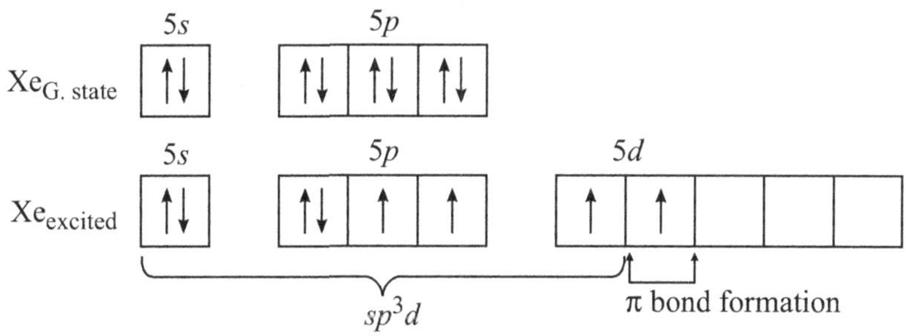

Accordingly one 5*s*, three 5*p* and 5*d* orbitals hybridise to give sp^3d hybridization. The three unpaired electrons form three σ bonds using two fluorine and one oxygen atom sp^3d hybridization proposes trigonal bipyramidal structure in which two positions are occupied by the two lone pairs. One unpaired electron in the 5*d* is not used in hybridization and therefore used in π-bond formation with the oxygen atom.

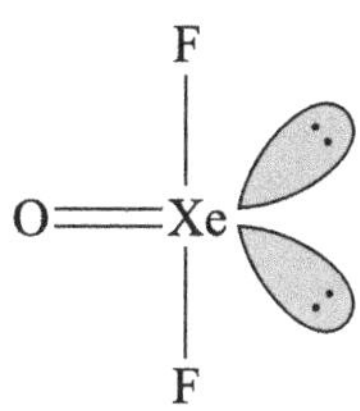

Oxides of Xenon

(*i*) Xenon Trioxide XeO_3 Preparation: It may be prepared by the hydrolysis of XeF_4 and XeF_6. Hydrolysis should be carried out carefully as the reaction is explosive.

$$3XeF_4 + 6H_2O \longrightarrow 2XeO_3 + 12HF + Xe$$
$$XeF_6 + 3H_2O \longrightarrow XeO_3 + 6HF$$

Properties

It is a colourless solid. It is a highly explosive substance.

$$XeO_3 \longrightarrow Xe + \frac{1}{2}O_2 \quad \Delta H = -402 \text{ kJ}$$

Structure

1. XeO_3 compound has a pyramidal structure.
2. Valence bond theory. The theory suggests that three 5*p* paired electrons get unpaired and promoted to high energy 5*d* orbitals.

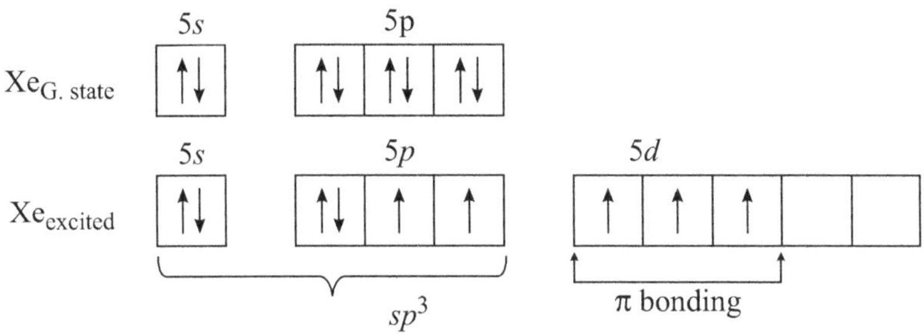

One paired 5*s* orbtial and three 5*p* orbitals hybridise to give sp^3 hybridization. The three 5*p* unpaired electrons form 3σ bonds with the three oxygen atoms. The three unpaired electron in the 5*d* orbitals form three π-bonds and do not participate in hybridization.

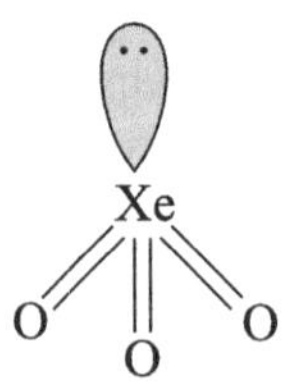

(*ii*) Xenon Tetraoxide XeO_4 Preparation: It is prepared by the action of concentrated H_2SO_4 on sodium or barium xenate at room temperature.

$$Na_4^{4+}XeO_6^{4-} + 4H^+ \longrightarrow X_4XeO_6 \xrightarrow{-2H_2O} XeO_4$$

Properties

It is quite unstable and decomposes to Xe and oxygen. At – 40°C it explodes to give.

$$XeO_4 \xrightarrow{-40^\circ C} Xe + 2O_2 \quad \Delta H = -643\ kJ$$

Structure

1. The XeO_4 compound shows tetrahedral structure.
2. Valence bond theory

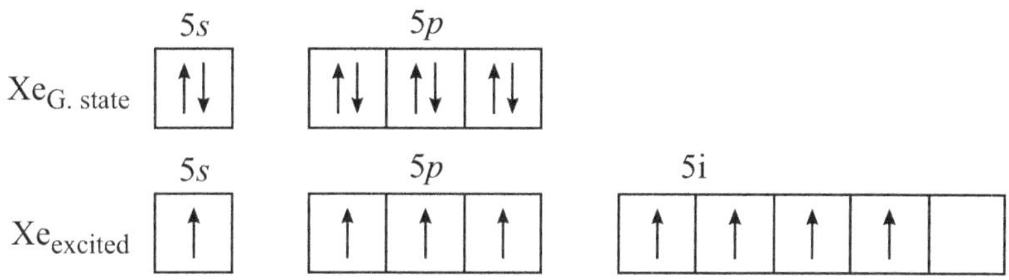

It this molecule all the electrons in one 5*s* and three 5*p* get promoted to the high energy 5*d* orbitals. One unpaired electron in 5*s* and three unpaired electrons in three 5*p* orbitals form 4σ bonds. With four oxygen atoms and hybridise giving rise to sp^3 hybridization and a tetrahedral structure. The four singlets in the 5*d* orbitals form four π bonds which do not participate in the hybridization.

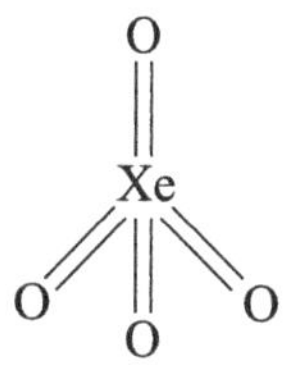

Tetrahedral

Other compounds of xenon *e.g.*, $XeCl_2$, $XeCl_4$ and XeO_2 find mention in the literature but they are not significant. $XeCl_2$ has a linear structure just like XeF_2.

EXERCISES

1. Explain why it is very difficult to form the compounds of Helium.
2. Write equations:
   (*i*) $XeF_6 + SiO_2 \longrightarrow$
   (*ii*) $XeF_4 + SF_4 \longrightarrow$
   (*iii*) $XeF_2 + H_2 \longrightarrow$
   (*iv*) $XeF_4 + Hg \longrightarrow$
   (*v*) $XeF_2 + SO_3 \longrightarrow$

(*vi*) $XeF_2 + SbF_5 \longrightarrow$
(*vii*) $XeF_6 + H_2O \longrightarrow$
(*viii*) $XeF_2 + 2SbF_5 \longrightarrow$

3. The first original compound obtained after the reaction of Xe and PtF_6 was:

   $Xe + PtF_6 \longrightarrow$

4. Give the reaction product of:

   $XeF_6 + AuF_3 + F_2 \longrightarrow$

5. Write the cations in the compounds $XeOF_4$ and XeO_2F_2.
6. Why the noble gases are assigned zero valency and are mono atomic gases? Comment.
7. The main constituents of the sum are:

   (*i*) H_2 and He (*ii*) H_2 and Ar (*iii*) H_2 and Xe (*iv*) Ar and Xe.

8. Which noble gas reacts with O_2 and F_2

   (*i*) He (*ii*) Ne (*iii*) Ar (*iv*) Xe

9. XeF_2 has structure:

   (*i*) Octahedral (*ii*) Tetrahedral (*iii*) Triangular (*iv*) Linear.

10. Which noble is not well known to form clothrate compounds.

    (*a*) Ne (*b*) Ar (*c*) Kr (*d*) Xe

11. Which of the follow shows sp3d hybridization

    (*i*) XeF_4 (*ii*) $XeOF_2$ (*iii*) $XeOF_4$ (*iv*) XeO_3

12. $XeOF_4$ compound can be prepared by the partial hydrolysis of:

    (*i*) XeF_6 (*ii*) $XeF + SiO_2$
    (*iii*) $Xe + F_2$ (1 : 4) in excess O_2 (*iv*) XeO_2F_2 in excess of F_2.

13. Discuss the factors which led to the formation of noble gas compounds.
14. (*a*) What are clathrate compounds of the noble gases? Discuss the type of bonding in these compounds

    (*b*) Why He and Ne do not form clathrates.

15. Discuss the preparation and structure of:

    (*i*) XeF_2 (*ii*) XeF_4 (*iii*) XeF_6 (*iv*) XeO_3
    (*v*) $XeOF_4$.

16. Complete the equations:

    (*i*) $XeO_2F_2 + H_2O \longrightarrow$
    (*ii*) $XeF_2 + H_2 \longrightarrow$
    (*iii*) $XeF_6 + BF_3 \longrightarrow$
    (*iv*) $XeF_6 + HCl \longrightarrow$
    (*v*) $XeF_4 + KI \longrightarrow$

17. Write the reaction of XeF_2 with NO_2, SO_2 and SbF_5.
18. Give the preparation of XeF_4. How does it react with (*i*) H_2O (*ii*) BCl_3. Draw the structure of XeF_3^+ ion.
19. How do XeO_3 and $XeOF_2$ differ in their structure. Explain.
20. Xenon forms the largest number of compounds out of the noble gases. Comment.
21. Why is it difficult to store XeF_6 in glass or quartz vessels?
22. Explain:
    (*i*) Melting points of noble gases increase with increase in atomic number.
    (*ii*) He_2 molecule is not formed.
23. Write short note on clathrates of noble gases.
24. Explain why Ar is used in the kroll process to extract Ti, for welding, and in electric bulles.
25. How do we prepare the compounds:
    (*i*) XeO_3 (*ii*) $XeOF_4$ (*iii*) Ba_2XeO_6.
    Discuss their structures.

Bibliography

Chapter 1

1. Anderson A.W. Domain representation of orbitals J. Chem. Edu., 1965, 42 (4)
2. Berry R.S. "Atomic orbitals" J. Chem. Edu., 1966, 43, 283
3. Cohen I. and Bustard. T "Atomic orbitals limitations and variation" J. Chem. Edu., 1966, 46, 187
4. Kiterchi, Q and Suzuki, K. "Orbital shape representations", J. Chem. Edu., 1985, 65, 206
5. Orgydo, E.A. and Porter G.B. "Contour surfaces for atomic and molecular orbitals", J. Chem. Edu., 1963, 46, 428
6. Powell, R.E. "The five equivalent d-orbitals". J Chem. Edu., 1968, 45, 1
7. Price W.C., Chissick, S.S. and Ravensdale, T. Edu., Wave mechanics, the first 50 years, Butterworth's, London, 1973
8. Pritzhard, H.O. and Skinner, H.A. "Electronegativity scales" Chem. Rew, 1955, 55, 745.

Chapter 2

1. Burdett, J.K. New ways to look at solids, Acc. Chem, Res. (1982) 15.34
2. Cos, P.A. The electronic structure and chemistry of solids (1987) Oxford Univ. Press.,
3. Jenkins, H.D.B. The calculation of lattice energy some problems and some solution revue do Chimie Mineral (1979) 16, 134–139
4. Ladd, M.F.C. Structure and Bonding in solid state chemistry (1974), Wiley. London.
5. Rao, C.N.R. and Gopala Krishnan, J. New Directions in solid state chemistry (1986), comp bridge university press. Comp bridge.
6. West, A.R. Solid state chemistry and its application, (1984) Wiley. New York,

7. Wells, A.F. Structural Inorganic Chemistry (1984), 5th edition. Oxford University Press. Oxford.
8. Shaman, R.D Revised effective ionic radii (1976) Acta. Cryst, A32, 751–767.
9. Greenwood, N.N. Ionic Crystals, Lattice Defects and Non-stoichiometry (1968) Butterworth's, London.
10. Galasso, F.S. Structure and Properties of Inorganic solids (1970) Pergamone, Oxford.
11. Galway, A.K. Chemistry of solids (1976) Chapman and Hall, London.
12. Dasent, W.E. Inorganic Energetics; an Introduction (1982) Comp bridge, University Press.

Chapter 3

1. Atkins, P.W. Molecular Quantum Mechanics (1983), Oxford University Press, Oxford.
2. Baltihausen, C.J. and Gray, H.B. Molecular electronic structures. Benjamin-Cummins, Menlo Park, California.
3. Coulson, C.A. revised by McWeiny, R, the shapes and structure of molecules (1982), 2nd ed. Clasendon Press, Oxford.
4. Douglas, B, McDanid D.H. and Alexander J.J. Concepts and Models in Inorganic chemistry (1983), 2nd ed., Wiley, New York.
5. Ferguson, J.E. Stereochemistry and Bonding in Inorganic Chemistry (1974), Prentice Hall Engle wood cliffs N.J.
6. Gillespie, R.J and Nyholm R.S. Q Rev. Chem. Soc., 11 (1957), 339 (Develops the Sidgewick-Powell theory into modern VSEPR theory)
7. Pauling, L. The Nature of Chemical bond (1961), 3rd ed. Oxford University Press. Oxford.
8. Sidgwick, N.V. and Powell H.M (The original paper on electron pair repulsion theory) 1940, Proc. R. Soc. 176A, 153
9. Kettle, S.F.A. Symmetry and structure (1985). Wiley, London
10. Concise Inorganic Chemistry, fifth Edition.
11. Inorganic chemistry, fourth ed., Huheey, Keiter and Medhi.
12. Basic Inorganic chemistry by cotton, Wilkinson and Gaus.

Chapter 4

1. Alfred, A.L. More on electronegativity values (1961) J. Inorganic Nuclear Chemistry 17, 215.
2. Emeleuces, H.J. and Sharpe, A.G. Moder Aspects of Inorganic chemistry 4th ed. (973) chapter 5 structure and energetics of inorganic molecules. (Chapter 6 Inorganic chemistry in aqueous midial Routbdge and Kangan Paul, Landon).

3. Huheey, J.E. Inorganic chemistry (1972) Harper and Row, New York (Discussion on electronegativity)
4. Mulliken, R.S. Mulliken electronegativity scale (1934) J. Chem. Physics 2, 782, 3, 573.
5. Pauling, L. The nature of the chemical bond (1960) 3rd ed. Oxford University Press, London.
6. Jolly, W.L. Inorganic chemistry (1976) McGraw Hill, New York.
7. Sharpe, A.G. Inorganic chemistry (1981) (Chapter 7 Inorganic chemistry in aqueous media), Longmans, London.
8. Rosotti, H. The study of Ionic equilibrium in aqueous solution, Longmans, London.
9. Inorganic chemistry by Moeller,
10. Inorganic chemistry by Kotz
11. Inorganic chemistry by J.D. fifth edition.
12. Inorganic chemistry 4th edition by Huheey.
13. Inorganic chemistry 3rd edition by Meissler.

Chapter 5

1. Fiacco, D.L. A. Toro and K.R. Leopold, Inorganic chemistry (2000) 39, 37
2. G. Olah and G.K.S. Prakash, super acids (1985) John Wiley and sons, New York.
3. G. Olah G.K.S. Prakash and J. Sommer science (1979), 206, 13.
4. R.J. Gillespie Acc. Chem. Res. (1968), I. 202.
5. L.P. Hammen and A.J. Dertup, J. Am. Chem. Soc. (1932), 54, 2721.
6. W.V. Gicha and F. Aubke, J. Am. Chem. Soc. (1989), 111, 4328.
7. D. Zhang, S.J. Retting, J. Trotter and F. Aubke, Inorganic chemistry (1996), 36m 6113.
8. R. Minkwitz, A. Knmath, W. Sawoday and J. Hahn, Inorganic chemistry (1996) 35, 3622.
9. Advanced inorganic chemistry by K.L. Kapoor and Vashisht 4th edition.
10. N. Barford, A.D. Phillips, R.W. Schurko, R.E. Wasilishen and J.F. Richardson. Chem. Communities (1997) Camp bridge, 2363.
11. The Lewis Acid-Base concept by W.B. Jensen (1980), Wiley Inter science, New York.
12. H.L. Finston and Allen C. Rychtman A new view of current Acid-Base theories (1982), John Wiley and sons, New York.
13. Advanced inorganic chemistry Satya Parkas, G.D. Tuli, S.K. Basu and R. D. Madan volume I, 1997.

Chapter 6

1. Duboil, P; Lilieur, J.P; Lepoutre, G Inorg. Chem. 1988, 27 (1883 – 1892), 1989, 28 (195 – 200)
2. Edward, P.P. Adv. Inorganic chemistry Radio chem. 1982, 25 (135 – 185)
3. Edwards P.P; Sienko, M.J; J. Am. Chem. Soc. 1981, 103 (2967)
4. Dye, J.L; De Backer, H.G; Nicely, V.A; J. Am. Chem. Soc. 1976, 92 (5226 – 5228)
5. Issa, D; Dye, J.L; J. Am. Chem. Soc. 1982, 104, (3781)
6. Hall, N.F; Conant J.B; J. Am. Chem. Soc. 1927, 49 (3047 – 3061)
7. Gillespie, R.J; Peel, T.E. Adv. Phys. Org. Chem. 1972, 9 (1 – 24)
8. Olah, G.A. Angew, chem. Int. ed. Engle 1973, 72 (173 – 212)
9. Olah G.A; Prakash, G.K.S; Sommer, J. Super acids, Wiley; New York, 1985

Chapter 7

1. Adv. Inorganic chemistry by Satya Prakash; Tuli, G.D; Basu, S.K; Madan, R.D. j 1997, (590 – 6556)
2. Inorganic chemistry by Ramesh Kapoor Vasisht, S.K; Chopra, R.S. 1998, (257 – 301)

Chapter 8

1. D.J. Adam, Early industrial electrolysis, Education in chemistry (1980), 17, 13–14, 16.
2. H.U. Borgetedts and C.K. Mathews Applied chemistry of alkali metals (1987), Plenum, London.
3. R.S. Boynton chemistry and technology of line and limestone (1980), 2nd ed. John Wiley, chicterter.
4. R.M. Stephenson introduction to the chemical process industries (1966) Van Nastrand Reinhold, New York
5. R.W. Purcell, the chlor-alkali industry chapter by Campbell, A; 31, the chemical society, London.
6. S. Venkatesh and S. Tilak Chlor-alkali technology (1983), J chem. Ed. 68, 276–278.

Chapter 9

1. N.A. Bell; Beryllium halides and complexes, Adv. Inorg. Radio chem. 14, 225
2. T.P. Hanusa; Re-examining the diagonal relationship, (1987) J. Chem. Ed. 64, 686 – 688.

3. W.E.C. Waker, Magnesium and Man, Harvasd University Press, London.
4. D. Parker, Alkali and Alkaline earth cryptstes, (1983), Adv. Inorg. Radio Chem., 27, 1 – 26
5. M.N. Hughes, The inorganic chemistry of Biological processes (1972), (chapter 8), John Wiley, London
6. B. Dietrich, Coordination chemistry of alkali and alkaline earth cations macro cyclic ligands, (1985), J. Chem. ed. 65, 954 – 964 (Crowns and Crypts)

Chapter 10

1. C.L. Christ, and J.R. Clark A crystal classification of borate structure with emphasis on hydrated borates (1977), phys. Chem. Minerals, 2, 59
2. N.N. Greenwood, Boron Pergamon press, Oford and Elms ford, NY.
3. N.N. Greenwood and I.M. Ward, metillobornaes and metal-boron bonding (1974), Chem. Soc. Rev. 3, 231 – 271
4. W.E. Haupin, Electro chemistry of the Hall-Heroult process for aluminum, J. Chem. ed. 60, 279 – 282.
5. A. G. Lee, The chemistry of Thallium, (1971), Elsevier, Amsterdam.
6. A.G. Lee, Coordination chemistry of Thallium (1), coordination Chem. Rev. 8, 289
7. A.G. Massey, the sub-halides of boron, (1983), Adv. Inorg. Chemc. Radio Chem. 26, 1 – 54.
8. E.L. Muettertion (ed.) Boron Hydride chemistry, (1975), Academic Press, New York.
9. K. Wade, Electron deficient compound, (1971), Nelson, London.
10. K. Wade and A.J. Bannister, Comprehensive Inorganic chemistry, vol. 1 (chapter 12) Aluminum, gallium induces and Thallium, (1973), Pergamon Press, Oxford.
11. R.T. Sanderson, More on complex hydrides (1997), Chem. Eng. News, 29, 3 ($LiAlH_4$) etc.

Chapter 12

1. H. Bode; Lead-Acid Batteries (1977), Wiley, New York.
2. B. Bolin; the carbon cycle, (1970), scientific American, September.
3. S. Elliott, F.S. Rowland chlorofluorocarbon and atmospheric ozone (1987) J. Chem. Ed. 64, 387 – 390.
4. L.S.D. Glosser sodium silicates, (1982), Chemistry in Britain, 33 – 39
5. G. Raabe and J, Michi, Multiple bonding to silicon (1985), Chem. Rev. 85, 419 – 509.
6. W. Noll, O. Clenz and G. Hecht, chemistry and technology of silicones, (1968), Academic press, New York.

7. J.S. Pizey (ed), synthetic reagents (1977), chapter 4 by R.N. Butter, lead retractile, John Wiley, chichester.
8. E.G. Rochow, Comprehensive inorganic chemistry (1973), Vol. I (Chapter 15, silicon) Pergamon Press, Oxford.

Chapter 13

1. C.C. Addison, Di-nitrogen tetra oxide, nitric acid and their mixtures as media for inorganic reactions (1980), Chem. Rev. 80, 21–39
2. M. Braudler, polyphosphorus compounds new results and new perspectives (1987), Angewandie chemic, international edition, 26, 419–44.
3. E.W. Schmidt, hydrazine and its derivatives. Preparation, Properties, Application (1984) John Wiley, New York.
4. J.D. Smith comprehensive inorganic chemistry vol. 11 (1973) (chapter 21) nitrogen, Pergamon Press, Oxford.
5. T, Kanazawa (ed.) Inorganic Phosphorus Materials (1989), Kodansha, Tokyo.
6. J. Lagowski The chemistry of Non-aqueous solvent chapter 4 by W.H. Lee (ed.) 1967 Nitric Acid chapter 7 by J.J. Lagowski and G.A. Moczyemba liquid ammonia academic press, N.Y.
7. A.N. Wright and C.A. Winkler Active Nitrogen (1968) Academic Press New York.
8. K. Jones comprehensive Inorganic chemistry vol. 11 chapter 19, Nitrogen (1973), Pergamon Press. Oxford.
9. W.L. Jolly, Metal Ammonia solutions (1972), Dowden, Hutchinson and Row Stroudsburg, P.A.

Chapter 14

1. K.W. Bagnall, Comprehensive inorganic chemistry, vol. 11 (1973) Pergamon Press, Oxford.
2. D.J.M. Bevan, comprehensive inorganic chemistry vol. 3 (1973) Pergamon Press, Oxford.
3. W.J, Bland sulhpuric acid-modern manufacture and uses (1984), Education in chemistry, 21, 7 – 10
4. A, Cocks and T. Kallend the chemistry of atmospheric pollution (1988), chemistry in Britain, 24, 884 – 885
5. H.J. Emelues and A.G. Sharpe, Modern aspects of inorganic chemistry (1973) 4th ed. (chapter 12) Perxodies and peroxy-acids Routledge and Kegan Paul, London.
6. M. Horvath, I. Bilitzky and J. Huttner, Ozone (topics in inorganic and General chemistry (1983), series No. 20) Ebsevier.

7. G. Nickless, Inorganic sulphur chemistry (ed.) (1968), Elsevier, Amsterdam.
8. A, Philips the modern sulphuric acid process (1977), chemistry in Britain, 13, 471.
9. M. Schimdt and W. Siebert, comprehensive inorganic chemistry, volume 11 (1973) (Chapter 23: Oxyacids of sulphur) pergamon press, Oxford)

Chapter 15

1. A.J. Downs and C.J. Adams comprehensive inorganic chemistry Vol. 11 (1973) chapter on chlorine, bromine iodine and astatine Pergamon Press, Oxford
2. R.J. Gillespie and M.J. Morton, Halogens and interhalogen cations (1971) Q. Rev. Chem. Soc., 25, 553.
3. R.H. Langley and I, Welch, Fluorine (chemical of the month) (1983), J. Chem. Ed. 60, 759 – 761.
4. J.S. Sconce, Chlorine its manufacture properties and uses (1962), Remhold, New York.
5. J. Shamir. Poly-halogen cations, structure and bonding (1979), 37, 141 – 210.
6. J.C. Tallow et. al. (eds.), advances in Fluorine chemistry, Butterworth, London.

Chapter 16

1. N. Bartlett, Xenon hexafluoro palatinate (V) $Xe^+ (PtF_6)^-$ (1962), Proc. Chem. Soc. 218 (The original paper on the noble gas compound).
2. N. Bartlett the chemistry of the noble gases (1971), Elsevier, Amterdam.
3. E. Berecz, and S.M. Balla ach. Gas Hydrates (studies in inorganic chemistry (1984) Vol. 4, Elsevier Oxford.)
4. D.T. Hawkins; W.E. Falconer and N. Bartlett Noble gas compounds (1978) Plenum Press, London.
5. J.L. Huston, chemical and physical properties of some xenon compounds (1982) Inorganic Chemistry 21, 685 – 688.
6. K. Sepplet, and D. Lentz Novel developments in noble gas chemistry Program Inorganic Chemistry 29, 167 – 202

Index

For Product Safety Concerns and Information please contact our EU representative GPSR@taylorandfrancis.com Taylor & Francis Verlag GmbH, Kaufingerstraße 24, 80331 München, Germany

Printed and bound by CPI Group (UK) Ltd, Croydon, CR0 4YY
14/09/2026
02216240-0004